depo
give
a pa
anim
of Cl
each
the in
specia
Un
water
ions i
being
in Ch
proces
also a
Tog
essenti
on hy
of lime
knowl
us to c
The
appeal
deposit
minera
discuss
althoug
reactio
are disc
detect
number
more t
foundat
Life
environ
There is
constant
loss of
an abilit
Study o
and bac
known,
mented

1. Cave Surveys

B. M. Ellis

Cave surveys are an essential tool for anyone carrying out scientific work underground because in addition to giving an impression of the overall layout of a cave system, they can be a source of information and a useful form on which to record scientific observations and to portray results. Geologists studying caves need surveys on which to plot dip and strike of bedding, the orientation of joints and the disposition of faults; geomorphologists need to plot erosional features to obtain a full understanding of a cave's morphological evolution; biologists need surveys to record the distribution of life forms underground; meteorologists may require a good survey from which to calculate the volume of cave air liable to movements; hydrologists need surveys to record their observations of water movement; chemists use surveys to record the movement of ions in solution; and archaeologists must have detailed surveys on which to record their finds. Surveys may also be useful historical documents, showing the researcher what used to exist at the time of the survey, even if since that time part has been lost by quarrying or other causes. Some uses of a survey are non-scientific; they can provide a route map enabling a caver to find his way through a system, and also inform him of other facts such as the ladders and tackle that he will require.

In preparing such a survey the cave surveyor has a much more difficult task than his topographical counterpart. Not only are the conditions under which he is working considerably more difficult (darkness, water, mud, confined spaces, and so on), but he has a much more difficult subject to portray. The topographical surveyor only has to depict what exists on one plane, even though the topography exists in three dimensions; but a cave exists fully in three dimensions—passages have height as well as given in Fig. 1.1. Because of the relative accuracies of the various instruments used, any cave survey should always be more accurate in its vertical measurements than in the horizontal ones. Further, due to the nature of the possible errors involved the resultant accuracy expressed as a percentage should be better the longer the distance surveyed, though error will reach a minimum of about 0·5% as is shown in Fig. 1.1. One technique that is being used more and more frequently to improve the accuracy of key points on the centre line of the survey is electromagnetic

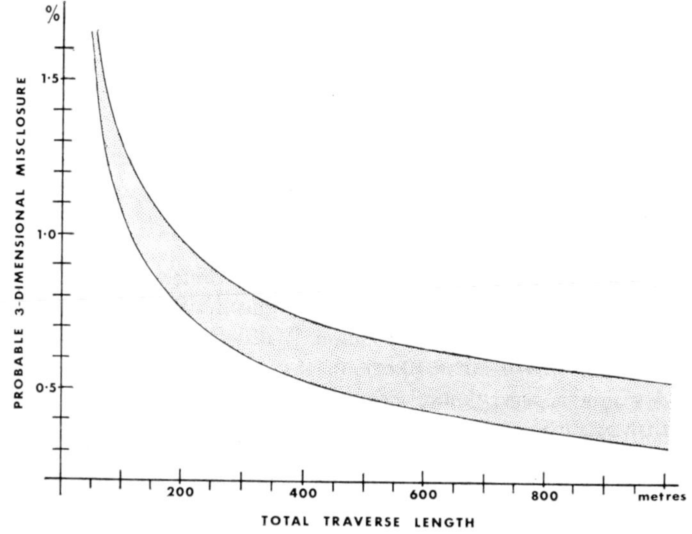

FIG. 1.1. The probable accuracy of a grade 5 survey.

location. This consists of using a "radio" transmitter and directional receiver to locate the position of key points in the cave in relation to the surface above. These points on the surface can then be tied together using conventional surveying techniques which are considerably more accurate than the techniques practical underground. If such a technique has been used in the preparation of a survey it is indicated by adding "/e" to the grading and classification, e.g. B.C.R.A. grade 5A/e. Surveyors are recommended to publish with every survey an article giving details of how the survey was made, and so on. If the user wishes to have a better indication of the survey's probable accuracy he should refer to this work, as it will probably include notes on any errors that were obtained, in addition to other information. Because a survey is neatly drawn and well presented it is not necessarily more accurate than one which is not.

Because of the complexity of the subject that the surveyor is attempting to portray, good cave surveys will normally consist of four separate parts.

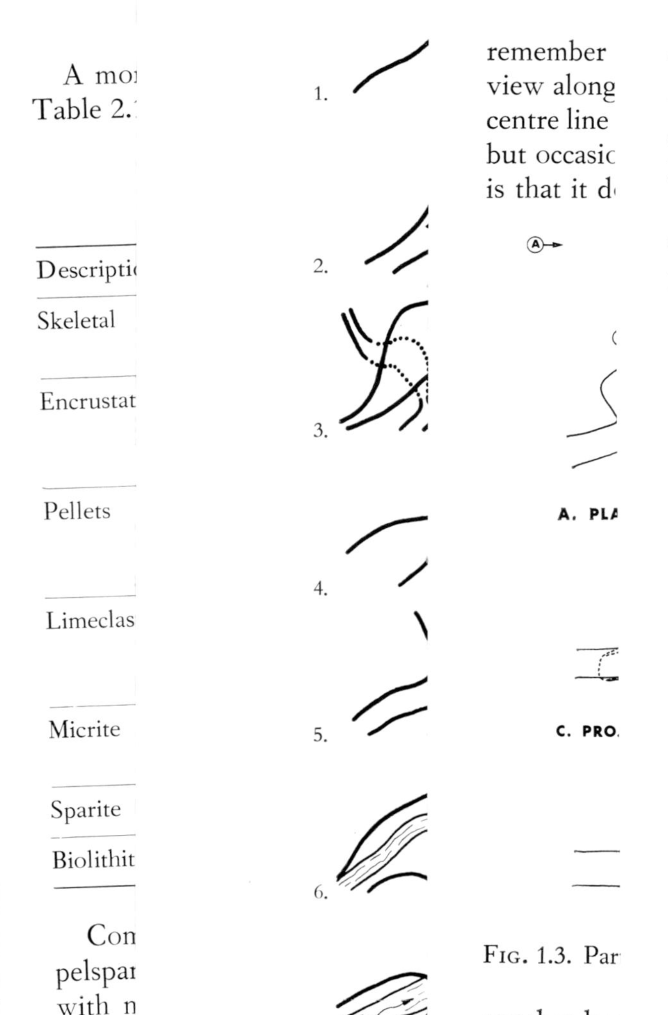

The first of these is the *plan view*. This is a view of the cave as seen from vertically above; generally this is the easiest to understand because it is similar to an ordinary map except that it is usually at a larger scale. It must be remembered when using a plan that the passages are not shown at their actual length unless they are horizontal. Because the plan is, as already stated, a view as seen from above, the passage length shown is the true passage length projected on to a horizontal plane, as shown in Fig. 1.2.

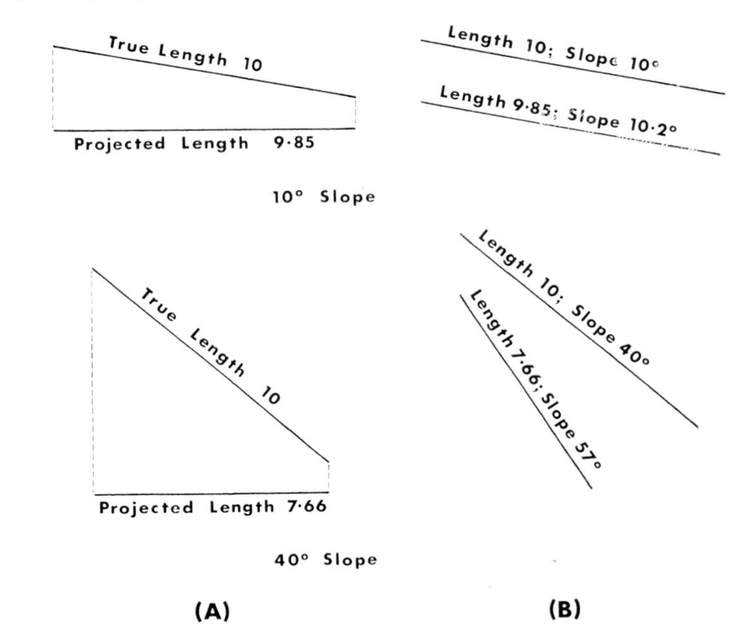

FIG. 1.2. The projection of a passage on to a plane.

This means that the steeper the passage slopes, the more it is foreshortened until in the extreme case of a vertical drop it has no length on the plan. The true passage length can be obtained from the extended elevation, which is described later. One problem that confronts the surveyor is the width of passage that should be shown on the plan view. If cave passages were always rectangular in shape there would be no problem, but in practice they are found to be far from regular. Therefore, which width should the surveyor show? In the past some very complex methods of showing passage shape have been devised, and the results can be seen on the plan views of some older surveys. These methods make the drawing very difficult to understand and are unnecessary. Surveyors are now recommended to show on the plan the typical width of the passage, roughly at a caver's eye height as he travels along the passage. If the user

PLATE 2.2. (b) Poorly sorted shelly calcarenite in a micrite matrix.

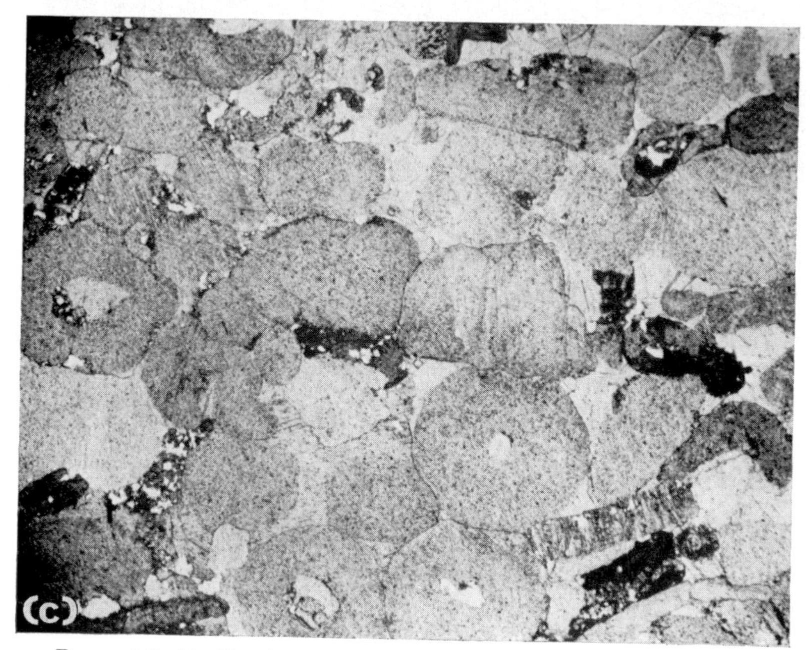

PLATE 2.2. (c) Closely packed crinoidal fragments in a calcirudite.

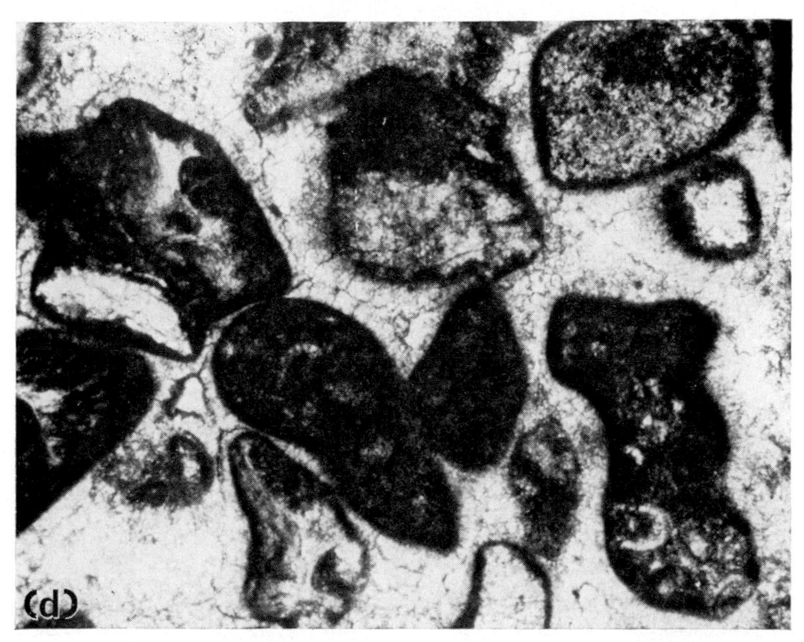

PLATE 2.2. (d) Pelletal calcarenite— probably faecal pellets, with thin dark envelopes of algal micrite, in a spar matrix.

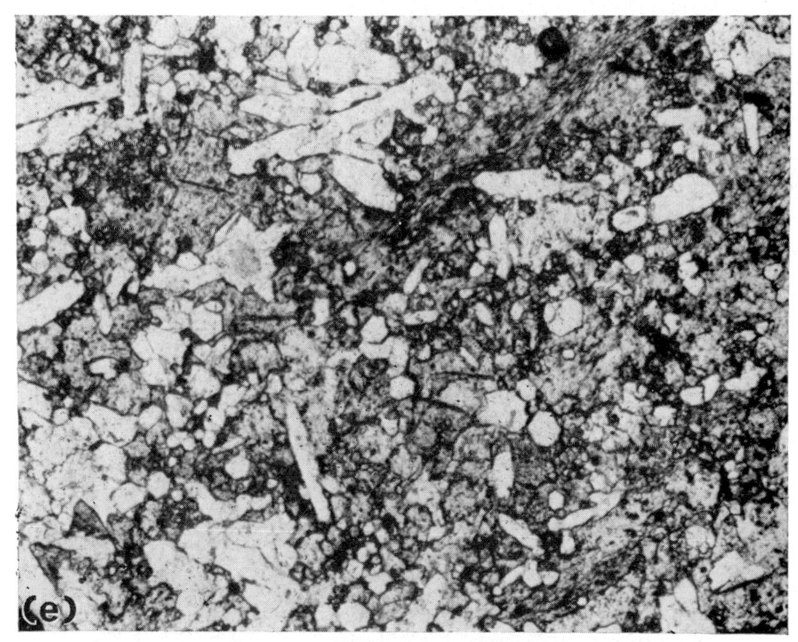

PLATE 2.2. (e) Authigenic quartz crystals replacing a recrystallized limestone with no surviving primary texture.

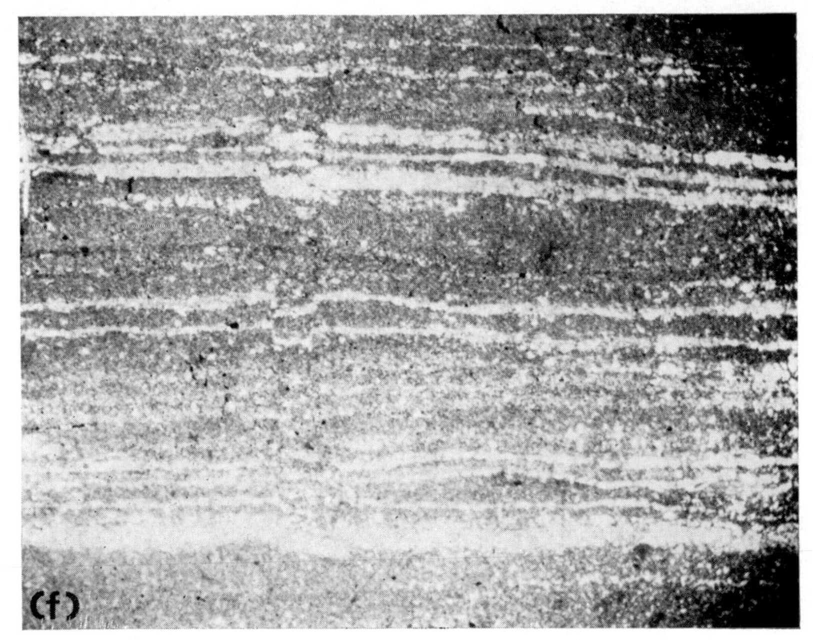

PLATE 2.2. (f) Laminated impure calcilutite.

LITHIFICATION

The process of lithification means the conversion of loose sediment into rock, in the case of limestones normally by calcium carbonate being deposited in the pore-spaces. This is carried to the site of cementation in solution both in connate water, that is the sea water trapped in the original sediment, and in later groundwater. The calcium carbonate may be derived directly from sea water, or it may result from the solution of some fragments of shells near by, notably those with the most porous structure and those made of aragonite.

The cementing calcium carbonate may be deposited in any one of three forms: coarsely crystalline spar, elongate fibres, or as very fine-grained micrite (Fig. 2.3). Taken together with the nature of the component grains these may be used to construct various compound limestone terms, such as the following proposed by Folk (1959). Some variations of these are illustrated in Fig. 2.3. (p. 15).

Intrasparite	intraclasts cemented by sparite
Intrasparrudite	large intraclasts cemented by sparite
Oosparite	ooliths cemented by sparite
Biosparite	fossil shells cemented by sparite

Pelsparite	pellets cemented by sparite
Biomicrite	shells cemented by micrite
Dismicrite	patches of spar in micrite
Biolithite	micrite with layering of biological origin

Fibrous cements normally encrust clasts and are themselves bound by spar or micrite cements, so that a qualifying adjective to indicate the fibrous character may be added.

The process of cementation may be accompanied by alteration of the enclosed grains in various ways, adding yet another factor to the already complex attempt to classify limestones. Such alteration may include *grain growth*—by the addition of a syntaxial rim, i.e. calcite added in crystallographic continuity with the grain, usually distinguishable under the microscope by the lack of enclosed carbonaceous matter, *fibrous calcite encrustation*, *micritic replacement* of the margins of the grain, covered by spar, fibrous or micritic cement and *fibrous recrystallization* of the grain, often spheroidally. Broadly, such changes may be considered as grain growth, grain diminution, or void filling. Voids in lime sediment are often the result of burial of soft organisms, for example, seaweeds, which rot away; or they may be the result of solution of an aragonitic shell. Low flat voids of the former sort often have a thin layer of fine sediment washed in to form a floor before the rest of the void is filled with spar. Such lenticular spar with micrite masses was once thought to be a fossil organism of unknown type and was named Stromatactis. This term is still used descriptively.

Voids partly filled with fine mud provide a "fossil spirit-level" by which the original attitude of the sedimentary layer may be determined. Such textures are known as geopetal fabrics.

DIAGENESIS

Diagenesis has a broader meaning than lithification, as it also includes those changes which take place in the rock owing to the migration of magnesium and silica, etc., which are discussed below. To some extent diagenetic changes in a limestone are an indication of age. Although much of the British Carboniferous Limestone was formed in similar environments to those of the Jurassic the limestones of the latter period have generally experienced much less diagenetic change and so are characterized by much of their original porosity, and thus by a lack of caves, that water can move through the whole body of the rock. Among the factors in the lesser diagenetic changes in the younger British limestones are the limited depth of burial under younger strata, and the relative lack of tectonic movements. In other parts of the world, both depth of burial and tectonism, like the Alpine earth movements, have contributed to highly lithified young limestones with extensive karst features, like some of the great caves of the Pyrenees developed in Mesozoic limestones, and those of Greece and Jamaica in Cainozoic limestones.

DOLOMITE

Both the mineral CaMg $(CO_3)_2$ and the rock composed of the mineral are usually called dolomite, though American geological literature has introduced dolostone for the latter. Dolomite results from the introduction of magnesium into the calcite molecule. The source of the magnesium may be either directly from sea water, or more likely from connate water enriched in magnesium by solution of some other mineral. Dolomitization may occur in the sediments below saline lakes or lagoons owing to the concentration of magnesium in the brine. The conversion from calcite to dolomite results in a theoretical reduction of molecular volume of 12%, which is often reflected in dolomite rock having a greatly increased porosity between the rhombic crystals of dolomite, and a loss of original textures. A dolomite layer within a limestone sequence will provide a preferential pathway for water movement; but as dolomite is generally less soluble than calcite, the initiation of caves may be slow or negligible. Often the process of dolomitization does not go to completion, and a mosaic of dolomite rhombs may still have interstitial calcite. Solution by moving waters will remove the calcite first and may leave behind an almost incoherent dolomite "sand", with little mechanical stability. Such solution is commonly along bedding and joints at first, so that bedding-joint networks may develop. Later the dolomite blocks may become unsupported and collapse ensues.

Sediments originally formed of dolomite are extremely rare and restricted to hypersaline lagoons, where they are associated with evaporite minerals such as gypsum, anhydrite and halite. Occasionally limestones, such as some of those in the Isle of Purbeck, result from calcite replacement of evaporites.

A reintroduction of solutions saturated in calcium carbonate may sometimes result in the growth of calcite crystals in the dolomite, often with spectacular radial and other patterns, such as the Cannon-Ball Limestone of Sunderland, or the "nodular" limestone of Agen Allwedd in South Wales.

The relative solubilities of dolomite and calcite still require full investigation, but some work has suggested that, if a little dolomite is present in an excess of calcite, then the dolomite is more soluble (Picknett, 1972). Many British Carboniferous Limestones contain a small proportion of dolomite, though whether this has been a factor in cave initiation is still unknown. Some also contain haematite (Fe_2O_3) and differential solution effects can be very striking (Plate 2.3).

PLATE 2.3. Differential solution in a partly dolomitized and partly haematized limestone. Forest of Dean. (*Photo*: D. M. Judson.)

ANHYDRITE

A recent discovery which may be of speleogenetic significance has been in two boreholes in Derbyshire where the Carboniferous Limestone has

unfavourable. The former probably have greater initial permeability or more pyrite, or, having clastic grains in a calcareous matrix, require less solution to allow disintegration of a given mass of rock. Shale partings are mechanically weak but less soluble, so that speleogenetic processes commonly start in the adjacent limestone but rapidly exploit the weakness of the shale. Bedding planes and the adjacent limestones are places where pyrite is concentrated, thus enhancing the possibilities of cave formation.

PLATE 2.4. Phreatic tubes developed along shale partings between beds of differing lithology; the roof is formed in a silicified shell bed. Carlswark Cavern, Derbyshire.

Plotting the distribution and thicknesses of beds and bedding planes could be a useful adjunct to a cave survey. Plates 2.4, 2.5 and 2.6 illustrate differing effects of bedding planes on cave morphology.

Recent work in the Ingleborough region by Waltham (1971) has shown that groups of limestones rich in shale partings along the bedding planes have been preferred sites for phreatic tube development.

The contrasts in properties between adjacent limestones such as a coarse-grained bed with high porosity in contact with a fine-grained bed may give preferred hydraulic routes for water through a limestone mass.

The form and thickness of beds are factors in speleogenesis. Thin-bedded limestones, with beds usually not more than 25-50 cm thick,

PLATE 2.5. Phreatic tube developed along a tight bedding plane in massive beds of similar lithology. Peak Cavern, Derbyshire.

PLATE 2.6. Bedding plane anastomosis developed along a shale bed. Bar Pot, Gaping Gill, Yorkshire.

provide many bedding planes, and a poor concentration of flow, so that cave development may be precluded. Thick-bedded limestones have fewer bedding planes and thus a limited number of preferred flow systems. Figure 2.4 gives a hypothetical example of the effect of different beds and bedding planes on the features of cave systems.

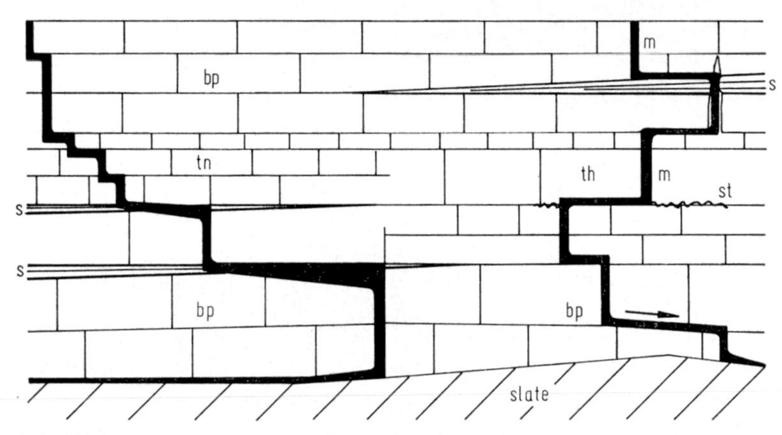

FIG. 2.4. Diagrammatic section to show the effect of minor stratigraphical variations in a limestone sequence on the profiles of two adjacent cave systems, with indications of the relative importance of correlation in controlling features. s = shale beds, locally wedging out or being replaced by well-developed bedding planes (bp) or by stylolites (st); th = thick-bedded limestones; tn = thin-bedded limestones; m = master joint.

Apart from simple beds there are reefs, or bioherms, banks and biostromes. The term "reef" has been so misused in both popular and geological literature that the other more easily definable terms are now in common use. A bioherm is a mound-like mass of lime sediment specifically built by the action of organisms, through the binding action of algal mats, or the trapping mechanism of bryozoan fans or coral skeletons (Fig. 2.5). A bank is a mound built by currents on the sea floor piling up detritus; binding organisms may occur sporadically, and a bank may become the nucleus of a bioherm. A biostrome is a horizontal bed built by the action of organisms such as a spread of corals across a lagoon floor. Bioherms and, to a lesser extent, banks may have vugs or internal cavities resulting from growth of organisms forming an umbrella. Such vugs may provide the voids which are later infilled with spar to form stromatactis textures. Flanking bioherms and banks there are commonly coarser, often crinoidal, limestones with increased permeability.

The presence of bioherms, banks or biostromes within a succession of limestone beds causes marked local variations in permeability, in the frequency of bedding planes, and in insoluble residues. In some reported cases, such as Dovedale in Derbyshire, bioherms have been preferential

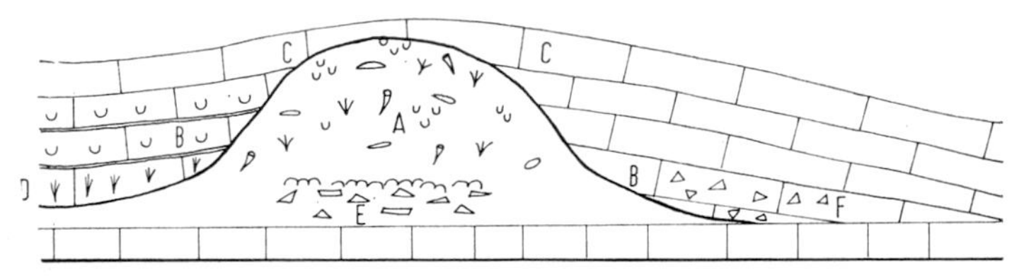

FIG. 2.5. Diagrammatic section through a bioherm and associated sediments. A = calcilutite bioherm, without bedding but containing stromatactis vugs, and framework builders such as corals and bryozoans with "pockets" of reef-dwelling shells. B = flanking coarse calcarenite or calcirudite, largely made of crinoid debris, thick-bedded on up-current (right) side, thin-bedded with shelly horizons on down-current (left) side in the lee of the bioherm. C = supra-reef limestones, mainly fine crinoidal calcarenite. D = a biostrome, being a limestone bed rich in corals representing a short-lived spread across the lagoon. E = a bank of lime-clasts swept up by currents and then encrusted by algae to form the nucleus of the bioherm. F = a current-piled bank of lime-clasts which failed to become the nucleus of a bioherm.

sites for cave development, perhaps owing to their vuggy character. Elsewhere, as in Giants Hole and some other Castleton caves, the uneven character of bedding planes around bioherms has given rise to sumps and other constrictions.

STYLOLITES

Many bedding planes in limestones exhibit pressure solution features known as stylolites (Fig. 2.3h). If a scatter of insoluble material lies along a bedding plane, the effect of the weight of younger superincumbent strata is to press the beds together. Under such pressure calcium carbonate in contact with quartz grains may dissolve, and such solution may be preferentially more on top of some grains and underneath in others. The net result is a sort of three-dimensional zig-zag graph. Seen on a joint face in two dimensions, it looks like the trace of a pen-recorder, and is therefore called a stylolite. Some stylolites start to form during lithification and so supply the calcium carbonate necessary for cementation of adjacent sediment. These surfaces of mutual pressure-solution vary from ones in which the projections from one bed to the next are only a millimetre or so, through jagged interprojections of 10 or 20 cm up to ones with lobes a metre long. Three different types of stylolite are recognized: denticular,

Underlying strata have a limiting effect on the depth of speleogenesis if they are impermeable. If the underlying strata are permeable, and a few cases are known, a cave passage may become impenetrable owing to the water seeping away through the pores, for example, in an underlying sandstone.

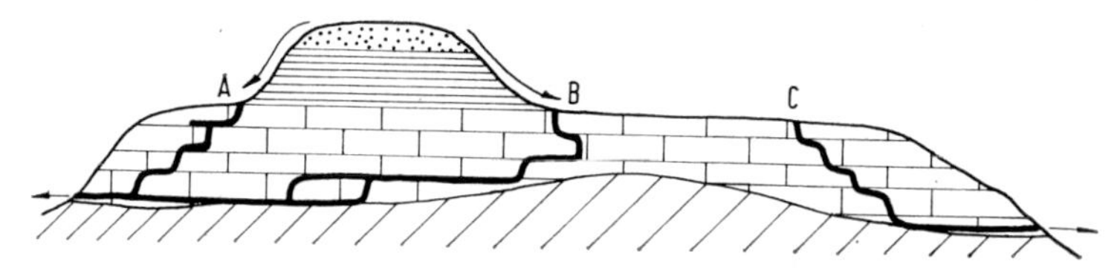

FIG. 2.6. Diagrammatic section to show the effects of overlying impervious strata and unconformable slate below. A = runoff from overlying strata sinks into jointed horizontal limestones and rises in adjacent valley. B = runoff sinks and at first follows a random path in horizontal upper beds; it then goes down the dip of beds above a buried hill of slate to follow an epi-phreatic path to the left. C = an abandoned system now receiving only local seepage, but once a precursor of system B before erosion of cap-rocks; its lower course was influenced by the inclination of the beds away from the buried hill.

If an area has been subject to folding, strata which underlie the limestone may rise higher topographically to provide a surface catchment, as is the case with the Old Red Sandstone cores of the Mendip anticlines (Fig. 2.7).

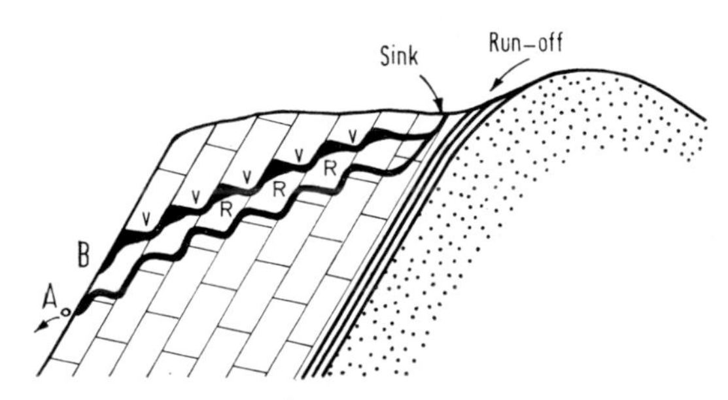

FIG. 2.7. Diagrammatic section to show the effect of an inclined series of limestone beds forming the flank of an anticline. A = an active drainage system with an undulating down-dip and up-joint profile. B = an abandoned system showing vadose trenching (V) of former high proportions, but still some intervening sumps, some with the roof eroded (R) and partly full of sediment.

Non-calcareous strata within a limestone sequence may be anything from a thick shale parting in a bedding plane up to interleaved massive sandstones or other rocks. A few feet of shale may form an impermeable barrier so that phreatic cave development takes place above it; but when rapid flow erodes through soft rock such as shale it may provide a mechanical weakness to be exploited by a vadose stream. A sandstone layer will generally resist downcutting, though it may allow seepage loss. An example is in Mossdale Caverns, in Yorkshire, where pools experience gradual seepage loss through joints in the underlying sandstone and dry up during dry periods.

In faulted strata, local barriers to water movement may be breached, allowing flow across the fault from above the impermeable layer on one side to below it on the other. Stepped cave profiles may result (Fig. 2.11).

Volcanic eruptions in a limestone sea may provide lava flows and ash-falls. Though generally of limited extent, these may provide discontinuous barriers to water movement as in Derbyshire. Chemical reactions not uncommonly give pyrite in limestones adjacent to volcanic layers. Faulting may again allow stepped profiles to develop.

Thus any cave system developed in limestones with non-calcareous strata in the vicinity may have special characteristics of its own. Most of the Northwest Yorkshire cave area has had its speleogenetic processes at least partly controlled by the Yoredale Shales and Millstone Grit, and they are thus not to be taken as a textbook generalization of cave development.

STRUCTURE

Once lithified, a limestone mass may be subjected to stresses and strains emanating from forces commonly called tectonic, within the Earth. Such stresses may result in tilting or folding, so that limestone beds are inclined and any inherent weaknesses inclined in the same direction are preferentially developed. The stresses may also result in fracturing of the limestone, producing joints and faults.

In many cases cave surveyors could usefully plot the disposition of joints, faults and the dips of inclined beds as the survey progressed. This would greatly assist the interpretation of the cave's origins at a later date and would remove the necessity for a subsequent geological survey, with the possible dangers of mislocation.

JOINTS

Joints and faults are both fractures, but joints show no displacement of the rocks on either side, whereas faults, by definition, are planes of displacement. Both are produced by stresses such as compression, tension and torsion, any of which may act in any direction. The mechanics of such stresses need not be discussed here. In the simple case of joints in more or a later chapter. Here it is only necessary to say that the inclination of limestone beds may serve to give preferential distributions of some joint and fault systems and of many of the lithological weaknesses in limestones as discussed in the preceding sections.

FAULTS

Faults are fractures with displacement, which may bring beds of different lithology in contact with each other, sometimes with a thoroughly porous breccia between them. The possibilities for speleogenetic initiation along faults are thus obvious. Furthermore, most faults are not perfect

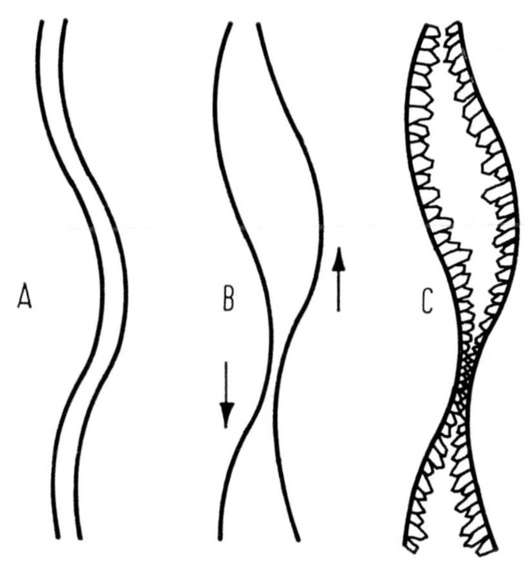

FIG. 2.9. Diagrams of a slightly irregular fault plane. A = before movement. B = after movement has brought bulge against bulge, with lens-shaped voids (rifts) in between. C = after mineralization has lined voids with crystals, and filled in narrow sectors.

geometrical planes, and displacement often brings the bulges on one side against the bulges on the other, giving tight contacts with little permeability, whilst the opposition of concavities may leave relatively open, or even breccia-filled, spaces (Fig. 2.9). "Rift" chambers developed along faults may thus show tight closures at each end.

Faults fall into three main groups: normal; wrench or tear; and reverse or thrust. Normal faults result from tensional stresses, and they are commonly inclined at about 45° to 60° from the horizontal though they may be vertical in massive limestones. They frequently occur in parallel or conjugate sets. In normal faults the displaced block is thrown down in the same direction as the inclination of the fault plane. Wrench or tear faults show lateral displacement, by which the opposing blocks of country have moved sideways relative to each other, sinistral if in a left-hand sense, and dextral if right-handed; in horizontal beds the lateral movements may be difficult to detect. Wrench fault planes are commonly

PLATE 2.8. A cave developed in a thrust fault. Traligill Swallet, Sutherland, Scotland.

nearly vertical. Reversed faults incline in the opposite direction to the displacement, indicating compressional stresses. Reversed faults of low angle, less than about 30° from horizontal, are known as thrusts, and they cause a duplication of the stratal sequence in a vertical profile.

Within the limits of a cave passage it is often impossible to match up beds across a suspected fault, so that vertical fracture planes may not be determinable as to which type of fault or joint is present.

An area which has been subjected to different stresses during its geological history may have a complex of any of the above classes of faults, providing a variety of pathways for water movement and complex cave profiles. Wrench faults have controlled the form of some caves in Yorkshire, such as Meregill Hole; they are more common in South Wales, where they have controlled much of the form of Dan-yr-Ogof, especially the Great North Road. Thrust faults have been utilized by cave streams in north-west Scotland, e.g. Cnoc nan Uamh and Traligill Caves (Plate 2.8).

Many fault planes, when bared by quarrying, show grooved or polished surfaces where the two walls have ground against each other; such surfaces are known as slickensides, and differential solution along them may have produced some vertically disposed anastomoses. Fine examples of slickensides may be seen in the Great North Road of Dan-yr-Ogof.

Fault movements often result in the crushing or grinding of rock fragments to form a zone or band of breccia rather than a clean-cut fault plane. Such breccias are commonly cemented by calcite, but sufficient permeability may remain to be an important factor in cave development.

One of the prime effects of faults is the displacement of beds of similar speleogenetic character away from each other (Fig. 2.10a). The fault

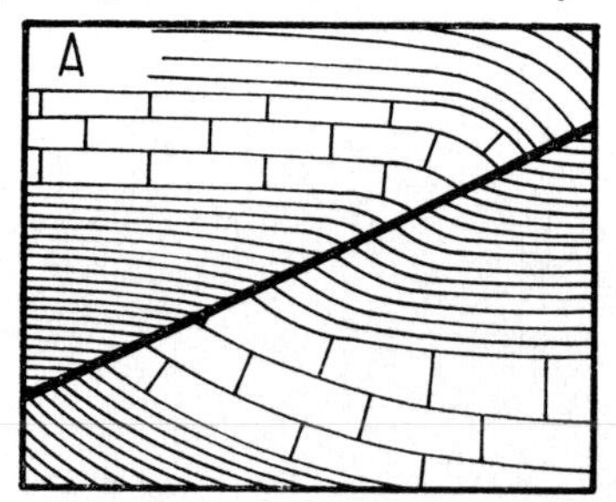

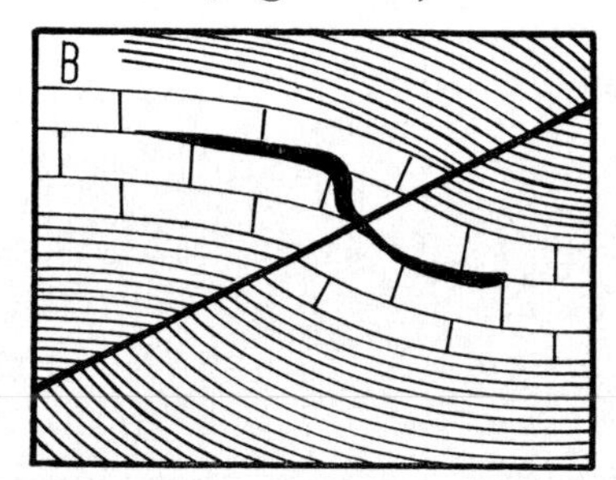

FIG. 2.10. Diagrams to show reversed (thrust) faulting and its effect on a limestone bed. A = with other strata forming a seal between limestones. B = with limestones in contact across the fault showing the possible cave profile.

plane provides a migration route to restore continuity in some cases, giving a stepped profile (Fig. 2.10b). Elsewhere faults may bring different beds, both with some speleogenetic characters, into opposition; the resultant cave may then have a marked change in size of detail of shape and features when the fault is crossed (Fig. 2.11). Faulting may so

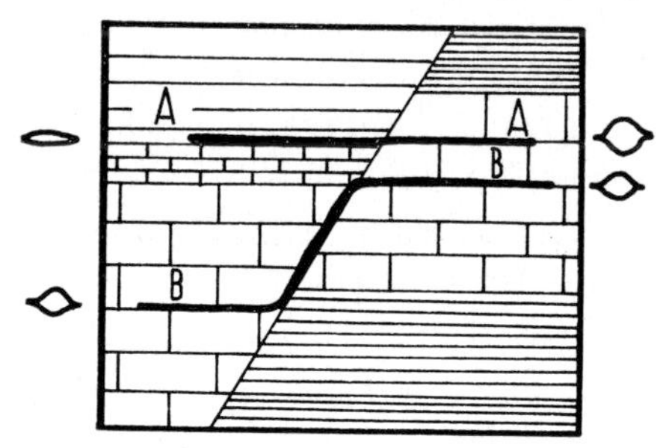

FIG. 2.11. Limestone beds with varying lithology affected by a normal fault with two possible cave profiles across it. A-A = the cave passage crosses the fault horizontally into beds of different thickness with a consequent change in the cross-sectional profile. B-B = the passage steps down the fault and continues in the same bed with the same cross-section, but an inclined rift develops in the fault plane.

affect the relationship of a limestone mass to adjacent shales that nearby cave systems exhibit marked differences in morphology. In Fig. 2.12 system A may show mainly vadose features while system B is chiefly phreatic, though both are active at the same time.

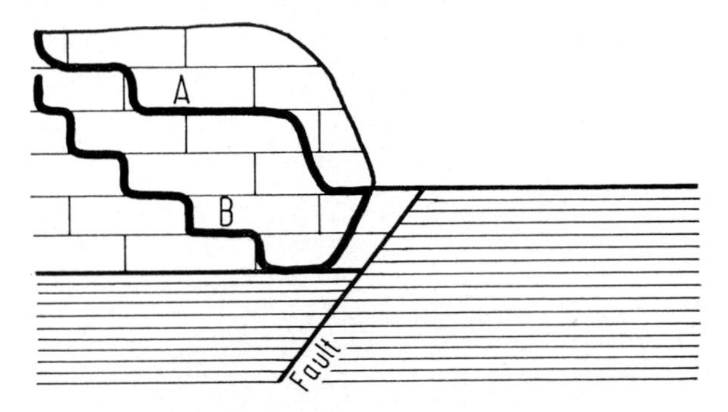

FIG. 2.12. Profiles of two cave systems with common catchments and resurgences, with their forms showing differences owing to fault control. More rapid erosion of the shale on the upthrown (right-hand) side of the fault has permitted draining of system A with consequent development of vadose features, while system B is still partly phreatic.

MINERAL VEINS

A fracture system in a limestone mass deeply buried within the Earth's crust may be utilized by mineralizing fluids, so that joints and faults become infilled with minerals such as quartz (SiO_2), galena (PbS), fluorite (CaF_2), baryte ($BaSO_4$) and calcite ($CaCO_3$) as the fluids rise and cool. Crystal growth into fault cavities frequently fails to fill them, and crystal-lined channelways may be left ready for take-over by speleogenetic streams at a later date (Fig. 2.9c; Plate 2.9). Solution may remove the calcite but the others are effectively insoluble and may remain in the conduits as residues, though they are easily abraded.

Both faults and mineral veins may provide conduits for water movement which penetrate far below a stratigraphically or lithologically controlled drainage system. Deep sumps and risings are to be expected in such systems and it is unlikely that they can be explored by diving (Fig. 2.12B). The very deep penetration over a long distance by such artesian waters may mean that they take several years to pass through a system and the depth may result in an increase in temperature so that a hot spring emerges at the surface, often with an excess of mineral matter in solution which is rapidly deposited as tufa, or sinter. One recent study has suggested that the Matlock warm spring waters have spent 15 years

PLATE 2.9. A resurgence in a mineral vein. The "boil-up" in Magpie Sough, Derbyshire.

passing through the system. Other warm waters may result from the exothermic reaction of the oxidation of pyrite heating the water, whilst others may be the surface expression of juvenile waters, i.e. those emerging for the first time from within the earth, or of waters heated by a dormant volcanic system. Though such waters, if near boiling temperatures, may be utilized for geothermal power plants, there have been few attempts to study such systems, past or present, from a speleogenetic point of view.

Mineral veins filled with insoluble minerals such as quartz may form impermeable barriers. Even calcite-filled veins in Derbyshire may sometimes be hydrological barriers; waterfalls over upstanding calcite veins occur in the Speedwell Cavern.

STRATIGRAPHY

The sequence of limestone beds, shale partings and other rock layers seen in a cave system is the stratigraphic record of the events which provided the rocks concerned. A greater understanding of that cave system in relation to its neighbours may be obtained by comparing the sequences of beds in adjacent caves (Fig. 2.4). This may provide important evidence of which beds were more favourable in speleogenesis, or of the sequence of events in development of the caves. Sometimes, comparison of the sequences in nearby caves may lead to the discovery of hitherto unsuspected linking passages. The sequences of beds in adjacent caves will not usually be exactly the same, and a study of the relationship may indicate why two caves differing in detail have developed side by side.

MECHANICAL PROPERTIES OF LIMESTONES

Masons commonly speak of soft or hard limestones when referring to their resistance to a saw; this is a reflection of the degree of cementation and the grain size. Architects often measure the crushing strength of limestones used in buildings—will the lower blocks withstand the load of the blocks above? Such properties have a bearing on the rate at which caves are enlarged by abrasion on the one hand, and on the degree to which the roof or walls of a cave can hold up overlying beds. Little attempt has been made to study either of these properties as far as they affect cave development, though breakdown of a cave roof due to failure under the latter stress is probably quite common. Breakdown of a roof is also controlled by joint distribution and bed thickness—the closer the joints and the thinner the beds the more likely is breakdown (Plate 2.10).

In cool temperate and cold climates a further feature of mechanical nature is the resistance to frost. Although most often seen on the surface, the effects of repeated freeze and thaw are common in cave entrances as shown by much "shattered" rock in the roof and walls and a pile of spalled-off fragments beneath. In an archaeological context such a pile may be an important climatic marker.

PLATE 2.10. Weathercote Cave, Yorkshire: an example of a collapse of well-jointed and thin-bedded limestones.

The resolution of stresses of a large cave passage may not always mean breakdown of the roof; instead, slabs may spall off the walls, occasionally leaving a conchoidal fracture surface.

Temporary loading by ice during the Pleistocene may have caused more breakdown than would be expected from present-day circumstances. An example is the widespread slab spalling off the walls of the entrance passages of Ogof Ffynnon Ddu II, where the present cover of strata is thin but the load of ice may have been some hundreds of metres in thickness. Deeper within the same cave, pillars in thin-bedded and closely jointed limestone show clear evidence of crushing under the superincumbent load, though whether this was initiated by the former cover of ice, or by the removal of hydraulic support when the transition from phreatic to vadose conditions took place, or whether it is simply due to the present lithostatic pressure, is not known.

CARBONATITES

An uncommon variety of volcanic rock is that in which carbonate minerals are an important, if not the dominant, component. Only one volcano is known to be erupting a carbonate lava today; it is Oldonyo Lengai in East Africa, with a lava of sodium carbonate, washing soda. Others in the past appear to have erupted lavas with calcium carbonate as an important constituent, and these carbonatite rocks form the eroded remnants of ancient volcanoes and intrusions, particularly in East Africa. Though no descriptions of cave systems in such rocks have been published, some of the hills exhibit surface karst features, and have drainage disappearing underground. From a study of the literature it appears that some archaeological sites, such as rock shelters, are in the entrances of solution caves in carbonatites, but speleological investigation is still required.

In some cases volcanic ashes, agglomerates or even lavas have been cemented or permeated by carbonate-rich waters soon after eruption. Subsequently karst processes have operated on these and have produced small solution caves. Other minerals sometimes form efflorescences on the walls, and in one case Glauber's Salt, sodium sulphate, is thought to be used by speleologically minded elephants as a purgative!

BURIED KARST

Most speleogenetic studies are concerned with present-day or at least geologically recent systems, but any limestone of appreciable geological age may have gone through more than one cycle of karstification, only to be buried again. For example, there is little doubt that much of the British Carboniferous Limestone was exposed during at least parts of Permian and Triassic times and suffered some karstification in a semi-arid climate. The recognition of ancient karst features in a modern karst region is difficult and controversial, but the possibility of the control of present-day karst drainage by fossilized systems should not be overlooked. Karstification is widespread in the United States at the base of the Pennsylvanian (Upper Carboniferous) where it rests on older limestones. As well as distinctive surface features, such as exhumed solution collapses, the subsurface existence of a buried karst has been proved in a number of oil-fields.

The iron ore bodies on the fringes of the Lake District are probably partly buried karstic features of Triassic age, with some of the detailed features obscured by metasomatic replacement of the limestone by haematite, and most of the evidence destroyed by mining.

CAVERNOUS CONGLOMERATES

Conglomerates are rocks composed of pebbles derived from older rocks and cemented together. Either some of the pebbles or some or all of the cement may be calcite and thus susceptible to solution. Void space between pebbles in conglomerates is often high, so that rapid movement of groundwater may take place and solution may be at a premium. This may result in solution throughout the whole conglomerate mass and gradual disaggregation and collapse, so that caves are not formed; but in some cases preferential channels may be opened, particularly if the conglomerate is jointed. Examples are in Wookey Hole, Somerset, of which the tourist cave is all developed in the basal Triassic Dolomitic Conglomerate, and in the Harpan River Caves of Nepal, as recently described by Waltham (1971).

GYPSUM

Whole rock formations may be composed of gypsum ($CaSO_4 . 2H_2O$) or to varying degree by its anhydrous equivalent anhydrite ($CaSO_4$) deposited by the evaporation of saline lagoons. Being soluble like limestone, these can develop gypsum karst landscapes, of which good examples are known in the Middle East. In wet temperate climates gypsum is so much more soluble that it rarely forms any topographic features, but solution caverns can be found beneath the surface during mining operations. Their features are not unlike those of phreatic solution caverns in limestone, though, as gypsum has less mechanical strength, collapse is more common. In hot climates, particularly those with strongly seasonal rainfall regimes, gypsum may occur as ranges of hills with topographic features very much like those of limestone massifs, and both vadose and phreatic cave systems may be found. Some of the characteristics of gypsum caves have recently been described by Hensler (1968) and by Kempe (1972).

ROCK SALT

The mineral halite (NaCl) is commonly deposited as an associate of gypsum by evaporation of saline lagoons. Being highly soluble, it does not survive for long at the surface in wet climates, but in arid areas like Iran salt plugs form prominent topographic features and may acquire a karst-like topography during the rare but heavy rain-storms of such desert areas. Man-made salt-karsts have been produced by long-continued pumping of brines from salt formations, as in Cheshire, where solution collapses are not unknown.

CONCLUSION

Many hypotheses of cave origin have taken as their starting point a homogeneous mass of limestone newly uplifted above sea-level. By assuming homogeneity, the facts of the variability of lime sediments and of diagenetic textures are ignored. The vagaries of grain and pore size, of differential permeabilities, of the nature of bedding planes, stylolites, joints, faults, mineral veins, buried karsts, may all have fundamental significance in controlling the place, time and rate of speleogenesis. No cave system can be fully understood until these factors have been analysed. Thin-section studies of limestone textures under the microscope have been all too few in relation to cave-forming processes. Studies of joint patterns and densities are similarly few and far between. Of the other factors, only a scatter of isolated studies is known.

Since no two geological situations are ever exactly the same, a single generalized theory of speleogenesis becomes unreal. Many of the factors will be common to many areas, but their relative importance and interplay will depend on the geological situation. Every area will thus have its own unique speleogenetic history.

At the time of initiation all speleogenetic pathways are in the phreatic zone. A hydraulic gradient is required to maintain flow; chemical reactions between moving waters and wall rocks are then required to allow enlargement of one or more pathways to cave proportions. The geological factors outlined above to a large extent control which pathway becomes a cave. The topography, particularly the erosion surfaces above the limestone mass, is significant only in providing input and outfall and a hydraulic gradient to the system. As recent studies of the concept of water-tables in limestones have shown, preferred pathways may develop into vadose cave systems, whilst others, less well endowed with speleogenetic properties, remain in the phreatic zone, sometimes in limestones directly above a preferentially developed vadose passage; this is normally a direct consequence of geological features, for a homogeneous limestone will not support such vagaries of cave development.

SEDIMENTS IN CAVES

Rock fragments in process of transport from their source to their ultimate resting place in the sea may have a temporary resting place in a cave and glacial, solifluction and other mass-movement effects. The causes of interruptions within stream cave systems are usually "accidents" of erosion of the cave itself: a roof collapse may form a dam with a resultant settling-tank effect on sediments carried in by the stream. Oscillations of flow stage may give a laminated alternating fine and coarse sedimentary deposit, such as that in the Sand Caverns of Gaping Gill (Plate 2.12).

PLATE 2.12. Layered silts and sands deposited as a result of the "settling tank" effect. Sand Caverns, Gaping Gill, Yorkshire.

Gradual removal of the dam by solution may allow the stream to cut a channel through the deposit, but then a further roof fall may restore the dam and allow the channel to be re-filled. Abandonment of a passage by a stream taking a lower route may allow flood waters to overflow periodically into the old passage there to deposit sediment, often progressively finer in grain upwards.

When a vadose stream becomes laden with sediment and enters phreatic passages its velocity is reduced and the coarsest fraction is dropped. This restricts the channel so that the next smaller grain sizes are carried further inwards, and so on. In an ideal phreatic tube system the grain sizes may all be sorted out downstream by this means. If a phreatic passage is drained or abandoned the sediments should reveal the former direction of flow, e.g. if a floor with pebble gravel is so exposed it may mean that the entry point of a vadose stream was not far away. Damming of the resurgence by ice may cause the return of a passage to epiphreatic conditions with similar sedimentary results.

It is not uncommon to find passages completely choked with fluvial sediments. In Ingleborough Cave, Yorkshire, several passages are completely full of bedded gravel except for the top few centimetres which have sand or silt. This type of deposit probably represents rapidly flowing streams heavily laden with pebbles, derived from glacial outwash on the plateau above. As the channel became full, perhaps owing to an obstruction downstream or to epi-phreatic conditions, the flow was diverted and the remaining passage became a backwater suitable for the settlement of fine-grained material. Pebbles frequently show imbricate stacking, i.e. each lying against the previous one, with an upstream dip, thereby indicating the direction of flow.

A common cave sediment, often of controversial origin in the past, is clay-fill, first described from Missouri caves by Bretz (1942). In general terms this represents the transition of an active cave to a condition of relative inactivity when it can become a settling-tank for fine-grained sediment. As Bretz argued, this may be during the period between the development of a phreatic "first-cycle" cave and a vadose "second-cycle" system. Equally it may be the settling-tank effect of the phreatic part of an integrated drainage system which receives only fine-grained input; or it may be due to outside climatic influences, such as a glacial episode cutting off all but a little undermelt flow beneath the ice so that the only available material is very fine-grained. Alternatively, a fine-grained fill may represent the washing through of a loess cover over the limestone mass. Sometimes caves close to a limestone-shale contact contain ochreous clay which is no more than the washed-in weathered shale subsoil. A detailed study of clay mineralogy may be necessary to distinguish the various types of clay-fill origin. Cave clays are commonly yellow-brown but may be red if residual iron-rich terra rossa soils are washed in from the surface. Blue-grey clays may result from the reduction of the iron to the ferrous state.

A special type of exogenetic material is of biological origin. Included are such deposits as bat and bird guano, the remains of insects and beetles which feed thereon, bone gravels washed in from the surface, bone-rich talus cones beneath pitfalls, bone accumulations from occupation by man or animals, and peat derived from surface deposits. These may occur as admixtures with other deposits. Peat and guano may take part in chemical occur beneath a hole in the roof, possibly caused by collapse. A mixture of roof blocks, soil and subsoil may form a heap beneath, which gradually builds up as further debris is washed or blown in, and as animals fall into the pit-fall trap. Temporary habitation by trapped but living animals may cause further mixing of the talus. A fine example is that in Joint Mitnor Cave in Devon, which formed the basis of that so ably described by Sutcliffe (1970) (see Chapter 13).

If a cave mouth opens in the bank of a river, the sediments may be affected by pouring in river-borne material, which will decrease in grain size inwards, in contrast to the downstream size-grading seen in phreatic tubes. Channelling and re-sorting of existing sediments may take place if a river breaches an existing cave mouth.

Some caves open directly on the sea shore and may thus contain marine deposits such as shell sands or beach cobbles. Other caves discharge their streams directly from the sea floor and when they become inactive they may be filled with marine sediments of various types.

The deposition of calcium carbonate cement in a cave sediment is an extension of the stalactite formation process and is discussed elsewhere in this book. It may be noted, however, that cemented cave sediments indicate a change in environmental conditions, which may be due to climatic hydrological or chemical factors. Such cementation must be taken into account in studying the history of a cave system.

Sedimentary deposits in caves can thus tell much about the history of a particular cave, such as the possible source areas of the inflowing streams, whether or not the area has been glaciated, submerged beneath the sea and so on. But there are many pitfalls in the study of such cave sediments. The possible modes of formation of the clay-fill have already been discussed. Laminated silts and clays may result from glacial ponding and so compare with the varved clays of proglacial lakes, or they may simply result from a roof fall and settling tank effect on a stream with variation in flow and sediment-carrying power. Boulder clay, periglacial solifluction deposits and aqueous sludge deposits can look very similar and may in fact be indistinguishable at times. The presence of palaeontological material such as scattered bones may help in interpretation but on the other hand, the bones may have been derived from an earlier deposit, transported along the cave and then buried again. Similarly vegetable debris in cave silts may be that of contemporary plants, or it may have been derived from an ancient peat deposit above the cave. Carbon-14 dates are meaningless in such cases.

In practice there has been little description or analysis of cave sediments (except for Sweeting, 1972) apart from those bearing archaeological remains. Much more descriptive work is needed before more accurate conclusions can be drawn and diagnostic tables set up. Size ranges of pebbles and sand, roundness tests and variations in stream flow stage need to be related to each other in as many areas as possible, and in turn they must be compared against a background of different source rocks in different climates, past and present.

REFERENCES

Bathurst, R. G. C. (1972). *Diagenesis of Limestone Sediments*. Elsevier, Amsterdam.

Blatt, H., Middleton, G. and Murray, R. (1972). *Origin of Sedimentary Rocks*. Prentice-Hall, New Jersey.

Bretz, J. H. (1942). Vadose and phreatic features of Limestone caverns. *Jour. Geol. 50*, 675-811.

Chilingar, G. V., Bissell, H. J. and Fairbridge, R. W. (1967). Carbonate rocks. Vols. 9A and 9B of *Developments in Sedimentology*. Elsevier, Amsterdam.

Coase, A. C. (1972). *Limestone Landforms* and *Caves*. Film-strips and notes. Diana Wyllie Ltd., London.

Doughty, P. S. (1968). Joint densities and their relation to lithology in the Great Scar Limestone. *Proc. Yorks. Geol. Soc. 36*, 479-512.

Drew, D. P. and Smith, D. I. (1972). An unconformity cave. *Proc. Univ. Bristol. Speleo. Soc. 13* (1), 89-103.

Folk, R. L. (1959). Practical petrographic classification of limestones. *Bull. Amer. Assoc. Petrol. Geol. 43*, 1-38.

Ford, T. D. (1971). Structures in limestones affecting the initiation of caves. *Trans. Cave Res. Grp. G.B. 13*, 65-71.

Friedmann, G. M. (1969). *Depositional Environments in Carbonate Rocks*. Soc. Econ. Pal. Min., Spec. Pubn. No. 14. Tulsa, Oklahoma.

Ham, W. E. (ed.) (1962). *Classification of Carbonate Rocks*. Memoir No. 1. Amer. Assoc. Petrol. Geol. Tulsa, Oklahoma.

Hatch, F. H., Rastall, R. H. and Greensmith, J. T. (1971). *Petrology of the Sedimentary Rocks* (5th edn). Murby, London.

Hensler, E. (1968). Some examples from the South Harz region of cavern formation in gypsum. *Trans. Cave Res. Grp. G.B. 1*, 33-44.

Kempe, S. (1972). Cave genesis in gypsum with particular reference to underwater conditions. *J. Brit. Speleo. Assoc. 6*, 1-6.

Newson, M. D. (1971). The role of abrasion in cavern development. *Trans. Cave Res. Grp. G.B. 13* (2), 101-108.

Picknett, R. G. (1972). The pH of calcite solution with and without magnesium carbonate present and the implications concerning rejuvenated aggressiveness. *Trans. Cave Res. Grp. G.B. 14* (2), 141-150.

Sutcliffe, A. J. (1970). A section of an imaginary bone cave. *Studies in Speleology*, *2* (2), 79-80.

Sweeting, M. M. (1972). *Karst Landforms*. Macmillan, London.

once the ground became permeable again. More recently Morgan (1971) has suggested that some of the smaller peripheral tributary valleys were initiated under these conditions. Another popular view is that dry valleys result from collapsed caves, Cheddar Gorge being the favourite example; but dry valleys outnumber caves, and there are few signs of progression from collapsed cave to dry valley. The plan of dry valleys also rarely resembles that of a cave system. A third possibility is that streams originating on a cover rock have been rejuvenated and cut down into the limestone. Where the cover has been largely removed water will sink into the

PLATE 3.5. Dry valley, Hartle Dale, Derbyshire.

limestone and many tributaries will cease to flow. As the main streams cut down the groundwater will become adjusted to the lowest streams and its level will lie below the floors of the side valleys. The writer has suggested this origin for many of the Derbyshire valleys (Warwick, 1964). A knick-point working upstream may also lower the water-table below the level of the upper part of a valley so that it may become dry above the point where the river first becomes adjusted to its new level (Warwick, 1960) (see Fig. 3.4). There are also transitional valleys where springs may break out in the valley floor in winter after heavy rains, known as *winterbournes*. Glennie (1960, 1962) has described the breaking out of the Hertfordshire Bourne. Some dry valleys terminate in very steep heads, suggesting a dried-up spring. Probably the origin of most dry valleys is complex, with the effects of successive water-table lowering being interrupted by touching up during glacial or periglacial conditions. Some collapsed caves occur, such as the Rak Polje in Yugoslavia, where each end of the depression is a rocky gorge, bridged in part by a natural bridge; Skoksjanske Jama in the same area is an even more spectacular example of collapse. But such features do not appear to be very common. For other recent contributions in this field see Small (1961, 1964), Kerney *et al.* (1964), Roglić (1964), Brown (1966) and Lewin (1969).

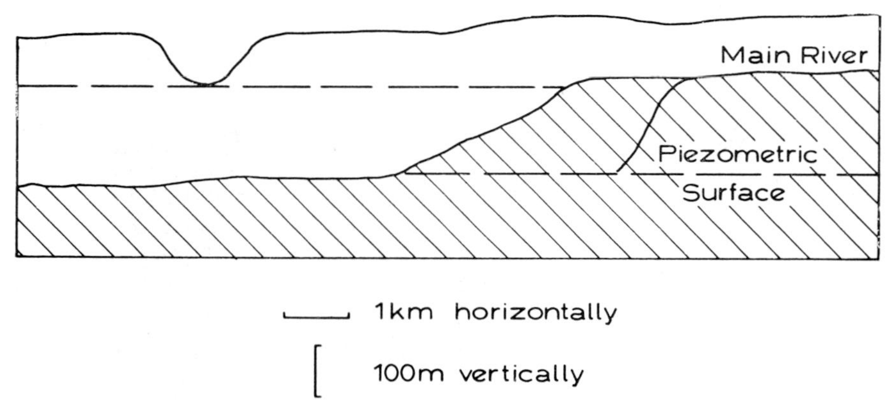

FIG. 3.4. The effects of knick-point recession upon the water-table (piezometric surface) and the disappearance of valley-floor streams.

LIMESTONE SCENERY—CLOSED DEPRESSIONS

Enclosed depressions of various types form other highly characteristic features of karstic regions, ranging in length and width from a few metres to kilometres and up to hundreds of metres in depth. The generally accepted terminology for these is based upon Serbo-Croat nouns which originally had a less specialized meaning. Gèze (1973) has provided an up-to-date glossary of the French terminology. The commonest depression is the *doline*, roughly circular in shape and like an inverted cone in form, usually a few metres in diameter but extending to a kilometre or so in width (see Fig. 3.5, Plates 3.6 and 3.7). These are called *sinkholes* in America and by some British authors, though *shakehole* (modified from an old Derbyshire lead-mining term) is more commonly used in caving circles. The term *swallowhole* has also been used (Thomas, 1954b), but this is best reserved for water sinking underground. The term shakehole implies an origin by collapse, yet some examples found on bare karst appear to be due to some other mechanism, such as solution. For such reasons doline, as a descriptive term, is to be preferred. Where an origin can be inferred from field evidence it is justified to speak of collapse and solution dolines. Dolines on covered karst most frequently show signs of instability on their sides, with small scars of bare soil above "sheepwalks", though occasionally bare rock may be visible (see Coleman and Balchin, 1959; T. D. Ford, 1967). At Crveno Jezero on the rocky side of Imotski polje in Yugoslavia almost vertical sides lead down to scree slopes and a dark pool over a hundred metres below the level of the plateau. Thomas (1954a) has described examples of Millstone Grit being let down into

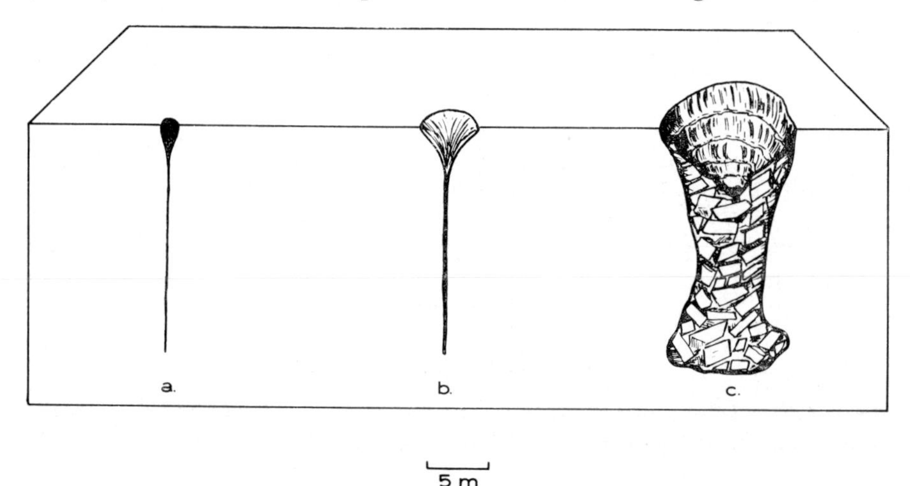

FIG. 3.5. The origin of dolines. (a) A pipe. (b) A solution doline. (c) A collapse doline.

cavities in the limestone, which remains long after the surrounding grits have been removed by erosion. Other examples may be seen in quarry exposures (Thomas, 1973). Probably many examples are due to a combination of localized solution and collapse (Gams, 1965). Dolines often occur in lines, at the edge of a till sheet or over a cave, though the surface pattern does not always coincide with an underlying cave system. Care has to be taken to avoid mistaking old mine-excavations or walling-stone quarries for dolines, though the irregularity of outline provides some guidance. Small shallow depressions may become filled with debris and can only be seen in cross-section in excavations; these are relatively common in chalk districts where they are known as *pipes*. Dolines may develop and combine, and some large ones may have relatively flat floors. Cvijić used the name *uvala* to describe large complex dolines, but Yugoslavian karst morphologists find it difficult to agree on a precise

PLATE 3.6. Collapse doline, Orton Fell, Cumbria.

PLATE 3.7. Solution dolines, Velebit Mountains, Yugoslavia.

definition so that this term should perhaps be abandoned. The larger dolines grade into the large flat-floored depressions known as *polja* (singular, *polje*) the base being covered with alluvium of non-limestone origin (see Plate 3.8). The sides are usually steeply sloping, but rarely cliffed, and may be breached by inflowing streams which sink into the polje floor (see Plate 3.9). Most of them are elongated, but there is a wide variety of plan. One sub-type is the *Randpolje*, where one side is composed of non-limestone rock. Some contain permanent streams which pass into cave systems, or the streams may be intermittent. Many of them are liable to flooding, some near the coast being permanently in this condition, yet others being always dry. Examples abound in Yugoslavia, but they are not uncommon in France and Italy. They are rare in Britain, but examples have been described from Ireland (the Carran polje (Sweeting, 1953)) and around the head of Morecambe Bay (Ashmead, 1974).

CLASSIFICATION OF CAVES

Caves form the subterranean class of karstic phenomena; but a brief acquaintance soon reveals a multitude of types. To a caver or potholer the distinction between a wet cave with its associated stream to be negotiated and a dry one fed only by drops and trickles of water is a very important one. The relationship of the cave to the surface hydrology is of more interest for the hydrologist. Caves which have streams flowing into them are often called influent caves; correspondingly, effluent caves have streams flowing from them. There are surprisingly few caves where one can enter with a stream and return to the open air at a lower level. These could be called through caves, or occasionally are referred to as tunnel caves; Mas d'Azil in the French Pyrenees is a good example, which is also utilized by a main road. There are also systems which do not have streams flowing into or out of them and which have been discovered by miners or quarrymen, though some contain water. All of these classes have their dry equivalents which have ceased to be conduits for the passage of water. In fact, most caves may be regarded as active or dry natural drains. In popular terminology a *cave* is essentially a horizontal cavity and a *pothole* or *pot* is one which contains vertical drops or pitches, often, but not necessarily, at the entrance. Such cavities are commonest in NW Yorkshire, where the term originated, but now it is in widespread usage within the British Isles. Several caves such as Minchin Hole include the word "hole" in their name to mean a cave, and the Irish *Poll* (*Pol* or *Poul*) has the same meaning (J. C. Coleman in Tratman, 1969, p. 220). The Welsh *Pwll* (lit. a pool) is also used in the same way though *Ogof*,

Plate 3.8. Planina Polje in flood, 17 September 1965, Yugoslavia.

Plate 3.9. Swallets (ponor), south-eastern part of Lika Polje, Yugoslavia.

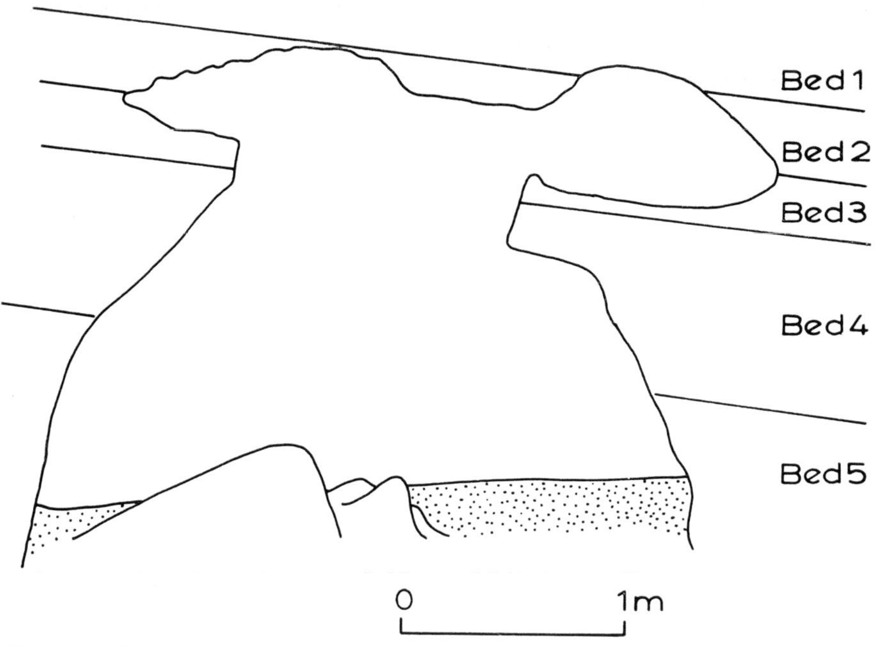

FIG. 3.9. The development of a cave passage from three separate tubes (after E. A. Glennie, 1948).

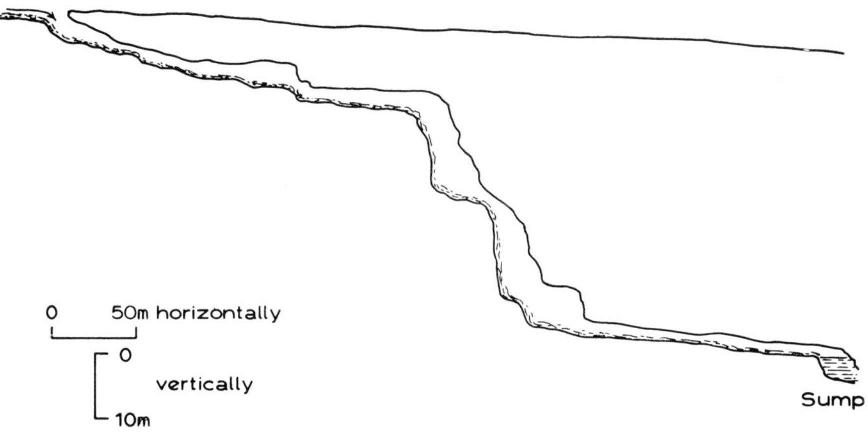

FIG. 3.10. Diagrammatic long profile of a simple influent cave.

in Goyden Pot in Nidderdale, Yorkshire, and in the upper part of Ogof Ffynnon Ddu I in Breconshire. Some cross-sections show an upper bedding plane development cut into by a stream passage.

The plan of a cave may be modified by its stream having formed a more efficient downward path, leaving its old course dry. This may be in the form of a simple "ox-bow" passage which has become by-passed or it may consist of a network in which the most favourable route has been adopted by a cave stream. There are other cases where there is no flowing water. For whatever reason a cave passage has been abandoned

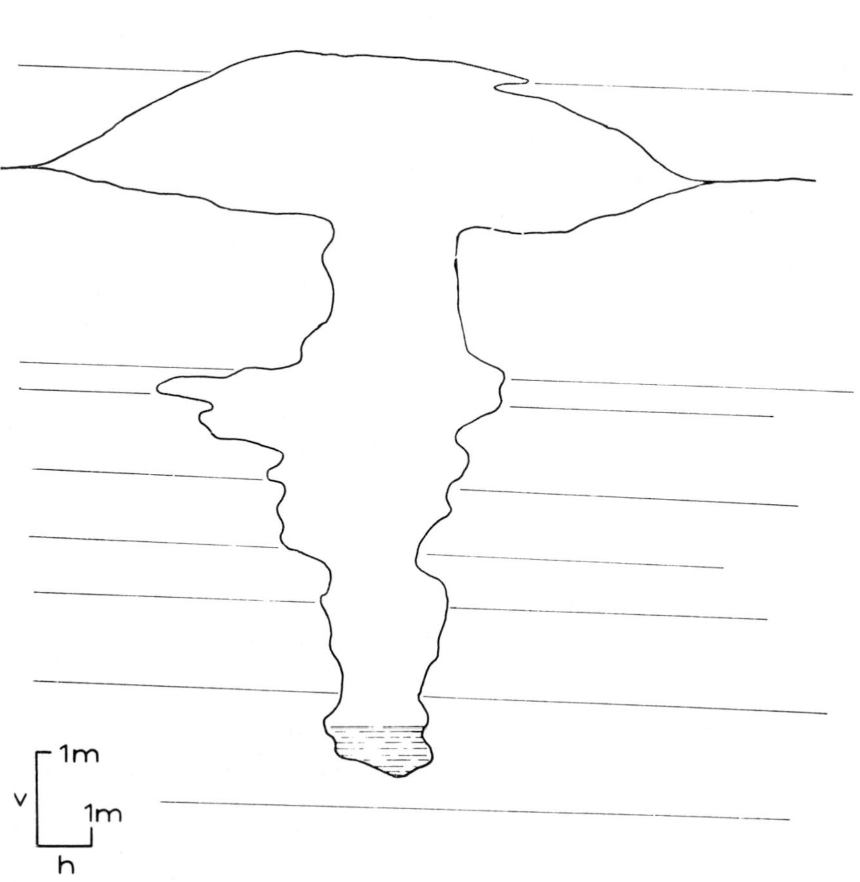

Fig. 3.11. Cross-section of a bedding plane passage which has been developed by a stream cutting into its floor.

it becomes a zone of accumulation. Fine particles may reach it by seepage along cracks and down walls, or in times of flood, water may back up into it and mud settle out. Some of the deposits may even have been laid down when the cave was occupied by water. Often these muddy accumulations, usually reddish in colour, are overlain by crystalline deposits of calcite or other minerals in the form of flowstone, stalactites and stalagmites—in other words, cave formations (see Chapter 8). Such formations may also cover blocks of limestone fallen from the roof and walls. Sometimes these abandoned sections are referred to as fossil passages, though true fossils are very rare in their contained sediments, being largely confined to cave entrances, especially of old effluent caves.

On the above basis caves may be classified in the following way, though the classes are not mutually exclusive and the list is not intended to be exhaustive:

(*a*) Simple influent caves, some of which possess a dendritic plan, becoming more simple downstream.
(*b*) Complex influent caves where a stream enters a pre-existing solutional network of passages and chambers.
(*c*) Cave systems with little or no direct connection with the surface.
(*d*) Simple effluent cave systems, including limestone springs.
(*e*) Complex effluent caves.
(*f*) Through caves.
(*g*) Flooded systems which may only be inferred indirectly or be explored by diving.

In all except the last case it is possible to have both dry and active caves, indicated by the presence or absence of running water. There are others which are only active seasonally.

Influent Caves

A surface stream flowing over limestone may gradually lose water until it ceases to flow, though more usually it disappears at a specific point, usually among boulders. Others pass abruptly underground via the mouth of an open cave, though this may have been dug out by cave explorers. If the cave takes all of the flow, the valley ceases to be deepened below the *swallow hole* or *swallet* and often the gradient increases and the stream passes underground beneath a steep slope or even a small cliff of bare limestone at the head of a *blind valley* (see Plate 3.10). The cliff may be only a metre or so high, but some reach tens or hundreds of metres. Such sites have often proved to be lucky, revealing an extensive cave system after a short dig, as at Swildon's Hole on the Mendips. Yet others, like Waun-fignen-felen, involving the construction of a timbered shaft through 20 m or more of loose rock, did not reveal their secrets despite great efforts. Influent caves are frequently most complicated near to their entrances with many minor feeders from the surface. Downstream their plans are usually simple, terminating in a narrow bedding plane or a pool.

PLATE 3.10. Blind-headed valley and swallet, Perryfoot sink, Derbyshire.

Their streams may also sink into an accumulation or "ruckle" of boulders. In NW Yorkshire influent caves usually begin with a gentle gradient and then plunge vertically down a steep pitch, resuming their gentle gradient once more (see Fig. 3.10). Sometimes the roof may collapse over a cave and afford an alternative entrance, e.g. to Long Churn Cave and more spectacularly at Alum Pot, over the vertical part of the system. Influent caves are usually situated well above the main valley floors and absorb streams coming off impermeable rocks overlying the limestone. Ingleborough is ringed with such caves, another series lies below Rushup Edge in Derbyshire.

Another group of influent caves occurs in the valley floors, but usually above a knick-point or steepening of the valley profile, as at Redhurst Swallet in the Manifold Valley, N Staffordshire, or in the Nidd Valley, Yorkshire, where Manchester Hole, Goyden Pot and New Goyden Pot absorb the river (Ford, 1963a). When the capacity of such influent caves is exceeded, the surface channel takes the excess flow, but in dry weather the bed remains empty. Rivers may even take permanent short-cuts through incised meander cores, as the River Lesse does at Han-sur-Lesse via the Trou d'Enfaule, reappearing inside the famous Grotte de Han. On smaller streams such a traverse may be only a few metres in length, as in the Little Dale Beck above Chapel-le-Dale.

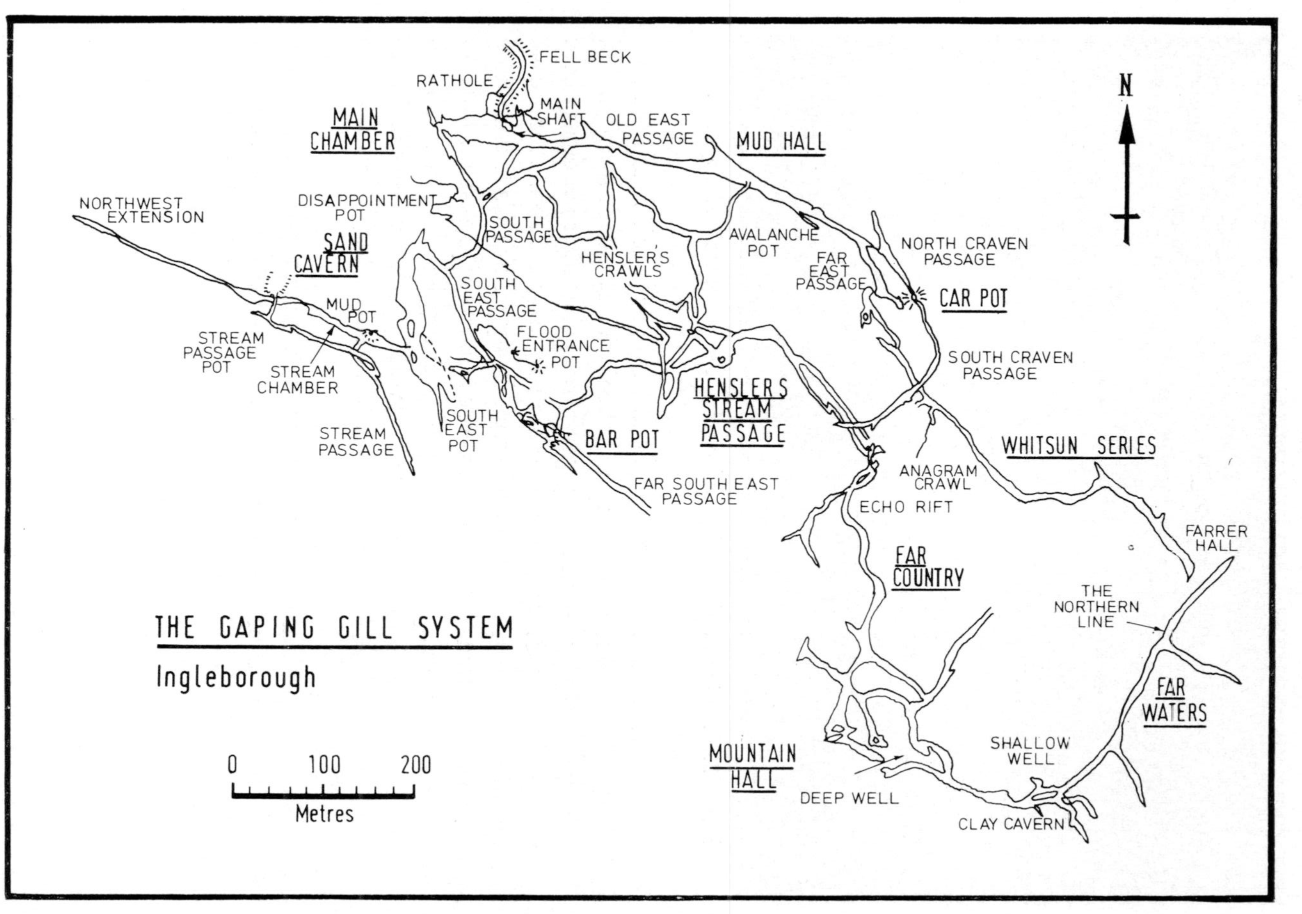

Fig. 3.12. The plan of a complex cave, Gaping Gill Hole, Yorkshire.

One of the most complex influent caves in the British Isles is Gaping Gill Hole, Yorkshire, where Fell Beck plunges 120 m into a large chamber where it sinks in the alluvial floor (see Fig. 3.12). Its plan reveals a series of passages radiating off from the Main Chamber, away from the point of entry. These passages leave at different heights and not all of them have stream-controlled floors. They are interrupted by vertical chambers which reach above and below the general passage level. Other caves with complex plans include St Cuthbert's Swallet and Swildon's Hole in Somerset, and Giant's Hole in Derbyshire. The Ease Gill-Lancaster Hole system of NW Yorkshire shows some features of a dendritic pattern becoming simpler, but its lower reaches are complicated again in three dimensions.

Caves with Little Apparent Surface Connection

Most of this group have been discovered by quarrying and mining. Agen Allwedd, Breconshire, is perhaps one of the most complex, being entered by a narrow passage revealed by quarrying, and now connected to the neighbouring Ogof Gam, the connection having been enlarged by cavers. There is a stream within the cave, but it appears to be fed by seepage through the overlying Millstone Grit and the water drains away, via an impassable pool, and feeds a spring in the Clydach Gorge which cuts back into the escarpment parallel to the northern outcrop of limestone. Tunnel Cave, Powys, is very similar, though the first and lower entrance was forced from a small effluent cave, through a heap of fallen blocks into a cave of much larger dimensions. Probably more such concealed caves remain to be entered. Another type of cave was encountered by miners seeking lead and other minerals, not infrequently full of water when first encountered. In cutting the Milwr drainage tunnel in North Wales a Z-shaped chamber was found which emptied into the adit and affected the flow of nearby springs. In the Halkyn Mine in the same district an even larger chamber 67 × 43 m and 30 m high was pumped clear of water. In Mill Close Lead Mine, Derbyshire, a still greater cavity was pumped dry at a depth of 225 m below the surface and 150 m below sea-level. Unfortunately this has been subsequently flooded and can no longer be inspected.

Effluent Caves

The simplest form of effluent cave is a spring, often rising from a round pool and fed from narrow openings in the rock floor (Fig. 3.13a). These springs are often much larger than springs from other porous rocks and are frequently called *kelds* in Yorkshire. These may occur at the side of the valley, and the outflows form short tributaries to the main stream such as Turn Dubs in the upper Ribble valley, or again they may lie at the foot of cliff at the head of a pocket valley, of which Malham Cove is a good example, though there water comes from a flooded bedding plane cave. Even simpler springs, flowing from a bedding plane, may be found, for example at Austwick Beck Head; alternatively joints may provide a channel for water to escape (see Fig. 3.13b). Cave openings with water

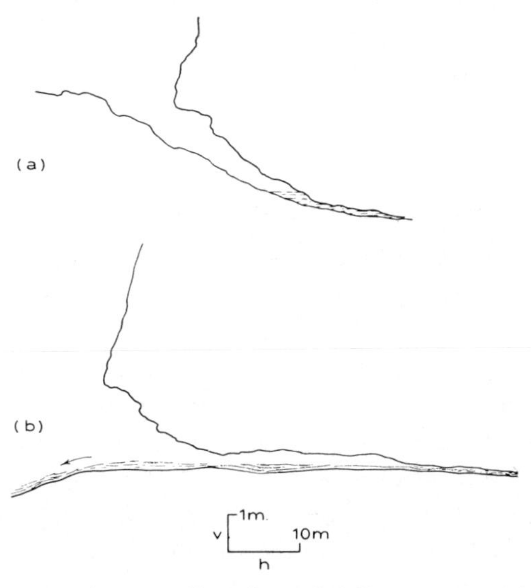

FIG. 3.13. Diagrammatic long profiles of simple effluent caves. (a) a dry resurgence. (b) An active cave with a free-air surface stream.

issuing as a stream with a free air surface are much rarer, as at Dan-yr-Ogof in the River Tawe Valley, Bown Scar Cave in Littondale, and Browgill Cave in Ribblesdale. Some caves of this type become tight and impassable after a short distance and may be subject to flooding, whilst others only flow occasionally, like Lathkill Head Cave in Derbyshire. Some effluent caves have a more complex plan, Dan-yr-Ogof for instance, where the open entrance leads back to a sump from which the water wells up, though an upper level can be explored from the stream passage. The stream is encountered again at a higher level and leads back into a more complex cave. Behind the apparently simple rising of Ffynnon Ddu, on the opposite side of the Tawe valley from Dan-yr-Ogof, lies Britain's most complex and longest cave system, Ogof Ffynnon Ddu, though access was only gained by digging (Railton, 1953; O'Reilly 1969). At Wookey Hole in Somerset the entrance leads back to a series of flooded chambers only accessible to divers, though at the time of writing a tunnel is being made to connect with one of the chambers which has a free air surface. On the other hand, in Ingleborough Cave (Clapham Cave), Yorkshire, an abandoned and much decorated passage leads back to an active stream which disappears into the cave floor to reappear at a lower level at Clapham Beck Head. This reaches a partially flooded bedding plane section which receives water by unknown passages from Gaping Gill Hole.

In addition to caves formed by marine erosion, risings and small effluent caves also occur around coastlines. Off the Yugoslavian coast one may find upwellings of cool, fresh water rising up through the sea water from springs on the sea bed. Many limestone coasts have old dry, effluent caves well above the present sea-level, usually in old sea-cliffs rising above old wave-cut platforms. Some of them formed the home of early man, such as Paviland Cave on the southern coast of the Gower peninsula (Bowen, 1970). Similar abandoned effluent caves occur above the valley floors of inland valleys cut in limestone, like Llanarmon Cave, Denbighshire, which is only about 12 m long but has a very large mouth. Such caves are usually simple in plan, though two and three-dimensional complications may be found in some cases, like Thor's Cave in the Manifold Valley of Staffordshire, or Cefn Cave in Clwydd.

Through Caves

Most caves fit into the influent or effluent category but there are a few where one can enter by an upper entrance and leave at a lower. There are only some four of these in NW Yorkshire amongst hundreds of influent and effluent caves. However, most geomorphological textbooks regard them as the norm. The simplest occur where a stream by-passes a meander, as at the Pont d'Arc in the Ardèche valley of France. Calf Holes-Browgill Cave is a fairly simple through system in Ribblesdale, Yorkshire. A more complicated and hazardous system is the Providence Pot-Dow Cave system, which cuts through the divide between two small tributaries of Park Gill Beck along a fault line; the exit lies in the valley of the upstream tributary.

Tectonic Caves

A few caves have developed as a result of the relative movement of rocks (see Fig. 3.14). Where a slab of rock is dipping steeply and slips downhill, it may rotate slightly and leave a cavity. In other circumstances limestones or other rocks may slide on lubricated clays on valley sides, leaving fissures, known as *gulls*. In Britain these are most common in former periglacial areas, especially in Northamptonshire, and the process is termed *cambering* (see Fig. 3.14c). Many gulls were filled with solifluxion debris, but open cavities are found in Ryedale in the south eastern part of the North York Moors (Fitton and Mitchell, 1950). These can be penetrated downwards for 30 m; the walls are sharp and angular and show little solutional modification. Similar features occur in the Cotswold Oolites, and have also been described from Montana (Campbell, 1968). Gèze (1955) has also claimed that cavities may form as a result of mountain-

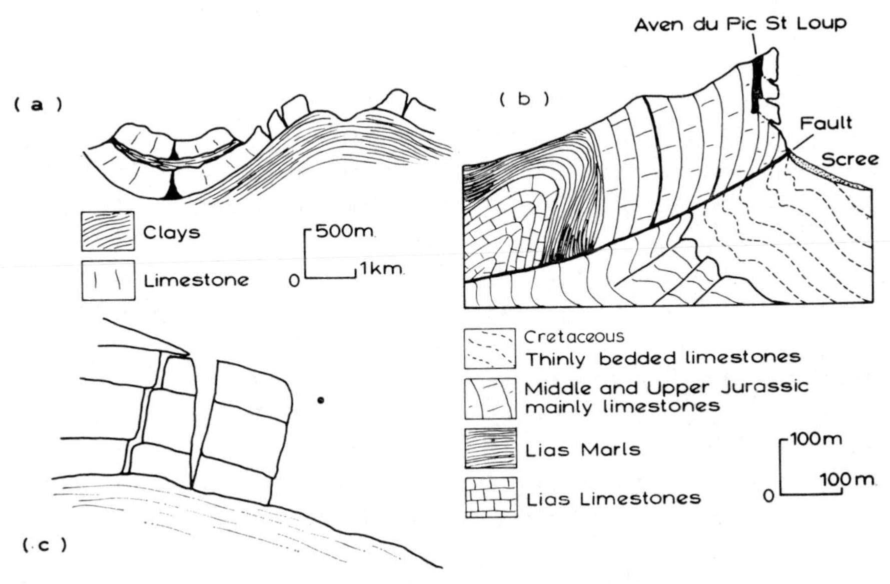

Fig. 3.14. Tectonic caves. (a) Diagram of sharply folded limestones with opened joints (after Gèze, 1955). (b) Section through Pic de Loup, Herault, France, showing gravity spread of vertical beds (after Gèze, 1955). (c) Cambered limestone blocks which have slid over an impervious clayey rock, producing "gulls", which may become filled with solifluxion debris.

building processes (see Fig. 3.14a and b). Arnberger (1954) put forward the hypothesis that in the Mammuthhöhle, in the Dachstein massif, a series of beds had slid out, leaving a rectangular-shaped passage (summarized in Groom and Coleman, 1958). Inspection of his main example reveals rounded corners and no signs of limestone being displaced on the face of a nearby cliff. Tectonic action causes rocks to crack and form joints which may facilitate water movement and solution. However, intense pressure is likely to cause voids to be filled with broken rock and under tension many cracks would be likely to form rather than one large void into which overlying rocks could collapse. Solution may develop much larger spaces which may be filled by steady collapse of the roof and walls or by a more catastrophic and sudden collapse of a mass of rock, more or less intact (see Thomas, 1973).

CAVES AND HYDROLOGY

This subject is dealt with more fully in Chapter 6 and by Newson (1971a), but it is necessary to draw together some of the threads already mentioned in passing in order to deal with problems concerning the origin and development of caves. Present-day British limestone regions show a general lack of surface water, indicating that rain water percolates downwards through the vegetation and soil and then into the joints and bedding planes of the underlying rocks. Streams flowing over impermeable rocks, on reaching limestone, often pass underground into caves, where they can be observed until they disappear into a bedding plane, a fissure, a collection of boulders or a more or less quiet pool. Some of the percolation water also enters caves as drips or trickles, but most of it moves downwards until it reaches the flooded or *phreatic* zone. The upper zone is called the *vadose* zone, where voids may contain as much air as water, or even more. In the phreatic zone most of the water storage lies within the larger openings; only a small proportion occupies the pore-spaces, and this does not flow very readily. If the macro-openings are well integrated, the theoretical surface or envelope formed by connecting the rest-levels in the joints and other openings will be level and of a low gradient. In a poorly integrated system of solutionally enlarged joints there may be greater variations in level. This surface is referred to as the *water-table* or *isopiestic surface*, and differences in altitude of this surface provide a hydraulic gradient under which water will flow towards the lower levels, though under pressure and not necessarily by a direct path in either the horizontal or vertical planes. The *groundwater*, i.e. the water of the phreatic zone, discharges via springs, effluent caves and seepages from the channel sides of surface streams. A fall of rain ultimately causes the water-table to be raised, so increasing the hydrostatic pressure and the rate of discharge until a steady state is reached. Such percolation water movement is hindered by friction on the sides of the fissures. However, water flowing through wider conduits moves much more quickly, as can be demonstrated by putting dyes in the water which may reappear within hours or a day or two. In more porous rocks with fewer solutional openings, such as chalk, the resurgence may take years.

Most valleys in upland areas in Britain provide evidence of repeated downcutting of the valley floors by the rivers flowing on them. In porous rocks one would expect the water-table to be lowered with each rejuvenation. In limestone regions springs would become dry and new ones develop at lower levels. Furthermore, since some caves are also developed at or near to river level, some relationship between older valley levels and dry caves should result. In Yorkshire, Marjorie Sweeting found a concentration of large chambers at similar heights to the erosion levels she established in that region (Sweeting, 1950). Martel (1921) also found many examples of large chambers near former river levels and beneath them restricted passages which he termed *oubliettes*. The writer found that some of the caves in the valley of the River Meuse near Dinant, notably the Grotte de Montfat, also exhibited these characteristics. In the U.S.A. Davies (1960) demonstrated that many cave systems in folded limestones were formed just below the water-table under geomorphologically stable conditions. On the other hand, passages and chambers in caves may show a strong correlation with the occurrence of impervious rock such as shale bands within the limestone. Waltham (1971) showed that in the Leck Fell Area vadose cave elements exhibited a strong relationship to the bedding planes and shale bands that was statistically significant. There is a need for more statistical analysis of the relationship between different types of caves, the geological structure and the external landscape. Valley deepening can also be achieved by the glacial scouring of river valleys, as Waltham (1970) has invoked for NW Yorkshire. It is generally assumed that such action would be quicker than normal river erosion and that the de-watering of the limestone would be also achieved more rapidly.

THE ORIGIN AND DEVELOPMENT OF CAVES

In deciphering the history of a cave the most difficult phase to identify is the *initial* one, and it can be argued that all that has occurred subsequently is *development* (Davies, 1960; Warwick, 1962b). In practice it is often impossible to separate the evidence for these phases, and hypotheses can only be advanced for testing. However, White and Longyear (1962) have produced some quantitative measure of distinction between phases. Many authorities agree that the initial stage for cave formation started with a widespread mesh of minutely open, randomly distributed, solutional courses (D. C. Ford, 1965b, p. 125). In such a network certain channels will offer an advantage to water flow in a certain direction. In soluble rocks this will induce an increased discharge compared with that along less-favoured routes, and in a given time the fissures taking most water will suffer most solution. This initial concentration of flow is probably very slow, and the rate of increase exponential. White and Longyear (1962) suggested that the critical diameter of a tube or width of an opening, when turbulent flow may commence, is 5 mm, and that subsequent flow would constitute a developed stage. The incidence of turbulence will be affected by roughness of the channel due to minute differences in solubility of its walls. D. C. Ford (1965b) considered that, in the Mendips, such an early developmental stage was also accompanied by a high water-table, with a well-developed surface drainage, and that complications developed with successive falls in the external base-level (see Fig. 3.15). The field evidence

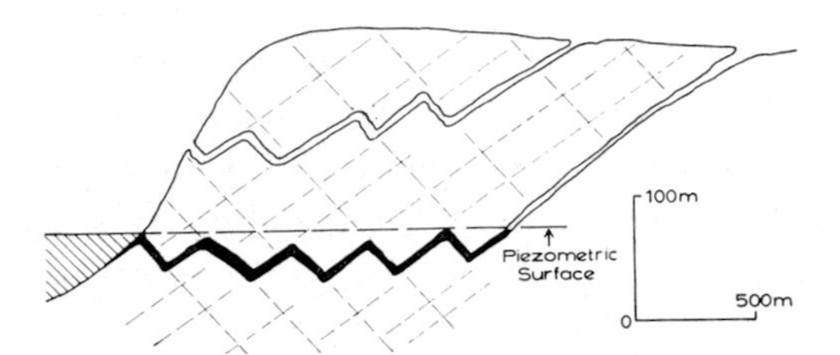

FIG. 3.15. Phreatic loops in steeply dipping rocks (after D. C. Ford, 1971).

from other areas suggests a similar history, so that this may be taken as a reasonable working hypothesis. The effectiveness of base-level lowering upon the phreatic zone is also bound up with the degree of removal of non-limestone cover rocks from the limestones. In the Mendips this process is virtually complete, but in the Peak District the limestone is flanked by impervious rocks to the east, west and north which rise above it; only to the south does the limestone stand higher than the surrounding rocks. In Yorkshire much of the surface of the soluble rocks is still covered and only part of the southern flank is exposed.

Most emphasis will be placed upon development in the post-initial phase, but rather than summarize the various theories that have been advanced on cave formation (as in Warwick, 1962b), discussion will concentrate upon influent caves, effluent caves and phreatic caves. This tacitly assumes that some solution can occur both at the water surface and below that level.

Caves within the Vadose Zone

This group of caves is the largest class in the British Isles. By definition such caves act as channels for water movement above a waterlogged zone. They are in this position because in the upland areas the water-table lies well below the surface. Many of this class of caves carry a surface stream underground in what is virtually a roofed channel. Most of them have very simple plans and long sections with a few complications. Typical examples of the simplest form of such caves are the Cullaun Caves on the western side of Poulcappa, Co Clare. But even here there is evidence of early development along bedding planes with wide, flat roofs, interrupted in places by half-tubes and rock pendants, some of which have been almost dissolved away (Ollier and Tratman, 1969). Most of the passages show entrenchment into their upper bedding-plane section by cave streams. The upper part presumably represents the initial phase and the trench secondary development after the local water-table had been lowered. Similar cross-sections may be found in NW Yorkshire, another region of gently dipping rocks. Where the dip of the rocks is steeper, caves tend to be orientated down-dip, though stream channels may be reduced to gentler gradients. D. C. Ford has analysed several of the Mendip caves and has found there is a tendency for many "dip-passages" occasionally to "jump" up a convenient joint into a higher bedding plane. He inferred that similar loops occur in the phreatic zone.

The most common morphological feature of caves in the vadose zone is the presence of *scallops*. These faceted forms cover the bottom and sides of stream channels with a series of asymmetric, concave hollows with rounded edges on their upstream sides and more pointed forms down-stream, and in section along their long axes with a steeper slope at their upstream ends (Plate 3.11). There is considerable variation in size of these scallops, though in a given locality the size is more or less uniform. Bretz (1942) took them to be indicators of the direction of stream flow and they are specially useful in deciphering the past history of passages that are now dry. Curl (1966b) has made a very sophisticated statistical study of these forms and confirms that they may be relied upon to act as indicators of direction of flow. He found some evidence to indicate that the size of the scallop is inversely proportional to the velocity. He also identified *flutes*, an allied form, produced under constant, turbulent flow conditions. These have parallel, rounded crests at right angles to the line of flow, but possess the typical asymmetric profiles of scallops, with a constant steeper slope of *c.* 71° from the horizontal. Similar forms are found on ice surfaces in caves, as in Eisriesenwelt. Curl regarded both scallops and flutes as solutional forms which are absent where abrasion is active or sediment is being deposited. Scallops are very common in the British Isles, but in some countries like Yugoslavia they are comparatively rare. Coleman (1949) first introduced the term scallop into the British literature; formerly they were often called flutes (Maxson, 1940).

Unfortunately this latter term is also used for rillenkarren on cave walls. In the Grotte de Niaux, in the French Pyrenees, in a large tunnel-like cave, scallops give way to a fluted form parallel to the direction of flow where the passage is suddenly reduced in size, producing a *venturi* through which flow was concentrated and discharge increased. The term flute is best left as a descriptive term, unless defined closely in the usage given to it by Curl.

PLATE 3.11. Scallops on a detached flake of limestone (*Photo*: G. P. Dowling).

The walls of many caves are damp and they are covered by a thin film of water which may move slowly downwards over the surface, producing rounded channels and sharp divides like surface rillenkarren. Cusp-like forms, intersecting in angular divides and a centimetre or two across, may also be seen on the roofs of caves, especially near to the entrance. These are symmetrical in cross-section and may be due to condensation. Possibly as the film breaks with conditions becoming drier, solution is concentrated in the water patches remaining, until they become embedded and the hollow is perpetuated. They are allied to the forms described above from lake margins, and possibly also to vermiculations. Where temperature changes at a cave entrance are greater than those associated with the formation of dew, freezing may occur and ice-crystals form in cracks which, when melted, may cause a flake of limestone to fall or at least to become loosened. This produces a very rough surface with minor angularities a few centimetres long and wide. High mountain caves often show this and many cave entrances in Derbyshire have layers of fine angular gravel, containing few animal bones; those which do occur

with frost action. In the past this process has been more active than today and the surface beds of limestone are often very much cracked as a result. There may also be more localized and severe damage due to the former presence of ice-bodies in areas subject to periglacial conditions. A similar process occurs where pyrite breaks down to form gypsum which replaces the limestone, especially in the vicinity of cave passages. The gypsum molecule is larger than that of the replaced calcium carbonate, and this sets up strains within the limestone. Often this causes the wedging off of chips from the roof, as reported from many caves in the eastern United States. On the other hand saturated solutions of calcium carbonate may lose carbon dioxide near a cave and deposit calcite in the fissures, acting as a cement to reinforce the roof.

Active caves within the vadose zone are subject to enlargement by the abrading action of debris transported by the streams, which, over a period of time, can be measured (Hanna, 1966). This wear, or corrasion, depends upon the nature of the load and the duration of the contact between particles and the bed of the stream (White and White, 1968; Newson, 1971b). In glaciated areas the till supplies a wide range from fine clay to coarse boulders. Surface streams may also carry underground material derived from non-limestone areas. Limestones themselves generally produce small amounts of fine, insoluble residues, in the clay and silt grades. This is often reddish in colour and usually does not have a marked corrading action. Sand grains are more effective and larger stones may produce highly localized erosion in the form of potholes or rock mills. Ford (1965a) has suggested that these are most common in young caves with steep profiles and a good volume of water, though with decreasing discharge this form of bed may become eliminated. The forms are associated with highly turbulent water and may occur in caves which are occupied by floodwater flowing uphill, as in the Hölloch, Switzerland (Bögli, 1970, p. 48). These *marmites de géant* have a characteristic form, usually roughly circular or oval in plan and undercut in a rounded fashion in cross-section. They are most probably caused by the rotation of a boulder which may leave a central conical rise in the floor. They are also quite common in the beds of surface streams, especially on limestones. This points to solution also being a factor in their formation. Excellent examples occur in the Upper Neath (Nedd) valley below Cwm Pwll y Rhydd. Similar forms also occur on limestone coasts in Co. Clare, where shingle provides "pestles" for these natural "mortars".

Not all cave streams have rough, long profiles. Some of the lowest parts of large cave systems, especially in Yorkshire, run for considerable distance at a very low gradient and are probably at or near to the general level of the water-table. Simpson (1935) described these as *master caves*, and they usually are of ample proportions. A good example is to be found in Lost John's Cave. Unfortunately Simpson did not elaborate his concept very clearly and the term is best used in a descriptive sense. The lower stream passages all experience flooding, resulting in the backing up of stagnant water, at least temporarily. If this water is not completely saturated, solution may occur resulting in irregular solutional fretting, especially where the limestones are not homogeneous. Just before the IVth International Speleological Congress visited the cave of Skoksk-janske Jama in September 1965 the flood level reached 80 m above the normal river level. Such floods may also do extraordinary damage to caves by forcing up flowstone floors and sweeping out sediments deposited during centuries of lesser flows (Hanwell and Newson, 1970).

Most cave streams in the vadose zone terminate in a pool, sink into loose rock debris, or disappear into very small passages. Upstream from such terminal points, the cave stream is more or less adjusted to them as a sort of "base-level", especially a master cave. Waltham (1974) has suggested that this process is helped by the junction of tributaries, making the trunk stream more efficient. This concept stresses the headward enlargement as the system progresses to achieve a steady state. Sometimes tributaries may be followed back to tight bedding planes, as in Rumbling Hole, Yorkshire.

Caves act as sediment traps, and many caves contain deposits several metres in thickness of gravels, sand, silts and clays. The size-grading tends to become finer as one proceeds underground, as wear reduces particle size and the dropping of the coarser material. Subsequent re-sorting may make this process of sorting more effective as the finer grades are washed out of mixed deposits. During the glaciation of the British Isles, meltwaters must have washed considerable quantities of debris underground. Later, as the supply of load diminished, streams could begin to cut down into these old deposits. The latter would remain as terraces on either side of the stream. If this downcutting proceeded far enough the river could encounter the original rock floor and start to cut into that, but not in the same places as the original stream (Warwick, 1968b; W. K. Jones, 1971).

Considerable differences of opinion exist about the role of solution in the enlargement of the channel of a cave stream. A turbulent stream tends to lose dissolved carbon dioxide and becomes less aggressive. However, Thrailkill (1968) considers that though percolation water may be more or less saturated when it reaches the groundwater or a cave stream, the chilling effect of the groundwater renders such water capable of absorbing more carbon dioxide and of dissolving more limestone.

It is generally agreed that for the initiation of vadose influent caves the external streams must cut down below the level of the top of the limestone. This would cause the water to be drained from those open joints, bedding planes and faults situated above the water-table. This primitive stage would follow the removal of impervious cap rocks from the limestone. So long as the streams did not cut deeply into the limestone, the surface network would remain fairly dense. The solutional widening of surface cracks would be minimal, and the discharge of most streams would be greater than the capacity of the joints in their rocky beds to swallow water. With the development of surface openings, especially in the top-most 10 m or so, some of the minor streams would lose water underground, and in times of drought dry up altogether. With further rejuvenation the water-table would be lowered and vertical movement of percolation water encouraged. The main streams would tend to entrench themselves, but side streams would probably find it difficult to match this as the water-table became adjusted to the main streams; the water-table might well pass below the level of small tributaries. This would assist stream sinking and the development of underground channels down to the water-table. With water removed from the high-lying tributary valleys, they would become dry and "hang" above the main valley. The path of the engulfed streams would depend upon the local structure. Gently bedded rocks would tend to produce gently graded caves developed along bedding planes, though major joints would encourage vertical passages, especially in massively bedded rocks. On the whole, caves tend to favour down-dip directions, especially where joints are tight. On steeper dipping limestones the dip-orientation is greater, though cave streams may "jump" into higher beds via connecting joints; but this is under waterlogged conditions (Weaver, 1973; Waltham, 1974). Cave breakdown is probably most common after caves have been dewatered and support withdrawn from cave roofs. After stability has been reached it will become unimportant, though as a system ages and solutional development becomes more efficient, it may become more important and even cause successive collapse to the surface. The relative dating of such developments is hard to assess on the basis of current knowledge. Youthful features may be recognizable in the form of poorly graded profiles, initial collapse, and a majority of caves occupied by active streams. When this passes to maturity and old age dating is more difficult and will depend on many circumstances. The stream routing should become simpler, leaving abandoned sections which may contain beautiful stalactite formations. Rejuvenation of the external drainage could lead to further dewatering and a greater simplification of surface drainage and fewer streams surviving above ground near the limestone boundary. Ultimately, as karstification proceeds, all surface drainage will disappear unless water flow in the major streams is fed from a large gathering ground on non-permeable rocks.

Complex Caves within the Vadose Zone

There are few caves in the vadose zone that do not show some complications. Often these take the form of captured passages which permit a more direct flow of conduit water. In other cases there is a complicated three-dimensional network of passages, though most of these will be relatively dry. Gaping Gill Hole, Yorkshire, is one of our best examples. Here Fell Beck approaches an open gulf via a blind-headed valley cut in till and literally falls 110 m into the enormous Main Chamber 145 m ×25 m ×35 m high; there it promptly disappears in the boulder-strewn floor. Other passages lead off from this chamber at varying heights, some being interrupted by enlarged chambers and shafts which penetrate above and below the general level of the connecting passages, like South-East Pot. The lower parts form low bedding planes or tubes with several long pools which become impassable in times of high water. The system has at least five other entrances; one of them, Car Pot, cannot be penetrated though a visual connection has been established (Glover, 1974). There is now widespread acceptance that such a complicated network can only be explained by postulating a fairly long period of initiation and development *below* the water-table and subsequent modification by running water with a free air surface afterwards. These views stem from Grund (1903) and were developed independently by Davis (1930, 1931) and tested in the field by Bretz (1942). Water movement under such conditions is controlled by hydrostatic pressure and the geometry of the solutionally developed network. Water movement is not continuously downwards as in the vadose zone and may occur upwards. Laterally, movement is towards major springs and surface rivers.

If the surface of a fallen slab in a complex system is examined, a network of small tubes developed along a bedding plane or, more rarely, along a joint plane, will probably be found. These are *anastomoses*, a term derived from anatomy and also applied to the interconnecting channels of certain large river deltas. Bretz first drew attention to their importance, and

Ewers (1966) has made additional observations in caves and also in simulation experiments in the laboratory, using blocks of salt. Ewers found that anastomoses develop under very gentle flow conditions and that solution tends to occur upwards and sideways. Where a joint permits easier movement and with greater flow velocities, a tube may form which takes most of the flow and inhibits further development in the lateral channels. If anastomosing networks develop between successive beds this favours cave breakdown. Sometimes the presence of anastomosing channels can be seen at the head of a "blind" passage such as Waltham described from Rumbling Hole. Most anastomoses are only 1 cm or so in diameter. A more complex development is the *spongework* (Bretz, 1942), with openings up to 10 cm or more and often a high percentage of limestone removed. In the Grotte de Lascaux at Montignac, France, certain parts of the wall beneath the painted surface are reduced to a condition resembling an intricately carved chinese ivory puzzle. Glennie (1950) found other examples of spongework in Ogof Ffynnon Ddu I (see Cullingford, 1962, Plate IIb). The tubes which may develop from anastomoses are larger still, up to a few metres in diameter. Some are more or less circular in cross-section, while others are oval. Sometimes two tubes developed close together in a bedding plane may coalesce into a complex form, which in turn may become a place of weakness for breakdown. Some caves have one-half of a tube etched into their ceilings, often following a meandering course. These usually occur under the undersides of bedding planes. They are often flanked by what appear at first sight to be rather fat stalactites up to half a metre long, but composed of living rock. Sometimes the form is blade-like or even reduced to small inverted pyramids. Bretz (1942) christened these *rock-pendants*; they might well be called inverted clints. The Germans refer to them as *Deckenkarren*, which could be roughly translated as roof grikes. A much larger development of connected passages is the *network* (Bretz, 1942) with passage widths measured in metres and intervals between junctions in tens or even hundreds of metres. This would fit the Far Country series in Gaping Gill Hole, and the Waterfall Series of Ogof Ffynnon Ddu I. Bretz found even larger networks in Missouri, such as the Mark Twain Cave which had been dissected by a huge surface valley cut into it. Bretz considered that these caves formed below the water-table would accumulate within them masses of fine clay, representing the insoluble residues from the limestone. This would inhibit solution downwards and concentrate it on the roof. Many of the examples he quoted from the U.S.A. have subsequently been found to be partially filled with debris of a much later age; sometimes stream-gravels occur beneath the clays. There is little doubt that flooded systems act as sediment traps, but much of it is of surface origin.

We can normally only enter caves formed in the phreatic zone after they have been drained, though cave divers have penetrated some that are still flooded, and others have been emptied in the course of metal-mining operations. As soon as this occurs they are subject to the laws of water movement in the vadose zone. They receive percolation water which may coalesce to form a stream, but many more systems have been invaded by surface streams which have developed along suitable joints and bedding planes. Where such streams have adapted an old network, the lowest and usually the simplest course is followed, though some sections may remain flooded. Some Pennine streams have invaded cave systems which they played little part in initiating (Warwick, 1965). To take an example, at Old Ing Cave a stream plunges 10 m to join a stream flowing underground beneath the surface brook. A more complex system was entered by Fell Beck at Gaping Gill Hole, as described above. Surface streams underground develop their channels, frequently cutting deep passages with typical solution features remaining near the roof (Malott, 1937; Bretz, 1942).

Collapse of the surface sometimes reveals large cavities with no stream in them, often with more vertical development than lateral, as in Eldon Hole, Derbyshire, where there is another covered chamber parallel to the one that has been exposed by breakdown. The roof of this inner chamber reaches up towards the surface, shaped like an inverted V in cross-section. Such features which occur in other caves nearby are termed *avens* and usually do not reach to the surface (Fig. 3.17). This term seems to be an adaptation of the French term for a pothole and it was used by Martel to describe the high chimneys in Peak Cavern (Martel, 1897). These are known as *dome-pits* in America and *foibe* (singular *foiba*) or *fusi* in Italy. Several French and Italian authors have suggested that they are caused by upward solution by the backing-up of floodwaters in chambers developed along master joints (Renault, 1952a and b). Maucci (1952, 1958) called it *inverse erosion*. More recently Pitty (1971) has supported this hypothesis with measurements of calcium carbonate solution in Peak Cavern after flooding. Pohl (1955) favours percolation water, especially below dolines, as the dominant mechanism for initiating and developing dome-pits. Such water may be reinforced by water descending through porous cap rocks, especially where they have collapsed into underlying cavities. In South Wales Burke and Bird (1966) found small circular pits 2·5-10 cm in diameter and 2·5-17 cm depth, below a thin layer of quartz conglomerate over carboniferous limestone and these they attributed to solution by acid water penetrating through the cap rock. These, however, had flat roofs in contrast to the dome-shape noted by Pohl, and are much smaller in size and are vadose phenomena.

The drainage of a cave system does not necessarily mean that it will be invaded by a surface stream. Some remain dry and undiscovered since there is little connection with the surface, and are only found by accident during quarrying or mining. They may, however, act as drains for

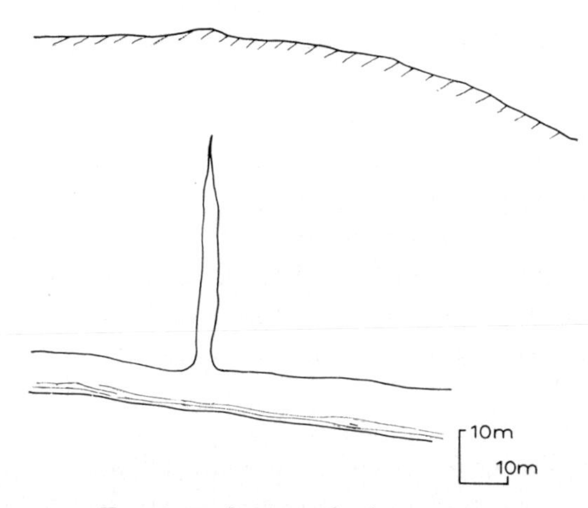

FIG. 3.17. Section of a dome-pit or aven.

percolation water and even collect sufficient to maintain a stream, as in Agen Allwedd. Some of the shallow cavities encountered in mining are associated with one chamber and little else. Eldon Hole would appear to belong to this category too, the roof having caved in; but there are no signs of a surface stream having entered this way.

Caves developed beneath the Water-table

These caves can only be entered by divers or after drainage, for example by mining. Direct conduits can be inferred by the rapid movement of flood pulses or of water containing dyes, from sink to rising. It is assumed that the operative mechanism for the water movement is hydrostatic pressure due to the difference in height between the resurgence and the highest part of the waterlogged network. It is a reasonable hypothesis that such movements may be lateral or downwards and then upwards. Our knowledge of past water movement in abandoned systems now above the water-table supports this view. Theoretical models based upon flow in porous media indicate that under hydrostatic pressure flow would take place along downward curving paths, rising steeply towards the rising. Limestones are non-porous, but it is presumed that flow would follow the bedding planes and joints nearest to the theoretical curves. Davis (1930) assumed a very deep movement, but Swinnerton (1932) proposed that the shorter paths, nearer to the top of the phreatic zone, take more water because they offered the most efficient route, and this would ultimately lead to more solutional widening (Fig. 3.18). Thrailkill (1968) favoured

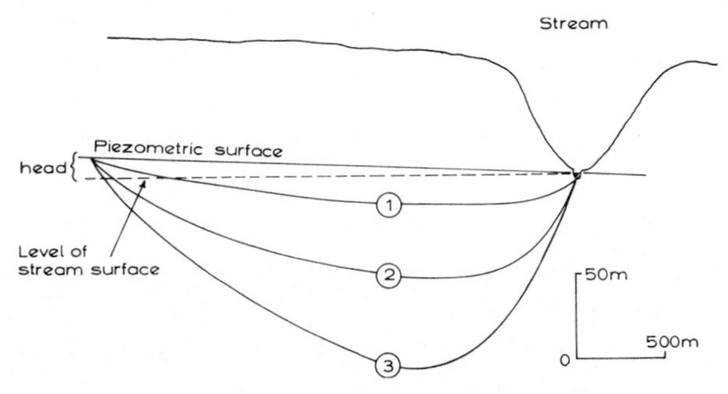

FIG. 3.18. Potential paths for water movement in the phreatic zone (after A. C. Swinnerton, 1932).

the shorter paths on the grounds that solution would be more likely here; below the surface zone the water would be more or less saturated. He considered that water movement would be concentrated in the upper 100 m of the phreatic zone and mainly in the uppermost 10 m. In the Central Kentucky karst, Watson (1966) made a series of measurements of the inputs and outputs of both water and sediment into the flooded zone. He found that inputs exceeded the outputs from Green River, and inferred that there was movement of water and sediment in complex cavities well below the river level. E. L. and W. B. White (1968) also demonstrated the movement of sediment in the upper part of the phreatic zone.

There is some divergence of opinion about the efficacy of phreatic water to dissolve limestone. Bögli (1960, 1971) has suggested that water percolating downwards in more or less independent joints may contain differing amounts of carbon dioxide in solution, and, when in equilibrium with limestone, would contain differing amounts of the rock in solution. Where such waters mixed with each other or a body of water in a flooded conduit, the resultant mixture would be capable of dissolving a certain amount more of the limestone, since the curve relating carbon dioxide in solution and dissolved calcium carbonate is curved upwards above the straight line representing the value of calcium carbonate in solution by mixing the two solvents (see Chapter 7 on solution). This process Bögli terms mixed corrosion. Thrailkill considers on theoretical grounds that such a process would be relatively unimportant, though Howard (1966) has produced evidence in support of the hypothesis. However, Bögli claims that many of the small, wide-mouthed fissures are produced by this effect in the roofs and walls of caves.

Direct observation in Ogof Hesp Alyn, a cave that was recently flooded and acted as a resurgence to the River Alyn in North Wales, indicates a rising cave with a water velocity sufficient to produce scalloping. This cave was deprived of its water as a result of a mine drainage adit at a lower level. Typical solutional rounding can be seen on the walls, picking out the bedding planes, and many passages are in the form of large circular tubes which in other parts of the cave are modified by stream trenches. This cave, which is only partially explored and surveyed, promises to be a good example of a shallow phreatic cave (Appleton, 1974). Moneymaker (1941) also reported on solutional cavities up to 80 m below river level in the Tennessee Valley.

Evidence for deep phreatic cavities has been produced from the cores extracted from exploratory boreholes in search of oil. Cavernous limestone has been encountered at depths of several thousands of metres. In some of the deeper mines in favourable geological structures it has proved possible to pump out the water from chambers at a faster rate than it can seep through narrow connecting joints. In some cases water movement occurs in veins where the cavities are as old as the original mineralization, though in other parts such veins may also act as impediments to water flow and permit isolated areas to be kept dry by pumping. Some connection may be sought between these deep chambers and the occurrence of hot springs flowing out at the surface within a metre or so of a much colder one. The writer (Warwick, 1965) has suggested that geothermal gradients induce vertical convection. Certainly convection currents have been observed in boreholes within 24 hours of the cessation of drilling, and they have to be stopped artificially with plugs if determinations of the temperature of the rocks are to be made with any accuracy. It is difficult to postulate an initiation for such movement naturally, unless by way of vein cavities, and even more difficult to find a mechanism to induce solution. Additional carbon dioxide from deep-seated igneous activity is also possible but not easy to establish. Perhaps a more likely solvent is sulphuric acid formed from the breakdown of sulphides (Morehouse, 1968).

Water may also flow under artesian or confined conditions within limestones (Fuller, 1908). This may be produced by the constraining action of impervious beds above and below the limestones, a situation which is repeated several times in NW Yorkshire. This may produce a perched water-table within the limestones. It is also possible to have artesian flow through conduits where the joints and bedding planes are very tight or where certain beds are specially favourable to solution. Glennie (1950) found in Ogof Ffynnon Ddu I that the lowest part of the cave was developed in only twelve individual beds. He later suggested that conduit flow in the upper part of the phreatic zone was also a form of artesian movement (Glennie, 1954a). He thought that at greater depths such flow would be very slow. He coined the terms *epi-phreatic* flow for water-table streams as defined by Swinnerton and Simpson; *bathy-* or *endo-phreatic flow* for stream flow beneath the water-table; and *hypo-phreatic flow* for artesian flow into the groundwater zone from a non-cavernous aquifer tapped by phreatic action. For true siphons carrying water above the general water level he suggested *pro-phreatic flow*. Howard (1964a) has made a special study of the caves of the Black Hills, South Dakota, to diagnose criteria of past artesian flow. He stressed the clustering of passages within narrow zones and the migration upwards of joint-determined passages between levels. He also found that most major passages follow the dip, and that in the levels favoured by solution there was easy intercommunication and a general absence of sediments. The presence of delicately etched-out mineral veins known as *boxwork* in closely jointed rock indicated gentle flows, as these solution features are very fragile. Perhaps most telling of all was the presence of ceiling pits of up to 30 m above the general cavern level. In Britain the limestones underlying the syncline of the South Wales Coalfield offer the most favourable structure for artesian flow. There is however little direct evidence of water following such a route. One pointer is the inrush of water encountered in the Elliott Pit in the Rhymney Valley to the north of the fold axis. This happened in 1891. By 1893 the calcium carbonate concentration was only 13 ppm, but this rose markedly to 130 and 147 ppm by January 1898, though subsequently declining to 120 ppm (Hann, 1899). It has been suggested that these flows were fed from the underlying limestones. There are also springs on the southern outcrop which show a steady discharge (Knox, 1933).

Effluent Systems

Groundwater discharges from springs, and many of these in limestone areas, are very large, such as the "boiling springs" of Missouri, which

also Warwick, 1960). A similar system of sinks and risings above a valley steepening in profile occurs in the Manifold Valley. A comparable situation occurs where a stream takes a short-cut through the spur of an incised meander, as at the Grotte de Han, though here the river cannot be followed through and there is an underground rising in a series of passages and large chambers; these seem to have been invaded by the River Lesse having originated independently when the river was flowing at a much higher level and the cave system was in the phreatic zone.

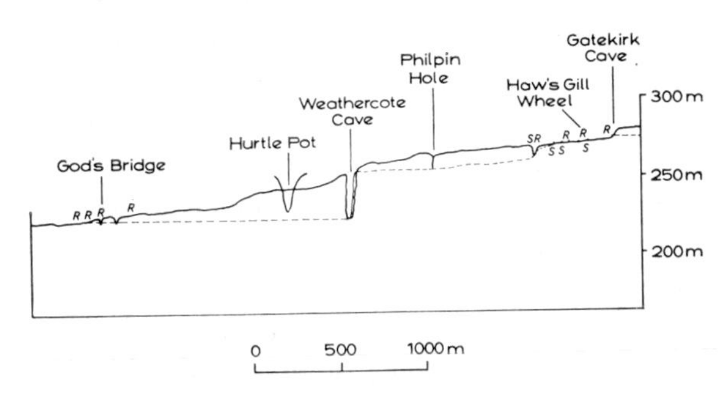

FIG. 3.19. Long profile of central Chapel-le-Dale, NW Yorkshire.

THE ORIGIN OF CAVES

It should be clear that there is not one universal origin for all caves. The type of cave in relation to the hydrology, the relation to the external drainage, and the relation to the geological structure are all important (T. D. Ford, 1971; Waltham, 1971). Few now would deny the importance of slow, solutional development below the water-table, and the modification of the network of enlarged joints and bedding planes following upon downcutting by the external drainage which controls discharge from the limestone. Drainage of underground passages and chambers permits vadose development or infilling with sediment and/or calcite formations, though this is rarely simple for there is evidence of cave formations being destroyed by re-solution and even patched up by further deposition. Variations in climate may affect these processes, affecting both the amount of water entering a system and the acidity of those waters. A detailed study of one system provides a means of testing various general theories of the origin of caves, though as many problems will be raised as are solved.

This account has been largely based upon experience of caves in the British Isles and to a lesser extent in Europe, mainly in the mild, temperate zone of NW Europe, with excursions to high mountain karst and Mediterranean regions. In the British Isles our caves and limestone scenery have been developed in fairly low hill country, well dissected by rivers, under a mild, rainy climate. Over half of the area has been glaciated and the rest subjected to periglacial conditions. It has also been a stable group of islands since mid-Tertiary times, subject to only mild flexures and a repeated lowering of base-level as the sea-level has fallen. No region is far from the sea and sea-level changes have worked well up most of our rivers. Reference has been made in passing to landforms and caves to be found in other climatic and tectonic environments, but there has been no attempt to make it all-embracing and worldwide in cover. A reading list covering mainly English and American sources is given below with some references to the French and German literature. The serious student of speleogenesis will find further enlightenment from this bibliography over a wider range of caves and countries than can be covered in this chapter.

NON-LIMESTONE CAVES AND PSEUDO-KARST

The commonest non-limestone caves occur around the coast as a result of wave action attacking joints and bedding planes and loosening blocks and slabs of rock, making them unstable. For this reason most of these caves are ephemeral. Where the coast is formed from soluble rocks solution may be added to the essentially mechanical agents. Coastal caves in limestone may combine features due to marine erosion with those associated with effluent caves. A fall in sea-level may leave an old cliff with caves at its foot, though this quickly becomes degraded through the formation of scree slopes. Some caves near Oban have been excavated from such debris and have yielded human artifacts.

Gypsum and salt form the next most common of the soluble rocks after limestones and dolomites in the British Isles. A few caves have been found in gypsum; above gypsiferous rocks the surface quickly becomes fretted in small cusps or with karren, and locally dolines may be found. The literature is not extensive and the deposits are not very thick. The gypsum is often confined between impervious clayey rocks and the area of outcrop small. The thickest deposits occur in the Vale of Eden, in Cumbria. In a mine at Kirkby Thore isolated chambers, filled with water, were found during mining operations using pillar and stall techniques. There appears to be little connection between these cone-shaped chambers, which are up to 3 m high and 1-2 m in diameter at the base. They are allowed to drain gradually into the workings and when dry they show a corbelled structure with overlapping thin beds of gypsum supporting the next highest bed, which extends out in a cantilever fashion. In open pits near by, irregularly shaped rundkarren are found beneath a cover of till. Further reports of the solution of gypsum have come from Nottinghamshire (Elliott, 1961; Firman and Dickson, 1968) where the mineral is found in beds up to 20 m thick in the Keuper Marl. Near the surface this is subject to considerable solution, and the beds have become reduced to large nodules known as "cakes". The upper surface is frequently convex and covered with fibrous gypsum, but the flatter undersurface is usually free of this fringe, and circular holes some 5 cm in diameter penetrate upwards for distances of 20 cm or more. No such holes are found on the upper surfaces, though some smaller nodules are pierced completely through. Firman suggested that these features indicate solution by water flowing upwards under artesian pressure, probably during one of the later phases of the Pleistocene glaciations. As far as the writer is aware, no extensive cave systems have been reported from the British gypsum deposits, though no doubt a close inspection of gypsum mines in Sussex and the Midlands would find many more cases of minor solution features. Much larger caves and other karstic phenomena are found in the Harz region of Germany (Hensler, 1968). Kammholz (1966) and Priesnitz (1969a and b), have made a special study of this gypsum karst. Only one British salt mine is now operating, but large quantities of salt are removed by brine extraction, whether by "wild" brine pumping or by supplementary injection of water. This results in the formation of cavities which collapse and let down the surface, especially in Cheshire, where the "meres" that form are the result of subsidence hollows filling with surface water. Other examples occur near Droitwich, where some of the resultant dolines are flooded whilst others remain dry. In one case, to the south of the town, there is a trench some 500 m long and 10 m wide running across the grain of the landscape where the ground has subsided 1-2 m. Some of the natural brine springs have been known for centuries and these no doubt have developed underground connections. In some cases the extensive pumping of brine has caused these springs to dry up. In one case at Stafford, a lycopodium spore introduced into a borehole in another part of the town was traced to the brine pump, proving a direct connection between the sites. Surface exposures of natural salt are rare, but they occasionally reveal solution, especially beneath superficial deposits.

Lava caves are a special class, with little to do with solution, and are dealt with in Chapter 4.

THE AGE OF CAVES

"A cave is a space rather than an object" (Curl, 1964) and is difficult to define or date, though some limits may be established. A lower limit is obviously the age of the rock in which the cave is developed. An upper limit is the present or the date of destruction of the rock surrounding the cave, as by quarrying.

The history of a cave may be amplified by the dating of deposits found within it, which clearly post-date the surface upon which they are lying. A third method is to establish a connection between the cave and some external geomorphological feature, such as a river terrace or an event such as a glaciation. A combination of these techniques is being increasingly used.

The relative age of the containing sedimentary rocks has long been established on the basis of the law of superimposition which states that in a sequence of bedded rocks the younger rocks normally lie upon the older. Certain key fossils may be used to subdivide these rocks into zones and stages. More recently an absolute chronology has been established by the use of radioactive isotope measurements (see below). The age range of the periods during which the major limestones were laid down in the British Isles is given below:

Period	Age in millions of years
Cretaceous	65-136
Jurassic	136-190
Triassic	190-225
Permian	225-280
Carboniferous	280-345
Devonian	345-395
Silurian	395-435
Cambrian	500-570

(After Harland *et al.*, 1964 and Kirkaldy, 1971)

Where a cave system is developed in two limestones of different age, the lower limit is moved upwards to the age of the younger limestone. This situation is found on the flanks of the Mendips where the Carboniferous Limestone is overlain by the Dolomitic Conglomerate, a breccia of Triassic age. In a few localities, mainly concerning Carboniferous and Triassic rocks, evidence has been found of *fossil-* or *palaeo-karst* (Walkden, 1972). The Carboniferous examples from Derbyshire (T. D. Ford, 1952; King, 1966), and South Wales (Dixon, 1910) occur beneath unconformities where solutional features of a land surface have been covered by a later deposit (T. D. Ford, 1972). In the Mendips and on the southern outcrop of Carboniferous Limestone in South Wales true caves have been found infilled with Triassic deposits, some containing animal remains (Robinson, 1957; Halstead and Nicoll, 1971). In southern Dyfed red-stained breccias are found among grey Carboniferous limestones. These were thought to be Triassic deposits in fissures produced by solution, but Thomas (1973) has shown them to be of tectonic origin and some may even be Tertiary in age. Apart from these palaeo-karstic features it is difficult to say when cave formation began, but in most cases it would be unlikely to have commenced in earnest until the limestones were stripped of their cover. This probably means an Upper Tertiary age, extending as far back as about 26 million years.

The collapse of surface rocks presumes a subjacent cavity. In South Wales Thomas (1954, 1963) has found many masses of Millstone Grit let down into the limestone, but all that can be inferred is that this is post-Namurian in age. However, continued collapse in the Carboniferous Limestone of the Peak District was matched by the deposition of silica sand, known as the Brassington Formation, which was presumably widespread upon the limestone surface but is now confined to the former depressions (Walsh *et al.*, 1972). These are worked commercially and the pits at Brassington and Friden in Derbyshire have revealed fossil pollen of Upper Miocene-Lower Pliocene age (*c.* 7 million years) (Boulter, 1971). The uppermost deposit in these pits is a thin layer of glacial till.

The mouths of old effluent caves often provide deep deposits of angular limestone rubble and clay, among which one may find the bones and teeth of animals, including Man. Man also left behind some of his more durable artefacts such as flints, metal objects, pottery, etc. The law of superimposition can be applied to these deposits and a relative dating applied (Cailleux, 1963). Care has to be taken in interpreting artefacts deposited on a slope and also where animals and human excavators have disturbed the sequence. Most of the artefacts are of post-glacial (Flandrian) age. However, the animal record is usually more complete, and many caves preserve records of animals that could tolerate cold conditions and these deposits may be equated with the Last Glaciation (Devensian). In a few instances a "warm" fauna may be recovered from beneath the "cold" layers including animals such as hippopotamus. Influent caves may also contain animal remains, but these are rarely stratified. The oldest British cave fauna comes from the Victoria Fissure at Doveholes, near Buxton, Derbyshire, where remains of sabre-toothed cat (*Homotherium latidens*) were discovered, probably Lower Pleistocene in age. Over the period of the

Pleistocene many of the upper cave mouths were probably destroyed by the processes of slope retreat or by glacial scouring.

In the interior of caves may be found deposits of clay, sand, gravel and calcite. These may show a relative succession; but this is difficult to correlate with external conditions, though radiometric dating of carbonaceous cave formations is now being established.

Table 3.1

DIVISIONS OF THE PLEISTOCENE

General Terminology (Zeuner)		Geological Soc. Recommended Stages		Approximate Time Range (B.P.)	N. European Sequence	Alpine Sequence
Post-Glacial		Flandrian		Post 10 000	Holocene	Post-Glacial
Last Glacial	III	Devensian	L	8-26 000	Weichselian	Würm Glaciation
	II		M	26 000 – 50 000		
	I		E	50 000+		
Last Interglacial		Ipswichian		105 000	Eemian	Riss/Würm Igl.
Penultimate Glaciation		Wolstonian		140 000	Saale	Riss Glaciation
Penultimate Igl		Hoxnian		195 000	Holstein	Mindel/Riss Igl.
Antepenultimate Glaciation		Anglian		275 000	Elster	Mindel Glaciation
Antepenultimate Interglacial		Cromerian		325 000		Gunz/Mindel Igl.
Early Glaciations		Beestonian Pastonian Baventian Antian Thurnian Ludhamian Waltonian				Gunz Glaciation

It is usually difficult to establish a direct connection between caves and their deposits and the external world, but in a few fortunate localities it is possible to correlate cave morphology with external morphological features and also to relate them to external superficial deposits such as river terrace gravels, glacial till or solifluxion "flows" (head). Near the coast there is the possibility of connecting caves with old sea-levels. In Britain it is now acknowledged that the Pleistocene Period was one of alternating cold and warm periods, and on the basis of existing knowledge the general succession (based on Mitchell *et al.*, 1973) is given in Table 3.1.

The correlations shown above are extremely tenuous before the Last Interglacial and are only indicated for general guidance. The single dates of the British stages should be taken as a mid-stage date. The effects of increased cold would be felt in different places at differing times, and are said to be diachronous. The commonest connection with caves is the presence of glacial till within a cave, which indicates that the particular cave is older than the ice-sheet which deposited the till (Warwick, 1971). In Co. Clare few caves received boulder clay, Pol-an-Ionain being exceptional. It has been suggested that the present cave streams carry sufficient limestone in solution to have formed the present caves in post-glacial times. In a few localities till overlies sediments containing a warm fauna, or even human artefacts. Raygill fissure in Lothersdale, Yorkshire, is an example of the first kind (Davis, 1886) and the North Welsh caves of Cae Gwyn and Ffynnon Beuno of the second (Hughes, 1887, 1888). In Yorkshire, pollen grains have been found in peat deposits (dating back to the second interstadial of Devensian age) which rest upon till, which in turn is found in the local caves. Waltham (1974) has summarized the evidence for NW England and produced a tentative correlation, though it is difficult to identify the individual glacial tills. In S Devon, at the Joint Mitnor Cave, Sutcliffe (1960) has established a correlation with a terrace of the River Dart. This cave was formed below the level of this terrace and after the river had cut down and lowered the local water table, the roof of the cave collapsed and let in debris, including gravels and a warm fauna of Ipswichian age. The nearby Pridhamsleigh Cave also bears witness to a rise in sea-level, since divers found stalactites at 13 m below the water level in a flooded passage, and eels were seen at 22m. This rise is known as the Flandrian transgression caused by the last melting of the ice-sheets. Old sea-caves and fossil effluent caves are to be found at the back of old marine erosion surfaces though they may become covered by scree. Small caves of this type have been found in the Oban area, containing remains of Mesolithic age. Many of these beaches have been dated, providing a late date for the associated caves. There have been several attempts to establish a connection between certain dominant cave levels and the external landscape. One of the most successful was that of Droppa (1966), who found nine river terraces in the neighbourhood of Demänová Cave, Czechoslovakia, and the same number of levels inside the cave. The highest terrace, number IX, has been tentatively dated to the uppermost Pliocene, and number VII with the Donau period (pre-Gunz). This would make this large system 1-2 million years old. Waltham (1974) proposed a similar younger age as one proceeds down some of the Yorkshire cave systems, but there are no datable terraces there. Sweeting (1950) has also put forward a relationship between some of the large chambers in Yorkshire caves and a denudation chronology which she established outside. The writer has also attempted to relate the morphology of a by-passed swallow hole cut in the side of the Meuse gorge at Dinant with local terrace levels (Warwick, 1959).

All of the above methods of dating are essentially relative, but with the development of radioactive methods, absolute dates can be assigned to objects and deposits in special circumstances, though subject to a considerable margin of error that may be as much as 25%. These new techniques depend upon the measurement of the rate of decay of naturally occurring radioactive elements to a more stable form. It has been established that such elements decay at an increasingly slower rate such that the amount of radioactivity is halved after a period of years, known as the "half-life", and is halved again after another such period. The radioactive elements, or isotopes, possess the same chemical properties as the stable form, but have extra neutrons. The most important of these elements for dating geologically recent events is carbon-14, abbreviated to ^{14}C (or in older literature, C^{14}) with a half-life of 5730 years. Dates using this method are quoted from 1950 and referred to as Before Present (BP). The method works best with materials containing carbon of organic origin, which is derived ultimately from atmospheric radioactive carbon dioxide. After impurities have been removed the material is burnt in pure oxygen and converted into carbon dioxide for counting (Burleigh, 1972; Suess, 1973). It is used chiefly to date wood and peat, though the collagen from bones can also be used (Sellstedt *et al.*, 1966). Derived peat debris in a cave sediment is of little use for dating since it can only give the date of the peat bog on the surface. The accuracy falls off with age and is of little use beyond 65 000 BP. Up to 20 000 years it is accurate to $\pm 1\%$, falling off to $\pm 5\%$ at 30 000 years. The method has been used to date collagen from mammoth bones found beneath boulder clay in Cae Gwyn Caves, Treimerchion, North Wales (see above), producing an age of 18 000 BP (Shotton and Williams, 1971). Recently corrections have had to be made to the accepted half life of ^{14}C from the older value of 5570 ± 30 years, but since earlier results were published on the older scale, new determinations are still given on this scale and a standard correction applied to all values to avoid confusion. Results are published in a special journal, *Radiocarbon*, and prior to 1959 as supplements in the *American Journal of Science*. Complications have also been introduced by the formation of radioactive carbon dioxide in the atmosphere from nuclear explosions. The application of the ^{14}C technique to the dating of shells and cave formations is more difficult since part of the carbon comes from "dead" carbonates dissolved from limestone whose radioactivity has decayed to imperceptibility. Such dates appear to be older than one would expect. Broecker and Olson (1959) found that modern, organically derived carbon dioxide accounts for *c*. 70% of very recent speleothems and Hendy and Wilson

(1968) observed figures of *c.* 50%. Pearson (1965) has found a more accurate method of estimating the inactive carbon pollutant by measuring the ratio of $^{13}C/^{12}C$ in the specimens (see also Hendy, 1970, 1971; Harmon, 1971a and b, 1972; Howard and Howard, 1972). There are a number of laboratories working on cave deposits including one in Yugoslavia (Slipčević and Planninick, 1974).

For older speleothems, measurements of Thorium isotopes derived from Uranium are required. From N Canada Ford and his collaborators (1972) have studied stalagmites formed in caves of phreatic origin some 1500 m above present sea-level. One stalagmite produced ages using the $^{230}Th/^{232}Th$ ratio, of 90 000 to 270 000 proceeding inwards. In Italy Fornaca-Rinaldi (1968), using similar methods, found that where a stalagmite had stopped growing and then growth was resumed, that such formations could not be used for $^{230}Th/^{238}U$ determinations. On the positive side he measured the age of a flowstone in the famous Romanelli cave of Apulia at 40 000 ± 3250 years BP and the calcite sealing a breccia on a nearby raised beach gave a figure of 69 000 BP.

Perhaps the nearest that one may come to dating a cave is where it occurs in very recent material, such as calcareous sand (calcarenite) or coral reefs. Jennings (1968) has referred to such phenomena as syngenetic karst. In the Riyukuyu Islands, Sakanoue *et al.* (1967) found that a stalagmite in a cave dissolved out of reef limestones was of a similar age to the shells forming the walls of the cave, though some difficulty was encountered in dating the stalagmite. For the deposits within the ^{14}C range of ages it is possible to cross-check some of these measurements such as Kaufman and Broecker (1965) made from carbonate materials in a former glacial lake. These results showed a reasonable degree of agreement, though where a discrepancy occurred the lower age appeared to be in error. Stuiver (1970) has compared more recent ^{14}C dates with those from tree rings (dendrochronology) and varves and found some discrepancies over the period 5000-7400 years BP of the order of 600-800 years.

Another technique is to measure the $^{18}O/^{16}O$ ratio which provides an indication of the mean annual temperature at the time of formation. This was first used on marine deposits but has now been used on cave formations in conjunction with radioactive thorium measurements (Duplessy *et al.*, 1969). M. Duplessy and his colleagues (1970) have carried out such a series of measurements on a stalagmite in Aven Orgnac, Ardèche, France. Their results show a slow increase in temperature of *c.* 4°C between 130 000 and 120 000 BP after which it remained virtually constant until 97 000 BP. The next 1000 years saw a sharp fall of 2·5°C with minor fluctuations to a minimum in 95 000 BP. Between 93 000 and 92 000 BP a further sudden rise of *c.* 4° was observed. Duplessy considers that these results mark the period of the Last (Ipswichian) Interglacial.

Radioactive hydrogen (Tritium or ^{3}H) can also be used to date groundwater (Int. Atomic Energy Agency, 1968), but chiefly for very recent ages since the introduction of atmospheric testing of the nuclear devices since 1952. Older waters require the use of ^{14}C methods. The results indicate that in some finely porous rocks water may take many years to reach the surface since movement is so slow. With well-developed karstic conduits, output from springs and floods is matched with inputs into cave systems, especially where radioactive tracers are used. There is also some indication that diffusion of water below the surface of the water-table is very slow, which supports the views of Swinnerton (1932).

One method of potential use in the vicinity of volcanoes is the dating of volcanic ash deposits by Potassium-Argon techniques. Where these deposits penetrate a cave entrance, they provide a useful marker horizon. For further details of the methods and their application the reader is referred to Hamilton (1965), Harper (1973), Int. Atomic Energy Agency (1967a and b, 1968) and Shotton (1966).

These new methods promise much greater precision for correlations, especially for closely related phenomena. There are still difficulties to be ironed out, but the chief practical one is of cost, since a ^{14}C determination costs something of the order of £100 and close supervision is required in the collection and processing of material. Radiocarbon is the most popular method and already the task of recording results on punched cards has become unworkable and results are being transferred to magnetic tape. For the student of speleogenesis the most exciting results will come from the use of thorium and uranium isotopes which can measure much further back into time. More accurate correlations between caves and changes in climate and between caves and external geomorphological events will also become possible. Perhaps it will be possible to answer the question "how old is this cave?" with a greater degree of certainty.

REFERENCES

Appleton, P. (1974). Subterranean courses of the River Alyn including Ogof Hesp Alyn, North Wales. *Trans. Br. Cave Res. Ass. 1* (1), 29-42.

Arnberger, E. (1954). Neue Ergebnisse morphotektonischen Unterschungen in der Dachstein-Mammuthöhle. *Mitt. der Höhlenkommission, Wien* (for 1953) *2* (3), 43-48.

Ashmead, P. (1974). The caves and karst of the Morecambe Bay area. *In* Waltham, A. C. (ed.), 201-226.

Bedinger, M. S. (1966). Electric-analog study of cave formation. *Bull. Nat. Speleo. Soc. 28* (3), 127-132.

Birot, P. (1966). *Le Relief Calcaire*. Paris.

Bögli, A. (1960). Kalklösung and Karrenbildung. *Zeit. für Geom. Supplementband, 2*, 4-21.

Bögli, A. (1965). The role of corrosion by mixed water in cave forming. *In* Štelcl, O. (ed.), 1965, 125-131.

Bögli, A. (1970). *Le Hölloch et son Karst*. Neuchâtel (in German and French).

Bögli, A. (1971). Corrosion by mixing of karst waters. *Trans. Cave Res. Grp. G.B. 13* (2), 109-114.

Boulter, M. C. (1971). A palynological study of two of the neogene plant beds in Derbyshire. *Bull. Br. Museum (Nat. Hist.), Geology, 19* (7), 361-410.

Bowen, D. Q. (1970). The palaeoenvironment of the "Red Lady" of Paviland. *Antiquity, 44*, 134-136.

Bramwell, D. (1949). Physical properties of some Derbyshire and Staffordshire cave earths. *Cave Res. Grp. N/L, 21*, 5-8.

Bretz, J. Harlen (1942). Vadose and phreatic features of limestone caverns. *J. Geol. 50* 675-811.

Bretz, J. Harlen (1953). Genetic relations of caves to Peneplains and Big Springs in the Ozarks. *Am. J. Sci. 251*, 1-24.

Bretz, J. Harlen (1954). Caves of phreatic origin. *Scientia, 89*, 13-18.

Broecker, W. S. and Olson, E. Z. (1959). C^{14} dating of cave formations. *Bull. Nat. Speleo. Soc. 21* (1), 43.

Brown, E. H. (1966). Dry valleys in the chalk scarps of South East England. *Biul. Peryglacjalny, 15*, 75-78.

Brucker, R. W. (1966). Truncated cave passages and terminal breakdown in the central Kentucky karst. *Bull. Nat. Speleo. Soc. 28* (4), 171-178.

Burke, A. R. and Bird, P. F. (1966). A new mechanism for the formation of vertical shafts in carboniferous limestone. *Nature, 210*, 831-832.

Burleigh, R. (1972). Carbon-14 dating with application to dating of remains from caves. *Stud. in Spel. 2* (5), 176-190.

Cailleux, A. (1963). Datation absolue des principales industries préhistoriques. *Bull. Soc. géol. de France.* (Sér. 7), *5*, 409-413.

Campbell, N. P. (1968). The role of gravity sliding in the development of some Montana caves. *Bull. Nat. Speleo. Soc. 30* (2), 25-29.

Coleman, A. M. and Balchin, W. G. V. (1959). The origin and development of surface depressions in the Mendip Hills. *Proc. Geol. Assoc. 70*, 291-309.

Coleman, J. C. (1949). An indicator of water-flow in caves. *Proc. Univ. Bristol Speleo. Soc. 6* (1), 57-67.

Common, R. (1955). Les Formes littorales dans les calcaires en Northumberland septentrionale. *Ann. de Géog., 64*, 126-128.

Corbel, J. (1959). Erosion en terrain calcaire (vitesse d'érosion et morphologie). *Ann. de Géog. 68*, 97-120.

Chorley, R. J. (1972). *Spatial Analysis in Geomorphology*. London.

Cullingford, C. H. D. (ed.) (1962). *British Caving. An Introduction to Speleology.* 2nd edn. Routledge and Kegan Paul, London.

Curl, R. N. (1960). Stochastic models of cavern development. *Bull. Nat. Speleo. Soc. 22* (1), 66-76.

Curl, R. N. (1964). On the definition of a cave. *Bull. Nat. Speleo. Soc. 26* (1), 1-6.

Curl, R. N. (1966a). Caves as a measure of karst. *J. Geol. 74*, 798-829.

Curl, R. N. (1966b). Scallops and flutes. *Trans. Cave Res. Grp. G.B.* 7 (2), 121-160.

Cvijić, J. (1960). *La Géographie des Terrains calcaires.* Belgrade.

Davies, W. E. (1949). Features of cavern breakdown. *Bull. Nat. Speleo. Soc. 11*, 171-178.

Davies, W. E. (1951). Mechanics of cavern breakdown. *Bull. Nat. Speleo. Soc. 13*, 36-43.

Davies, W. E. (1960). Origin of caves in folded limestones. *Bull. Nat. Speleo. Soc.* 22 (1), 5-18.

Davis, J. W. (1886). On the exploration of the Raygill Fissure in Lothersdale. *Proc. Yorks. Geol. and Polytechnic Soc. 9*, 280-281.

Davis, S. N. (1966). Initiation of ground-water flow in jointed limestone. *Bull. Nat. Speleo. Soc. 28* (3), 111-118.

Davis, W. M. (1930). Origin of limestone caverns. *Bull. geol. Soc. Am. 41*, 475-628.

Davis, W. M. (1931). The origin of limestone caverns. *Science, 73*, 327-333.

Dawkins, W. B. (1873). Observations on the rate at which stalagmite is being accumulated in the Ingleborough Cavern. *Rep. Br. Assoc. for Adv. Sc.* 80.

Deike, G. H. and White, W. B. (1969). Sinuosity in limestone solution conduits. *Am. J. Sci. 267*, 230-241.

Deike, R. G. (1969). Relations of jointing to orientation of solution cavities in limestones of Central Pennsylvania. *Am. J. Sci.* 267, 1230-1248.

Dixon, E. E. L. (1910). Unconformities on limestone and their contemporaneous pipes and swallow holes. *Rep. Br. Ass. for Adv. Sci.* for 1901, 477-479.

Droppa, A. (1966). The correlation of some horizontal caves with river terraces. *Stud. in Spel. 1* (3), 186-192.

Duplessy, J. C., Labeyrie, J., Lalou, C. and Nguyen, H. V. (1970). Continental climatic variations between 130 000 and 90 000 years BP. *Nature, 226*, 631-633.

Duplessy, J. C., Lalou, C. and Gomes de Azevedo, A. E. (1969). Étude des conditions de concrétionnement dans les grottes au moyen des isotopes stables de l'oxygène et du carbone. *C.R. Ac. Sc. Paris, 268*, 2327-2330.

Dury, G. H. (ed.) (1966). *Essays in Geomorphology.* London.

Elliott, R. E. (1961). The stratigraphy of the Keuper Series in Southern Nottinghamshire. *Proc. Yorks. Geol. Soc. 33*, 197-234.

Ewers, R. O. (1966). Bedding-plane anastomoses and their relation to cavern passages. *Bull. Nat. Speleo. Soc. 28* (3), 133-140.

Firman, R. J. and Dickson, J. A. D. (1968). The solution of gypsum and limestone by upward flowing water. *Mercian Geologist, 2* (4), 401-409.

Fitton, E. P. and Mitchell, D. (1950). The Ryedale Windypits. *Cave Sci.* 162-184.

Ford, D. C. (1965a). Stream potholes as indicators of erosion phases in limestone caves. *Bull. Nat. Speleo. Soc. 27* (1), 27-32.

Ford, D. C. (1965b). The origin of limestone caves: a model from the Central Mendip Hills, England. *Bull. Nat. Speleo. Soc.* 27 (4), 109-132.

Ford, D. C. (1968). Features of cavern development in Central Mendip. *Trans. Cave Res. Grp. G.B. 10* (1), 11-25.

Ford, D. C. (1971a). Research methods in karst geomorphology. In *Research Methods in Geomorphology, Proc. 1st Guelph Symposium on Geomorphology, 1969*, 23-47.

Ford, D. C. (1971b). Geological structure and a new explanation of limestone cavern genesis. *Trans. Cave Res. Grp. G.B. 13* (2), 81-94.

Ford, D. C., Thompson, P. T. and Schwarz, H. P. (1972). Dating calcite deposits by the uranium disequilibrium method: some preliminary results from Crowsnest Pass, Alberta. *Research Methods in Pleistocene Geomorphology, Proc. 2nd Guelph Symposium on Geomorphology, 1971*, 247-255.

Ford, T. D. (1963a) The Goyden Pot drainage system, Nidderdale, *Trans. Cave Res. Grp. G.B. 6* (2), 79-89.

Ford, T. D. (1963b). The Dolomite Tors of Derbyshire. *E. Midland Geog. 3* (3), 148-153.

Ford, T. D. (1967). A quartz-rock-filled sink-hole on the Carboniferous Limestone near Castleton, Derbyshire. *Mercian Geologist, 2* (1), 57-62.

Ford, T. D. (1969). Dolomite tors and sandfilled sink holes in the Carboniferous Limestone of Derbyshire, England, pp. 387-397 in Péwé, T. L., *The Periglacial Environment, Past and Present*. Montreal.

Ford, T. D. (1971). Structures in limestone affecting the initiation of caves. *Trans. Cave Res. Grp. G.B. 13* (2), 65-72.

Ford, T. D. (1972). Evidence of early stages in the evolution of the Derbyshire karst. *Trans. Cave Res. Grp. G.B. 14* (2), 73-77.

Ford, T. D. and King, R. J. (1966). The Golconda Caverns, Brassington, Derbyshire. *Trans. Cave Res. Grp. G.B.* 7 (2), 91-114.

Fornaca-Rinaldi, G. (1968). ^{230}Th/^{234}Th dating of cave concretions. *Earth and Planetary Sc. Letters, 5*, 120-122.

Frye, J. C. and Swineford, A. (1947). Solution features on Cretaceous sandstone in Central Kansas. *Am. J. Sci. 245*, 366-379.

Fuller, M. L. (1908). Summary of the Controlling Factors of Artesian Flows. *Bull. 319, U.S. Geol. Survey, Washington.*

Gams, I. (1965). Types of accelerated corrosion. *In* O. Štelcl (1965), 133-139.

Gèze, B. (1955). La genèse des gouffres. *Communications du 1er Congrès Int. de Spél. (Paris, 1953), 2*, 11-23.

Gèze, B. (1975). *La Spéléologie Scientifique*. Paris.

Gèze, B. (1973). Lexique des termes français de spéléologie. *Ann. de Spél. 28* (1), 1-20.

Glennie, E. A. (1950). Further notes on Ogof Ffynnon Ddu. *Trans. Cave Res. Grp. G.B. 1* (3), 1-47.

Glennie, E. A. (1951). Some observations in Long Kin Cave. *Cave Res. Grp. G.B. N/L. 35*, 10-14.

Glennie, E. A. (1952). Vertical development in caves. *Trans. Cave Res. Grp. G.B. 2* (2), 73-93.

Glennie, E. A. (1954a). Artesian flow and cave formation. *Trans. Cave Res. Grp. G.B. 3* (1), 55-71.

Glennie, E. A. (1954b). The origin and development of cave systems in limestone. *Trans. Cave Res. Grp. G.B. 3* (2), 73-83.

Glennie, E. A. (1960). The Hertfordshire Bourne in 1959. *Trans. Herts Nat. Hist. Soc. 25* (3), 105-107.

Glennie, E. A. (1962). The Hertfordshire Bourne, 1960-61. *Trans. Herts Nat. Hist. Soc. 25* (5), 200-204.

Glover, R. R. (1974). Cave development in the Gaping Gill System. *In* Waltham, A. C. (1974), 343-384.

Goldie, H. (1973). The limestone pavements of Craven. *Trans. Cave Res. Grp. G.B. 15* (3), 175-190.

Goodchild, D. M. F. and Ford, D. C. (1971). Analysis of scallop patterns by simulation under controlled conditions. *J. Geol. 79*, 52-62.

Groom, G. E. and Coleman, A. (1958). *The Geomorphology and Speleologenesis of the Dachstein Caves*. Cave Res. Grp. G.B. Occ. Pub., No. 2, Berkhamstead.

Grund, A. (1903). Die Karsthydrographie. Studien aus Westbosnian. *Pencks. geogr. Abhandl.* 7 (3), 103-200.

Guilcher, A. (1953). Essai sur la zonation et la distribution des formes littorales de dissolution du calcaire. *Ann. de Géog. 62*, 161-179.

Halstead, L. B. and Nicoll, P. G. (1971). Fossilised caves of Mendip. *Stud. in Spel. 2* (3/4), 93, 102.

Hamilton, E. I. (1965). *Applied Geochronology*. London.

Hann, E. M. (1899). Deep pumpings at the Elliott Colliery. *Proc. S. Wales Inst. o, Eng. 21* (5), 248-256.

Hanna, F. K. (1966). A technique for measuring the rate of erosion of cave passages. *Proc. Univ. Bristol Speleo. Soc. 11* (1), 83-88.

Hanwell, J. D. and Newson, M. D. (1970). *The Great Storms and Floods of July 1968 on Mendip*. Meare, Somerset.

Harland, W. B. *et. al.* (eds) (1964). *The Phanerozoic Time-Scale*. Geographical Society, London.

Harmon, R. S. (1971a). The application of stable carbon isotope studies to karst research: Part I, Background and theory. *Caves and Karst, 13* (3), 17-28.

Harmon, R. S. (1971b). The application of stable carbon isotope studies to karst research: Part II, An example from central Pennsylvania. *Caves and Karst, 13* (4), 29-35.

Harmon, R. S. (1972). The application of stable carbon isotope studies to karst research: reply. *Caves and Karst, 14* (2), 13-16.

Harper, C. T. (1973). *Geochronology Radiometric Dating of Minerals and Rocks*. Chichester.

Hendy, C. H. (1970). The use of ^{14}C in the study of cave processes. *Proc. Nobel Symposium*, Uppsala, 1969, 419-433.

Hendy, C. H. (1971). The isotopic geochemistry of speleothems—I. The calculation of the effects of different modes of formation on the isotopic composition of speleothems and their applicability as palaeoclimatic indicators. *Geochimica et Cosmochimica Acta, 35*, 801-824.

Hendy, C. H. and Wilson, A. T. (1968). Palaeoclimatic data from speleothems. *Nature, 219*, 48-51.

Hensler, E. (1968). Some examples from the Southern Harz Region of cavern formation in gypsum. *Trans. Cave Res. Grp. G.B. 10* (1), 33-44.

Herak, M. and Stringfield, V. T. (eds) (1972). *Important Karst Regions of the Northern Hemisphere*. Amsterdam.

Howard, A. D. (1964a). Model for cavern development under artesian ground water flow. With special reference to the Black Hills. *Bull. Nat. Speleo. Soc. 26* (1), 7-16.

Howard, A. D. (1964b). Process of limestone cave development. *Int. J. Speleo. 1* (1/2), 47-60.

Howard, A. D. (1966). Verification of the Mischungskorrosion effect. *Cave Notes*, *8* (2), 9-12.

Howard, A. D. and Howard, B. Y. (1972). The application of stable isotope studies to karst research—Discussion. *Caves and Karst*, *14* (2), 9-13.

Hughes, T. McK. (1887). On the drifts of the Vale of Clywyd and their relation to the caves and cave deposits. *Q. J. Geol. Soc. 43*, 73-120.

Hughes, T. McK. (1888). On the Cae Gwyn Cave. *Q. J. Geol. Soc. 44*, 112-147.

International Atomic Energy Agency (I.A.E.A.) (1967a). *Radioactive Dating and Methods of Low-Level Counting*. Vienna.

International Atomic Energy Agency (1967b). *Isotopes in Hydrology*. Vienna.

International Atomic Energy Agency (1968). *Guidebook on Nuclear Techniques in Hydrology*. Vienna.

Jennings, J. N. (1968). Syngenetic karst in Australia. *Contributions to the Study of Karst*. Res. Schl. of Pacific Stud., Dept. of Geog. Publication G/5, Canberra, 41-110.

Jennings, J. N. (1971). *Karst*. Cambridge, U.S.A.

Jones, R. J. (1965). Aspects of biological weathering of limestone pavement. *Proc. Geol. Assoc. 76*, 421-433.

Jones, W. K. (1971). Characteristics of the underground floodplain. *Bull. Nat. Speleo. Soc. 33* (3), 105-114.

Kammholz, H. (1966). On the problems relating to the covered karst in the south-eastern foreland of the Harz Mountains. In Štelcl, O. (1966), 51-62.

Kaufman, A. and Broecker, W. (1965). Comparison of Th^{230} and C^{14} ages for carbonate materials from Lakes Lahontan and Bonneville. *J. Geophys. Res. 70*, 4039-4055.

Kerney, M. P., Brown, E. H. and Chandler, T. J. (1964). The late-Glacial and post-Glacial history of the chalk escarpment near Brook, Kent. *Proc. Roy. Soc.* Ser. B, *248*, 135-204.

King, R. J. (1966). Epi-syngenetic mineralization in the English Midlands. *Mercian Geologist*, *1* (4), 291-301.

Kirkaldy, J. F. (1971). *Geological Time*. Edinburgh.

Knox, G. (1933). The Mid-Glamorgan water supply. *Proc. S. Wales Inst. of Eng. 49*, 250-271.

Lange, A. L. (1959). Introductory notes on the changing geometry of cave structures. *Cave Studies*, *11*, 69-90.

Lewin, J. (1969). The formation of chalk dry valleys: the Stonehill Valley, Dorset. *Biul. Peryglacjalny*, *19*, 345-349.

Malott, C. A. (1937). The invasion theory of cavern development (abstract). *Proc. Geol. Soc. Am.* 323.

Martel, E. A. (1897). *Irlande et Cavernes Anglaises*. Paris.

Martel, E. A. (1921). *Nouveau Traité des Eaux Souterraines*. Paris.

Maucci, W. (1952). L'ipotesi del' 'erosione inversa' come contribute allo studio della speleogenesi. *Boll. Soc. Adriatica di Scienze Naturali*, Trieste, *46*, 1-60.

Maucci, W. (1958). Il fenomeno della retroversione nella morfogenesi degli inghiottitoi. *Atti del VII Congresso Naz. di Spel.* (Sardegna, 1955), 221-237.

McConnell, H. and Horn, J. M. (1972). Probabilities of surface karst. *In* R. J. Chorley (ed.), 1972, 111-133.

Mitchell, G. F. *et al.* (1973). *A Correlation of Quaternary Deposits in the British Isles.* Geol Soc. London, Spec. Rept. No. 4.

Moneymaker, B. C. (1941). Subriver solution cavities in the Tennessee Valley. *J. Geol. 49*, 74-86.

Moneymaker, B. C. (1949). Limestone cavities in the Papaloapan Basin. *J. Tennessee Acad. Sc. 24*, 117-122.

Morehouse, D. F. (1968). Cave development via sulfuric acid. *Bull. Nat. Speleo. Soc. 30* (1), 10-1.

Morgan, R. P. C. (1971). A morphometric study of some valley systems on the English chalklands. *Trans. Inst. Br. Geog. 54*, 33-44.

Mowat, G. D. (1962). Progressive changes of shape by solution in the laboratory. *Cave Notes*, *4*, 45-49.

Myers, J. O. (1948). The formation of Yorkshire caves and potholes. *Trans. Cave Res. Grp. G.B. 1* (1), 26-29.

Myers, J. O. (1955). Cavern formation in the Northern Pennines. *Trans. Cave Res. Grp. G.B. 4* (1), 29-49.

Newson, M. D. (1971a). A model of subterranean limestone erosion in the British Isles based on hydrology. *Trans. Inst. Br. Geog. 54*, 55-70.

Newson, M. D. (1971b). The role of abrasion in cavern development. *Trans. Cave Res. Grp. G.B. 13* (2), 101-108.

Ollier, C. D. and Tratman, E. K. (1969). Geomorphology of the caves. Chapt. 4 of Tratman, E. K. (1969).

O'Reilly, P. M., O'Reilly, S. E. and Fairbairn, C. M. (1969). *Ogof Ffynnon Ddu. Penwyllt, Breconshire.* S. Wales C.C., Swansea.

Pasquini, G. (1973). Forced flow passages in karst massifs. *Trans. Cave Res. Grp. G.B. 15* (2), 89-90.

Pearson, F. J. (1965). Use of C^{13}/C^{12} ratios to correct radiocarbon ages of materials initially diluted by limestone. *Proc. VIth Int. Conf. on C^{14} and H^3 Dating*, 357-366.

Pearson, F. J. and Hanshaw, B. B. (1970). Sources of dissolved carbonate species in groundwater and their effects on carbon-14 dating. In *Isotope Hydrology*, 1970, Int. Atomic Energy Comm., Vienna, 271-286.

Pigott, C. D. (1962). Soil formation and development on the Carboniferous Limestone of Derbyshire: parent materials. *J. Ecol. 50*, 145-156.

Pigott, C. D. (1965). The structure of limestone surfaces in Derbyshire. *Geog. J. 131*, 41-44.

Piper, A. M. (1932). *Ground Water in North Central Tennessee.* Water Supply Paper 640, U.S. Geol. Svy.

Pitty, A. F. (1971). Evidence related to the development of avens, from karst water studies in Peak Cavern, Derbyshire. *Trans. Cave Res. Grp. G.B. 13* (1), 53-55.

Pohl, E. R. (1955). *Vertical Shafts in Limestone Caves*. Occ. Paper No. 2,, Nat. Speleo. Soc. Arlington, Va.

Priesnitz, K. (1969a). Kurze Übersicht über den Karstformenschatz des südwestlichen Harzrandes. *Jh. Karst-u. Höhlenkunde*, *16*, (9), 11-23.

Priesnitz, K. (1969b). Das Nixseebecken, ein Polje im Gipkarst des südwestlichen Harzvorlandes. *Ibid*. 73-82.

Railton, C. L. (1953). The Ogof Ffynnon Ddu system. *Cave Res. Grp. G.B*. Pub. No. 6. Leamington.

Reid, C. (1887). On the origin of the dry chalk valleys and of the coombe rock. *Q. J. Geol. Soc*. London, *43*, 364-373.

Renault, P. (1952a). Distinction de deux types d'avens sur les Plans de Canjuers (Var). *C.R. Acad. Sci. Paris*, *234*, 1519-1520.

Renault, P. (1952b). Influence du sens se circulations aquifères, sur le creusement des avens des Plans de Canjuers (Var). *C.R. Acad. Sci. Paris*, 1672-1673.

Renault, P. (1957). Sur deux processes d'effrondrement karstique. *Ann. de Spél*. *12*, 19-46.

Renault, P. (1970). *La Formation des Cavernes*. Paris.

Reynolds, D. L. (1961). Lapiés and solution pits in Olivine-Dolerite Sills at Slieve Gullion, Northern Ireland. *J. Geol*. *6* , 110-117.

Rhoades, R. and Sinacori, M. N. (1941). The pattern of ground-water flow and solution. *J. Geol*. *49*, 785-794.

Robinson, P. L. (1957). The Mesozoic fissures of the Bristol Channel area and their vertebrate faunas. *J. Linnean Soc*. (Zool.), *43*, 260-282.

Roglić, J. (1957). Quelques problèmes fondamentaux du karst. *L'Information Géog*. *21* (1), 1-12.

Roglić, J. (1964). "Karst Valleys" in the Dinaric karst. *Erdkunde*, *18* (2), 113-116.

Sakanoue, M., Konishi, K. and Komura, K. (1967). Stepwise determinations of thorium, protactinium and uranium isotopes and their applications in geochronological studies. *In* Int. Atomic Energy Agency, 1967, 313-329.

Sellstedt, H., Engstrand, L. and Gejvall, N. G. (1966). New application of radiocarbon dating to collagen residue in bones. *Nature*, *212*, 572-574.

Semmel, A. (ed.) (1973). Neue Ergebnisse de Karstforschung in den Tropen und im Mittelmeeraum. *Geog. Zeit. Beihefte*, *32*, Wiesbaden.

Shotton, F. W. (1966). The problems and contributions of methods of absolute dating within the Pleistocene period. *Q. J. geol. Soc. London*, *122*, 356-383.

Shotton, F. W. and Williams, R. E. G. (1971). Birmingham University radiocarbon dates, V. *Radiocarbon*, *13* (2), 141-156.

Simpson, E. (1935). Notes on the formation of Yorkshire caves and potholes. *Proc. Univ. Bristol Speleo. Soc*., *4*, 224-232.

Slipčević, A. and Planninick, A. (1974). The age determination of the secondary limestone sediments by the method of radioactive carbon. *Naše jame*, *15* (for 1973), 71-75. (In Serbo-Croat, with English summary.)

Small, R. J. (1961). The morphology of chalk escarpments: a critical discussion. *Trans. Inst. Br. Geog*. *29*, 71-90.

Small, R. J. (1964). Escarpment dry valleys of the Wiltshire Chalk. *Trans. Inst. Br. Geog*. *34*, 33-52.

Smith, D. I. (1971). The concepts of water flow and water tables in limestone. *Trans. Cave Res. Grp. G.B. 13* (2), 95-99.

Štelcl, O. (ed.) (1965). *Problems of the Speleological Research*. I. Prague.

Štelcl, O. (ed.) (1966). *Problems of the Speleological Research*. II. Brno.

Štelcl, O. (ed.) (1969). *Problems of the Karst Denudation*. Brno.

Stuiver, M. (1970). Tree ring, varve, and carbon-14 chronologies. *Nature*, *228*, 454-455.

Suess, H. E. (1973). Natural radiocarbon. *Endeavour*, *32*, 34-38.

Sutcliffe, A. J. (1960). Joint-Mitnor Cave, Buckfastleigh. *Trans. Torquay Nat. Hist. Soc. 13* (1) (for 1958-59), 3-28.

Sweeting, M. M. (1950). Erosion cycles and limestone caverns in the Ingleborough district. *Geog. J. 115*, 63-78.

Sweeting, M. M. (1953). The enclosed Depression of Carran, Co Clare. *Irish Geography*, *2*, 218-224.

Sweeting, M. M. (1965). Denudation in limestone regions: A symposium. I—Introduction. *Geog. J. 131*, 34-37.

Sweeting, M. M. (1966). The weathering of limestones, with particular reference to the Carboniferous Limestones of Northern England. In Dury, G. H. (1966), 177-210.

Sweeting, M. M. (1968). The solution of limestones. *Research Methods in Geomorphology*, 1st Guelph Symposium on Geomorphology, 1-21.

Sweeting, M. M. (1972). *Karst Landforms*. London.

Sweeting, M. M. (1973). Some results and applications of karst hydrology: A symposium. I. Some aspects of karst hydrology. *Geog. J. 139*, 280-285.

Swinnerton, A. C. (1932). Origin of limestone caverns. *Bull. geol. Soc. Am. 43*, 662-693.

Swinnerton, A. C. (1942). Hydrology of limestone terranes. *In* Meinzer, O. E., *Physics of the Earth: IX Hydrology*. New York.

Thomas, T. M. (1954a). Solution subsidence outliers of Millstone Grit on the Carboniferous Limestone of the North Crop of the South Wales Coalfield. *Geol. Mag. 90*, 73-82.

Thomas, T. M. (1954b). Swallow holes on the Millstone Grit and Carboniferous Limestone of the South Wales Coalfield. *Geog. J. 120*, 468-475.

Thomas, T. M. (1963). Solution subsidence in South-East Carmarthenshire and South-West Breconshire. *Trans. Inst. Br. Geog. 33*, 45-50.

Thomas, T. M. (1970). The limestone pavements of the North Crop of the South Wales Coalfield. *Trans. Inst. Br. Geog. 50*, 87-105.

Thomas, T. M. (1971). Gash-breccias of South Pembrokeshire: fossil karstic phenomena? *Trans. Inst. Br. Geog. 54*, 89-100.

Thomas, T. M. (1973). Solution subsidence mechanisms and end-products in south-east Breconshire. *Trans. Inst. Br. Geog. 60*, 69-86.

Thornbury, W. D. (1969). *Principles of Geomorphology*, 2nd edn. New York.

Thrailkill, J. (1968). Chemical and hydrological factors in the excavation of limestone caves. *Bull. Geol. Soc. Am. 79*, 19-46.

Tratman, E. K. (1955). Vertical development in some Irish Caves. *Trans. Cave Res. Grp. G.B. 4* (1), 1-27.

Tratman, E. K. (1957). Some problems of solution in caves under vadose conditions. *Trans. Cave Res. Grp. G.B. 5* (1), 53-59.

Tratman, E. K. (ed.) (1969). *The Caves of North-West Clare, Ireland.* Newton Abbot.

Trudgill, S. T. (1972). The influence of drifts and soils on limestone weathering in N.W. Clare. *Proc. Univ. Bristol Speleo. Soc. 13* (1), 113-118.

Walkden, G. (1972). Karstic features in the geological record. *Trans. Cave Res. Grp. G.B. 14* (2), 180-183.

Walsh, P. T., Boulter, M. C., Ijtaba, M. and Urbani, D. M. (1972). The preservation of the Neogene Brassington Formation of the southern Pennines and its bearing on the evolution of Upland Britain. *J. Geol. Soc. Lond. 128* (6), 519-559.

Walsh, P. T. and Brown, E. H. (1971). Solution subsidence outliers containing probable Tertiary sediment in north-east Wales. *Geol. J.* 7 (2), 299-320.

Waltham, A. C. (1970). Cave development in the limestone of the Ingleborough District. *Geog. J. 136*, 574-585.

Waltham, A. C. (1971). Controlling factors in the initiation of caves. *Trans. Cave Res. Grp. G.B. 13* (2), 73-80.

Waltham, A. C. (ed.) (1974). *The Limestones and Caves of North-West England.* David and Charles, Newton Abbot.

Warwick, G. T. (1955). Polycyclic swallow holes in the Manifold Valley, Staffordshire. *Proc. 1er Congrès Int. de Spél.* (Paris, 1953), *2*, 59-68.

Warwick, G. T. (1956). Caves and glaciation: I—Central and Southern Pennines and adjacent area. *Trans. Cave Res. Grp. G.B. 4* (2), 125-160.

Warwick, G. T. (1959). Some observations on by-passed swallow holes in the Meuse and Lesse Valleys. *Mém. du Colleq. Int. de Spél.* (*Bruxelles, 1958*), 66-72.

Warwick, G. T. (1960). The effect of knick-point recession on the water-table and associated features in limestone regions, with special reference to England and Wales. *Zeit. für Geom. Supplementband*, 292-299.

Warwick, G. T. (1962a) The characteristics and development of limestone regions in the British Isles with special reference to England and Wales. *Actes de 2me Congrès Int. de Spél.* (Bari-Lecce-Salerno, 1958), *1*, 79-105.

Warwick, G. T. (1962b). The origin of limestone caves. *In* Cullingford, C. H. D. (1962), Chapt. 3, 55-82.

Warwick, G. T. (1964). Dry valleys of the southern Pennines, England. *Erdkunde*, *18*, 116-123.

Warwick, G. T. (1965). Influent streams of the southern and central Pennines. *Geog. J. 131*, 49-53.

Warwick, G. T. (1968a). Some primitive features in British Caves. *Proc. 4th Int. Spel. Congress Yugoslavia*, (*1965*), *3*, 239-252.

Warwick, G. T. (1968b). A subterranean knickpoint and associated gravels in Clapham Cave, Yorkshire, England. *Proc. 4th Int. Speleo. Cong. Yugoslavia* (*1965*), *3*, 637-641.

Warwick, G. T. (1971). Caves and the Ice Age. *Trans. Cave Res. Grp. G.B. 13* (2), 123-130.

Warwick, G. T. (1974a). River Karren (Flusskarren). *Proc. 6th Int. Spel. Cong.* (Olomouc, 1973). *In the press.*

Warwick, G. T. (1974b). The metamorphosis of Karren in the North of England. Ibid.

Watson, R. A. (1966a). Underground solution canyons in the Central Kentucky karst, U.S.A. *Int. J. Speleo. 2*, 369-376.

Watson, R. A. (1966b). Central Kentucky karst hydrology. *Bull. Nat. Speleo. Soc. 28* (3), 1959-1966.

Weaver, J. D. (1973). The relationship between jointing and cave passage frequency at the head of the Tawe Valley, South Wales. *Trans. Cave Res. Grp. G.B. 15* (3), 169-173.

White, E. L. and White, W. B. (1968). Dynamics of sediment transport in limestone caves. *Bull. Nat. Speleo. Soc. 30* (4), 115-129.

White, E. L. and White, W. B. (1969). Processes of cavern breakdown. *Bull. Nat. Speleo. Soc. 31* (4), 83-96.

White, W. B. (1960). Termination of passages in Appalachian caves as evidence for a shallow phreatic origin. *Bull. Nat. Speleo. Soc. 22* (1), 43-53.

White, W. B. (1969). Conceptual models for carbonate aquifers. *Ground Water*, 7 (3), 15-21.

White, W. B. and Longyear, J. (1962). Some limitations on speleogenetic speculation imposed by the hydraulics of groundwater flow in limestones. *Nittany Grotto Newsletter*, *10*, (9), 155-167.

Williams, P. W. (1966). Morphometric analysis of temperate karst landforms. *J. Irish Spel. 1*, 23-32.

Williams, P. W. (1967). Limestone pavements with special reference to Western Ireland. *Trans. Inst. Br. Geog. 40* (for 1966), 155-172.

Williams, P. W. (1968). An evaluation of the rate and distribution of limestone solution and deposition in the River Fergus Basin, Western Ireland. *Contributions to the Study of Karst*. Research School of Pacific Studies, Dept. of Geog. Pub. G/5 (1968), Canberra, 1-40.

Williams, P. W. (1972a). Morphometric analysis of polygonal karst in New Guinea. *Bull. Geol. Soc. Am. 83*, 761-796.

Williams, P. W. (1972b). The analysis of spatial characteristics of karst terrains. In Chorley, R. J. (ed.), 1972, Chapt. 5, 135-163.

Woodward, H. P. (1961). A stream piracy theory of cave formation. *Bull. Nat. Speleo. Soc. 23* (2), 39-58.

Zotov, V. D. (1941). Pot-holing of limestone by Development of solution cups. *J. Geomorph. 4* (1), 71-73.

4. Caves in Rocks of Volcanic Origin

C. Wood

The superficial resemblance between the forms of limestone and of volcanic rocks is often striking. Features characteristic of karst—lapiés, solution and collapse dolines, swallow holes, extensive cave systems, gorges and natural bridges—may all be seen in regions of volcanic rocks. Solution is not, however, the dominant process of landform evolution in volcanic rocks, and the geomorphology of some volcanic tracts has been termed "pseudokarst" (Kosack, 1952; Halliday, 1954, 1960). Without examining the concept of pseudokarst in its entirety, caves in volcanic rocks will be discussed here within this context of karst-like landforms of non-solutional origin.

The investigation of caves in rocks of volcanic origin is known as "vulcanospeleology". Volcanic caves present as wide a diversity of size, morphology and origin as limestone caves, though they are somewhat restricted in occurrence to rocks which are of Tertiary age, or younger, because weathering agents destroy them relatively rapidly. Speleogenesis may occur both in lavas and in pyroclastic rocks.

In lavas, the special flow characteristics resulting in cave formation are a function of fluidity, which is determined in turn by the gas content, the chemical composition and the temperature of the lava. "Acid" lava rich in silica, such as rhyolite or dacite, is too viscous to form caves. Basaltic lava, on the other hand, has greater mobility and it is in this rock that volcanic caves attain their greatest significance. Even here, basaltic lava flows fall into three general types—*pahoehoe* (or "ropy"), *aa* (or "cindery") and *block lava*—depending upon the physical properties of the magma and the resulting forms produced by the flow, each type differing in importance as a cave-former.

In pahoehoe lava the volatiles remain trapped in solution, which results in slower congelation and a greater mobility than in aa or block lava.

Farther from the vent, however, the loss of gas and progressive cooling induces a loss in fluidity and a change to aa or block varieties. Pahoehoe is therefore regarded as the fundamental form of basaltic lava (Macdonald, 1967, p. 2). Flow continues beneath an elastic skin which becomes twisted into ropy wrinkles and congeals with a smooth glistening surface. The front advances by the protrusion of tongues of mobile magma through ruptures in its still half-consolidated skin.

Aa and block lava have a surface form and mechanism of advance which is distinct from pahoehoe. In aa, the rapid loss of volatiles leads to a sudden increase in viscosity and rapid consolidation. Spiny-surfaced masses of solidifying lava are ejected by the mobile magma, giving to the flow a characteristic rough, jagged, clinkery surface. The front of the flow advances with a rolling motion which has been described as resembling the front of the endless track of a tractor, sluggishly burying the blocks of clinker which avalanche down from the top of the flow. Block lava has a surface made up of fragments which are more regular than aa, lacking its rough, spinose character.

Although not common, lava tube caves and other minor cavities may form in aa lava flows (Powers, 1920). The rarity of such features may be due to the greater viscosity of aa and its inability to vacate tubes after cessation of activity at the vent. In pahoehoe, because lava tube caves are regarded as characteristic features, and because a whole variety of smaller cavities are associated with crustal forms upon its surface, discussion of lava caves will be restricted principally to this type of lava.

Accumulations of fragmentary rocks and minerals blown from the vent by the explosive discharge of volcanic gases are known as pyroclastic deposits. Larger and coarser fragments, including volcanic bombs, lumps of scoria or pumice, or blocks of older rocks, pile up near the vent to form roughly bedded volcanic breccia or agglomerate. Finer material that showers down from volcanic clouds as ash travels farther from the vent and, when more or less indurated, is known as tuff. Often beds of agglomerate and tuff alternate with lava flows, and it is in such situations that speleogenesis is most likely in pyroclastic rocks.

CATEGORIES OF VOLCANIC CAVES

The literature reveals few attempts to classify caves in volcanic rocks. In his classic work Kyrle (1923) recognized speleogenesis in lavas and in pyroclastic rocks, and listed two important types of lava caves: "blasenhohlen" and "lavahohlen", corresponding to gas pocket caves and lava tube caves respectively. An attempt was made by Poli (1959) to classify genetically the wide variety of caves he examined on the flanks of Mount Etna in Sicily. A more sophisticated classification was advocated by Montoriol-Pous and de Mier (1969). Caves formed during the period of consolidation of the rock in which they are found were termed "syngenetic", while caves formed after consolidation of the enclosing rock by weathering and erosion were termed "epigenetic". These two major categories were then subdivided into minor groups. Similarly Szentes (1971) classified the caves in the volcanic rocks of Hungary into two main groups, according to whether they formed though primary processes or secondary processes. Concerned solely with lava caves, Lindsley (1966) listed the more important categories of lava caves, while Harter and Harter (1970) devised a classification of lava tube caves.

There is a wide diversity of speleological forms to be found within volcanic rocks. To avoid confusion the classification of the more important volcanic cavities will be divided into those landforms formed by processes which originate within the earth's crust (endogenetic), and those formed by processes which originate on the outside of the Earth's crust (exogenetic). Caves formed in volcanic rocks fall into both of these categories:

Endogenetic: (*a*) cavities beneath pressure ridges; (*b*) cavities beneath spatter cones; (*c*) blister caves; (*d*) vent caves; (*e*) fissure caves; (*f*) lava tube caves.

Exogenetic: (*a*) caves in bedded tuffs and agglomerate.

MINOR VOLCANIC CAVITIES OF ENDOGENETIC ORIGIN

Because the thickened pahoehoe crust is buckled by the movement of the mobile lava beneath, or by gases escaping from the depths, the surface of a lava flow is commonly diversified by features with a relatively strong relief. Many of these features, when ultimately vacated of liquid lava, contain cavities of variable sizes making the flow surface very cavernous in places.

Of larger order, ranging from 1 to 12 m in height, are long narrow ridges which lie transversely to the direction of flow, termed pressure ridges. They result from compression of the crust by the movement of the mobile magma beneath, and cavities are present in many. Caves are commonly 1-2 m in height, with a roughly triangular cross-section, extending unbroken for lengths of up to 800 m. Examples of cavities in pressure ridges were described by Russell (1902) from Cinder Buttes, Idaho.

Sometimes gases escaping from the lava carry liquid clots of lava through cracks in the crusted surface of the flow to form steep-sided heaps of welded spatter known as spatter cones. Frequently withdrawal of the magma results in the formation of a dome-shaped chamber, and sometimes spatter cones afford entrance to lava tube caves. In the Craters of the Moon National Monument, Idaho, some cavities are 9 m in diameter and up to 20 m in depth (Peck, 1962). Ollier (1967) described a cavity in a spatter cone near Mt Eccles, Victoria, Australia, which was shaped like an inverted wineglass, with a depth of 30 m, and walls festooned with lava stalactites.

Some cavities develop from the true blister-like lifting of a lava sheet by pockets of steam or other gases trapped within the flow. Small cavities in the surface of McCarty's flow are only 1 m in length and 13 cm in depth, and clearly formed through the coalescence of vesicles (entrapped bubbles of gas) (Nichols, 1946). Larger features were described from Iceland (Mercer, 1966), though from their description they may have been confused with tumuli (domical upwellings of the flow surface by differential pressure on the crust from the mobile flow beneath) or spatter cones. Many of these domes were ruptured at the top revealing bell-chambers of large size, some of which had circular tunnels leading off. Some other interesting blister caves formed in ignimbrite (welded ash) on the lower slopes of the Fantalle volcano, Ethiopia (Sutcliffe, 1970; Gibson, 1974). Many blisters held circular domed chambers 18-30 m across and about 3-5 m high. It was thought that the layered ignimbrite had become sufficiently plastic to dome up locally where gases had built up.

At the vent it is usual for the level of the magma chamber to subside after cessation of extrusive activity. A cavity formed in this way described by Lindsley (1966) from the crater of Mt Pisgah, California, consisted of a narrow passageway angling down for 3-5 m. A rather different feature was described from Auckland, New Zealand (Anon., 1957), with a depth of 14 m terminating in a choke. From its description this feature may be similar to the "mortars" examined by Stearns (1924) in the Craters of the Moon National Monument, Idaho, and developed by jets of steam or gas. Some mortars had several branches which united to form a main tube, though most were filled with debris 3-5 m below the surface.

Open fissures are common in lava fields and some formerly carried molten lava to the surface. They are mostly narrow cracks, only a few feet wide and not more than 6 m deep, though they may extend for considerable distances. One enormous fissure system, however, contributed much of the lava of the Snake River Plain, Idaho, and has a depth of several hundred feet (Halliday, 1959, 1966; Ross, 1969). The famous

Grjotagjá fissure at Myvatn, NE Iceland, is celebrated for its caverns which contain a hot spring at a temperature of 38·0°C (105·5°F). Some 260 m of cave has been surveyed in this fissure (England, 1959).

LAVA TUBE CAVES

The feeding rivers of basaltic lava flows, particularly pahoehoe, are extremely complicated. Flow patterns frequently consist of an internal network of interconnecting conduits which sometimes attain considerable vertical and horizontal complexity. The speleologist has an interest in these conduits, or lava tubes, for many are emptied of liquid lava, either wholly or partly, to form lava tube caves. These are speleological features which must rank in importance with limestone caves for size, beauty of internal decoration, complexity of form and detail of formation. They are not merely curious geological features of limited occurrence, for it seems the development of tubes is a characteristic of pahoehoe, and they play an important role in moving lava away from the vent during periods of extrusive activity. A few lava tube caves have been discovered in andesitic lavas.

In Europe, lava tube caves are found in Iceland, the Azores, the Canary Islands, and on some of the Mediterranean volcanoes, notably Mt Etna in Sicily. The Middle East has examples in Syria and Israel, while in Africa lava tube caves have been reported in the Cameroun Mountains, Malagassy, Ethiopia, Uganda, Kenya and are most probably to be found in other countries dissected by the African rift valley system, for example, Zaire. Australia and New Zealand have important locations, and some of the longest lava tube caves in the world are found in Korea. Japan has lava tube caves and so also have many of the Pacific islands, such as Easter Island, Galapagos and, of course, the Hawaiian Islands. References are coming to light of lava tube caves in Argentina, and undoubtedly other locations on that sub-continent will be discovered soon. The largest number of explored lava tube caves are in the western U.S.A. and in Mexico. There are notable exceptions to this list, but it is not regarded that such areas are devoid of lava tube caves, rather that the features are difficult to locate, or there is as yet insufficient local interest.

Even though long conduits are initially developed, the thin fragile roof of drained lava tubes readily collapse (see Plate 4.1), and continuous cave segments are difficult to find. To date, the longest lava tube cave is the Cueva del Viento, on the island of Tenerife, Canary Islands, but even here the total of 10 km is broken into two segments of 7·9 km and 2·1 km respectively. These lengths are short, however, in terms of partly

PLATE 4.1. A roof collapse at Surtshellir, Iceland, showing typical lava structures.

collapsed tube systems in the U.S.A. and Australia, which have lengths of nearly 40 km in some cases. There is a great diversity too in the nature of lava tube caves (see Fig. 4.1). The Cueva del Viento, for example, comprises a fantastic complex of small interconnecting passages on three different levels. It contrasts with some Icelandic caves, such as the Surtshellir/Stephánshellir system of interconnecting spacious passages formed at one level, or the beautiful Viðgelmír cave with its vast, single, meandering passage. It is interesting also that some caves may be entered through the wall of the crater at the vent, such as one of the Gullborgarrhaun caves in Iceland; while other caves seem to originate considerable distances from the vent, such as the Surtshellir/Stephánshellir system and Raufarhólshellir which lie 26 km and 10 km respectively below the vent.

In terms of speleogenesis, it is possible to distinguish in lava tube caves forms and features which have developed in two quite separate phases of development. The first phase involves the development of a conduit beneath the congealed surface of the flow, through which liquid lava is transmitted from the vent to the advancing flow front. In the second, activity at the vent ceases, the conduit drains and is modified by the adherence of cooled lava to the walls. Both phases are responsible for complexities in the morphology of the lava tube cave, and it is important not to confuse the two.

The cross-section of a lava tube cave is developed both by the mobile lava stream entirely filling the conduit, and by the degree of meandering which takes place within the flow. The diameter of caves is variable and

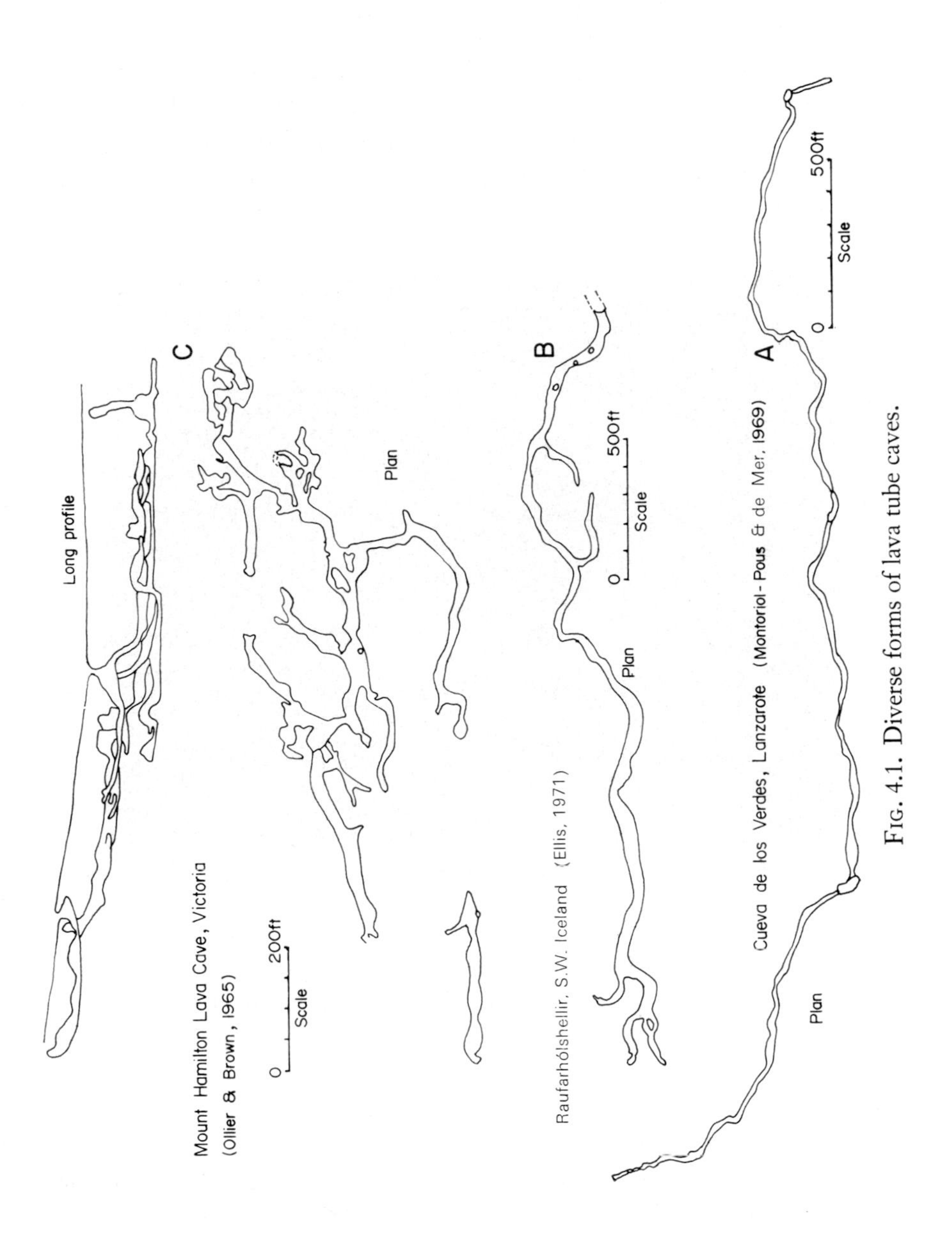

FIG. 4.1. Diverse forms of lava tube caves.

may range from 1 to 25 m. Commonly, in straight tube sections the cross-section is almost circular, though for a variety of reasons the perfect form is rarely attained. At meander bends the cross-section is asymmetrical with the highest part of the tube situated on the outside of the bend, a direct analogy with the form of a river meander. These cross-sectional forms, however, are developments of the first phase of speleogenesis, and the majority are subsequently modified from the moment draining of the tube commences.

PLATE 4.2. Lateral benches and roof collapse, Stephánshellir, Iceland (*Photo*: D. J. Wilkinson).

Lowering of the lava level within the tube may proceed spasmodically, with individual stands being marked by longitudinal deposits known as lateral benches (Plate 4.2). If the lowering of the lava level is arrested, it sometimes happens that a crust will form over the lava stream, so that with subsequent lowering of the lava level a horizontal partition, or false floor, may remain across the cave. This process may be repeated on numerous occasions, the partitions remaining whole to give the tube what appears to be superimposed tube levels, or they may collapse to leave shelf-like ledges that protrude from the walls. Occasionally, as at Surtshellir (Mills and Wood, 1971), the crust was still plastic when the lava stream left it, so that it sagged beneath its own weight.

Apart from modifying the cross-section of a tube, flow features ornament the walls and floors of the cave. Blocks of solid lava which have become incorporated in the liquid flow may gouge the walls of the conduit to produce flow grooves and flow ridges. In multi-level caves solidified cascades of lava, termed lava falls, may be the result of lava passing from level to level. The floor of a lava tube cave may show a wealth of interesting flow features. The more common rough, clinkery floor is caused by gases frothing to the surface as the lava cools, while flows broken into polygonal slabs are the result of jointing due to contraction upon solidification. Sometimes blocks of solid crust may fall from the roof of the cave while the floor is still plastic, and the resulting ripples may be "frozen" and preserved as a floor formation known as a splash concentric.

During the lowering of the lava level, burning gases maintain the temperature in the interior of the lava tube at 1200°C. Jagger (1947) observed this phenomenon through a "window" in the roof of the Postal Rift Tube, Hawaii, and attributed the maintenance of these high temperatures to a blast furnace effect. As a result, lava tube caves are lined with a black vitreous glaze that was sufficiently fluid at the time of its formation to trickle and produce decorations on the walls and ceiling. Usually cave walls exhibit a rippled structure, but if the glaze was highly fluid and flowed rapidly down the walls, vertical ridges or corrugations result, producing minute gour-like formations where each corrugation passes across a flow ridge. Commonly remelting of the ceiling of the tube causes blistering, though blisters may have burst before solidification to leave circular ridges a few inches across, from which hang tiny glazed stalactites.

In most lava tube caves lava stalactites and stalagmites are common features (Plate 4.3), though it is strange that some tubes are completely devoid of them. The larger forms usually hang from the outer edges of wall protuberances and are the result of the initial draining of the tube, or of lava flung as spatter on to the walls and ceiling. Smaller forms are made of glaze and possess a greater variety of shapes. Many are delicate rod or straw-like features, 6-13 mm in diameter, while others may be tapered or tear-drop in shape. Some straw stalactites have an erratic form consisting of contorted spiral structures, and others are partly crushed like a pipe-stem. Some are delicately ornamented with flow patterns. In the Viðgelmír cave in Iceland forests of stalagmites lie beneath walls festooned with rod and straw stalactites (Fig. 4.2). These stalagmites consist of tiny globules of glaze which are piled one above the other, so that heights of 30 cm are sometimes reached. In another Icelandic cave, Borgarhellir, straw stalactites are united with globular stalagmites to form erratic columns over 1·2 m high and individual straws exceed lengths of 1 m (Thorarinsson, 1957). Some interesting speleothems described by Jagger (1931) were "barnacle stalactites" which were described as extrusions of molten lava through pores in the walls of the cave. Similar extrusions which oozed through cracks formed thin,

PLATE 4.3. Straw stalactites and globular stalagmites of lava, Viðgelmír, Iceland. The stalagmites average 20 cm high (*Photo*: D. J. Wilkinson).

papery, ribbon stalactites. The examination of the chemistry and structure of lava stalactites has been carried out by several workers (Hjelmquist, 1932; Williams, 1963; Ollier and Brown, 1965). On many speleothems the outer skin was found to be coated with silvery magnetic oxide of iron. Internally the vesicles were elongated vertically and lined with crystals of augite and feldspar. In Government Cave, Arizona, Harter (1971a) distinguished between "lava drip pendants" and "lavacicle stalactites". He noted that lava drip pendants were only rarely responsible for the formation of lava stalagmites, and he separated the two stalactitic forms on the basis of vesicle arrangement.

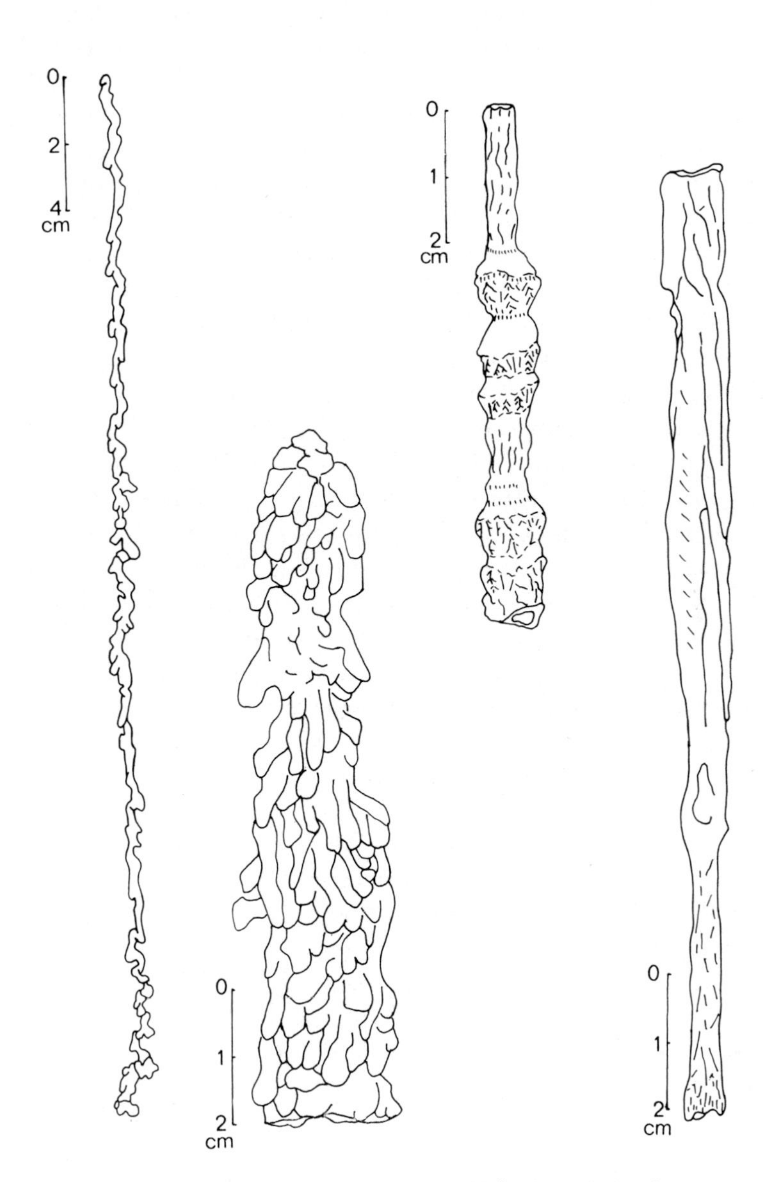

FIG. 4.2. Lava speleothems from Viðgelmír, Iceland.

The secondary features in lava tube caves obscure those details of the lava structures which provide evidence of the genesis of the original conduit. Only where roof segments or wall linings have collapsed are the important lava structures visible, but in many caves collapse is of infrequent occurrence. This meant in the past that models of tube genesis were highly speculative, and some even controversial. Today, new observations of lava tube systems in ancient lavas, and of tube formation during periods of active volcanicity, have contributed greatly to the discussion.

In Europe, an early interest in lava caves was shown by Victorian travellers in countries like Iceland and Sicily. Many formed an opinion of the origin of a particular cave, though the conception of some theories was often as fantastic as the weird landscape through which they travelled. Two popular explanations are notable. An earlier view held that lava caves were formed by blistering as gas escaped from the lava during its consolidation, and some authors believed the famous Surtshellir to consist of a chain of linked bubbles. The theory was based upon the apparently billowed and blistered surface of pahoehoe, but a little later some observers noted horizontal striations on the walls of the caves and regarded these features as evidence of flow of hot magma through the cave. A new theory then became popular and involved the draining of fluid magma from beneath the congealed crust of the lava flow during the later stages of its emplacement. This idea persists to the present day as an explanation for the genesis of lava tube caves, though it has taken many different forms.

In his study of the lava caves of Mt Etna, Sicily, Poli (1959), for example, believed that the wall structures in lava tube caves indicated laminar flow. The lava flows were thought to be constructed of many successively enclosed cylinders of lava whose viscosity increased externally. In the caves this was observed where the walls were composed of layered concentric structures (interpreted here as lava tube crusts). At the cessation of activity the more fluid internal cylinders drained of lava, leaving a tube-like cave.

Another variation of the traditional theme was suggested by Bravo (1964) as an explanation of the origin of the Cueva de los Verdes, Lanzarote. He thought the cross-sectional form of the cave indicated two phases of formation. In the first phase extensive flooding of lava erupted from the volcano "la Corona" formed the lava field "Malpais de la Corona". During the second phase, lava continued to be erupted, but in diminished quantity, so that it was confined to a channel eroded across the pre-existing lava field. Levees were developed along the borders of the channel by the ejection of scoriaceous material, and during a momentary cessation of flow a crust was formed over the lava. Bravo envisaged a deepening of the conduit by lava melting the floor, until total cessation of the flow led to the draining of the remaining fluid magma.

In many ways earlier observations of actively forming lava tubes in newly erupted pahoehoe supported the simple traditional concept. The Icelandic geologist Kjartansson, for example, was fortunate to observe the formation of a lava tube during the 1947-48 eruption of the Icelandic volcano "Hekla" (Kjartansson, 1949). He described how the lava was confined to a narrow channel which became partially crusted over, and how marginal levees developed by progressive welding of crustal fragments towards the centre of the channel until, ultimately, the channel became completely covered over. A similar process was also described by Wentworth and Macdonald (1953). Citing Stearn's observations of the 1935 eruption of Mauna Loa, Hawaii, and Macdonald's observations of the 1942 eruption, Wentworth and Macdonald showed how the flow near the vent was confined to a narrow channel, where levee construction proceeded by spattering and overflow, and where a roof could be formed across the channel by the jamming of crustal slabs carried along by the lava stream. Farther from the vent, they envisaged the margins of the flow to be fed by a myriad of small distributary tubes which branched from the main tube. They offered an alternative explanation of tube formation by describing how small flow units, or pahoehoe toes, could be elongated by repeated outbursts of lava from the front, and how each toe subsequently developed a shell by chilling, which gave, if drained, a small lava tube cave.

As new lava fields were discovered and explored by speleologists and geologists, however, it was realized that lava tube caves were much more common and possessed a greater diversity of form and size than was formerly held, and many workers became discontented with the traditional explanation. In his *Caves of Washington*, one of the most important regional surveys of lava tube caves, Halliday (1963, p. 5), for example, noted "as a group, these caves do not seem entirely in accord with the traditional concept of these caves as simple conduits with distal ramifications". Similarly, Ollier and Brown (1965, p. 225) noted that the traditional concept "does not account for all the observed shapes and structural features encountered in lava tubes". In order to explain the wide diversity of form met with in the lava tube caves of Victoria, Australia, Ollier and Brown advocated a more elaborate explanation of laminar flow, which was based upon discernible structures within the flow and the cave. They recognized in lava flows layers up to several feet thick which lay parallel with the flow surface. The layers were of compact basalt separated by trains of vesicles, buckles and partings. This "layered lava" was said to result from differential movement within one thick flow along shear planes. In consideration of the caves and the lava structure, Ollier and Brown suggested the following mechanism of internal flow. The liquid lava concentrated between laminae during the formation of the layered lava became segregated and came to occupy tubes running through the lava. This mobile lava eventually became concentrated in a few major channels that were a continuing source of heat, so that the earlier layered lava could be eroded. The end-result was cylinders of liquid lava flowing through tubes cut in virtually solid rock.

In later years this theory found a great deal of support, particularly from North American vulcanospeleologists, though it has become a controversial topic even there. Also, geologists and speleologists were still expounding on hypotheses that followed traditional lines (Kermode, 1970; Macdonald and Abbot, 1970) (Fig. 4.3). There was a reluctance to reject the traditional explanation, and Wood (1971) found in a later study of the Icelandic lava tube cave Raufarhólshellir that traditional concepts could still apply to more complicated cave systems. Like Ollier and Brown, Wood based his conclusions upon the relationship between flow structure and tube morphology. Cross-sections of the lava flow were identified in the main passage of Raufarhólshellir, and at the collapse entrance. The structures were similar to those recognized by Ollier and Brown, but at Raufarhólshellir it was felt the structural characteristics of the flow resulted from the superimposition of small flow units. In terms of tube morphology, Wood recognized four general passage types based upon detailed cross-sectional measurements. The smallest tube type was generally found above lavafalls in the extremities of the cave and had a form consisting of a flat floor and an arched roof. A second tube type, into which the first type passed at the lavafalls, carried three lateral benches, one of which was found above a lateral shelf and was continuous with the single lateral bench of the first tube type (Fig. 4.4). A third type of tube was the joint controlled, rectangular, breakdown tube, and a fourth was composed of a series of irregular forms of larger size that constituted the main tube. Cross-sections of the lava were rarely observed in the smaller tubes, but it was observed that because of the natural weaknesses at flow unit contacts, ropy surfaces were sometimes exposed. By the identification of successive contacts, therefore, the relationship between the flow structure and tube form could be understood. It appeared that the smallest

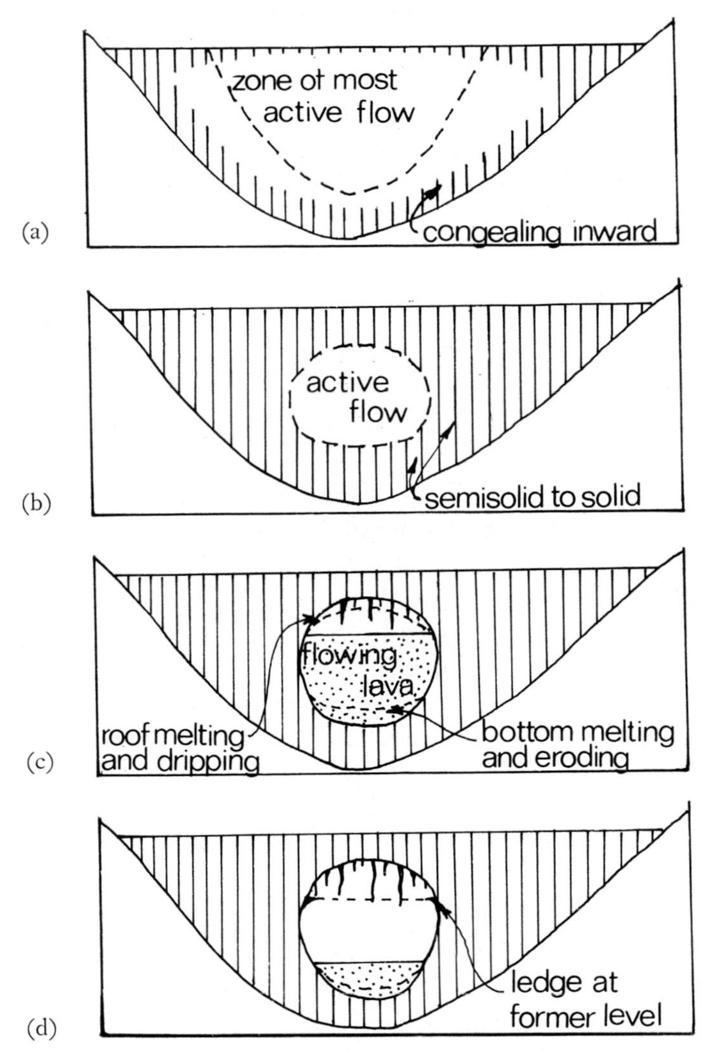

FIG. 4.3. Model of speleogenesis in lava (after Macdonald, 1970). (a) A lava flow (confined to a valley) develops a thin crust and starts to freeze inward from the edges, but the centre remains hot and continues to flow. (b) The active movement becomes restricted to cylindrical pipelike zone near axis of flow. (c) Supply of liquid lava diminishes and no longer entirely fills pipe. Burning gases heat roof and cause it to drip. (d) Further diminution of supply lowers level of liquid, which congeals to form tube floor.

tube passages represented the drained cores of single flow units, and that each flow unit represented a potential "primary tube unit". Larger tubes were seen to be constructed of multiples of this single unit, due to erosion and remelting of the crusts of the original flow units by the lava stream. In one case, where one unit was seen to diverge and then re-enter the main conduit, a small loop passage had been established.

In latter years, the development of lunar geology and the problems of lunar landform evolution has given an impetus to the study of terrestrial lava tubes. The study has leapt from almost total insignificance to great importance, because of the believed close analogy between lava channels and tubes and lunar sinuous rilles. Vulcanologists are also taking interest and have been impressed at the considerable role lava tubes play in the growth and importance of Hawaiian type volcanoes (Peterson and Swanson, 1974).

One of the first important studies of this nature was on the Bandera lava tubes of New Mexico by Hatheway and Herring (1970). The study was of an impressive complex of lava channels and lava tubes situated in eleven distinct basalt formations. All the tubes appeared to have formed in single flow units, except the Bandera Crater tube, traceable for 28·5 km, which was common to three flow units and had a more complicated origin. Hatheway and Herring discussed three alternative modes of formation of the Bandera Crater tube, and quoted methods of tube formation after Wentworth and Macdonald and Ollier and Brown. They concluded that the model proposed by Ollier and Brown, if slightly modified, adequately explained the formation of very long tube systems, while they regarded the processes recorded by Wentworth and Macdonald to apply only to shorter tubes. In consideration of the lunar implications, Hatheway and Herring analysed bend sinuosity, gradient and roof collapse. On the basis of their study of gradients, they suggested that the production of long tubes in olivine basalts would take place if gradients lay between 0°21′ and 0°35′.

Following this study, Greeley (1971a and b; Greeley and Hyde, 1971) have recently made a large contribution to the discussion on tube genesis, as a result of interest in lunar sinuous rilles. In his study of the lava tube caves of the Bend area of Oregon (Greeley, 1971a), the fieldwork in general confirmed the "layered lava" hypothesis of Ollier and Brown. Greeley also agreed with Hatheway and Herring in that two tube types were recognizable: minor and major lava tube caves. Minor lava tube caves were described as less than 10 m wide and a few hundred metres long, which formed in small, single flow units and often occupied the entire flow.

They were often feeders for larger tubes, or they formed in discrete lava flows which emanated straight from the vent. Most caves in the Bend area, however, were said to be major lava tube caves "of the type described by Ollier and Brown", and these were found in flows several kilometres long. Like Hatheway and Herring, Greeley also discussed sinuosity and gradient, and he concluded that the greater degree of complexity of the Horse system was attributable to a lesser gradient. Greeley (1971b) was also fortunate to observe actively forming lava channels and tubes during

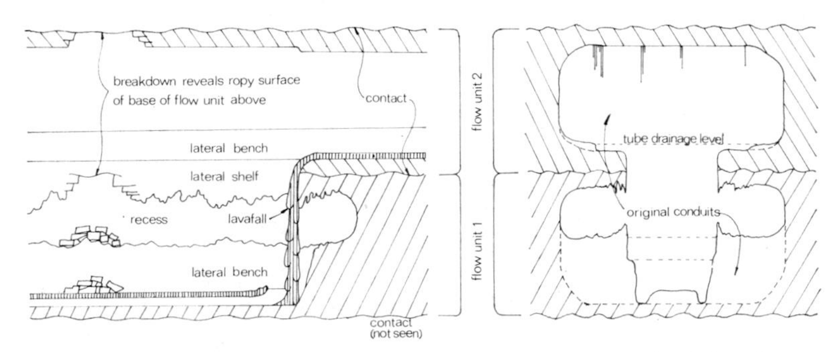

FIG. 4.4. Evidence of stacked conduits in Raufarhólshellir (after Wood, 1970).

the 1970 eruption of Kilauea, Hawaii. His observations showed that roofs over open channels were constructed by simple crusting, by the jamming and fusing together of crustal slabs, and by levee formation through overflow and spatter. He also noted multiple flow along rifts and suggested that this may be a mechanism leading to the formation of unusual cross-sections seen in some lava tube caves (Fig. 4.5). Greeley noted that a single channel could display braided channel flow, open flow, mobile crustal plates and roofed channel along its length. These observations were of later use in the study of 833 m of lava tube in the Cave Basalt, Mount St Helens (Greeley and Hyde, 1971), where they believed the caves to have two modes of formation. Some tube segments were developed in "layered lava" and resulted from laminar flow. Other segments were thought to be due to spatter accretion leading to the formation of arched levees and eventually a complete roof. They thought that laminar flow was a product of lesser gradients, while spatter accretion was the result of more turbulent flow on steeper gradients. Greeley and Hyde made the important point that subsequent lava flows could modify quite extensively the first formed tube by filling or partially filling, remelting the tube roof to form vertically elongated tubes, reshaping and eroding the tube walls, stacking additional tube levels above the first, or any combination of these (Fig. 4.5).

Cruikshank and Wood (1972) also observed the development of lava conduits on Kilauea in 1969, as part of a study concerned with the terrestrial analogues of lunar sinuous rilles. They examined lava channels left at various stages of development, and they were able to recognize stages in the roofing of channels to form lava tubes. They provided two examples of channel closure. Common to both examples, they proposed, a thin flow developed marginal levees by spatter along the flow boundary

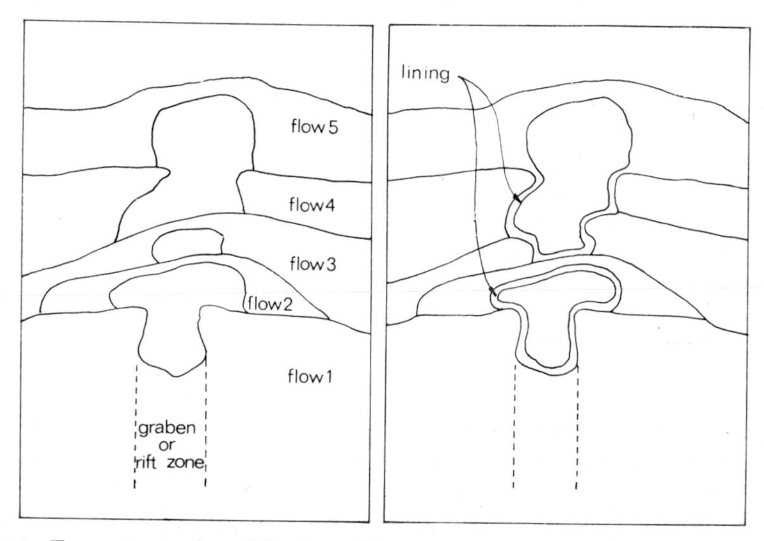

FIG. 4.5. Formation and modification of lava structures as a result of subsequent lava flows (after Greeley, 1971b).

and by slowly undercutting the channel walls. Later stages of development, however, either involved the formation of roofs by marginal spatter, so that arched levees fused and became reinforced by surface flows or, alternatively, a surface crust on the flowing lava in the channel fused and grew, later to be reinforced by surface flows. Cruikshank and Wood recognized that channels closed by spatter and overflow lay on topographic highs orientated along the tube axis caused by repeated lateral flow of lava overflowing the channel before closure. They also noted several factors that governed the method of closure. Proximity to the vent, for example, was held to be important with regard to the loss of gases to the atmosphere, so that once the gas was removed from the magma, spatter was reduced and levee formation was lost. Thus this method of roof construction would not be expected farther from the vent. Turbulence also caused exsolution of gases, and they suggested that roof construction by lateral spatter could only occur in narrow channels. An important observation that confirmed earlier speculation (e.g. Wood, 1971) was the deepening of lava channels, with the result that older routes could be captured by newer ones. Cruikshank and Wood could not apply the explanation of Ollier and Brown and Hatheway and Herring to speleogenesis in the Hawaiian pahoehoe, but suggested instead that tubes developed by the crusting over of relatively fast-moving streams by fusion of floating crustal fragments and/or by the joining of spatter built laterally on to levees. Other observations also held to be important were bi-level conduits, fluvial-like processes of meandering, bank cutting, channel capture, and channel deepening in active tubes through melting and plucking of the country rock.

Yet other observations of the Kilauea activity were recorded by Peterson and Swanson (1974). Their observations agreed with those made by Wentworth and Macdonald, Greeley and Cruikshank and Wood, though they saw no evidence to support the proposal by Ollier and Brown that lava tubes are the result of the internal shearing of thick flows. Unlike the other observers, Peterson and Swanson described the budding of pahoehoe toes away from the vent, suggesting another tube-forming process operating in these finely anastomizing distributaries. They also observed through "skylights" spectacular glowing underground lavafalls, which were formed when lava from a high level tube plunged into a deeper tube. It seemed that this occurred when the young lava emptied into the skylight of an older tube, or when a weakened roof of a lower tube collapsed beneath a stream in an overlying tube. It was also seen at times that there was a tendency for a flowing lava stream to develop a lower level roof beneath skylights, and at some skylights, if this process was repeated at successively lower levels, three- or four-tiered tubes developed.

Although, as this review shows, new observational data of actively forming lava tubes is available to aid the interpretation of lava tube caves in ancient lava flows, controversy still reigns. In a recent review of the literature, Wood (1974) decided that the difficulties found by speleologists and geologists in interpreting lava cave development was the result of considering individual hypotheses as over-all explanations. In order to evaluate the past models, discussion should first centre upon the genesis of the primitive conduit, i.e. the simple tube passage, and then, with that knowledge to build upon, to construct a model of more complex tube varieties. The model proposed by Ollier and Brown was discounted by Wood as an explanation of more complex tube systems, and instead suggested two alternatives based upon the established observational evidence. These alternatives took account of individual tube segments as simple conduits which could have formed either by the roofing of an open channel or the chilling of a shell around a small flow unit or pahoehoe toe (Fig. 4.6). As a first alternative, Wood pointed to observations made in Hawaii which showed that lava streams assume intricate braided channel networks, any part of which, if crusted and drained, would form a complex lava tube cave. As his second alternative he pointed to the importance of coalescing drainage channels carried in stacked conduits or flow units (Fig. 4.4). He suggested that three-dimensional complexes like Mount Hamilton Lava Cave, or the Cueva del Viento, may be due to a combination of both processes.

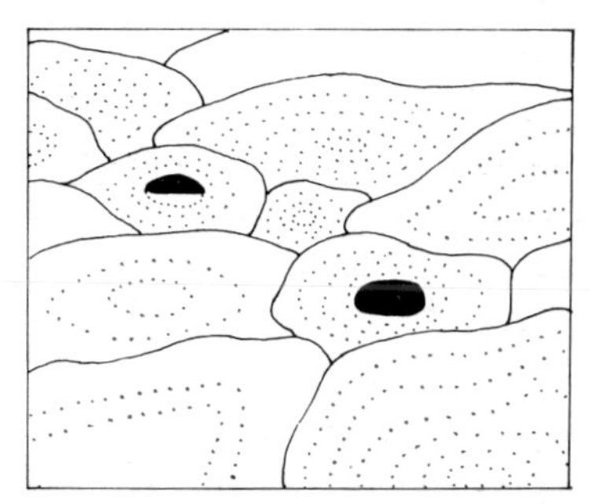

FIG. 4.6. Idealized section of pahoehoe toes (after MacDonald, 1967).

Even with detailed mapping of more lava tube caves, and even with more observation of actively forming lava tubes, it will be a long time before adequate explanations are forthcoming. In fact, it may be that because of the large variety of controlling factors governing speleogenesis in lavas, explanations may only account for individual lava tube caves. There remains great scope in this field of speleology.

LAVA CAVE MINERALIZATION

Some solution of magmatic rocks occurs and, in favoured localities, geomorphic features of solutional origin, such as lapiés (Palmer, 1927) and solutional depressions (Le Grand, 1952) have been found. In general, however, solution of basalt by normal weathering processes cannot be regarded as wholly responsible for mineralization in lava caves. The most common mineral deposits to be found are calcite, quartz, opal, chalcedony, gypsum and zeolites. Hydrated silica and calcite in particular are often found in stalactitic forms up to 15 cm long, while gypsum occasionally coats cave walls and lava stalactites. More complicated structures of opal of coralloidal form were discussed in detail by Swartzlow and Keller (1937). These authors believed that solution of the basalt was greatly facilitated by chemical alteration by hot, moist gases during the cooling of the flow.

CAVES OF EXOGENETIC ORIGIN IN PYROCLASTIC ROCKS

Less common are large-scale speleogenic features formed in pyroclastic rocks. The volcanic caves of Mount Elgon, Uganda, fall into this category and serve as an example, having been described in detail by Ollier and Harrop (1958). Mount Elgon consists of layers of volcanic agglomerate which is interbedded with subsidiary lava flows, and the majority of the caves are associated with ash bands located within the agglomerate. One cave at Sipi was reported to be 122 m in length, while another visited by Sutcliffe (1970) was said to have an entrance 60 m in width. The caves occur in rows and are frequently located in cliff faces. The agglomerate which underlies the ash band in which the caves had formed was reported by Ollier and Harrop to have had a hard, baked soil layer, and it was believed that water percolating through the upper agglomerate was held above this impervious soil horizon, finding an easy outlet along the band of ash which it attacked chemically. Cave formation was therefore partly attributed to the solution of the ash, which possessed a high concentration of sodium salts, and partly to normal erosional processes carried out by the subterranean streams. Similar features have been described elsewhere in the world. In the caves of South Georgia, U.S.S.R., for example, Dsavrivili (1965) has shown how these speleogenic processes can lead to the collapse of the overlying basalt, while other features in the Carpathian Mountains of Rumania hold interesting speleothems of limonite, resulting from solution of concretions in the rock above the cave (Naum and Butnaru, 1967). Szentes (1971) also listed this type of cave as present in Hungary. In the U.S.A., an interesting study by Parker, Shown and Ratzlaff (1964) of Officer's Cave, Oregon, showed that solution was not necessarily of great importance to cave formation in pyroclastic rocks. Officer's Cave occurs in altered montmorillonitic tuff and volcanic ash and constitutes the uppermost of four cave levels, with a cavern complex of 214 m depth. Although chemical analysis showed a high content of sodium in the tuff, solution was discounted as a major process of speleogenesis. Rather, it was because the nature of the rock was to become disaggregated when wet, offering little resistance to erosion, that cave formation occurred.

CONCLUSION

In the preface of his early bibliography Harter (1971b) wrote: "The infant science of . . . vulcanospeleology suffers greatly from fragmentation. Terms are not standardized. References are often difficult to acquire, and information may be unreliable." To this one must add that the study of volcanic caves has suffered most from a piecemeal approach which has lacked, until quite recently, any scientific method. To some extent these difficulties resulted because the areas of study are widely separated, and from the fact that cave research in volcanic rocks is only now being recognized as an exciting alternative to cave research in limestone. Over the past three years, for example, British cavers have contributed to the cumulative knowledge of lava tube caves by expeditions to Iceland and Tenerife, where detailed survey and geological work has shown remarkable contrasts in cave form and genesis. This type of speleological study from the world's active caving community, together with the new interest held by professional geologists, is certain to clarify many of the outstanding problems in vulcanospeleology in the near future.

REFERENCES

Anon. (1957). A lava vent. *New Zealand Speleo. Bull. 2* (22).

Bravo, T. (1964). *En volvan y el malpais de la Corona. La "Cueva de los Verdes" y "Los Jameos"*. Publicaciones del Cabildo Insular de Lanzarote, Arrecife.

Cruikshank, D. P. and Wood, C. A. (1972). Lunar rilles and Hawaiian volcanic features: possible analogues. *The Moon*, *3*, 412-447.

Dzavrisvili, K. V. (1968). De la génèse des grottes de lava. *Proc. 4th Internat. Cong. Speleo.*, Yugoslavia, 1965 (Pub. Ljubljana, 1968), 71-73.

England, P. (1959). Grjotagjá cave, Lake Myvatn, NE Iceland. *Cave and Crag Newsletter*, March/April, 9-11.

Gibson, I. L. (1974). Blister caves associated with an Ethiopian volcanic ash-flow tuff. *Studies in Speleo. 2* (6), 225-232.

Greeley, R. (1971a). *Geology of Selected Lava Tubes of the Bend Area, Oregon*. State of Oregon Dept. of Geol. and Min. Industries, Bull. 71, 1-46.

Greeley, R. (1971b). Observations of actively forming lava tubes and associated structures, Hawaii. *Modern Geology*, *2*, 207-223.

Greeley, R. and Hyde, J. H. (1971). *Lava Tubes of the Cave Basalt, Mount St. Helens.* Washington: NASA Technical Memorandum, NASA TM X-62 022, 1-33.

Halliday, W. R. (1954). Pseudokarst. *Technical Note* 25, *Salt Lake Grotto*, *Nat. Speleo. Soc.* 1-3.

Halliday, W. R. (1959). *Adventure is Underground.* Harper and Bros., New York.

Halliday, W. R. (1960). Pseudokarst in the United States. *Nat. Speleo. Soc. Bull. 22*, (2), 109-113.

Halliday, W. R. (1963). *Caves of Washington.* Washington Dept. of Conservation, Div. of Mines & Geol., Circ. 40, 1-132.

Halliday, W. R. (1966). *Depths of the Earth: Caves and Caving of the United States*. Harper and Row, New York and London, 1-398.

Harter, R. G. (1971a). Lava stalagmites in Government Cave. *Plateau* (Museum of N. Arizona), *44* (1), 14-18.

Harter, R. G. (1971g). Bibliography on lava tube caves. *Bull. 14, Misc. series, Western Speleo. Survey*, 1-52.

Harter, J. W. and Harter, R. G. (1970). Classification of lava tubes. *Northwest Caving*, *3* (1), 7-13.

Hatheway, A. W. and Herring, A. K. (1970). The Bandera lava tubes of New Mexico and lunar implications. *Commun. Lunar & Planetary Lab., Univ. Arizona*, *8* (4), 299-327.

Hjelmquist, S. (1932). Über lavastalaktiten aus einer lavahohle auf Süd-Island. *Kungl. Fysiogr.* I Lund, *2* (2), 6-8.

Jagger, T. A. (1931). Lava stalactites, stalagmites, toes and squeeze-ups. *Volc. Letter* (345), 1-3.

Jagger, T. A. (1947). Origin and development of Craters. *Geol. Soc. America, Memoir*, 21.

Kermode, L. (1970). Lava caves, their origins and features. *New Zealand Speleo. Bull.* *4* (76), 441-465.

Kjartansson, G. (1949). Nyr hellír i Hekluhrauni. *Natturufraedingurinn*, *19*, 175-184.

Kosack, H. P. (1952). Die verbreitung der karst—und Psuedokarsterscheinungen über die Erde. *Petermans Geog. Mitt.* *96* (1), 16-22.

Kyrle, G. (1923). *Grundiss der theoretischen Spelaologie*. Der Osterreichischen, Straatsdruckerei, Wien.

Lindsley, N. (1966). Lava caves, Parts 1 and 2; *Texas Caver*, July and Aug., 86-90 and 95-98.

Le Grand, H. E. (1952). Solution depressions in diorite in N. Carolina. *Amer. J. Sci.* *250*, 566-585.

Macdonald, G. A. (1967). *Extrusive Basaltic Rocks*, Chap. 1, in Hess, H. H. and Poldervaart, T. A., *Basalts: the Poldervaart Treatise on Rocks of Basaltic Composition*, 2 vols. Interscience, New York.

Macdonald, G. A. and Abbott, A. T. (1970). *Volcanoes in the Sea: The Geology of Hawaii*. Univ. Hawaii Press, Honolulu.

Mercer, D. C. (1966). Icelandic caves. *The Speleologist*, *2* (9), 11.

Mills, M. T. and Wood, C. (1971). A preliminary investigation of Surtshellir, West-central Iceland. *J. Shepton Mallet Caving Club*, *5* (1).

Montoriol-Pous, J. and De Mier, J. (1969). Estudio morfogenico de las cavidades volcanicas desarrolladas en las Malpais de la Corona. *Karst*, *6* (22), 524-562.

Naum, T. and Butnaru, E. (1967). Les volcano-karst des Calimani. *Ann. Speleologie*, *22* (4), 725-755.

Nichols, R. L. (1946). McCarty's basalt flow, Valencia Co., New Mexico. *Geol. Soc. Amer. Bull.* *57*, 1049-1086.

Ollier, C. D. (1967). *Landforms in the Newer Volcanic Province of Victoria*, Chap. 14, in Jennings, J. N. & Marbutt, J. A., *Landform Studies from Australia and New Guinea*. Cambridge Univ. Press.

Ollier, C. D. and Brown, M. C. (1965). Lava caves of Victoria. *Bull. Volcanologique*, *28*, 215-229.

Ollier, C. D. and Harrop, J. F. (1958). The caves of Mount Elgon. *Uganda Journal*, *22* (2), 158-163.

Palmer, H. S. (1927). Lapiés in Hawaiian basalts, *Geog. Review*, *19*, 627.

Parker, G., Shown, L. M. and Ratzlaff (1964). Officer's cave, a pseudokarst featured in altered tuff and volcanic ash of the John formation in Eastern Oregon. *Geol. Soc. Amer. Bull. 75*, 393-401.

Peck, S. (1962). Lava caves of Craters of the Moon National Monument. *Speleo Digest*, 27-35.

Peterson, D. W. and Swanson, D. A. (1974). Observed formation of lava tubes during 1970-71 at Kilauea volcano, Hawaii. *Studies in Speleo. 2* (6), 209-222.

Poli, E. (1959). Genesis e morfologia di aleure grotte del' Etna. *Bull. della Soc. Geog. Italiana*, *12*, 452-463.

Powers, S. (1920). A lava tube at Kilauea. *Bull. Hawaiian Volc. Obs. 8* (3), 1-4.

Ross, S. H. (1969). *Introduction to Idaho Caves and Caving*. Idaho Bureau of Mines and Geol, Earth Science series, No. 2.

Russell, I. C. (1902). Geology and water resources of the Snake River Plains of Idaho. *U.S. Geol. Surv. Bull. 199*, 1-94.

Stearns, H. T. (1924). Craters of the Moon National Monument, Idaho. *Geog. Review*, *14*, 362-372.

Sutcliffe, A. J. (1970). Caving in Africa. *William Pengelly Cave Studies Trust Newsletter*, *14*, 13-16.

Swartzlow, C. R. and Keller, W. D. (1937). Coralloidal opal. *J. Geol. 45* (1), 101-108.

Szentes, G. (1971). Caves formed in the volcanic rocks of Hungary. *Karszt-es Barlangkutatas*, *6*, 117-129.

Thorarinsson, S. (1957). Hellar i Gullborgarrhrauni. *Lesbok Morgunbladsins*, Sun. 6th Oct., 501-504.

Wentworth, C. K. and Macdonald, G. A. (1953). Structures and forms of basaltic rocks in Hawaii. *U.S. Geol. Surv. Bull.*, *994*, 1-98.

Williams, L. A. J. (1963). Lava tunnels on Suswa mountain, Kenya. *Nature*, *199*, 348-350.

Wood, C. (1971). The nature and origin of Raufarhólshellir. *Trans. Cave Res. Grp. G.B. 13* (4), 245-256.

Wood, C. (1974). The genesis and classification of lava tube caves. *Trans. Cave Res. Grp. G.B. 1* (1), 15-28.

5. The Erosion of Limestones

T. C. Atkinson and D. I. Smith

INTRODUCTION

All over the world limestones often stand up above the surrounding landscape, not because they are mechanically stronger than other rocks (Table 1 in Jennings, 1971), but because some of the erosional processes which would normally act at the surface are concentrated below ground in limestones. In particular, the debris produced by mechanical weathering cannot be transported away because there are no surface streams and little or no surface water on slopes during rain. The weathering processes are essentially similar to those acting on other rocks, but transport of the waste products is different. Because particulate waste, produced by mechanical weathering, cannot easily be removed from the surface of the landscape, most of the rock eroded from the surface is transported in solution. It is this fact which accounts for the distinctive nature of limestone landforms.

In Chapter 7, the chemistry of solution of limestone is described. Discussion is here restricted to rates and processes of erosion in limestone terrains. It is generally assumed that solutional erosion predominates over mechanical, though there are almost no quantitative estimates to justify this. However, the literature contains numerous estimates of solutional erosion rates. Rates of solutional erosion are normally assessed by measuring the discharge and dissolved load at a site in a stream or spring. The dissolved load is measured by taking water samples and analysing these for calcium and magnesium. The results are normally expressed as milligrams per litre of calcium carbonate and magnesium carbonate, hereafter referred to as mg/l $CaCO_3$ or $MgCO_3$ (sometimes noted as "parts per million"). The sum of the two carbonate figures is referred to as total hardness, often expressed in terms of the equivalent concentration of calcium carbonate. Some of the earlier studies in the literature were only concerned with calcium carbonate, and neglected the presence of magnesium.

The product of total hardness and stream discharge at any one time gives the instantaneous rate of removal of material from the catchment upstream of the sampling site. Because both discharge and hardness may vary within wide limits at a site, the instantaneous rate of erosion fluctuates wildly with time. For comparison with other sites, the mean annual erosion rate should be calculated. Ideally this should be done by collecting a long record of the discharge of the stream or spring, in order to estimate the mean annual flow or the duration of flows of a particular magnitude.

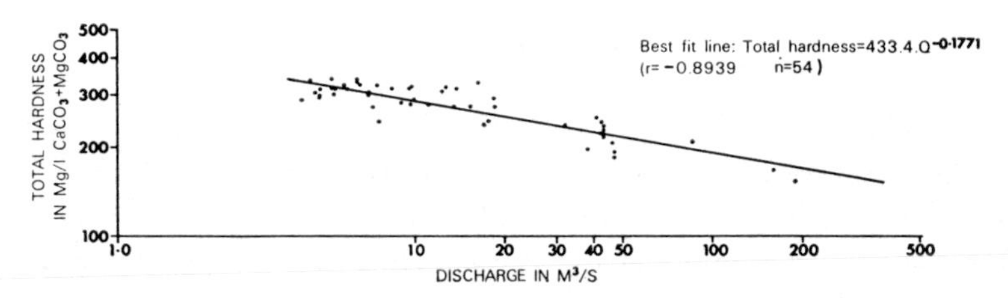

FIG. 5.1a. Graph of water hardness against discharge for the R. Avon at Melksham, Somerset.

For most streams a record of 30 years' length is really needed, though more approximate results can be obtained in 10 or even 5 years. In practice this is only possible by using records from existing river-gauging stations. Water hardness frequently varies with discharge, and at least 50 samples should be collected over a period of at least one year, covering as wide a range of discharge conditions as possible. The hardness can then be plotted against the discharge and the relationship between the two variables established. Normally, the hardness falls off at higher discharges, and a typical relationship for a surface river is shown in Fig. 5.1a. Limestone springs which are fed by percolation water constitute the majority of springs and show little or no variation with discharge except in the most extreme flood conditions. Conduit-fed springs, depending upon their swallet water contribution, occupy an intermediate category.

From the graph of hardness (or concentration) against discharge in Fig. 5.1a, a second graph of instantaneous solutional load in the stream can be prepared, as shown in Fig. 5.1b. The instantaneous load is the rate of removal of material from the catchment area above the sampling site and is normally expressed in gm/sec. There are two methods of calculating the mean annual rate of erosion from the data shown in Fig. 5.1b. The first is to read off the load corresponding to the mean annual flow, and to calculate the total amount of material which would be removed in a year. Application of this method to the Bristol Avon at Melksham, whose mean flow is 7·00 m^3/s, gives a mean solutional erosion rate of 67 784 000 kg/yr (or 67 784 tonnes). The second method, which is the more rigorous, is to establish from the flow duration curve of the river the percentage of the time during which flows of a given magnitude occur. For example, in Fig. 5.2 flows of between 1·0 and 1·5 m^3/s occur for 5·8% of the time.

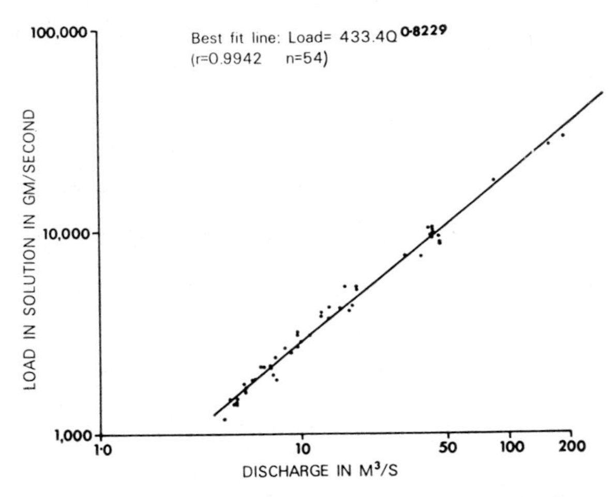

FIG. 5.1b. Solutional load and discharge of the R. Avon.

If the load corresponding to a flow of 1·25 m^3/s is L gm/s (521 gm/s for the Bristol Avon), the annual erosion accomplished by flows between 1·0 and 1·5 m^3/s is

$$\left(L \times \frac{5\cdot 8}{100} \times (\text{Number of seconds in a year}) \text{ all divided by } 1000\right) \text{kg}$$

or 952 800 kg for the Bristol Avon. By extension to all classes of discharge, the mean annual rate of erosion is:

$$\sum \frac{L_Q . P_Q .}{100} \frac{(\text{Number of seconds in a year})}{1000} \text{kg},$$

which is the sum from the smallest to largest class of discharge. L_Q is the load at discharge Q, and P_Q is the percentage of time occupied by flows of the class whose mid-point is Q. The mean annual erosion of the basin of the Bristol Avon above the measuring point by this method is 55 218 tonnes/yr over a catchment area of 666 km^2.

These examples show the calculation of the mean annual erosion rate in kg/yr for a particular site. For comparison with other sites, the erosion rate per unit area of catchment is usually used. This is normally given in kg/km^2/yr. By dividing this by the specific gravity of the rock, the more commonly used unit for limestone solution studies, of m^3/km^2/yr, is obtained. This is numerically equal to a lowering rate of mm/1000 years, which is also a frequently used unit, but the former (m^3/km^2/yr) is preferable, because the actual rate of surface lowering is not equal to the total erosion rate.

More often than not, adequate discharge records are not available and the discharge is approximated from the nearest available climatic records.

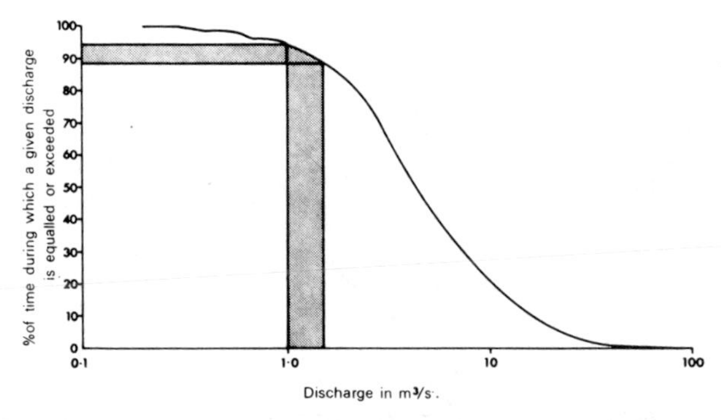

FIG. 5.2. Long term flow duration curve for the R. Avon at Melsham, Somerset.

Ideally these should include evapotranspiration figures, in which case the erosion rate is estimated by the formula:

$$\text{Erosion rate in m}^3/\text{km}^2/\text{yr} = \frac{(P-E)\,.\,\overline{H}}{1000\,.\,\rho},$$

where P is the mean annual precipitation in mm; E is the mean annual evapotranspiration in mm; $\overline{H}$ is the mean hardness of at least 50 samples in mg/l; ρ is the specific gravity of the rock. This formula is a variant on that proposed by Corbel (1959) and widely used since in the literature. Once again, adequate climatic data are not always available. Records of evapotranspiration are usually lacking, and one must often resort to estimates based upon temperature, such as those given in tables by Thornthwaite and Mather (1957).

Most of the estimates in the literature were made by the last of these methods, and were based upon water samples taken from springs and rivers draining areas of tens or hundreds of square kilometres. Data from such large areas represent an average rate of erosion over the catchment, and, if carefully collected, are useful for comparing erosion rates from one climatic zone to another. To establish the detailed distribution of erosion within an individual landscape samples should be taken from a variety of local sites. The former approach is discussed in the second section of this Chapter, the latter in the third.

FACTORS CONTROLLING SOLUTIONAL EROSION AND ITS WORLDWIDE VARIATION

There are only four factors which affect the rate at which a limestone is eroded, if the erosion is measured as an average rate over the whole of the drainage basin. They are the composition and resistance of the rock itself (the lithological factor), the amount of runoff from the basin, the temperature and the prevailing levels of CO_2. The interrelation of these variables and the effects of each, may now be discussed in turn.

LITHOLOGY

This is perhaps the least understood of the factors mentioned. While it is possible to draw a sharp distinction between carbonate limestones and soluble non-carbonate rocks such as gypsum or rock salt, the latter are comparatively rare, and so soluble that in humid climates they rarely crop out at the surface. In arid landscapes, the solution rate may be so low that gypsum survives at the surface, forming a landscape similar to that of limestones in more humid regions, and included in the general term "karst". Occasionally, where large thicknesses of gypsiferous rocks occur, karst features are formed in humid temperate regions, as in the Harz Mountains in Germany (Priesnitz, 1969).

The two principal minerals in carbonate rocks are calcite and dolomite, and the relative proportions of these affect the solubility of the rock. Calcite is the commonly occurring form of calcium carbonate, and may contain up to 7·75% of magnesium carbonate as a minor constituent. The mineral dolomite is the mixed carbonate (Ca, Mg)CO_3, and contains equal proportions of calcium and magnesium. Both minerals may, and often do, occur together in the same rock. There is, however, a tendency for individual beds to be composed mainly of either calcite or dolomite. Figure 5.3, which presents data from 209 analyses of various Scottish limestones (Robertson *et al.*, 1949), illustrates this tendency. Occasionally beds of dolomite alternate with beds of calcite limestone but more frequently large areas are underlain by uniform successions of calcite or dolomite limestones. In Britain the Chalk and the Jurassic limestones are chiefly calcite, the Carboniferous Limestone is locally dolomitized whereas the Magnesian Limestones of northern England and the lower units of the Durness Formation of NW Scotland are dolomite. The presence of magnesium carbonate as a minor constituent (up to about 15%) will affect the solubility of calcite in water. (This phenomenon is discussed in detail in Chapter 7 of this volume.) At present it is recognized in the laboratory but has not been investigated in detail in the field.

In theory, dolomite is slightly less soluble than calcite (Garrels and Christ, 1965) and the few laboratory studies that have been attempted

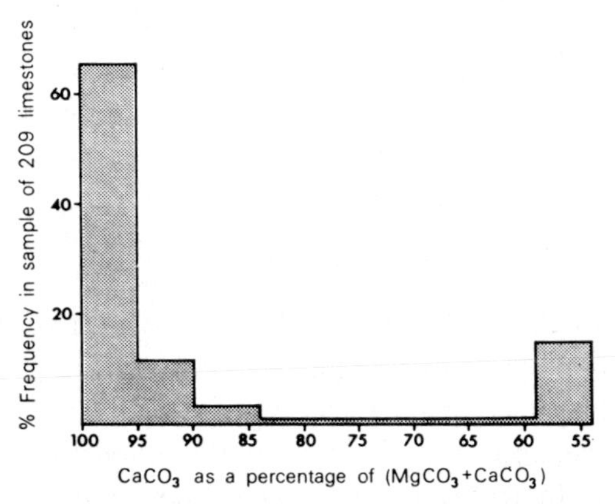

FIG. 5.3. Composition of Scottish limestones (data from Robertson *et al.*, 1949).

tend to confirm this (Priesnitz, 1967). In the field, however, differences in hardness due to the different solubilities of calcite and dolomite are swamped by variation due to other factors. On the Durness Formation, the total hardness of the Traligill River at Inchnadamph varies seasonally from 45 mg/l $CaCO_3+MgCO_3$ in winter to about 100 mg/l $CaCO_3+MgCO_3$ in summer. On the Permian Magnesian Limestone near Sheffield the amount of dolomite in solution in springs is around 300 mg/l $CaCO_3+MgCO_3$, with an equal quantity of gypsum ($CaSO_4$) also present. A similar range of variation is found in calcite limestones in Britain, for example 250 mg/l $CaCO_3$ in the Mendip Hills, 300 mg/l in the Cotswolds (Smith, 1965), 140 mg/l in NW Yorkshire (Sweeting, 1969), and 83 mg/l in South Wales (Newson, 1971a). Most of the variation between these areas is undoubtedly due to soil and hydrological factors, and is so great as to drown any small differences due to lithology.

A possible approach to the problem of the effects of lithology is to examine the detailed morphology of neighbouring strata within the same small area. This has been done for calcite limestones by Sweeting and

Sweeting (1969), who investigated the influence of textural differences between beds upon the scar, pavement and stream morphology in Littondale, Yorkshire. They found that coarse-grained limestones (sparites) tended to be more resistant to erosion, and to form barriers in stream beds, and to underlie limestone pavements and form scar edges. Fine-grained rocks (biomicrites and micrites) are less resistant. Whereas the Sweetings' study was of textural differences, a similar approach might be applied to investigate the effect of chemical composition, by studying the detailed morphology and analysing appropriate rock samples for calcium and magnesium.

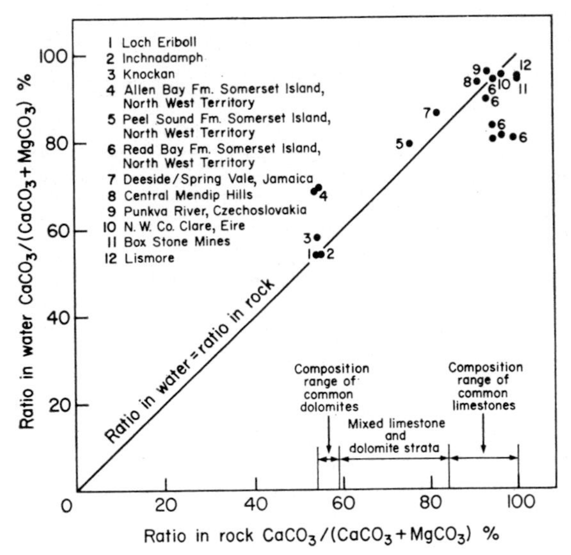

FIG. 5.4. Comparison of the composition of carbonates in solution with carbonates in rocks.

While the actual amounts of rock in solution are difficult to relate to lithology, the ratio between calcium and magnesium in solution is almost always very similar to that of the rock that is being dissolved. Figure 5.4 shows representative figures of rock analyses for a number of areas plotted against representative water analyses. The best line fit is very close to a 1 : 1 correspondence. This ratio in water samples at each site remains constant regardless of discharge, even though the total hardness may vary by several times. In areas where gypsum is present, waters will contain a considerable concentration of calcium sulphate ($CaSO_4$) in addition to calcium and magnesium bicarbonates. For example, the Permian Magnesian Limestone of northern England dissolves to produce hardnesses of about 600 mg/l $CaCO_3$ of which about half is sulphate.

TEMPERATURE AND CARBON DIOXIDE—EROSION ON BARE ROCK SURFACES

Limestone is scarcely soluble in water except in the presence of carbon dioxide. Atmospheric CO_2 levels deviate only slightly from the global mean of 0·03% by volume (Bolin and Keeling, 1963). At sea-level, water in equilibrium with the atmosphere can dissolve 74 mg/l $CaCO_3$ at 10°C. At high altitudes, where atmospheric pressure is lower, this figure will be correspondingly reduced. The solubility of CO_2, and thus of limestone, not only varies with the partial pressure of CO_2, but also with temperature, being greater at low temperatures. This variation is shown for a

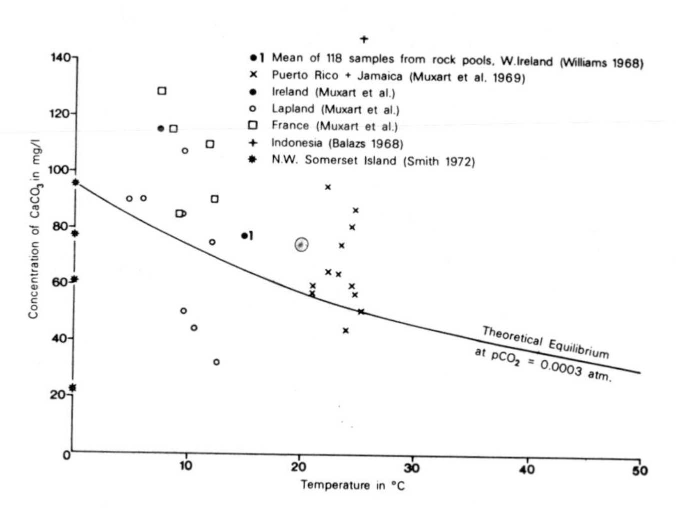

FIG. 5.5. Variation in the solubility of calcite with temperature, under a partial pressure of 0·0003 atm CO_2. Plotted points are waters from bare rock surfaces.

CO_2 pressure of 0·0003 atm (i.e. 0·03% by volume) in Fig. 5.5. It was claimed by Corbel (1957) that natural waters in the Arctic would be much harder than those in the tropics as a result of this effect. This is not so, however, because most of the carbon dioxide in the limestone solution process is derived from the soil air, where CO_2 concentrations are 10-100 times greater than in the atmosphere (Adams and Swinnerton, 1937; Smith and Mead, 1962). Field results bear this out, for they show that limestone waters often contain concentrations of up to 350 mg/l $CaCO_3$, which is far in excess of the amount dissolved by waters in contact with the atmosphere (Sweeting, 1964).

In theory, it should be possible to study the effect of temperature alone on limestone solubility in the field by comparing waters which have trickled over bare rock surfaces at different temperatures or under different climates. Several workers, including Bögli (1960), Balazs (1968), Miotke (1968), Williams (1968) and Muxart *et al.* (1969) have collected samples of this type. In almost all cases, individual samples do not plot on the line shown in Fig. 5.5, and they usually plot above it. The reason for this seems to be that very few limestone surfaces are completely devoid of vegetation, lichens in particular being almost ubiquitous. The effect of vegetation and organic debris is to supply additional CO_2 and increase the equilibrium hardness. Only in the high arctic and high mountain regions are there large areas of limestone which are truly free of vegetation. Data from vegetation-free arctic areas near sea-level plot below or on the line in Fig. 5.5, never above it (Smith, 1972). Typical points are shown on the figure. No data from mountain areas are shown because at high altitudes the partial pressure of CO_2 is less than 0·0003 atm, the pressure for which the diagram is constructed.

The influence of organic debris and vegetation on the surface solution of limestone was illustrated by a detailed study of rock pools on the limestone pavements of western Ireland (Williams, 1966, 1968). The mean total hardness for 118 samples was 77 mg/l $CaCO_3$ at a mean temperature of 15°C. This point is shown on Fig. 5.5, plotting somewhat above the theoretical line. The pools were extensively colonized by blue-green algae. The data show a wide range, 35-190 mg/l $CaCO_3$ and 6-24°C. A study of an individual pool throughout the year showed the effect of lower temperatures in raising solubility in so far as total hardness was 30 mg/l greater in winter than in summer, with an annual range from 60 to 90 mg/l.

In areas where snow lies for a considerable portion of the year it is possible that CO_2 levels that are slightly higher than the normal atmospheric value may occur in and beneath snowbanks. This was first suggested by Williams (1949) but subsequent literature has been confused. A review of the problem is given by Smith (1972), who shows that for arctic snowbanks there is no evidence of enhanced CO_2 concentrations. However, it is possible that the repeated freezing and thawing that occurs within alpine snowbanks may expel CO_2 from solution in the snow and meltwater and thus increase the CO_2 content of the air within the bank.

CARBON DIOXIDE IN THE SOIL ATMOSPHERE—EROSION IN SOIL COVERED AREAS

Except in areas of extreme climatic conditions, most rocks including limestones are covered by soil. The soil supports not only vegetation but a teeming microfauna, as well as larger soil animals. The production of carbon dioxide by these organisms is considerable, and the respired gas is concentrated in the small, air-filled pores of the soil. For this reason, the CO_2 content of soil air is much greater than that of the ordinary atmosphere. Typical values for tropical and temperate soils are given in Table 5.1. Some of the values given are from fairly recent studies in which measurements were made at a variety of depths at different seasons. Earlier studies are often incomplete in this respect.

In general, carbon dioxide concentrations increase with depth from the surface, and in limestone solution it is the CO_2 concentration at the base of the soil profile which controls how much limestone is dissolved. From the few figures available (Table 5.1) it would appear that CO_2 values are higher in the deeply weathered soils of the tropics. However, when individual measurements made at the same depth in different soils are compared, it can be seen that there is little difference between the CO_2 values in temperate and tropical soils. Soils which are formed solely by the weathering of the underlying limestone are usually thin, both in tropical and temperate zones. Thus, it is the CO_2 values for the shallow depths that are generally applicable in studies of limestone solution, and at these depths the figures in Table 5.1 show almost no difference between the two regions. The interpretation of the figures is at variance with the suggestions of Bögli (1960), Gerstenhauer (1960), Lehmann (1960) and Balazs (1968) that CO_2 values would be higher in tropical limestone soils because of higher temperatures and allegedly higher rates of humus production.

The only published study dealing specifically with soil CO_2 in tropical limestone soils is that by Nicholson and Nicholson (1969). The other figures in Table 5.1 from the tropics are from soils underlain by other rocks. The Nicholsons compared detailed results from several sites in Jamaica with those from a single site in Britain. They concluded that the Jamaican values are generally higher; but the measurements in Britain were made in winter. More recent work by the present authors has shown that summer values in Britain are three to four times greater than those of winter. The seasonal variation may not be so marked in the tropics, where temperatures are high all the year round. Summer values from limestone soils in the temperate zone show similar values to those of the tropics, viz. 0·5 – 1·5% CO_2 at about 30 cm depth.

The importance to limestone solution of varying levels of soil CO_2 can be seen from Fig. 5.6. This diagram shows the hardness which a solution

Table 5.1

CARBON DIOXIDE CONCENTRATIONS IN SOIL AIR

Source	Soil/Vegetation		Usual % CO_2	Summer	Winter	Extreme Values
Tropical						
Zonn and Li, 1960	Evergreen Forest	10 cm	0·5- 1·0	—	—	—
		200 cm	3·4- 6·3	—	—	—
	Bamboo Forest	10 cm	0·2- 3·5	—	—	—
		200 cm	4·1-10·8	—	—	—
				Wet Season	*Dry Season*	
Vine, Thompson and Hardy, 1943	Cacao Plantation, Trinidad	10 cm	—	2·8- 6·5	0·2- 0·8	—
		25 cm	—	3·0- 8·5	0·8- 1·7	—
		45 cm	—	4·2- 9·7	1·4- 3·8	—
		90 cm	—	4·5-14·3	3·4- 7·6	—
		120 cm	—	3·6-17·5	3·7- 6·8	—
Nicholson and Nicholson, 1969	Limestone Soils, Jamaica	15 cm	0·3- 1·6	—	—	—
		30 cm	0·4- 3·0	—	—	—
		60 cm		—	—	—
Temperate						
Russell, 1961	Arable		0·9	—	—	—
	Pasture		0·5 - 1·5	—	—	0·5 -11·5
	Sandy arable		0·16	—	—	0·05- 3·0
	Arable loam		0·23	—	—	0·07-0·55
	Moorland		0·65	—	—	0·28- 1·4
	Arable		0·1 - 0·2	—	—	0·01- 1·4
	Manured arable		0·4	—	—	0·03- 3·2
	Grassland		1·6	—	—	0·3 - 3·3
Chulakov, 1959	Dark Chestnut	7 cm	0·1	—	—	—
		300 cm	1·7	—	—	—
Matskevitch, 1957	Steppe: tree coenoses		2·5 - 3·4	—	—	—
	herbaceous		1·2 - 2·0	—	—	—
Gerstenhauer, 1969	Sandy loam	30 cm	—	2·5	0·3	0·2 - 3·6
		30 cm	—	1·5	0·1	0·1 - 1·9
	Loamy sand	50 cm	—	0·8	0·2	0·2 - 1·1
		20 cm	—	0·9	0·1	0·05- 2·0
Nicholson and Nicholson, 1969	Brown earth on limestone		0·27- 0·41	—	—	0·08- 0·7
Boynton and Compton, 1944	Orchard/grass	30 cm	—	1·5-2·5	0·1-1·0	—
		90 cm	—	2-5	1-3	—
		150 cm	—	4-9	2-6	—
Sheikh, 1969	Valley bog	5 cm	—	1-3·5	—	—
Boussingault and Levy, 1852	"Sandy Soil"		1·06	—	—	—
	"Manured sandy soil"		9·74	—	—	—
	"Black clay"		0·66	—	—	—
	"Fertile moist soil"		1·79	—	—	—

All values in % CO_2 by volume.

of calcite would attain at equilibrium with differing CO_2 levels in the atmosphere. The upper curve ("equilibrium solubility") refers to the case where the solution, gaseous CO_2 and solid limestone are all in contact, whereas the lower ("anaerobic") case refers to solutions brought

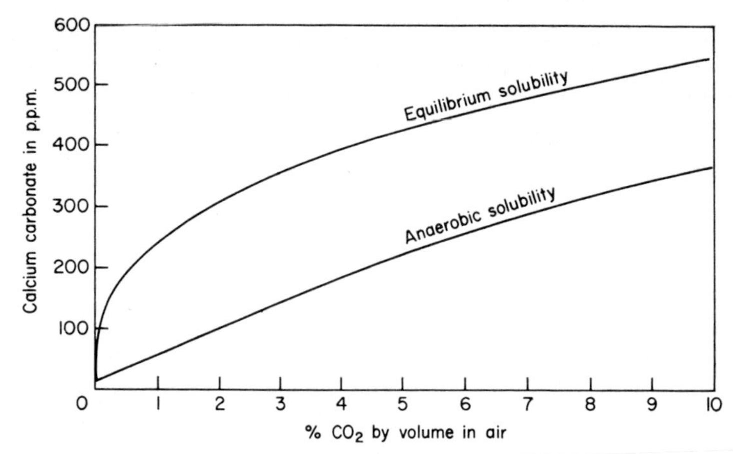

FIG. 5.6. Variation in calcite solubility with CO_2 composition of the gas phase, at 10°C (after Smith and Mead, 1962).

to equilibrium with gaseous CO_2, and then brought into contact with limestone in the absence of the gas phase. In the former case, therefore, as dissolved CO_2 is used up in forming calcium bicarbonate, it is replenished by further solution of CO_2 from the atmosphere.

Thus:

$$\left.\begin{array}{c} CO_2(\text{gas}) + H_2O \rightarrow H_2CO_3 \\ + \\ CaCO_3 \end{array}\right\} \rightarrow Ca(HCO_3).$$

In the "anaerobic" case, only the amount of CO_2 initially dissolved in the water is available to react with calcite, and the total amount of the latter dissolved is thus less. The reaction is a two-stage one:

$$1.\quad CO_2(\text{gas}) + H_2O \rightarrow H_2CO_3$$
$$2.\quad H_2CO_3 + CaCO_3 \rightarrow Ca(HCO_3)_2.$$

A full explanation of this effect will be found in Chapter 7.

Both "equilibrium" and "anaerobic" hardness vary greatly over the range 0-3% CO_2, which roughly corresponds to the range found in soil air. The scale of this variation is much greater than that produced by variations in temperature at a constant CO_2 level (cf. Fig. 5.5). Figure 5.6

shows the relationships of hardness to CO_2 level at 10°C; it may be corrected for other temperatures by applying the following factors to the hardness scale (Table 5.2).

Table 5.2

CORRECTION FACTORS FOR THE HARDNESS SCALE

Temperature	0°	10°C	20°C	30°C	40°C
Correction factor	1·28	1·00	0·78	0·64	0·43

Under field conditions there is sometimes some difficulty in deciding whether the "equilibrium" or "anaerobic" relationship applies. In the soil, however, and in the uppermost few metres of bedrock, water is percolating slowly enough for the "equilibrium" condition to apply with the prevailing level of CO_2 being that at the base of the profile.

In temperate areas, the usual CO_2 levels lie in the range 0·1-2·5%. A figure of rather under 1% is probably representative as an annual mean. This would give a mean hardness of about 240 mg/l $CaCO_3$, and a range from 125 mg/l to 325 mg/l. Figure 5.7a shows the frequency distribution of average hardness of waters from 44 springs and rivers in temperate soil-covered areas, taken from the literature on denudation rates and covering a wide geographical distribution. The mean is 189 mg/l $CaCO_3$, and the standard deviation 48·5 mg/l. Considering that the CO_2 data are of very poor quality and the portion concerned of the "equilibrium" curve in Fig. 5.6 is of extremely steep gradient, there is a reasonable agreement between mean CO_2 levels and this mean hardness. The extreme values on Fig. 5.7a are 137 mg/l and 349 mg/l, which agree very well with the range of hardness predicted from the extreme CO_2 concentrations.

From the literature searched, only 9 values of mean hardness could be found for tropical springs and rivers. Their mean is 163 mg/l $CaCO_3$, and their range 145-193 mg/l. These values are lower than the mean for temperate areas, though not greatly so. Since Table 5.1 shows no significant difference in CO_2 levels between the tropics and temperate areas, it may be that this discrepancy in hardness is due to the difference in mean annual temperatures under the two types of climate. This hypothesis may be tested by specific reference to data collected in the Maroon Town area of Jamaica (Nicholson and Nicholson, 1969; Smith, 1969). In this area the mean CO_2 content of soil air (47 observations) is 1·0%, which at 10°C would give a hardness of 240 mg/l in the spring water. The actual temperature of the spring water was 23°C, and this, being the temperature of groundwater, is the best estimate of the mean annual temperature. If a temperature correction is applied, the expected hardness is reduced to 170 mg/l. The hardness of Dromilly Clear Spring, a percolation-fed spring draining the area, is 172 mg/l, and this value is typical for much of Jamaica.

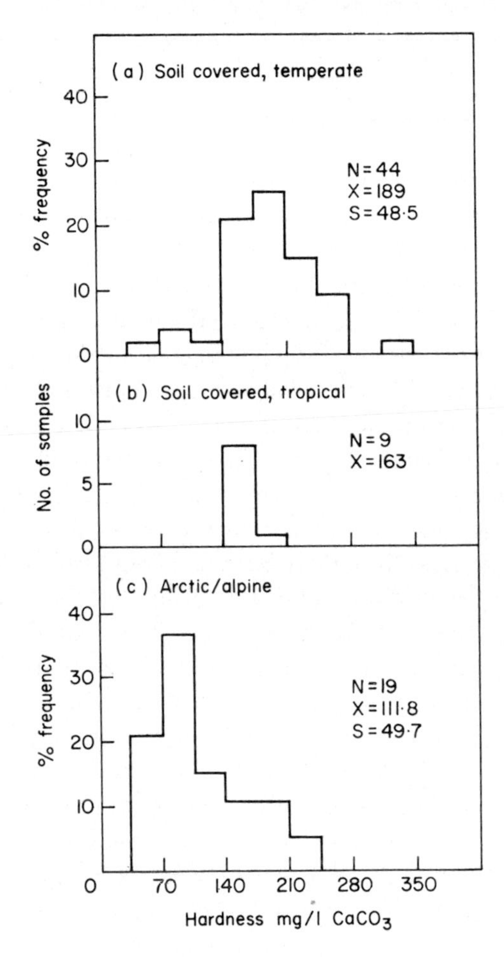

FIG. 5.7. Distributions of average hardness of natural waters from various climatic zones.

The hardness of waters from temperate soil-covered areas is compared in Fig. 5.7 with that from arctic and alpine areas with a sparse and discontinuous soil cover. It is clear that the hardnesses from the two areas are different, those from arctic and alpine environments having a mean hardness of 112 mg/l $CaCO_3$, against 189 for temperate areas and 186 for temperate and tropical areas combined. These differences were confirmed by a Kolmogorov-Smirnov statistical test, which was significant at the 99·9% level. The hardness of arctic and alpine waters may be raised somewhat by the temperature effect, but the main reason for their lower hardness is the paucity of soil cover, and consequent low rates of production of biogenic CO_2. This is confirmed by studies in mountain areas such as those by Kotarba (1971) in the Tatra Mountains and Ford (1971) in the Rocky Mountains. Figure 5·8a is reproduced from Ford

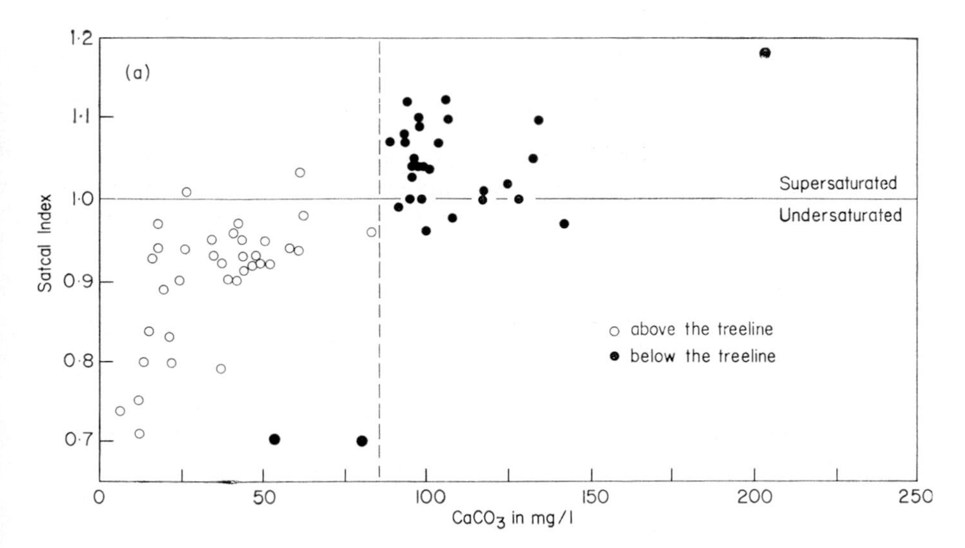

FIG. 5.8a. Chemical equilibrium of water from above and below the tree line in the area around Crow's Nest Pass and Mt Castleguard, Rock Mountains, Canada (redrawn after D. C. Ford, 1971).

(1971, Fig. 7) and shows clear differences between waters above and below the tree-line. Above the tree-line there is only a tundra vegetation and runoff is rapid. The hardness is 10-90 mg/l $CaCO_3$, and these waters are generally undersaturated. Below the tree-line the slower circulation of runoff through the soil and the higher CO_2 levels in the soil lead to a hardness of 95-150 mg/l, and generally these waters are saturated. Reference to Fig. 5.8b shows that the concentrations of CO_2 with which the waters are in equilibrium is 10-100 times smaller above the tree-line than below it.

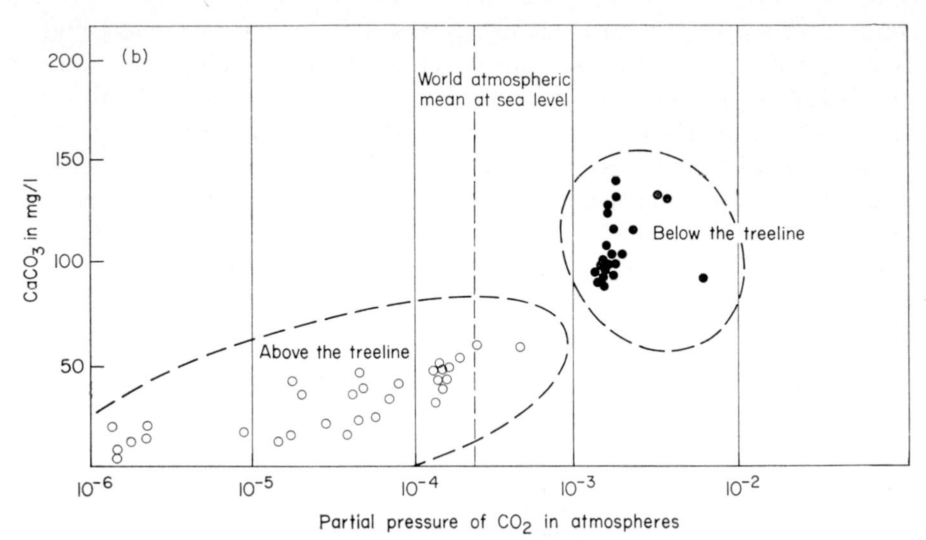

FIG. 5.8b. Hardness of waters from the same area plotted against the partial pressure of CO_2 in the gas phase in which they are in equilibrium (re-drawn after D. C. Ford, 1971).

RUNOFF AND EROSION RATES

The best method for calculating erosion rates involves the use of long-term discharge and hardness records, as shown earlier; but estimates based on these are rarely to be found in the literature. Most authors employ the Corbel formula:

$$\text{Erosion rate} = \frac{(P-E)\,.\,\bar{H}}{1000\,.\,\rho},$$

which was quoted above. Twenty-five results from various climatic regions are plotted in Fig. 5.9. They are taken from a variety of authors, all of whom used the Corbel approach. The foregoing discussion has shown that hardness varies little between tropical and temperate areas. Variation within the temperate regions is much greater, but the highest values on Fig. 5.7a are only four times greater than the lowest. The runoff, in contrast, varies by a factor of 10-40 times in soil-covered temperate areas. Even greater variations occur amongst the nearly soil-free high altitude and high latitude environments, within which hardness varies by a factor of only 2-3. This accounts for the close relationship between erosion rate and runoff displayed by the data in Fig. 5.9. The upper group of data is from soil-covered areas, whereas the lower group is from soil-free arctic and alpine regions. The difference in gradient of these two lines reflects the different hardness of waters from each, the ratio of gradients being approximately equal to the ratio of mean hardnesses in Fig. 5.7.

It has been suggested by Balazs (1968) among others that the highest erosion rates would generally be found in the humid tropics because of the very high rainfall there. In fact, the high rainfall is often accompanied by

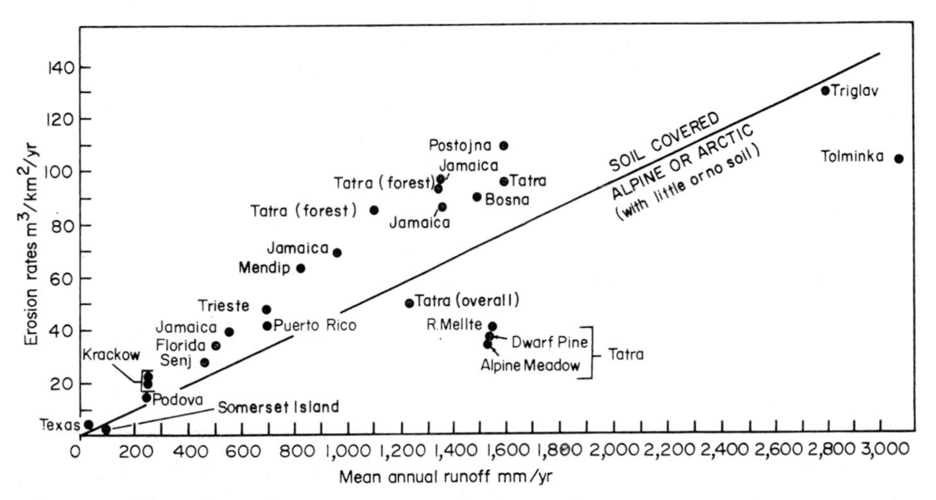

FIG. 5.9. The relationship between overall solutional erosion rate and mean annual runoff.

high rates of evapotranspiration, and the range of runoff and consequently of rate of erosion is not as great for tropical areas as for the temperate zone. The highest solutional erosion rates of all are found in high mountains where there is very high precipitation and relatively low evapotranspiration.

THE DISTRIBUTION OF EROSION WITHIN THE LANDSCAPE

The discussion above has dealt with values of overall erosion rate under different climates. There have only been a handful of studies in which the erosion rates have been measured at several sites within a single drainage system. Indeed, Beckinsale (1972) has criticized workers in this field for concentrating on overall rates, and has pointed out that much erosion occurs beneath the surface within the mass of limestone and therefore should not be thought of as contributing to the denudation of the land surface. Those reports in which some breakdown of distribution is made all indicate that the most important site is at the base of the soil profile or in the uppermost part of the bedrock, as Table 5.3 shows.

Table 5.3

RATES OF EROSION

Area	Overall Rate	Remarks	Source
Fergus R., Ireland	55 $m^3/km^2/yr$	60% at surface, up to 80% in the top 8 m	Williams, 1963, 1968
Derbyshire	83 $m^3/km^2/yr$	Mostly at surface	Pitty, 1968
North-west Yorkshire	83 $m^3/km^2/yr$	50% at surface	Sweeting, 1966
Jura Mountains	98 $m^3/km^2/yr$	58% at surface, 37% in percolation zone, 5% in conduits	Aubert, 1969
Cooleman Plains, N.S.W., Australia	24 $m^3/km^2/yr$	75% from surface and percolation zone, 20% from conduit and river channels, 5% from covered karst	Jennings, 1972
Somerset Island, N.W.T., Canada	2 $m^3/km^2/yr$	100% above permafrost layer	Smith, 1972

The unit of erosion rate, $m^3/km^2/yr$, is numerically equal to the rate of surface lowering in mm/1000 yrs. Thus, in the five temperate soil-covered areas listed in Table 5.3, about 60% of the total erosion took place close

PLATE 5.1. Limestone surface exposed by marine erosion in Co. Clare from beneath a cover of calcareous till. There is no enlargement of the joints on the freshly exposed limestone surface (*Photo:* S. T. Trudgill).

PLATE 5.2. Solutionally enlarged joints beneath an acid soil cover in Co. Clare (*Photo:* S. T. Trudgill).

to the surface, and could be thought of as surface lowering. Whether or not it is the actual bedrock which is lowered depends upon the nature of the soil or superficial deposits overlying the limestone. Normally soils which develop *in situ* by the accumulation of insoluble residue from the limestone have nearly neutral pH values (pH 6·5-7·5) and, provided they are not thin rendzinas, they contain little or no fragmented bedrock. Water percolating through them does not pick up much calcium, because there is little available, and most of the near-surface solution contributes

PLATE 5.3. A pedestal at Fanore, Co. Clare, formed beneath a glacial erratic demonstrating the surface lowering of the limestone pavement (*Photo*: D. I. Smith).

directly to lowering of the bedrock. In contrast, some thin soils over rubbly bedrock surfaces contain a high proportion of limestone fragments, and in these the percolating waters are almost saturated by the time they reach the bedrock proper.

In areas which have undergone glaciation (including most of the British Isles north of a line from Bristol to London) the limestone is often overlain by glacial deposits. Where these contain no limestone fragments, the underlying surface is usually deeply fretted and the joints opened by solution. Beneath calcareous drifts, the bedrock surface is smooth, joints are usually closed, and glacial striae are often perfectly preserved,

Plates 5.1 and 5.2 illustrate bedrock surfaces beneath calcareous and acid boulder clays in western Ireland.

The simplest case of surface lowering is on bare limestone pavements, such as those in north-west Yorkshire and western Ireland. In the former area, Sweeting (1966) found that rain water from pools on the pavements contained 120-140 mg/l $CaCO_3$. This is higher than the 74 mg/l which might be expected from consideration of the CO_2 content of the atmosphere (see above), but few rock surfaces are truly unvegetated; most are covered by lichens or mosses, and algae are common in rock pools. From these figures for hardness, Sweeting calculated the rate of lowering of the pavements as 40 mm/1000 yrs, and suggested a total lowering of 40-50 cm since the last phase of the Pleistocene glaciations. This agrees remarkably well with the average height of the Norber boulders, which are large glacial erratics resting on a pavement surface. Beneath each one is a pedestal, where the limestone surface has been locally protected by the boulder from erosion. Similar results have been recorded from western Ireland by Williams (1968) and Trudgill (1972b) (see Plate 5.3).

Probably the most detailed breakdown of erosion within a single area is in work by one of the present authors in the Mendip Hills, Somerset. The geology, caves and landforms of this area have been described by Smith and Drew (1975) who also reviewed the available solution data. Water samples were taken for analysis from three springs, at Cheddar, Rodney Stoke and Wookey Hole, and most of the swallets feeding them, and also from soil water at 36 sites and from percolation waters in caves. Discharges were measured continuously at the springs and as often as possible at the swallets. The results corroborate those of Smith and Mead (1962) who worked in the same area.

The combined mean hardness of the three springs (weighted by discharge) was 261 mg/l (total hardness as $CaCO_3$), of which 243 mg/l represents dissolved limestone, the remainder being present in the rainfall. Combining this with the runoff figure, the overall erosion rate is found to be 81 $m^3/km^2/yr$ on the parts of the area underlain by limestone. The swallet streams showed variable hardness, both from one site to another and with varying discharge at a single site. The mean for each site varies from 31 to 229 mg/l $CaCO_3$ with an overall mean of 126 mg/l (weighted by discharge). However, swallets contribute only 2·9% of spring discharge, and only 1·3% of the total erosion occurs on the limestones in the eleven swallet catchments.

The swallet streams are still slightly aggressive where they vanish below ground, and on average are capable of dissolving a further 3 to 6 mg/l $CaCO_3$, depending on the season. Much greater increases of hardness than this have been recorded by sampling downstream from cave entrances (Ford, 1966), but these increases are usually due to additions of saturated percolation water to the cave stream (Stenner, 1971). Probably the actual amount of limestone dissolved from cave walls is no greater than the 3-6 mg/l mentioned above. If so, the solutional erosion of cave passages amounts to only 0·05% of the total erosion. This figure for erosion in caves is startlingly low when compared with the amount of erosion accomplished by percolation waters. Water from the soil profile, which is not calcareous, contains an average of only 49 mg/l $CaCO_3$ in summer and 57 mg/l in winter, in spite of soil CO_2 levels of up to 2·5% in summer. Erosion within the soil profile amounts to only 10·1% of the total. However, the soil water is markedly aggressive to calcite, though the aggressiveness varies with season. Most of this aggressiveness is probably expended in the uppermost 10 m of bedrock, a supposition which is confirmed by the hardness of drips in shallow caves. Fifty-seven per cent of the total erosion takes place in this uppermost 10 m. Thus 67% of the total erosion contributes to the surface lowering of the soil and bedrock. The amount of limestone dissolved by percolation waters from the mass of the bedrock is found by difference between the sums of the swallet, cave passage, and superficial erosion and the total; it amounts to 31·4%.

The distribution of erosion in the Mendip Hills area is illustrated in Fig. 5.10. It is similar to that found by Jennings (1972) for the grass-covered uplands of the Cooleman Plains, New South Wales, where the limestone forms an outlier among forested crystalline rocks. Aubert (1969) describes a similar pattern of erosion in the Jura Mountains (see Table 5.3). Apart from the fact that over 50% of the erosion takes place close to the surface, the most significant feature is the low rate of erosion in cave passages. Far from being the principal expression of the effects of limestone solution, caves are merely the gutters which carry away the wastes produced by the degradation of the landscape.

In conclusion, it is of interest to enquire how caves are eroded. Newson (1971a and b) has measured and discussed the concentrations of suspended sediment in swallet, cave stream and spring waters in the Mendips. The absolute values of sediment concentration are low, rising above 10 mg/l at Cheddar springs only in flood. However, the sediment entering the swallets feeding Cheddar springs contains only 4% of calcareous matter, whereas 98% of that at the springs is calcareous. This clearly indicates that calcareous sediment is formed within the system. Samples collected from cave systems at varying distances downstream from the swallet show a rise in the proportion of calcareous material in suspension, and an increasing roundness of the particles. Newson gave no figures for absolute rates of suspended sediment formation and removal, for it is extremely difficult to quantify the process. There can be little doubt, however, that the majority of the calcareous sediment is derived from turbulent streams in cave passages (see Chapter 6), and that corrasion is a major process in the enlargement of caves in areas like the Mendips where there is an input of siliceous sediment at the swallets.

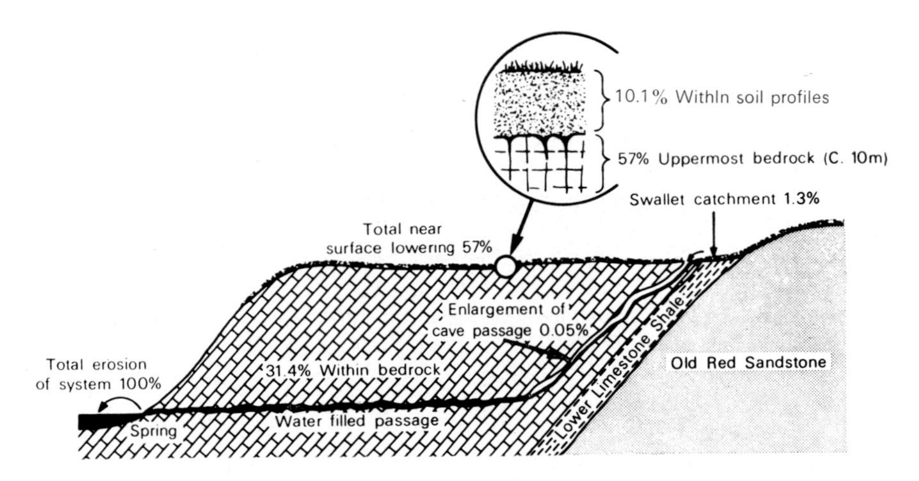

FIG. 5.10. Diagram to show the distribution of erosion with the landscape. Generalized from data for the Mendip Hills.

The actual rate of increase of solutional load along cave streamways is usually too small to measure, especially as tributaries and drips of hard percolation water add to the hardness of the stream and mask the small increase in load due to solution of the stream bed. A review of this and other difficulties is given by Smith, Nicholson and High (1969). To attempt a direct measurement of the rate of pick-up of load, a site must be selected in a long streamway free from tributaries and drips. On two isolated occasions, measurements by the authors in Ogof Agen Allwedd, South Wales, showed a pick-up of 1-2 mg/l $CaCO_3$ per 100 m of passage. Similar though lower values of 0·08-0·8 mg/l per 100 m were obtained from measurements of the absolute rate of lowering of the passage, using a micro-erosion meter. (A description of this instrument and its method of use was given by High and Hanna, 1969.)

Absolute rates of cave passage erosion have been measured over several years by High (1970) using the micro-erosion meter technique. Working in Co. Clare, western Ireland, High found that the rates at which cave floors were lowered by streams were highest at the swallets, where the waters were highly aggressive. Lowering rates of 0·25 to 0·5 mm per *year* occur at the swallets, compared to the value of 0·17 mm per year at a major rising, where the waters are nearly saturated. In the upstream parts of the swallet caves this rate of erosion corresponds to a solutional pick-up of 0·25 mg/l $CaCO_3$ per 100 m of streamway. This value is very close to the pick-up estimated by direct measurement.

REFERENCES

Adams, C. S. and Swinnerton, A. C. (1937). The solubility of limestone. *Trans. Am. geophys. Un. 11*, 504-508.

Aubert, D. (1969). Phénomènes et formes du karst jurassien. *Eclogae geol. Helv. 62*, 325-399.

Balazs, D. (1968). Karst regions in Indonesia. *Karszt-és Barlangkutatás*, *5*, 3-61. Budapest.

Beckinsale, R. P. (1972). The limestone bugaboo: surface lowering or denudation or amount of solution? *Trans. Cave Res. Grp. G.B. 14* (2), 55-58.

Bolin, B. and Keeling, C. D. (1963). Large-scale atmospheric mixing as deduced from the seasonal and meridional variations of carbon dioxide. *J. Geophys. Res. 68*, 3899-3920.

Bögli, A. (1960). Kalklösung und Karrenbildung. *Zeit. für Geom. Supplementband 2*, 4-21.

Boussingault, J. and Levy, B. (1852). Sur la composition de l'air confiné dans la terre végétale. *C.R. Acad. Sci., Paris*, *35*, 765.

Boynton, D. and Compton, O. C. (1944). Normal seasonal changes of oxygen and carbon dioxide percentages in gas from the larger pores of three orchard soils. *Soil Sci. 57*, 107-117.

Chulakov, Sh. A. (1959). The problem of the formation of soil structures. *Soil Fertil. Abstr.* 1960 (134).

Corbel, J. (1957). *Les Karsts du Nord-Ouest de l'Europe. Inst. Études Rhodaniennes de l'Univ. Lyons Mem. et Doc., 12.*

Corbel, J. (1959). Erosion en terrain calcaire. *Ann. Geogr. 68*, 97-120.

Ede, D. P. (1972). Comment on "Seasonal fluctuations in the chemistry of limestone springs" by E. T. Shuster and W. B. White. *J. Hydrol. 16* (1), 53-56.

Ford, D. C. (1966). Calcium carbonate solution in some central Mendip caves, Somerset. *Proc. Univ. Bristol Speleo. Soc. 11* (1), 46-53.

Ford, D. C. (1971). Characteristics of limestone solution in the southern Rocky Mountains and Selkirk Mountains, Alberta and British Columbia. *Canadian J. Earth Sci. 8* (6), 585-609.

Garrels, R. M. and Christ, C. L. (1965). *Solutions, Minerals and Equilibria.* Harper and Row, New York.

Gerstenhauer, A. (1960). Der tropische Kegelkarst im Tabasco (Mexico). *Zeit. für Geom. Supplementband 2*, 22-48.

Gerstenhauer, A. (1969). Offene Fragen der klimagenetischen karstgeomorphologie der Einfluss der CO_2 Konzentration in der Bodenluft auf die Landformung. In *Problems of the Karst Denudation*, *Studia Geographica*, Brno, 43-52.

High, C. (1970). *Aspects of the solutional erosion of limestone: with a special consideration of lithological factors*. Unpublished Ph.D. thesis, University of Bristol.

High, C. and Hanna, F. K. (1970). A method for the direct measurement of erosion on rock surfaces. *Tech. Bull. Br. Geomorph. Res. Gp. 5*, 24.

Jennings, J. N. (1971). *Karst. Introduction to Systematic Geomorphology v.7*. Australian National University Press, Canberra.

Jennings, J. N. (1972). The Blue Waterholes, Cooleman Plain, N.S.W., and the problem of karst denudation rate determination. *Trans. Cave Res. Grp. G.B. 14* (2), 109-117.

Kotarba, A. (1971). The course and intensity of present-day superficial chemical denudation in the western Tatra Mountains. *Studia Geomorphologica Carpatho-Balkanica*, *5*, 111-127. Krakow.

Lehmann, H. (1960). Introduction to meeting on karst phenomena, Vienna, 1959. *Zeit. für Geom. Supplementband 2*, 1-3.

Matskevitch, V. B. (1957). Carbon dioxide regime in soil air of steppe and semi-desert under tree and herbaceous coenoses. *Soil Fertil. Abstr.* 1961 (718).

Miotke, F. D. (1968). Karstmorphologische Studien in der glazialüberformtem Höhenstufe der "Picos de Europa", Nordspanien. *Jahrbuch der Geographischen Gesellschaft zu Hannover*, Sonderheft 4.

Muxart, R., Stchouzkoy, T. and Frank, J. (1969). Contribution a l'étude de la dissolution des calcaires par les eaux de ruissellement et les eaux stagnantes. *Problems of the Karst denudation*, *Studia Geographica* (5), 21-42. Brno.

Newson, M. D. (1971a). A model of subterranean limestone erosion in the British Isles. *Trans. Inst. Br. Geogr. 54*, 55-70.

Newson, M. D. (1971b). The role of abrasion in cavern development. *Trans. Cave Res. Grp. G.B. 13* (2), 101-107.

Nicholson, F. H. and Nicholson, H. M. (1969). A new method of measuring soil carbon dioxide for limestone solution studies with results from Jamaica and the United Kingdom. *In* Limestone geomorphology: a study in Jamaica. *J. Brit. Speleo. Ass. 6* (43/44), 136-148.

Pitty, A. F. (1968). The scale and significance of solutional loss from the limestone tract of the southern Pennines. *Proc. Geol. Ass. 79* (2), 153-177.

Priesnitz, K. (1967). Zur Fräge der Lösungsfreudigkeit von Kalkgesteinen in Abhängigkeit von der Lösungsfläche und ihrem Gehalt an Magnesiumkarbonat. *Zeit. für Geom. 11* (4), 491-498

Priesnitz, K. (1969). Über die Vergleichbarkeit von Lösungsformen auf Chlorid-Sulfat- und Karbonatgestein—Überlegungen zu Fragen der Nomenklatur und Methodik der Karstmorphologie. *Geologische Rundschaus*, *58*, 427-438.

Robertson, T., Simpson, J. B. and Anderson, J. G. C. (1949). The limestones of Scotland. *Mem. Geol. Surv. Spec. Rep. Miner. Resour. Gt. Br.*, no. 30.

Russell, E. W. (1961). *Soil Conditions and Plant Growth*. London.

Sheikh, K. H. (1969). The responses of *Molina caerulea* and *Erica* tetralix to soil aeration and related factors: gas concentrations in soil air and soil water. *J. Ecol. 57* (3), 727.

Smith, D. I. (1965). Some aspects of limestone solution in the Bristol region. *Geog. J. 131*, 44-49.

Smith, D. I. (1969). The solutional erosion of limestone in the area around Maldon and Maroon Town, St James, Jamaica. *In* Limestone geomorphology: a study from Jamaica. *J. Brit. Speleo. Ass. 6* (43/44), 120-135.

Smith, D. I. (1972). The solution of limestone in an arctic environment. *Spec. Pub., Inst. Brit. Geog. 4*, 187-200.

Smith, D. I. and Drew, D. P. (1975). *Limestones and Caves of the Mendip Hills.* David and Charles, Newton Abbot.

Smith, D. I. and Mead, D. G. (1962). The solution of limestone with special reference to Mendip. *Proc. Univ. Bristol Speleo. Soc. 9* (3), 188-211.

Smith, D. I., Nicholson, F. H. and High, C. (1969). Limestone solution and the caves. In *The Caves of North West Clare, Ireland*, 96-123. David and Charles, Newton Abbot.

Štelcl, O., Vlček, V. and Pišе, J. (1969). Limestone solution intensity in the Moravian Karst. *Problems of the Karst Denudation, Studia Geographica* (5), 71-86. Brno.

Stenner, R. D. (1971). The measurement of aggressiveness of water to calcium carbonate, Pts II and III. *Trans. Cave Res. Grp. G.B. 13* (4), 283-296.

Sweeting, M. M. (1964). Some factors in the absolute denudation of limestone terrains. *Erdkunde, 18* (2), 92-95.

Sweeting, M. M. (1966). The weathering of limestones, with particular reference to the Carboniferous Limestones of northern England, pp. 177-210. In *Essays in Geomorphology*, Dury, G. H. (ed.). Arnold, London.

Sweeting, M. M. and Sweeting, G. S. (1969). Some aspects of the Carboniferous Limestone in relation to its landforms with particular reference to northwest Yorkshire and Co Clare. *Revue Géographique des Pays Méditerranéens*, 7, 201-209.

Thornthwaite, C. W. and Mather, J. R. (1957). Instructions and tables for computing the potential evapotranspiration and the water balance. *Drexel Inst. Techn. Pubn. in Climatology, 10* (6), 185-311.

Trudgill, S. T. (1972a). The influence of drifts and soils on limestone weathering in N.W. Clare, Ireland. *Proc. Univ. Bristol Speleo. Soc. 13* (1), 113-118.

Trudgill, S. T. (1972b). Process studies of limestone erosion in littoral and terrestrial environments, with special reference to Aldabra Atoll, Indian Ocean. *Unpublished Ph.D. Thesis, University of Bristol.*

Vine, H., Thompson, H. A. and Hardy, F. (1943). Studies on aeration of cacao soils in Trinidad. 2. Soil air composition of certain cacao soil types in Trinidad. *Trop. Agric. Trin. 19* (11), 215-223.

Williams, J. E. (1949). Chemical weathering at low temperatures. *Geogr. Rev. 4*, 432-444.

Williams, P. W. (1963). An initial estimate of the speed of limestone solution in Co. Clare. *Irish Geog. 4* (6), 432-441.

Williams, P. W. (1966). Limestone pavements with special reference to western Ireland. *Trans. Inst. Br. Geogr. 40*, 155-172.

Williams, P. W. (1968). An evaluation of the rate and distribution of limestone deposition and solution in the R. Fergus basin, western Ireland. In *Contributions to the study of karst. Dept. Geog. Publ.* G/5 (1968), *Australian Nat. Univ. Canberra*, 1-40.

Zonn, S. F. and Li, G. K. (1960). Characteristics of the energy relations of biological processes of tropical forest soils. *Soil Fertil. Abstr.* 1961 (716).

6. The Hydrology of Limestone Terrains

D. I. Smith, T. C. Atkinson and D. P. Drew

INTRODUCTION

Hydrology is the study of how water precipitated on to the land surface is carried away. The natural landscape evolves a system of gutters and conduits, both on and below the surface, adjusted to evacuate surplus water most efficiently in the same way as the storm drains of a city. The nature and density of this network is primarily a function of local climatic, geological and vegetational factors.

The disposal of water at, or close to, the surface of the land, by direct runoff into streams and lakes and by evaporation, is termed *surface water hydrology*, whereas the study of the underground movement of water is called *groundwater hydrology*. Ultimately all the water will reach the major rivers in an area and thence flow to the sea or, in some cases, drain by subterranean routes directly into the sea. The process of water drainage from the land may be visualized as a series of transfers at differing rates between storages of differing duration. For example, storage occurs in lakes, transfer as river flow, storage as deep groundwater and transfer by cave stream flow to springs and rivers. Figure 6.1 shows diagrammatically this cyclic sequence of water movement for a limestone area.

The chemical and structural characteristics of crystalline limestones are such that their drainage is largely subterranean and the landforms of karst areas result largely from this.

POROSITY AND PERMEABILITY

The two characteristics of a particular rock type that primarily determine the behaviour of groundwater within it are porosity and permeability.

Porosity is defined as the total proportion of the rock which is made up of voids, and is commonly expressed as a percentage. For hydrological purposes it is only the interconnected voids which are important, because completely isolated pores are unable to exchange water with the surrounding rock. The simplest and most common method of measuring porosity is to weigh an oven-dried sample of the rock and then to soak it in water

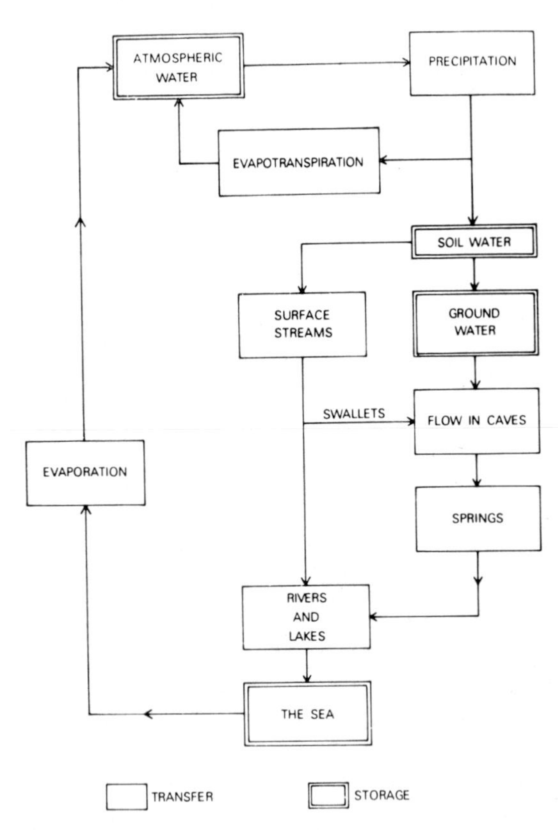

FIG. 6.1. Outline hydrological cycle for limestone areas.

for a period of days and reweigh it. The change in weight is a measure of the interconnected pore-space and the percentage porosity may be expressed as follows:

$$\text{Porosity } (\%) = \frac{(\text{Change in weight}) \times 100}{\text{Dry weight of rock}}.$$

Values of porosity found in commonly occurring rock types vary between zero and about 50%. Some typical values are shown in Table 6.1.

Whereas porosity is a measure of the total volume of pore-space in a rock, the size of individual interconnected pores may vary greatly (Table 6.1). In general, rocks which are composed of large grains, such as well-sorted gravels, have correspondingly large pores and transmit water easily, whereas rocks such as silts or clays, which have small pores, are less efficient at transferring groundwater. The efficiency of a rock in transmitting groundwater is a function of both porosity and pore size, as illustrated in Table 6.1. Greatest efficiency is achieved by a combination of large pore size and high porosity. However, the effect of pore size is rather greater than that of porosity; and clays, which are inefficient in transmitting water, have a high porosity but very small pore sizes.

Table 6.1

PRIMARY POROSITY, PORE SIZE AND PERMEABILITY OF NON-LIMESTONE ROCKS

Rock Types	Porosity %	Pore Size (mm)	Permeability (m/day)
Igneous			
Granite	0·3	0·001∼0·01	—
Basalt	7·7	0·001∼1·0	$1{\cdot}4\times10^{-5}$
Consolidated sedimentary			
Quarzite	0·5	0·001∼0·01	$1{\cdot}9\times10^{-6}$
Sandstone	10-30	0·01∼0·1	$5{\cdot}0\times10^{-1}$
Unconsolidated sedimentary			
Gravels	10-20	*c*. 1·0	$10^3\sim10^5$
Sands	35-50	0·01∼0·1	0·5 ∼50
Clay	30-60	0·001∼0·01	$1{\cdot}0\times10^{-5}\sim2{\cdot}5\times10^{-3}$

The efficiency with which a rock transmits water is known as its *permeability*, and is expressed as a flow rate. A rigorous definition of permeability was first given by Henri D'Arcy in 1856. He performed experiments which established the nature of groundwater flow through porous media. The apparatus used to measure permeability today does not differ basically from that employed by D'Arcy. It is illustrated diagrammatically in Fig. 6.2. Water is passed through a cylinder (C) containing the rock or porous medium under test. A constant head is applied to the system by means of the apparatus A, and the discharge, Q, produced by this head is measured at the outflow D. The head at various points in the cylinder is shown by the manometers (B_1-B_4), and it can be seen that the head decreases linearly from A to D. D'Arcy found that the discharge was proportional to the gradient of the head loss in the system.

Thus:

$$Q \alpha \frac{h}{l}$$

or

$$\frac{Q}{a} = K\frac{h}{l}.$$

expresses the fact that the discharge per unit cross-sectional area (i.e. the apparent velocity of the water) is proportional to the hydraulic gradient. The constant of proportionality K is the permeability. If h and l are expressed in metres, q in cubic metres per day and a in square metres, then K is in metres per day. In the technical literature a plethora of units for permeability is to be found but in this account metres per day (m/day)

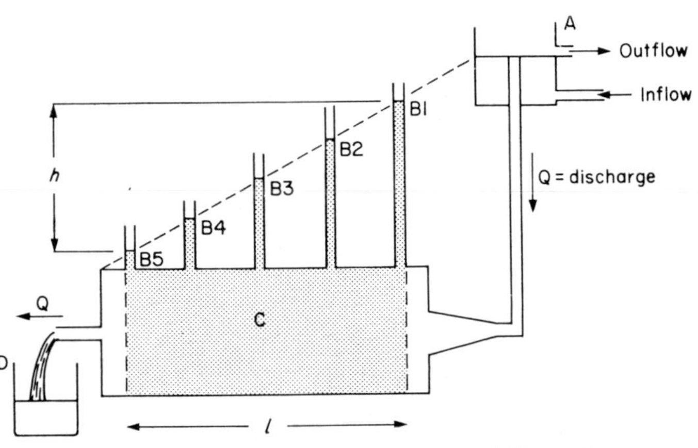

FIG. 6.2. Diagram of apparatus for measuring permeability of a porous medium. A—Constant head apparatus, B—Manometers. C—Porous medium to be measured, contained in cylinder of cross-sectional area, a. D—Outflow, with discharge Q.

will be used. Conversion factors to other units may be found in Lovelock (1970). Permeability varies by several orders of magnitude in common rock types from $1{\cdot}0 \times 10^{-5}$m/day in clays to 10^4m/day in well-sorted coarse gravels. Values are given in Table 6.1.

The methods of measuring porosity and permeability described above are normally performed on small samples in the laboratory. Strictly, they are determinations of *primary* porosity and *primary* permeability. These are the porosity and permeability resulting from the *intergranular* voids in the rock. Most rocks, particularly massive limestones, are transected by fractures such as joints, bedding planes and faults. These greatly enhance the permeability. For example, the primary permeability of the Chalk was measured as $1{\cdot}5 \times 10^{-4}$ to $3{\cdot}7 \times 10^{-3}$ m/day, whereas field determinations of overall permeability were 1·5 to 15 m/day (Ineson, 1962). The difference between these measurements is the value for the *secondary* permeability due to fractures. Likewise, *secondary* porosity is defined as the percentage volume of a very large mass of rock in the field occupied by such fractures. In massive limestones, primary porosity and permeability are usually very low, secondary porosity rarely exceeds 2% and secondary permeability is of overwhelming importance (Fig. 6.3). This point is illustrated by Fig. 6.2, in which the

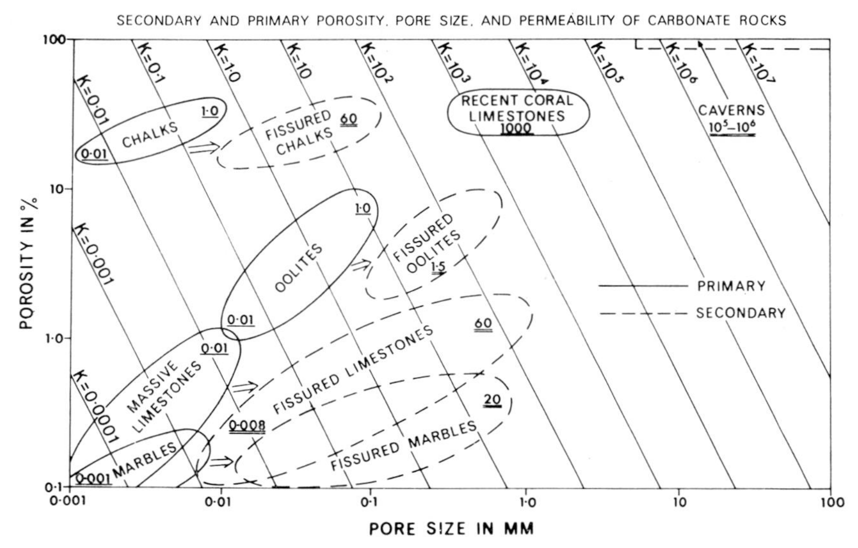

FIG. 6.3. Secondary and primary porosity, pore size, and permeability of carbonate rocks. Contour values of K, the theoretical permeability, are based upon the assumption that the rock behaves as a bundle of straight, parallel capillary tubes. Values shown underlined indicate the general range of permeability found in the lithology concerned. Double underlining indicates total permeability, single underlining primary permeability.

porosities and pore sizes of the principal limestone rock types are plotted. Each rock type is represented by two fields, of which the one to the left of the diagram shows the primary porosity and pore size. The second field shows the total rock porosity and the range of size of the secondary voids. There is very little change in porosity between primary and secondary fields, but the secondary pore size is 10 to 100 times greater than the primary pore size. For a given porosity, pore size is the principal factor controlling the permeability. On the diagram, Fig. 6.2, contours of permeability, in m/day, are shown, increasing from 0·0001 to 0·001 at pore sizes of less than 0·01 of a millimetre to 10 000 or more at pore sizes of 10-100 mm. These values are calculated on the basis of the following assumptions: that the flow is laminar; and that the rock behaves as if it were a bundle of straight, parallel tubes aligned in the direction of flow. These assumptions are not strictly valid, especially for very large pore sizes or very high or low porosities. The calculated values, however, agree quite well with the field and laboratory determinations of permeability available from the literature which are also plotted on the diagram. The effect of secondary porosity on the permeability is best shown by massive limestones in which the enlargement of joints and bedding planes leads to a 10 000-fold increase in permeability over that of the primary pores. The development of caverns leads to an increase in nominal permeability of over 10 million times.

A rock formation which can hold and transmit significant quantities of water through its voids is known as an *aquifer*, whereas a formation which allows little or no water movement is called an *aquiclude*. Aquifers may be classified into two principal types—*confined* and *unconfined* (Fig. 6.4). In an unconfined aquifer the water level is free to rise and fall following variations in output and input, and its upper boundary is a free surface at atmospheric pressure. This upper boundary, below which the rock is saturated with water, is termed the *water-table*. When a fully saturated aquifer is overlaid by an aquiclude and therefore has no free water surface, it is said to be confined. The level to which water rises in wells sunk into confined aquifers is called the *piezometric surface* (see Fig. 6.4). When the level of the piezometric surface is above the base of the confining bed, the well is said to be *artesian*. In some cases the well may overflow and this condition is known as *overflowing artesian*. A spectacular example of a confined aquifer is provided by the Severn Railway Tunnel, which during construction in 1879 intersected a major zone of secondary permeability with the Carboniferous Limestone beneath the Severn Estuary. This created within the tunnel itself a strong spring with a discharge of about 1000 l/s, flooding the tunnel, which has had to be pumped dry continuously ever since (Drew, Newson and Smith, 1969). Artesian flow is exemplified by wells sunk into the chalk in the central London area. The chalk here is confined by overlying London Clay. Originally the fountains in Trafalgar Square were simply an overflowing well, but the everincreasing use of groundwater supplies has lowered the piezometric surface, so that today the fountains are no longer supplied directly from groundwater. In the case of cavernous limestones it is valid to regard either the whole limestone mass as an aquifer or to visualize each individual cave system as a separate leaky aquifer. This is illustrated by Thrailkill (1968).

Some rock types (sands and gravels for example) have consistent values for porosity and permeability throughout their extent and are said

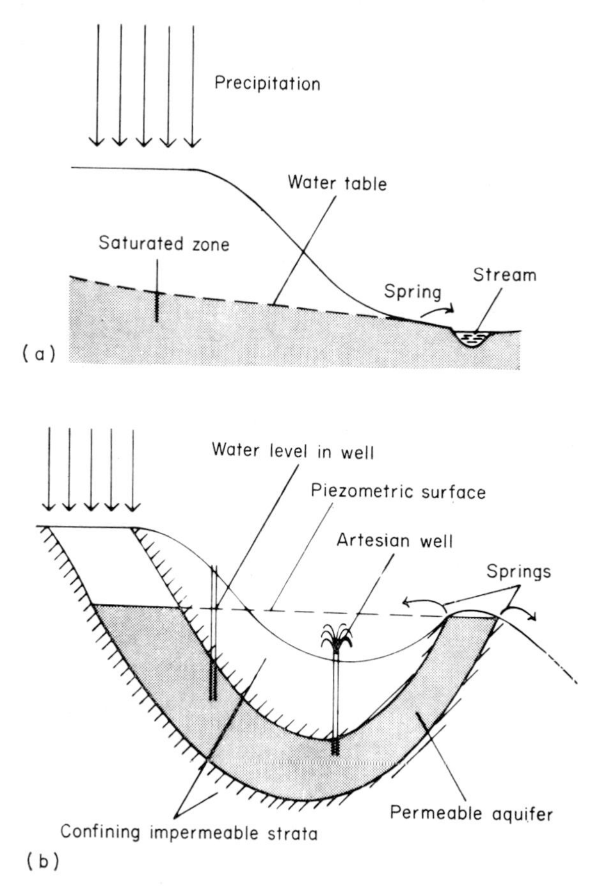

FIG. 6.4. Types of aquifer. (a) Unconfined aquifer, (b) confined aquifer.

to be hydraulically homogeneous, whereas in other rocks (lavas and chalks are examples) these values may vary greatly between different parts of the aquifer which is then termed hydraulically heterogeneous. Likewise, the values for porosity and permeability may be uniform in any direction from a point in an aquifer or may be markedly orientated in preferred zones. In the former case the aquifer is *isotropic*, in the latter, *anisotropic*. Crystalline limestones are markedly anisotropic, showing very high values of permeability along joints, faults and bedding planes, and virtually zero permeability and porosity within the undisturbed massive blocks of the rock itself. However, such limestones may also be hydraulically homogeneous over large areas if the bedding and jointing is an areally extensive and regular feature. Figure 6.5 illustrates these concepts.

Limestone aquifers, more so than any other rock types, are continually evolving. The flow of water through the aquifer alters and modifies the flow paths by preferential solution and thus the hydrological characteristics tend to change as the groundwater system develops. In general a limestone aquifer tends to become more markedly anisotropic and more permeable with time.

THEORETICAL CONSIDERATIONS OF GROUNDWATER FLOW

The laws governing groundwater flow, derived both empirically and theoretically, are well established. However, their relevance to the flow conditions within cavernous limestones is sometimes marginal and therefore only a brief summary of these will be made. A full account may be found in Hubbert (1940) and broader summaries in Todd (1959) and DeWeist (1965).

Most theoretical and practical treatments of groundwater motion assume that the flow is *laminar*. This means that individual particles of water move only in parallel layers in the direction of flow, with no transverse component to their motion. Laminar flow is generally confined to intergranular pores and very small pipes and channels with low velocities. *Turbulent flow*, which is common in cavernous limestones and ubiquitous in surface rivers, involves transverse mixing and eddying motions superimposed on the main flow direction. In pipes and channels greater than about 1 cm in diameter the flow is generally turbulent.

Within a single channel or pipe of a given size, the flow will be laminar when the velocity is very low and will become turbulent if the velocity is increased above a certain critical value, specific to that channel, which is defined by the Reynolds Number (Vennard, 1961). The Reynolds Number expresses the product of the velocity and the diameter of the pipe:

$$N_R = \frac{\rho v d}{\mu}$$

where N_R = Reynolds Number, v = average velocity, d = diameter of pipe, ρ is the density of the fluid and μ = viscosity. In general, the flow will be laminar when the Reynolds Number lies below a value of about 500 and fully turbulent at values greater than about 2000 (see Fig. 6.6). At intermediate values the flow is only partially turbulent, an occurrence

(a) Properties measured at a single point

A1

A2

(i)

(ii)

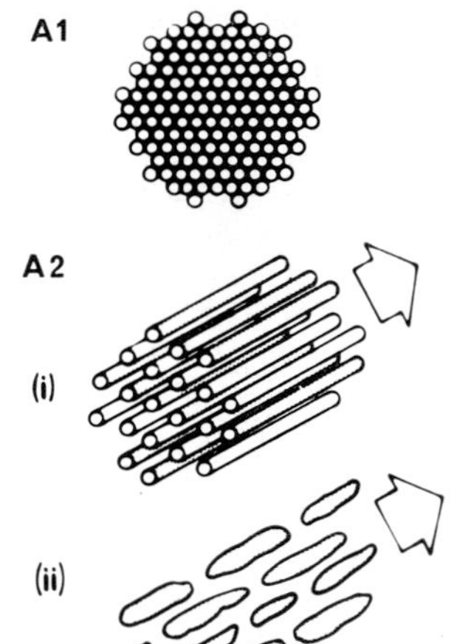

ISOTROPIC Permeability equal in all directions e.g. sand or gravel.

ANISOTROPIC : High permeability along the arrowed direction, low permeability at right angles to it. e.g.(i) Bundle of parallel tubes in the extreme case, with zero permeability at right angles to direction of tubes. (ii) Gravel composed of aligned, elongated pebbles in which pores are less tortuous in the direction of the long axes than across them.

(b) Properties measured over a wide area

B1

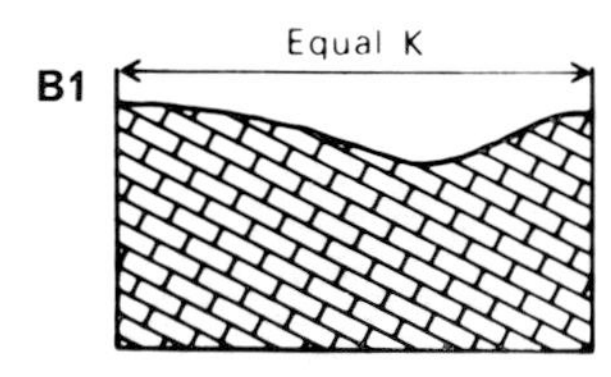

HOMOGENEOUS : Section through homogeneous aquifer with even joint spacing and porosity, and equal permeability over wide areas. e.g. Bunter Sandstone in parts of the Midlands.

B2

(i)

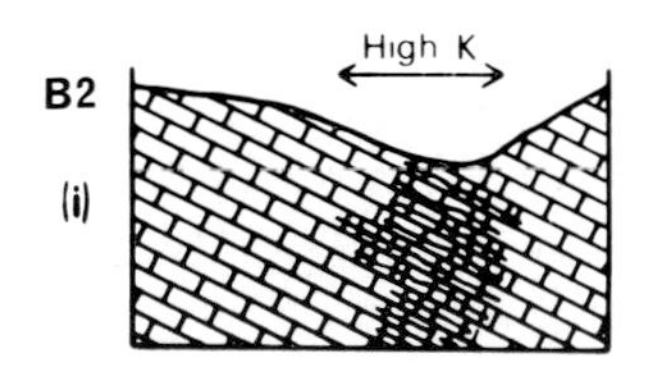

INHOMOGENEOUS Variable joint spacing, producing higher permeability in areas of greater fracture density. e.g. Chalk in Southern Britain.

(ii)

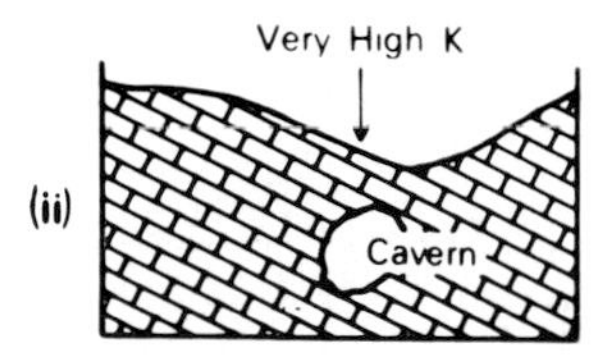

HIGHLY INHOMOGENEOUS : Cavern has extremely high nominal permeability and remainder of aquifer has most evenly distributed permeability as in B.1 e.g. Carboniferous Limestone in upland Britain

FIG. 6.5. Concepts of isotropy and homogeneity in aquifer properties.

G*

which, even under laboratory conditions, reflects the imperfections of the theory on which the concept of the Reynolds Number is based. The importance of the transition from laminar to turbulent flow during cave genesis is discussed by White and Longyear (1962) and Atkinson (1968).

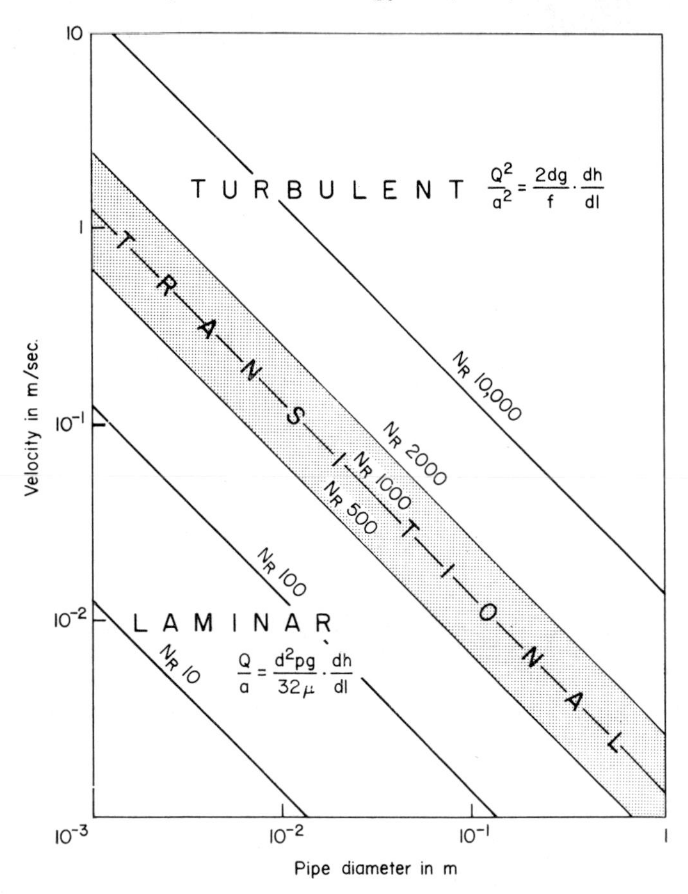

FIG. 6.6. Values of Reynolds Number at various velocities and pipe diameters, with fields of different flow regimes.

Where flow in a pipe or channel is turbulent the discharge may be related to the diameter of the pipe and the hydraulic gradient by the D'Arcy-Weisbach equation:

$$\frac{Q^2}{a^2} = \frac{2\,dg}{f} \cdot \frac{dh}{dl},$$

where Q = discharge, a = cross-sectional area, g = gravitational acceleration, f is a friction factor and dh/dl = the hydraulic gradient. The friction factor depends on the Reynolds Number and the roughness of the channel, except at very high values of Reynolds Number (about 20 000 for limestone solution conduits). Then the friction factor depends on roughness alone, as defined in the following equation:

$$\frac{1}{\sqrt{f}} = 1{\cdot}14 + 2{\cdot}0 \log \frac{d}{e},$$

where e is the relief of the projections in a pipe of diameter d.

In contrast, laminar flow in a pipe may be described by the Hagen-Poiseuille equation:

$$\frac{Q}{a} = \frac{d^2 \rho g}{32\mu} \frac{dh}{dl}$$

in which the notation is the same as that used above. This equation is of the same form as the D'Arcy's Law:

$$\frac{Q}{a} = \mathrm{K} \frac{dh}{dl}$$

and it is clear that D'Arcy's Law is the special case of the Hagen-Poiseuille equation applied to granular media. The application of these equations to groundwater flow in cavernous limestones is discussed by Thrailkill (1968), to whose paper the reader is especially recommended.

In both laminar and turbulent cases groundwater motion depends upon the viscosity which varies markedly with temperature. For example, the viscosity of water is approximately halved over the range of temperature from 5°C to 35°C. While variations in viscosity from one climatic region to another may be ascribed to this cause, variations in groundwater temperature within a region are generally minimal.

GROUNDWATER IN CRYSTALLINE LIMESTONES

SWALLET WATER AND PERCOLATION WATER

The chief characteristic of most limestone regions is the lack of surface water. In the early stages of landscape development a surface stream network may form. The dry valleys which are ubiquitous in limestone topography were eroded by such streams, the water having vanished below ground because of the gradual increase of the permeability with time. Eventually the only surface streams remaining will be those whose catchments are on less permeable strata and which disappear underground into swallets (stream sinks) when they reach the limestone outcrop.

The movement of groundwater through massive limestones is illustrated by Fig. 6.7. The process is depicted in the form of a diagrammatic flow chart in which the input of water to the system is precipitation,

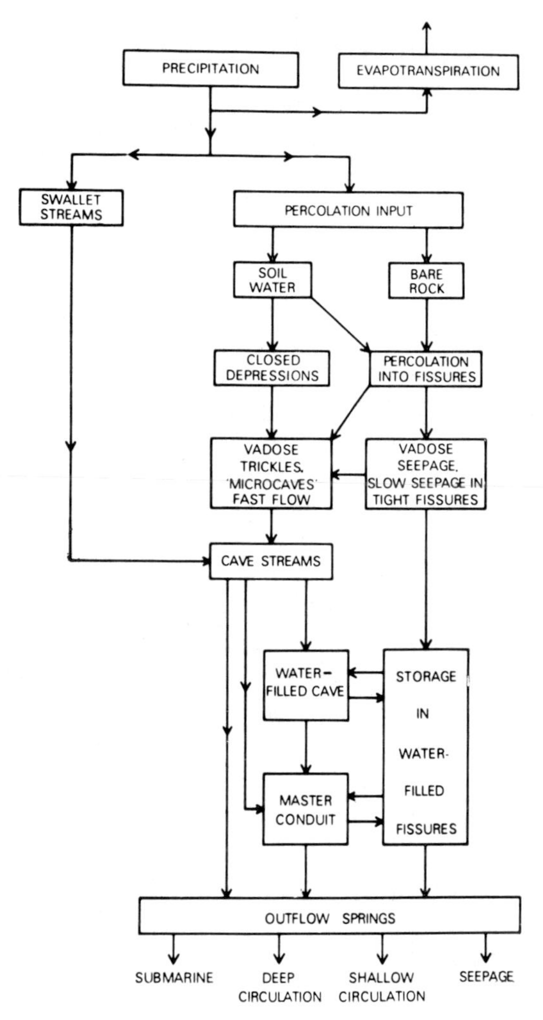

FIG. 6.7. Possible routes taken by flow in a limestone aquifer.

either directly as rainfall or indirectly as snowmelt or concentrated flow in streams. The output is in the form of discrete springs or indeterminate seepages at the surface or beneath the sea. Between surface and spring the water may traverse a variety of routes and be detained in storage for varying lengths of time, the particular condition of these two variables determining the nature of the outflow.

A proportion of the precipitation reaching the ground surface will be evaporated or transpired by plants; the remainder is termed *runoff*. Some of the runoff will reach surface streams by flow within the soil and a more limited proportion by overland flow; in a limestone region such streams will eventually sink underground. In Fig. 6.7 this component is termed *swallet streams*. The remainder of the runoff will infiltrate into the soil and thence into openings in the rock beneath, or in some cases into bare rock exposed at the surface; this is *percolation input*. Percolation water enters the bedrock through open fractures and planes of secondary permeability. In most areas not every fracture is sufficiently enlarged to permit water movement and therefore inputs tend to become concentrated, particularly at joint intersections. Similarly, water within the limestone tends to move along the more open fractures, and as these routes are enlarged by solution they become able to cope with increasing quantities of runoff. The spacing of these input points will vary with the geological and climatic conditions within a particular limestone area. The number per unit area may be very high, as for example in parts of the Indiana karst, where there may be as many as 360 per square km (Malott, 1945). The morphology of karst of this type has been discussed in detail by La Valle (1968), Williams (1972) and Palmer and Palmer (1975).

Over a period of time the inlet routes of the percolation water may become sufficiently enlarged near the surface to cause movement and gradual collapse of the soil and the uppermost layers of the underlying bedrock. The repeated subsidence of joint blocks towards a central solutional pipe creates a saucer or funnel-shaped *closed depression*. Typically, such depressions, when excavated, exhibit an upper layer of fine material underlain by a jumble of large boulders solutionally detached from the adjacent bedrock, and below this a vertical shaft tapering downwards into solid rock. In much of the European literature these depressions are known as *dolina*, and in North America as *sinkholes*.

If the precipitation falls directly on to a bare limestone surface the number of entry points available is usually greater and therefore the local catchments for individual fissures are smaller. Thus there will be less tendency for one point of engulfment to develop preferentially at the expense of those around it. The distinction made between percolation and swallet input is largely arbitrary, and swallets and soil-covered fracture intersections are opposite extremes in a continuum of concentration of runoff before it enters the ground. Whereas swallets may concentrate the runoff from many square kilometres, joint intersections beneath the soil and on limestone pavements have catchments of only a few square metres.

On rocks such as the chalk of southern England, which has no adjacent caprock and a very high fracture density, swallets are almost unknown, and there is little preferential enlargement of joint intersections. As a consequence closed depressions are uncommon, and the percolation water is almost entirely unconcentrated. The area around Inchnadamph, Sutherland, provides a complete contrast. Here large swallet streams collect on the quartzites and metamorphics which form the upper part of the Traligill River basin (Ford, 1959). On reaching the relatively restricted outcrop of Durness Limestone which underlies the lower basin they sink underground for about a kilometre. The discharge of the risings is little greater than the combined total of the swallets, and percolation waters from the limestone forms less than 10% of the basin outflow. A comparable situation occurs in South Wales where the Carboniferous Limestone outcrop is restricted in relation to the swallet catchment areas. Percolation water constitutes between 20 and 40% of the spring discharge (Newson, 1971). An intermediate example is provided by the Mendip Hills, Somerset, where swallet water contributes between zero and 50% of the discharge at various springs. The percolation water is partially concentrated into closed depressions, of which 566 have been mapped in about 36 km^2 of central Mendip (Ford and Stanton, 1968).

Swallet Water and Conduit Flow

By virtue of the larger discharges involved, the water entering the ground at swallets tends to form larger subterranean channels than percolation water. Most explorable caves are the active or fossil remnants of such channels. In passages which have not suffered collapse the dimensions of the active stream channel are related to the discharge of the stream that formed them (White and White, 1970). Where the passage is waterfilled, this may not necessarily apply.

Explorations of swallet caves have shown that the main swallet stream is normally joined by underground tributaries fed by other swallets and by percolation water. Figure 6.8 is a diagram of the stream network in the Pollnagollum-Poulelva cave system, Co. Clare, Ireland, which illustrates this well (Tratman, 1969). In all cases the stream remains confined within the cave passage to the downstream exploration limit even when the latter is a waterfilled passage which has been explored by diving. It does not in general disperse into the limestone (for a discussion of the behaviour in flood, see below). Similarly, diving at resurgences and springs has shown that underground streams are confined to conduits there too. From the evidence of direct exploration, therefore, it would appear that cave systems act in much the same way as surface stream networks, collecting tributaries from swallets and percolation water, sometimes joining to form the big river passages called *master caves*. The master caves in turn become waterfilled close to the level of the spring that they feed. It is rarely possible to explore such a system in its entirety, but this has almost been achieved in Pollnagollum in Ireland and in the Kingsdale System in NW Yorkshire (Brook, 1972, illustrated opposite p. 38).

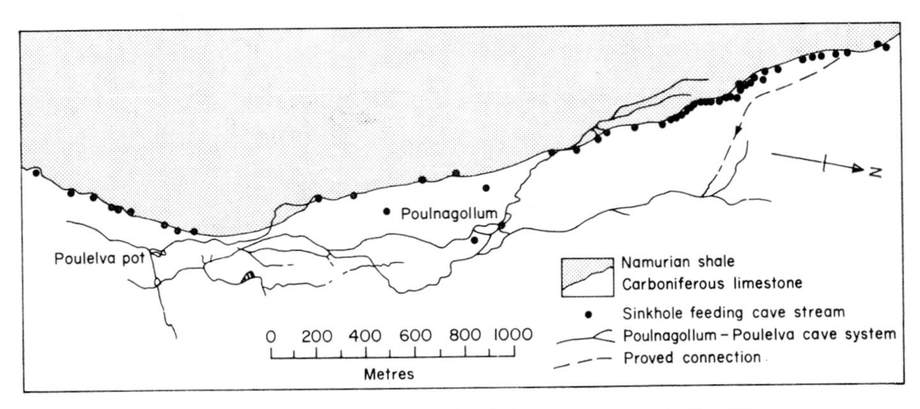

FIG. 6.8. The Poulnagollum-Poulelva cave system, Co. Clare.

A substitute for direct exploration is to conduct extensive water-tracing tests. There are a variety of methods for such tests (Drew and Smith, 1969), all of which depend upon tagging a swallet or cave stream with a distinctive material, either in suspension or solution. The presence of this material in samples of water taken subsequently from a spring establishes a positive connection between the spring and the point where the tracer was introduced. By taking frequent samples or by continuous monitoring of the spring water, it is possible to determine the average time of travel of the tracer, the mean straight-line velocity between the input point and the spring, and in some cases to deduce additional information about the nature of the underground flow. Using most techniques it is only possible to trace a single swallet to one or more springs at any one time. However, using *Lycopodium* spores, which can be dyed in up to five different colours, several separate tracing experiments can be performed simultaneously, which is a considerable advantage.

Chemical methods have been in use for many years to trace single swallets to springs. A classic example is that of the Malham Tarn area in

NW Yorkshire, where a stream overflows from a lake and sinks into the Carboniferous Limestone. This stream was traced by Carter and Dwerryhouse (1905) in 1899, using three tons of common salt. The water was found to reappear at Airehead Springs, 3·4 km away and 190 m lower than the swallet. Modern fluorescent dye tracers require

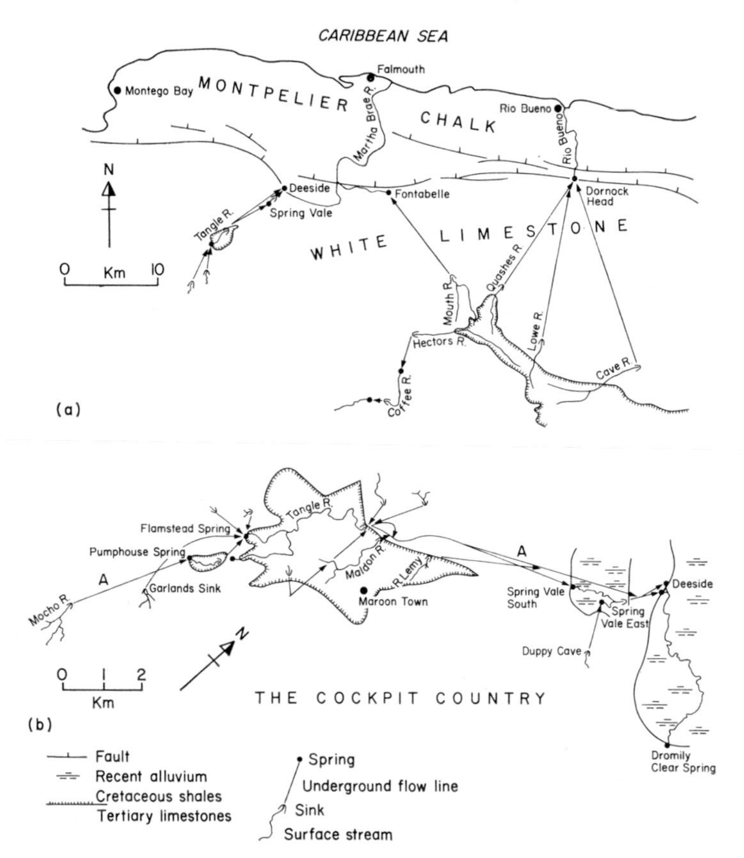

FIG. 6.9. Subterranean drainage in Jamaica
(a) The underground drainage of north central Jamaica (after Brown, 1966, and Drew, 1969).
(b) The underground drainage of the Maroon Town area.

the use of far smaller quantities, as exemplified by the work of Brown and Ford (1971) at Medicine Lake, Alberta. Here, a lake 8 km long and 15 m deep, with no surface overflow, was traced on three separate occasions to springs 15 km away and 430 m lower, The dosage of dye used was about 300 000 times smaller per unit discharge than that of common salt at Malham.

There are fewer studies describing the results of tracing programmes designed to elucidate the hydrology of a whole area. Three areas in which this has been done are the Dachstein Alps in Austria (Maurin and Zötl, 1959), the Mendip Hills, Somerset (Atkinson, 1971a), and Jamaica (Brown, 1966, Drew 1969). In all three cases, most of work was conducted using *Lycopodium* spores. Figure 6.9a shows the underground drainage established by Brown and by Drew in northern Jamaica. While on a regional scale the drainage pattern is fairly simple, Fig. 6.9b shows detailed results from the Maroon Town area. The complex pattern of this underground drainage remarkably resembles the network of a surface river, bearing out the conclusions drawn from explorable caves in this and other areas. However, there are important differences between the pattern of underground drainage and that of a river. In all three of the areas mentioned, positions are found such as those at the points "A" on Fig. 6.9b, where flow lines cross one another without mixing. This can only be achieved if flow is confined to discrete conduits. Also, it is not uncommon for a swallet to be traced to more than one spring, and the manner in which this might be achieved is illustrated by the downstream portion of the system shown in Fig. 6.9b. This phenomenon is particularly well illustrated by the Dachstein area. Conversely, most springs lie at the convergence of several conduits, for example the Dornock Head Rising shown in Fig. 6.9a. In some cases, conduits do not in fact meet before reaching the spring but resurge separately side by side. Thus, at St Dunstan's Well, east Mendip, there are two springs, only 2 m apart horizontally, each fed by a separate system of swallets and with its own distinctive water chemistry (Drew, 1969a, 1970a). Double or multiple risings of this kind are not uncommon, further examples being provided by the Killeany Rising, Co. Clare, where separate waters resurge from opposite sides of the same aperture (Smith and Nicholson, 1964), Airehead Springs, Yorkshire, and Deeside Springs, Jamaica.

In many of the experiments in these areas, measurements were taken of time of travel and mean velocity of flow. Table 6.2 shows the mean and standard deviations of velocity in Jamaica and Mendip together with the number of experiments on which the figures are based. Where measurements have been made in other areas they fall within the range shown. These figures, which are all from massive, crystalline limestones, show groundwater velocities greatly in excess of the few metres per day found in most other rocks (see Table 6.1). As discussed above, the flow is confined to conduits, and these are unlikely ever to exceed about 10 m in diameter. Reference to Fig. 6.6 will show that flow must be turbulent to support velocities of the order shown in Table 6.2. This confirms once again the similarities between the conduit systems of limestone areas and surface rivers, which have a velocity only slightly greater than that of conduits (Leopold, Wolman and Miller, 1964, p. 167). Like surface rivers, the velocity in conduits increases markedly with increasing discharge, as shown by the results of Brown and Ford (1971) from Maligne Lake and the present authors from work on the Mendip Hills.

Because the flow is confined to conduits it is possible to study the behaviour of natural or artificial flood pulses and deduce the volume of

Table 6.2

FLOW RATES FROM SWALLETS TO SPRINGS IN JAMAICA AND BRITAIN

District	Mean Flow Velocity (km/day)	Standard Deviation	Number of Traces
White Limestone, Jamaica	3·45	4·05	40
Carboniferous Limestone,			
Central Mendip Hills	7·36	5·91	23
Eastern Mendips	6·00	1·68	16

the waterfilled parts of the system. This technique was first described by Ashton (1966) and is as follows. A flood pulse is created either by rainfall, or artificially by building and breaching a dam at the entrance to a swallet. In the airfilled streamways of the system the flood wave thus produced travels only slightly faster than the actual water. However, on reaching the parts of the conduit which are completely waterfilled, the flood wave is transmitted almost instantaneously by displacing the water in a way similar to that observed in a U-tube. The floodwater takes a measurable amount of time to traverse the flooded conduit, and it is clear from Fig. 6.10 that this time is that between the moment of arrival of the flood wave at the spring and the arrival of the floodwater. The floodwater may be recognized either by tagging it with a tracer when it enters the swallet, or by a decline in hardness of the spring waters. The volume of the flooded parts of the system may be estimated by measuring the total volume discharged by the spring between the first arrival of the flood wave and the first decline in hardness or arrival of the tracer. Caution should be observed in interpreting the results, as the volume will be overestimated if there are tributaries in the waterfilled conduits, or if not all of the swallets have the same response to a natural storm. The method is probably of most use when applied to simple systems. For more complex systems the pulse patterns are interleaved, duplicated and superimposed. Computer analysis may be an advantage in these cases (Wilcock, 1968; see also Chapter 14 of this volume).

Artificial pulse waves were used as long ago as 1879 (Tate, 1879) to establish the connection between Malham Tarn sink and Airehead springs, as mentioned above. This, and subsequent experiments in 1899 (Carter and Dwerryhouse, 1905) and 1972, show that a flood always takes 90 minutes to traverse the system, regardless of discharge conditions. In

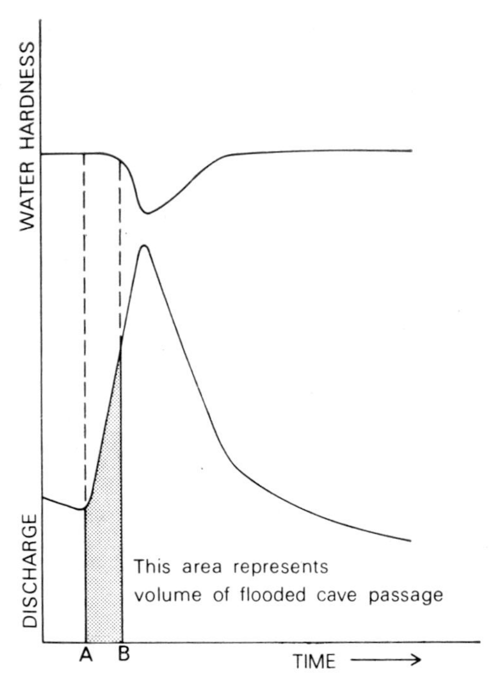

FIG. 6.10. Diagram to illustrate the principles of flood pulse experiments.

1972, a simultaneous tracer test showed that the volume of the flooded conduit was 42 000 m³. Similar experiments, employing natural flood pulses, have been conducted at springs in the Mendip Hills. Here the Cheddar spring, which has a mean discharge of 81 000 m³/day and drains an area of 9 km², has been shown to be fed through a flooded conduit whose volume is about 10 000 m³. Similarly, the smaller St Dunstan's Well spring (mean discharge 13 000 m³/day, catchment area 5·8 km²) has a conduit volume of about 1000 m³. These figures indicate that the volume of conduit is small in comparison with the area drained, a fact which reflects the efficiency of the subterranean drainage system.

Percolation Water

The behaviour of percolation water will depend to some extent upon whether the area has a soil cover. In the case of lapiés or bare limestone pavement, the input is through many small fissures. Runoff after rain is very rapid and water from each open fissure is quickly joined by tributaries from other fissures to form integrated cave streams. This type of system is commonly found in Alpine areas, for example the Trou du Glaz and the Gouffre de la Pierre St Martin, France (Chevalier, 1951; Queffélec, 1968). In these caves flash flooding is common, reflecting the rapid response and integration of the runoff from the bare lapiés above.

In comparison with bare limestone areas, those covered by soil have a much slower response to rainfall. One reason for this is that the water moves only slowly through the heavily textured soils which cover most limestone areas. In addition, water can only leave the soil and enter the limestone when the soil is saturated. In Britain, these conditions are usually found only in winter, and summer flooding of percolation-fed systems is comparatively rare, because of the large deficit of soil moisture in summer months.

When percolation water is seen entering caves, it takes a variety of forms, the two end-cases of which may be called *vadose trickles* and *vadose seepages*. Vadose trickles are small streams, without direct swallet feeders, which often originate in closed depressions. They flow in conduits which are usually smaller than swallet caves but of the same form. Where they can be followed, these trickles commonly receive tributaries from avens in their roofs, and appear to be composed of relatively fast flowing percolation water which is integrated into trickles and streams at no great depth. Vadose seepage is slow drips from narrow and relatively unopened joints in the cave roof. A study of St Cuthbert's Cave, Mendip, by Stenner (1966) shows that those waters which might be regarded as vadose trickles increase in discharge and change in temperature after rainfall (though more slowly than beneath a bare limestone area), whereas vadose seepages show little variation in discharge and none in temperature. It is tempting to equate vadose trickles to inputs of percolation water from closed depressions and vadose seepages to inputs from single fractures beneath the soil. In terms of the type of flow feeding springs, Schuster and White (1971) characterize springs mainly fed by vadose seepage as being fed by "*diffuse flow*", and those supplied by predominantly swallet and vadose trickles as having "*conduit flow*".

A few successful experiments in tracing percolation water have been performed (Drew, 1968b, 1969, 1970b), but all of these have been in vadose trickles. In one experiment, fluorescent dye was placed in a marshy depression above Contour Cavern in the Mendip Hills. Charcoal detectors were sited beneath all of the major drips in the cave, and, on recovery and analysis after one week, most showed a positive connection with the depression. In other experiments, dye was introduced in the floors of closed depressions in the Mendips and in cockpits in Jamaica, and in two cases on to bare rock in a quarry floor. Thus, these experiments refer only to vadose trickles. The results indicate that this type of percolation water moves at velocities in the range of 10-200 m/day. This is somewhat slower than swallet water but still well in excess of groundwater velocities in other rocks. Once again, the flow regime is almost certainly turbulent or transitional.

There are no direct measurements of the rate of movement of water in vadose seepages. Indirect evidence based on variations in water quality from cave sites in Derbyshire and Yorkshire have been discussed by Pitty (1966). By determining the lag necessary to produce the best correlations between surface temperatures and water quality in the cave, he suggested that some drips may take several weeks to penetrate to a depth of 30 m in the limestone.

Very little of the total quantity of vadose seepage water is likely to appear in caves, because the latter have a small areal extent compared with the total area of limestone. The majority of the seepage will eventually reach waterfilled fissures at levels close to that of the nearest spring or waterfilled cave passage. It is thought that in periods of baseflow there is a substantial contribution by these waterfilled fissures to the discharge in waterfilled conduits by means of water moving through the fissures which intersect with the conduit walls. In flood, the direction of this movement is reversed, and some of the floodwater in the conduit is stored in fissures close to it. As the flood recedes, the fissure once again begins to discharge into the conduit. Empirical evidence of this mechanism was obtained by Atkinson, Smith *et al.* (1973) at the Stoke Lane Slocker–St Dunstan's Well system. Dye introduced at the crest of a flood wave emerged from the spring as a strong pulse over a period of time marking direct flow through the system followed by a second, extended peak of dye at very low concentrations. This second peak corresponded to the release of tagged floodwater from fissure storage.

The magnitude of fissure storage and other types of storage have been determined by one of the authors, using methods of hydrograph analysis. The volumes of water in storage in the Cheddar System during a large flood are shown in Table 6.3. It can be seen that the volume of storage in waterfilled fissures is about 30 times greater than the storage in conduits. It is undoubtedly this large and important body of water which maintains the flow of springs in drought.

Table 6.3

DISCHARGE AND STORAGE DURING A FLOOD IN THE CHEDDAR SYSTEM

Total discharge	349 000 m³/day
Total storage	3·77 million m³
Fissure storage	3·27 million m³
Vadose storage	400 000 m³
Storage in conduits	100 000 m³

Summary of the Limestone Drainage System

It can be seen from the previous discussion that the different parts of the limestone drainage system behave in very different ways. The underground drainage in an area may be classified on the basis of the relative importance of different types of input and the form of the flow to the spring. A continuous spectrum exists from systems fed entirely by percolation water with predominantly diffuse routes to the spring, to those fed largely by swallets, in which conduit flow is most important. Both of these end members and an intermediate case are illustrated in Figs 6.11, 6.12 and 6.13. The flow diagrams are all based upon that shown in detail in Fig. 6.7. The active linkages of the flow diagram are shown as heavy lines, and shading in the boxes represents the relative importance of the various parts of the overall drainage system in each case. The diagrams also show a sketch map of two systems, together with spring hydrographs produced by actual storms of about 20 mm total rainfall in each case.

An example of a conduit system, the Traligill basin in NW Scotland, is shown in Fig. 6.11. Here the percolation component of the spring discharge is very small, most of the flow coming directly from the swallets. The response to rainfall is rapid, and the hydrograph is sharply peaked, with a rapid recession. The maximum discharge (only river level is shown in Fig. 6.11) is about 400% of the baseflow.

The opposite extreme, a percolation-fed system, is illustrated by the Dromilly spring, St James, Jamaica (Smith, 1969), in Fig. 6.12. The catchment area of this spring lies in the wilderness of the Cockpit Country,

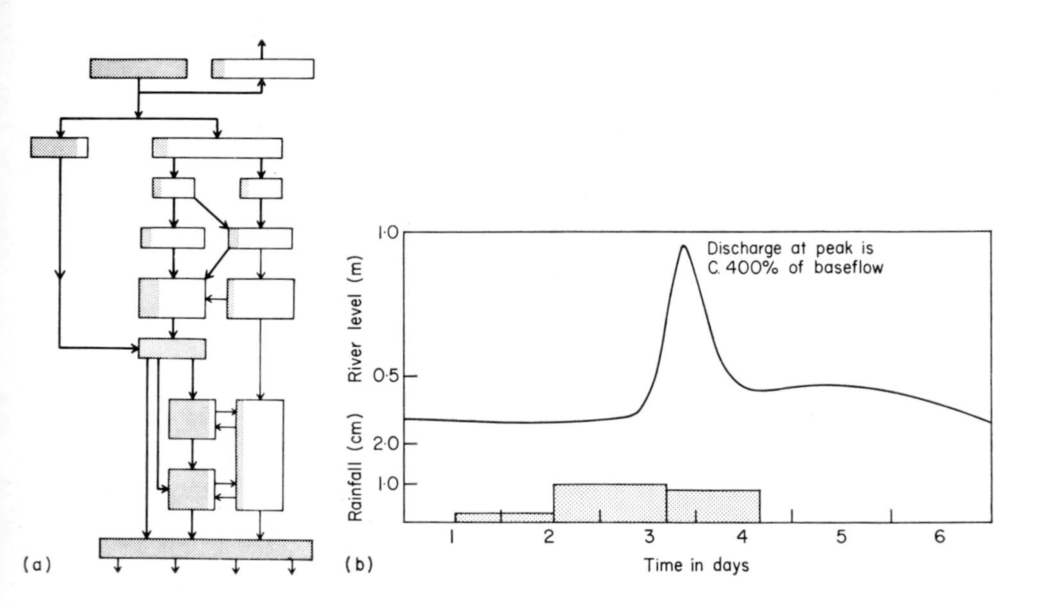

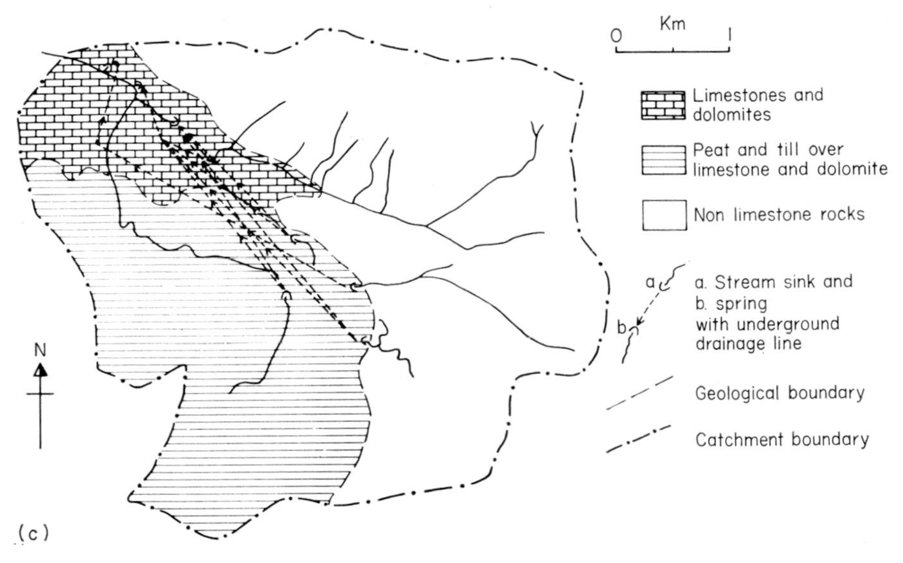

FIG. 6.11. Flow in a conduit aquifer.
(a) Diagram (based upon Fig. 6.7) showing relative importance of flow routes.
(b) Hydrograph at basin outlet of Traligill River, Sutherland, Scotland.
(c) Catchment of the Traligill River.

where surface streams are unknown. The spring, fed entirely by percolation water, has a delayed response to rainfall, with only a 20% increase in discharge, and a prolonged recession limb.

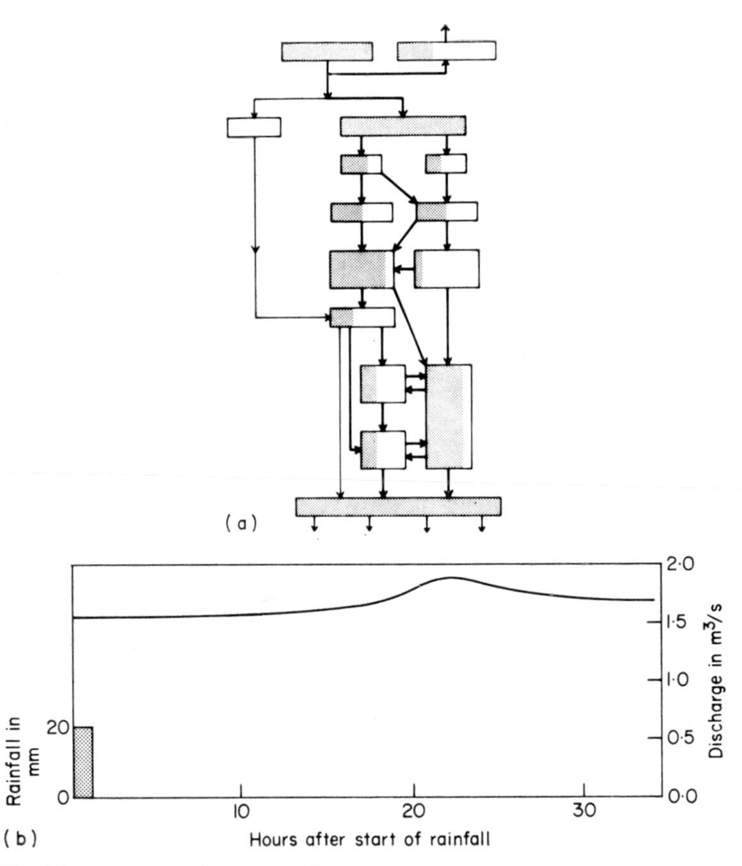

FIG. 6.12. Flow in a percolation aquifer.
(a) Diagram showing relative importance of flow routes.
(b) Hydrograph at Dromilly Bridge, Dromilly Clear Spring, Trelawny, Jamaica. (For location of spring and catchment, see Fig. 6.9a).

The Cheddar spring in the Mendip Hills is fed by a mixture of percolation and swallet waters (Fig. 6.13). The effect of the conduit portion of the system is seen in the rapid rise of the hydrograph, which is followed by a prolonged recession as percolation water reaches the spring. The maximum discharge is 100% greater than baseflow.

In a mixed system there are pronounced variations in the relative

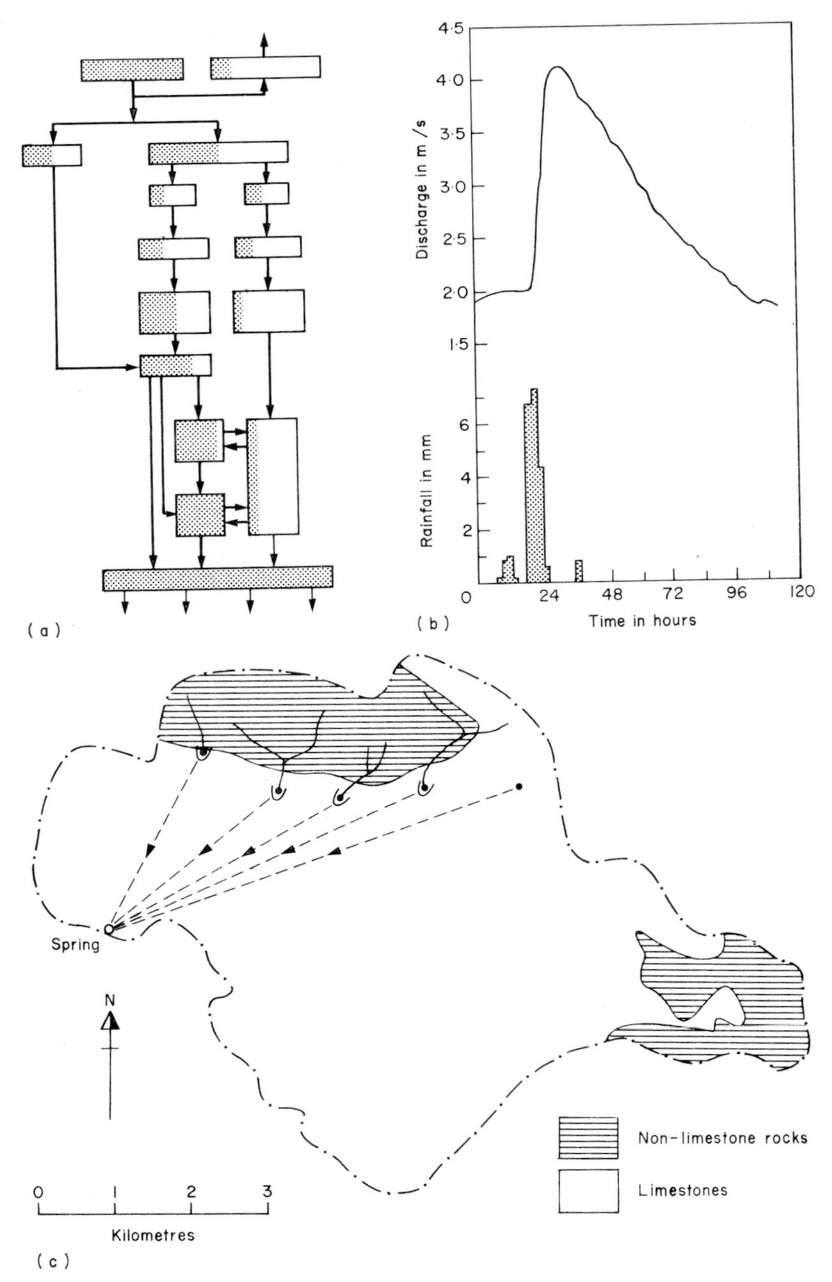

FIG. 6.13. Flow in a mixed aquifer.
(a) Diagram showing relative importance of flow routes.
(b) Hydrograph at Cheddar Spring, Somerset.
(c) The catchment area of the Cheddar Spring.

importance of conduit and diffuse flows at different times during and after the passage of a storm hydrograph. The Cheddar catchment is again used as an example and Fig. 6.14a shows its behaviour during a slightly idealized flood, with the discharge plotted as daily flows. The data shown on the accompanying flow diagrams are based upon detailed analysis of a year's flow records. The swallet and conduit parts of the system respond more quickly to rainfall than the percolation water, and swallet streams may increase immediately after or even during a storm. The flow diagram, Fig. 6.14b, shows a situation corresponding to point A on the rising limb

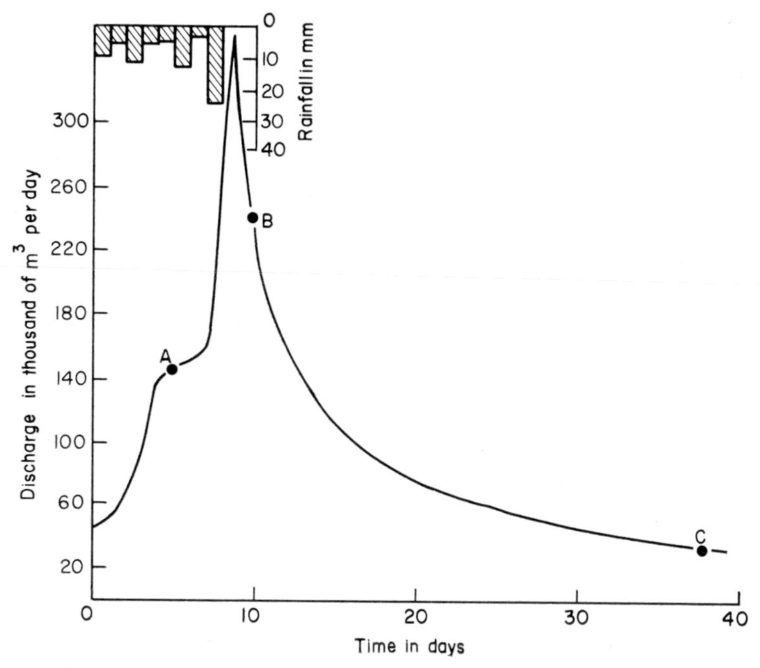

FIG. 6.14. Hydrology of a mixed aquifer during and after a flood.

(a) Hydrograph of an isolated flood in the Cheddar system (slightly idealized).

(b) Flow in the Cheddar system during and immediately after precipitation (see point "A" on Fig. 6.14a). Percentage values show proportions of inputs.

(c) Flow in the Cheddar system following precipitation (see point "B" on Fig. 6.14a). Percentage figures show proportions of total flux through system.

(d) Flow in the Cheddar system during drought (see point "C" on Fig. 6.14a). Percentage figures show proportions of total flux through system.

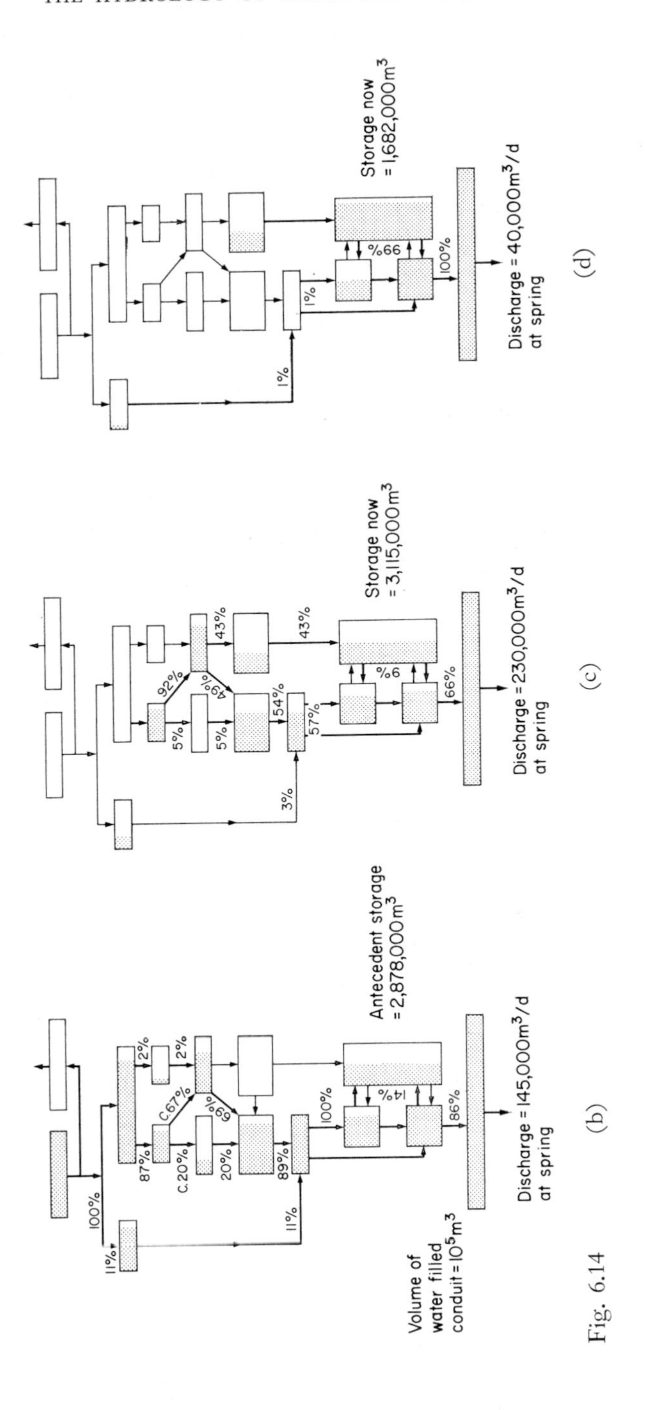

Volume of water filled conduit = 10⁵m³
Antecedent storage = 2,878,000m³
Discharge = 145,000m³/d at spring
(b)
Storage now = 3,115,000m³
Discharge = 230,000m³/d at spring
(c)
Storage now = 1,682,000m³
Discharge = 40,000m³/d at spring
(d)

Fig. 6.14

of the spring hydrograph. The total discharge is 145 000 m³/day of which 11% is contributed by swallets. The remainder of the effective rainfall is still replenishing the soil moisture, and the percolation part of the system has not yet responded to the rainfall. However, parts of the percolation system are still actively supplying water to the spring as a result of rainfall some days beforehand. Most of the remaining discharge of the spring will be supplied from this source until runoff from the storm under consideration enters the percolation system from the soil.

Because the swallet system responds more rapidly than the percolation system, the hydraulic head in the waterfilled Cheddar master cave will be higher at this stage than that in adjacent waterfilled fissures. This causes a net flow from the cave to the fissures, amounting to about 20 000 m³/day in this case. This temporary recharge of the fissures, which is analogous to bank storage in rivers, will continue until the discharge from the swallet declines after the end of the rainfall and the hydraulic gradient from waterfilled caves to fissures is reversed. By the time that this has happened, at point B on the hydrograph on Fig. 6.14a, the percolation system is fully active and streams entering swallets account for only 3% of the spring discharge of 230 000 m³/day. Rather more than half of the percolation water is in the form of vadose trickles, which eventually contribute to cave streams. The remainder is vadose seepage which will recharge the storage in waterfilled fissures. This storage has already increased from its pre-storm value of 2·88 million m³ to 3·12 million m³. Simultaneously, the waterfilled fissures are now contributing 20 000 m³/day to the conduit flow. Over the whole of this flood event 57% of the runoff was in the form of vadose trickles and swallet streams and 43% was contributed by vadose seepages via storage.

Finally, Fig. 6.14d shows the baseflow conditions following several weeks' drought. Swallet waters contribute only 1% of the spring discharge which has by now fallen to 40 000 m³/day. The storage is still sufficiently large to maintain flow, albeit diminishing, for over six months without further rainfall.

THE WATER-TABLE CONCEPT IN LIMESTONE AQUIFERS

One of the most long-standing controversies in the study of limestone hydrology is whether the water-table concept may be legitimately applied to limestone areas. It was first suggested by Martel (1910) that the importance of flow in caves and conduits was so great that it was irrelevant to regard a cavernous limestone as possessing a water-table, in the sense of a single surface below which the rock was entirely saturated. Conversely, Grund (1903) regarded cavernous limestones as being essentially similar to other aquifers in that they possess a network of waterfilled fissures, the rest level in which is the water-table. He regarded caves and waterfilled voids, though common, as exerting only a minor influence on the hydrology. Subsequently, there has been a tendency for English-speaking workers to favour the views of Grund, whereas some French and other European authors have inclined towards the ideas of Martel. Arguments against the importance of the water-table concept have centred around the conduit nature of much limestone drainage. Drew (1966) cited the evidence of crossing flow lines and rapid flow velocities, demonstrated by water tracing, to suggest the importance of conduits in massive limestones. He argued that the lack of mixing from one conduit to another demonstrates that they are watertight, and that the limestone surrounding them is almost dry and acts as an aquiclude. He also envisaged that percolation water was largely of the vadose trickle type, and could be considered as flowing in conduits too small to be entered. This can now be seen to be an overstatement of the case, for more recent work on the Mendip springs (some of the results are described in the preceding section) has shown that there are substantial amounts of water stored in flooded fissures close to the level of the spring. These fissures are narrow, unlike grikes opened by solution immediately beneath the soil, although at least some are large enough to drain freely under gravity. They constitute the secondary porosity of the limestone. In the absence of well records, there are no reliable estimates for the secondary porosity of the Mendip limestone, but Moore, Burchett and Bingham (1969) quote a value of 0·4% for the massive limestones of the Stones River basin, Tennessee. Only 58% of the total annual discharge from the Cheddar spring is contributed by direct conduit flow of combined swallet and percolation water. The remaining 42% is derived from fissure storage replenished mainly by vadose springs. When fully charged, this storage is equivalent to 105 mm of rainfall. Since the mean annual runoff from the area is about 600 mm, it can be inferred that the water is stored for an average period of 2 months. Although the flow-through time is undoubtedly less than this for some seepage water, the average time is so long that it is probable that laminar flow conditions obtain in many of the waterfilled fissures. The average permeability of the fissures has been determined by analysis of baseflow recession curves at the springs, and has a value of 89 m/day. This is a value similar to that of the Lincolnshire Limestone, which has been demonstrated by Downing and Williams

(1969) to possess rather ill-developed conduits, and rather higher than those in the Chalk (Ineson, 1962) which is not cavernous.

It is apparent that the conflict between the two schools of thought can be resolved by regarding limestone aquifers as being composed of both

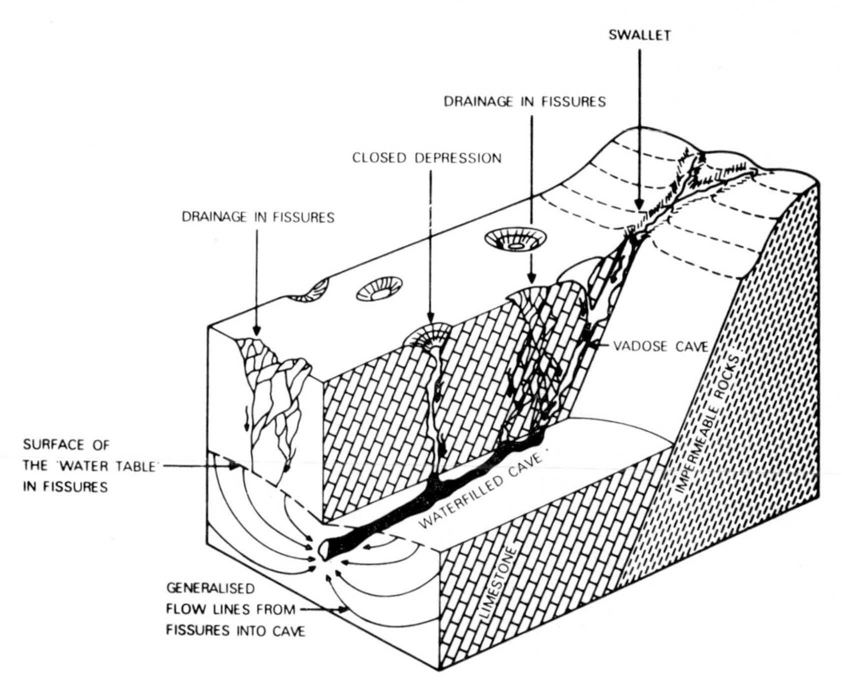

FIG. 6.15. Block diagram showing drainage of a limestone area via caves, closed depressions, and fissures.

conduit and diffuse flow systems. In the diffuse system in massive limestones water is confined to a network of interconnected fissures in which the air-water interface is a skeletal surface which is equivalent to the water-table in granular porous media. In this respect, the water-table in unconfined limestone aquifers is similar to that in other fractured rocks. The conduit system, on the other hand, is analogous to a surface river system or a town drainage scheme. Such systems are extremely localized and any concept of a general water surface is irrelevant to them. The two systems interact in a variety of ways. Firstly, there is an exchange of water between them during flood events (see Fig. 6.14). Secondly, the flow in the conduits is turbulent and reference to the D'Arcy-Weisbach equation earlier in this chapter will show that it is the square of the discharge which is proportional to the hydraulic gradient, whereas in the laminar regime of the diffuse flow, discharge is proportional to the first power of the gradient. Thus it is clear that the water-table above a submerged conduit will be depressed, and the course of the conduits will be marked by troughs or valleys in the water-table. This creates a net gradient towards the conduit, which is consequently fed by diffuse flow through fissures in its walls. This is illustrated in Fig. 6.15, which is in the form of a block diagram. When the conduit system is at shallow depths, severe flooding will cause backing-up and the water-table may rise as far as the ground surface. This occurs spectacularly in Co. Clare, Ireland, where Tratman (1968) has described how water spurts out along every joint line and closed depression in a severe flood.

CONCLUSION

Much heat and little light have been generated by the debate on the nature of limestone drainage which has spluttered and flared in the literature for the past sixty years. Much of the heat, but little of the light, has arisen from the dogmatic positions of the two main schools of thought. One school, taking its ideas from Grund (1903), has maintained that water circulation in limestones is essentially similar to that of any other fractured rock, and that caves develop as a consequence of the circulation without greatly influencing its pattern (see, for example, Davis, 1930; Swinnerton, 1932; Rhoades and Sinacori, 1941). On the other hand, the school following Martel (1910) tend to maintain that without caves and conduits there can be no underground circulation, or virtually none, and that the groundwater regime of limestone terrains is thus utterly different from that of other rocks (Trombe, 1952). A brief exposition of the details of this controversy is given in Jennings (1971, pp. 88-97). Both schools have generalized from inadequate data, and it is only in the past few decades that measurements have been made which allow some resolution of the problem and understanding of the processes involved. As we have seen, limestone drainage occurs both by conduit and diffuse flow, which interact to form a single integrated system rather like a river basin on the surface. The caves and conduits are analogous to the main stream and its tributaries, whereas the diffuse flow corresponds to the water in the soil which acts as a reservoir and maintains the flow of the river in drought. Because of this duality, and because so much of the system is hidden below ground, limestone drainage is often hard to understand in the field, and consequently difficult to quantify. There are very few numerical estimates of the way in which the drainage system operates, but in this chapter we have tried to review those that there are.

They indicate that the two apparently opposing models of Grund and Martel are but extreme cases of a spectrum of possible drainage systems. Most limestone regions exhibit features of both conduit and diffuse flow, while areas which show either to the exclusion of the other are rare.

REFERENCES

Ashton, K. (1966). The analyses of flow data from karst drainage systems. *Trans. Cave Res. Grp. G.B.* 7 (2), 161-203.

Atkinson, T. C. (1968). The earliest stages of underground drainage in limestone—a speculative discussion. *Proc. Brit. Speleo. Ass.* 6, 53-70.

Atkinson, T. C. (1971a). The dangers of pollution of limestone aquifers. *Proc. Univ. Bristol Speleo. Soc. 12* (3), 281-290.

Atkinson, T. C. (1971b). Hydrology and erosion in a limestone terrain. *Unpublished Ph.D. thesis, University of Bristol.*

Atkinson, T. C., Smith, D. I., Lavis, J. and Whitaker, R. J. (1973). Experiments in tracing underground waters in limestone. *J. Hydrol. 19*, 323-349.

Brook, D. (1972). Caves and caving in Kingsdale. *J. Brit. Speleo. Ass. 6* (48), 33-47.

Brown, M. C. (1966). *The 1965-6 Karst Hydrology Expedition to Jamaica.* Full report. (Private publication.)

Brown, M. C. and Ford, D. C. (1971). Quantitative tracer methods for investigation of karst hydrologic systems. *Trans. Cave Res. Grp. G.B. 13* (1), 37-51.

Carter, W. L. and Dwerryhouse, A. R. (1905). The underground waters of north-west Yorkshire. Pt. 1, *Proc. Yorks. Geol. Soc. 14*, 1-18; Pt. II, *15*, 248-292.

Chevalier, P. (1951). *Subterranean Climbers.* Faber and Faber, London.

Davis, W. M. (1930). Origin of limestone caverns. *Bull. Geol. Soc. Am.*, *41*, 475-628.

DeWeist, R. J. M. (1965). *Geohydrology.* Wiley, New York.

Downing, R. A. and Williams, B. P. J. (1969). The groundwater hydrology of the Lincolnshire Limestone, with special reference to the groundwater resources. *Water Res. Board Publ. 9.*

Drew, D. P. (1966). The water table concept in limestones. *Proc. Brit. Speleo. Ass. 4*, 57-67.

Drew, D. P. (1968a). A study of the limestone hydrology of the St. Dunstans Well, eastern Mendip, Somerset. *Proc. Univ. Bristol Speleo. Soc. 11* (3), 257-276.

Drew, D. P. (1968b). Tracing percolation waters in karst areas. *Trans. Cave Res. Grp. G.B. 10* (2), 107-114.

Drew, D. P. (1969). Water tracing (in Limestone Geomorphology, a study from Jamaica). *J. Brit. Speleo. Ass. 6* (43/44), 96-110.

Drew, D. P. (1970a). Limestone solution within the east Mendip area, Somerset. *Trans. Cave Res. Grp. G.B. 12* (4), 259-270.

Drew, D. P. (1970b). The importance of percolation water. *Groundwater*, *8* (5), 8-11.

Drew, D. P. and Smith, D. I. (1969). Techniques for the tracing of subterranean drainage. *Brit. Geomorph. Res. Grp. Tech. Bull. 2.*

Drew, D. P., Newson, M. D. and Smith, D. I. (1969). Water tracing of the Severn Tunnel Great Spring. *Proc. Univ. Bristol Speleo. Soc. 12* (2), 203-212.

Ford, D. C. and Stanton, W. I. (1968). The geomorphology of the south-central Mendip Hills. *Proc. Geol. Ass.* 79, 401-427.

Ford, T. D. (1959). The Sutherland caves. *Trans. Cave Res. Gp. G.B.* 5 (2), 139-190.

Grund, A. (1903). Die Karsthydrographie, Studien aus Westbosnien. *Geogr. Abh.* 7 (3), 103-200.

Hubbert, M. K. (1940). The theory of groundwater motion. *J. Geol.* 48, 785-944.

Ineson, J. (1962). A hydrogeological study of the permeability of the Chalk. *J. Instn Water Engns*, 16, 449-463.

Jennings, J. N. (1971). *Karst.* Australian National University Press, Canberra.

La Valle, P. (1968). Karst depression morphology in south central Kentucky. *Geografiska Ann.* 50A, 94-108.

Leopold, L. B., Wolman, M. G. and Miller, J. P. (1964). *Fluvial Processes in Geomorphology.* Freeman, London.

Lovelock, P. E. R. (1970). The laboratory measurement of soil and rock permeability. *Water Supply Papers, Instn geol. Sci., Tech. Communication,* 2.

Martel, E. A. (1910) La Théorie de la "Grundwasser" et les eaux souterraines du karst. *La Géogr.* 21, 126-130.

Maurin, V. and Zötl, J. (1959). Die Untersuchung der Zusammenhänge unterirdischer Wasser mit besonderer Berücksichtigung der Karstverhaltnisse. *Steirische Beitrage zur Hydrogeologie*, Neue Folge, 1-2.

Moore, G. K., Burchett, C. R. and Bingham, R. H. (1969). Limestone hydrology in the Upper Stones River basin, central Tennessee. *Tenn. Dept. Conservation and Dept. Water Resources*, 59.

Newson, M. D. (1971). A model of subterranean limestone solution in the British Isles. *Trans. Inst. Brit. Geog.* 54, 55-70.

Palmer, M. V. and Palmer, A. N. (1975). Landform development in the Mitchell Plain of Southern Indiana: origin of a partially karsted plain. *Zeit. für Geom. N.F.* 19, 1-39.

Pitty, A. F. (1966). An approach to the study of karst water. *Univ. Hull Occ. Paper in Geog.* 5.

Queffélec, C. (1968). *Jusqu'au Fond du Gouffre.* Stock, Paris.

Rhoades, R. and Sinacori, M. N. (1941). Pattern of groundwater flow and solution. *J. Geol.* 49, 785-794.

Schuster, E. T. and White, W. B. (1971). Seasonal fluctuations in the chemistry of limestone springs: a possible means of characterising carbonate aquifers. *J. Hydrol.* 14, 93-128.

Smith, D. I. and Nicholson, F. H. (1964). A study of limestone solution in northwest Co. Clare, Eire. *Proc. Univ. Bristol Speleo. Soc.* 10 (2), 119-138.

Smith, D. I. (ed.) (1969). Limestone geomorphology, a study from Jamaica. *J. Brit. Speleo. Ass.* 6, 43/44, 85-166.

Stenner, R. D. (1966). The variation of temperature and hardness in St. Cuthbert's Swallet, a progress report. *Belfry Bull.* 21 (1), 117-121.

Swinnerton, A. C. (1932). Origin of limestone caverns. *Bull. Geol. Soc. Am.* 43, 662-693.

Tate, T. (1879). The source of the R. Aire. *Proc. Yorks. Geol. Soc.* VII, 177-187.

Thrailkill, J. V. (1968). Chemical and hydrologic factors in the excavation of limestone caves. *Bull. Geol. Soc. Am.* 79, 19-45.

Tratman, E. K. (1968). A flash flood in the caves of northwest Clare, Ireland. *Proc. Univ. Bristol Speleo. Soc. 11* (3), 292-296.

Tratman, E. K. (1969). (ed.) *The Caves of Northwest Clare, Ireland.* David and Charles, Newton Abbot.

Trombe, F. (1952). *Traité de spéléologie*. Paris.

Vennard, J. K. (1961). *Elementary Fluid Mechanics*. Wiley, New York.

White, W. B. and Longyear, J. (1962). Some limitations on speleogenetic speculation imposed by the hydraulics of groundwater flow in limestones. *Nittany Grotto Newsletter* (Nat. Speleo. Soc. U.S.A.), *10* (9), 155-167.

White, W. B. and White, E. L. (1970). Channel hydraulics of free surface streams in caves. *Caves and Karst*, *12* (6), 41-48.

Wilcock, J. D. (1968). Some developments in pulse-train analysis. *Trans. Cave Res. Grp. G.B. 10* (2), 73-98.

Williams, P. W. (1972). Morphometric analysis of polygonal karst in New Guinea. *Bull. Geol. Soc. Am. 83*, 761-796.

7. The Chemistry of Cave Waters

R. G. Picknett, L. G. Bray and R. D. Stenner

This chapter is concerned with the solutional effects of natural waters on cave-bearing rocks (limestone, with gypsum and anhydrite), and with the deposition of solids from the resulting solutions. In Part I the subjects are covered in general terms, Part II contains more detailed treatments of certain aspects of solution and deposition, and Part III deals with practical techniques of analysis. A glossary of scientific terms and symbols is included.

PART I: GENERAL REVIEW

R. G. Picknett

INTRODUCTION

Pure water has little effect on limestone: the solvent action of natural waters depends on the acids which it contains, the most important being carbonic acid from dissolved carbon dioxide. Water absorbs carbon dioxide from the atmosphere, and the resulting solution can dissolve considerably more limestone than pure water. Even so, the solvent power provided by carbon dioxide from the open air is much less than is commonly found in natural waters, the extra carbon dioxide being supplied by biological processes in the soil and in rotting vegetation. Pitty (1966) has reviewed this subject. Figure 7.1 shows the effect of the carbon dioxide content of the water on the solubility of calcite, the form of calcium carbonate which is the main constituent of limestone. If water takes up carbon dioxide, its ability to dissolve calcium carbonate is enhanced; conversely, if the water loses some of its carbon dioxide, then part of the calcium carbonate in solution tends to be precipitated, perhaps forming flowstone, stalagmites and stalactites.

Also indicated in Fig. 7.1 is the relationship to the carbon dioxide content of the air. The normal atmosphere contains 0·03 to 0·035% (by volume) of carbon dioxide, and pure water in contact with the atmosphere can dissolve about 70 mg of calcium carbonate per litre (70 ppm). Soil air contains up to about 10% of carbon dioxide (Shoeller, 1962), corresponding to an upper limit of 500 mg of calcium carbonate per litre of solution (500 ppm). It is uncommon to find concentrations of calcium carbonate higher than this in natural waters.

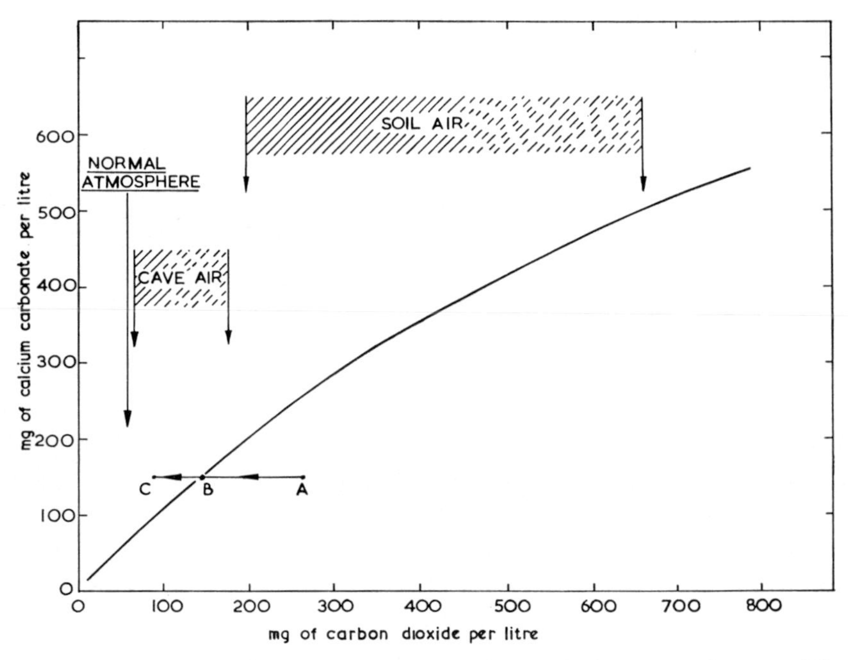

FIG. 7.1. The amount of calcium carbonate (calcite) which can dissolve at 10°C in water containing carbon dioxide. Also indicated are the ranges of carbon dioxide concentration to be expected when the water is in contact with air from various sources.

Other acids besides carbonic can be important for limestone solution. Sulphuric acid is often found in water from bogs (Gorham, 1958) and, although the concentration is low, a significant amount of limestone solution can result. This acid can also be produced underground by the oxidation of sulphide ore, thereby causing localized cavern formation (Caro, 1965). Organic acids such as formic, acetic and butyric are formed in plants both during life and, by decay, after death. Murray and Love (1929) have established that they are available in the soil and can be leached by water to dissolve limestone. Oxidation eventually converts them to carbon dioxide, but appreciable penetration of the limestone may occur before this process is complete (Howard, 1964; Bray, 1972). Although humic acids may not attack limestone directly (Fetzer, 1946), they certainly provide carbon dioxide on oxidation, and they may have an indirect influence by stabilizing the pH and by aiding the transport of certain metals which retard limestone solution. Industrial pollution may be a significant source of acids, even in remote regions (Rodhe, 1971).

LIMESTONE AND CALCIUM CARBONATE

The main chemical constituent of limestone is calcium carbonate: this sometimes is very pure, but more commonly it is associated with magnesium carbonate as a solid solution or as the mineral dolomite, which is the double carbonate of calcium and magnesium. Impurities usually

Table 7.1

THE SOLID VARIETIES OF CALCIUM CARBONATE

Name	Formula	Crystal Structure	Density
Calcite	$CaCO_3$	rhombohedral	2·710 g/cc
Aragonite	$CaCO_3$	orthorhombic	2·930
Vaterite	$CaCO_3$	hexagonal	2·51
Calcium carbonate monohydrate	$CaCO_3.H_2O$	hexagonal	2·4
(Calcium carbonate hexahydrate)	($CaCO_3.6H_2O$)	(monoclinic)	1·8

present include silica, clays, iron, manganese, lead, uranium, sulphide, sulphate and phosphate. The full chemical picture is complex, but fortunately a sound understanding can be obtained by studying the behaviour of calcium carbonate alone.

Pure calcium carbonate exists in three solid varieties differing from one another in the spatial arrangement of the calcium and carbonate ions in the crystal structure, and there is at least one crystalline form containing water (Table 7.1). Calcite is the stable form of calcium carbonate and is the main constituent of most limestones and speleothems. Aragonite is less stable, but occurs frequently in caves. Vaterite is much less stable, being transformed to calcite or aragonite very rapidly in the presence of water, and has not been found in caves. Calcium carbonate monohydrate is rare, although it has been reported in a small cave where the air was of low relative humidity (Fischbeck and Müller, 1971). Like vaterite, it is unstable. Calcium carbonate hexahydrate has been described many times in the literature (Brooks *et al.*, 1950; Warwick, 1962), but the X-ray diffraction pattern is omitted from the ASTM index, and some doubt must remain concerning both this substance and other hydrates which have been reported in the past.

CALCIUM CARBONATE SOLUTION

The process of dissolving calcium carbonate is traditionally represented by the chemical equation:

$$CaCO_3 + CO_2 + H_2O = Ca(HCO_3)_2,$$

accompanied by the statement that calcium bicarbonate, $Ca(HCO_3)_2$, is more soluble than calcium carbonate. This equation is incorrect. There is no evidence for the existence of calcium bicarbonate molecules in solution (Nakayama, 1968), and the ratio of $CaCO_3$ molecules dissolving to molecules of CO_2 in solution is not 1 : 1, as implied. A more accurate account of the chemistry is now given.

In an aqueous solution containing only calcium carbonate, several different species of ion and molecule exist which are capable of reacting with one another. The following representation of the reactions is a simplification of the facts, but is sufficient to illustrate the solution process:

$$\underset{\text{(solid)}}{CaCO_3} \underset{\text{slow}}{\overset{\text{slow}}{\rightleftharpoons}} Ca^{2+} + CO_3^{2-} \qquad 1$$

$$CO_3^{2-} + H^+ \underset{\text{fast}}{\overset{\text{fast}}{\rightleftharpoons}} HCO_3^- \qquad 2$$

$$HCO_3^- + H^+ \underset{\text{fast}}{\overset{\text{fast}}{\rightleftharpoons}} H_2CO_3 \qquad 3$$

$\longrightarrow$ forward direction
$\longleftarrow$ reverse direction.

When calcium carbonate is placed in pure water, calcium and carbonate ions are released by Reaction 1. Some of the carbonate ions combine with the hydrogen ions always present in water according to Reaction 2, and some of the resulting bicarbonate ions also combine with hydrogen ions to produce carbonic acid as in Reaction 3. The last two reactions are so fast as to be practically instantaneous, while the first is slow. In the early stages of solution the reverse process of Reaction 1 is much slower than the forward process, but as the concentrations of calcium and carbonate ions increase, so the rate of the reverse process increases. Eventually the forward and reverse rates become equal and the calcium carbonate stops dissolving. At this stage the other reactions, being faster, are also in equilibrium and the solution is saturated with calcium carbonate.

Any acid added to the solution will increase the concentration of hydrogen ions. The result is to displace the equilibria of Reactions 2 and 3 in the forward direction so as to reduce the concentrations of the carbonate and bicarbonate ions. The lowered carbonate concentration permits more calcium carbonate to dissolve by Reaction 1 because the rate of the reverse process is reduced, and eventually equilibrium is re-established with the resulting saturated solution having a higher calcium content. Thus the action of acids in increasing the solubility of calcium carbonate is explained.

Dissolved carbon dioxide is an acid and acts precisely as described, although its solution gives a further supply of carbonate and bicarbonate ions. However, it differs from most acidic substances found in natural waters in that it is gaseous and can be exchanged between air and water. The additional reactions involved are:

$$\underset{(air)}{CO_2} \underset{slow}{\overset{slow}{\rightleftharpoons}} CO_2^0 \qquad 4$$

$$CO_2^0 + H_2O \underset{slow}{\overset{slow}{\rightleftharpoons}} H_2CO_3. \qquad 5$$

CO_2^0 represents a carbon dioxide molecule in solution. Since in caves the air pressure is reasonably constant, the exchange between air and solution is governed by Henry's law. Thus the concentration of the species CO_2^0 in solution (not H_2CO_3 as has sometimes been claimed) is proportional to the concentration of carbon dioxide in the air. Increasing the concentration in the air causes carbon dioxide to dissolve by Reaction 4, and this causes more carbonic acid to form by Reaction 5. The result is an increased calcium carbonate solubility. Figure 7.2 shows the relationship between calcite solubility and the carbon dioxide content of air in contact with the solution for several temperatures. In terms of constant carbon dioxide content of the air the solubility decreases with increasing temperature, the rate of change at 10°C being about 1·3% per degree C.

Reactions 1 to 5 are not a complete description of the chemistry of calcium carbonate solution. Other reactions are known to occur but normally have little effect on the final equilibrium and will not be discussed now. However, two applications merit discussion. The first concerns

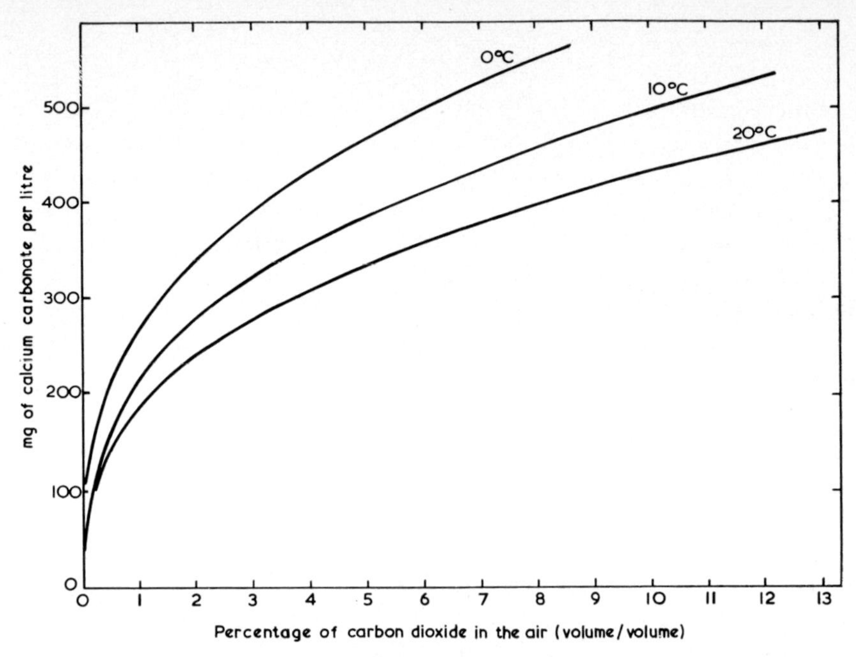

FIG. 7.2. The amount of calcium carbonate (calcite) which can dissolve in water which is in contact with air containing carbon dioxide. (Calculated values.)

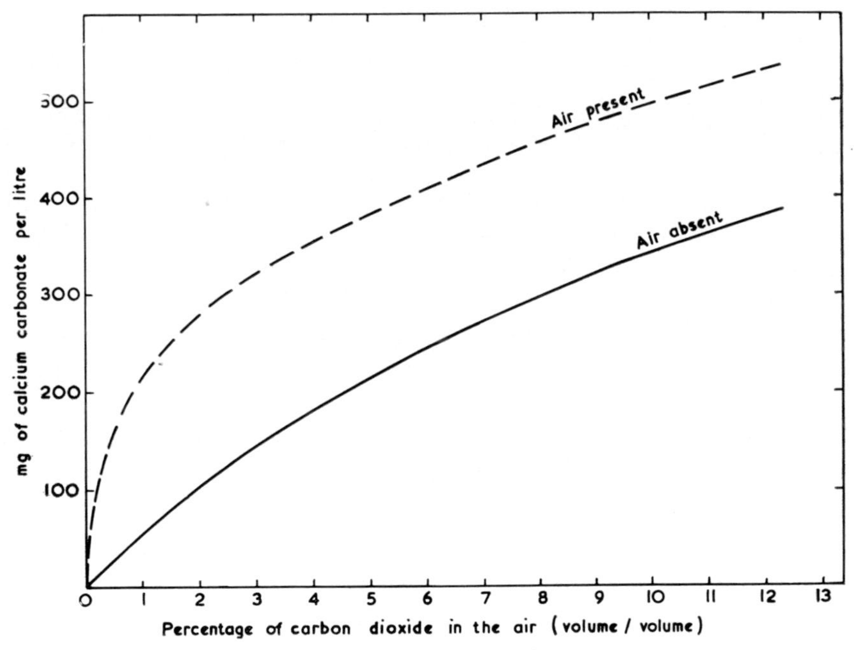

FIG. 7.3. Anaerobic solubility. Water at 10°C is brought to equilibrium with air containing carbon dioxide and then the air is removed. The solid curve shows the amount of calcium carbonate (calcite) which can dissolve in this water. The broken curve shows the amount which can dissolve when the air is always present (taken from Fig. 7.2).

anaerobic solubility, where water comes to an initial equilibrium with air, and then the air is cut off before calcium carbonate is introduced. This may occur, for example, when water enters a water-filled fissure in limestone. Whilst in contact with air the water will absorb carbon dioxide up to the equilibrium concentration given by Henry's law. Once the air is removed and calcium carbonate added, some of the carbonate ions from the dissolving solid combine with hydrogen ions according to the forward

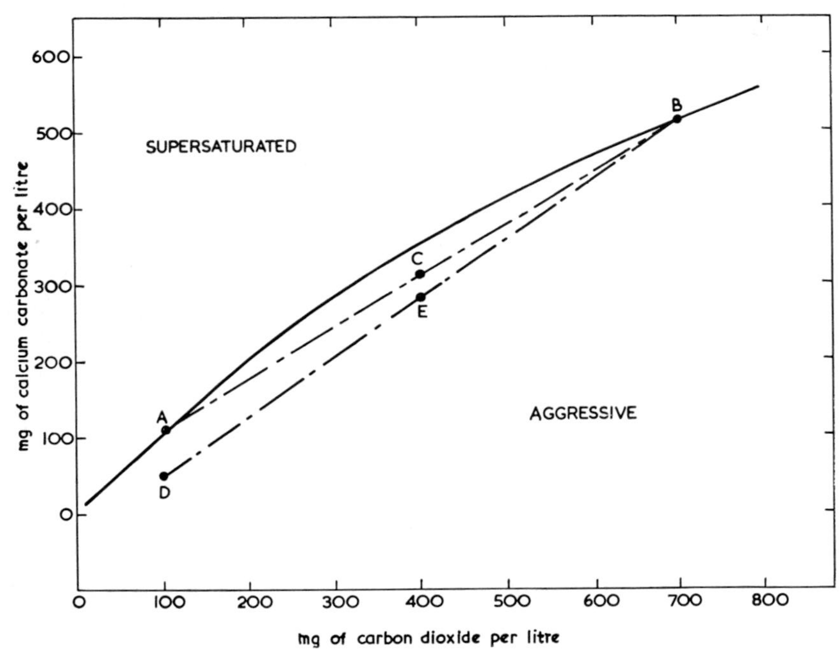

FIG. 7.4. Mischungskorrosion effect. The solid curve shows the solubility of calcium carbonate (calcite) at 10°C with respect to the total carbon dioxide in solution. The mixing of two saturated solutions A and B results in increased aggressiveness, as at C.

direction of Reaction 2. The reduced hydrogen ion concentration in turn affects the balance of Reaction 3, with the result that some CO_2^0 is removed by the forward process of Reaction 5. Normally this carbon dioxide would be replaced from the air, but this is not possible under the present conditions and so less calcium carbonate can dissolve than would be the case with air constantly present. Figure 7.3 illustrates the result. The solid curve shows the anaerobic solubility of calcium carbonate, while the broken curve shows the full solubility obtained with air constantly present. The difference can be large with percolation water in limestone covered by soil. While passing through the soil, water usually gains much carbon dioxide, but insufficient calcium carbonate for saturation. If it then travels in limestone fissures without air space, the water will fail to utilize the high potential for calcium carbonate solution that the soil air provides.

The second application concerns aggressiveness, or the ability of water to dissolve further calcium carbonate. If two saturated solutions of calcium carbonate are mixed it is possible for the resulting water to be aggressive, the phenomenon being called *mischungskorrosion* by the discoverer, Bögli (1964, 1971). Consider the solubility curve of Fig. 7.4, where carbon dioxide concentration is plotted against calcium carbonate solubility. A and B represent saturated solutions of calcium carbonate:

	Solution A	Solution B
Total CO_2 in solution (mg per litre)	100	700
$CaCO_3$ in solution (mg per litre)	110	510

On mixing equal volumes of these solutions the resulting composition will be 400 mg per litre of CO_2 and 310 mg per litre of $CaCO_3$. This is represented by C on the diagram; C is below the solubility curve and must represent an aggressive solution. In fact, C lies on the line joining A and B, as do all mixtures of solutions A and B. Since the line is on the aggressive side of the curve, Bögli's rule becomes apparent: the mixing of different saturated solutions of calcium carbonate always produces aggressive solutions. In the example discussed, the solution C is capable of dissolving 20% more calcium carbonate, but this figure is excessive for natural waters, where 1% or 2% would be more usual. Mixture of different proportions of the two saturated solutions result in different degrees of aggressiveness, and Bögli's rule can even be extended to solutions which are not saturated. In Fig. 7.4 an aggressive solution D mixed with a saturated solution B may result in increased aggressiveness as at E. The mixing effect is usually of little significance with vadose water because the degree of saturation is dominated by carbon dioxide transfer between solution and air. However, phreatic water, moving slowly deep underground, has had ample time to attain limestone saturation and temperature equilibrium; cavern formation under such conditions would be hard to explain without this concept of rejuvenated aggressiveness. The principle has now been extended to the mixing of waters which differ in the concentration of dissolved materials such as magnesium carbonate (Caro, 1965; Runnells, 1969; Picknett, 1972): either aggressiveness or supersaturation can result.

SOLUTIONAL EFFECTS IN CAVES

The soil is an important source of acids, and water percolating through it is usually capable of dissolving much limestone, i.e. it has high solutional power. The same applies to water which has passed through rotting vegetation or organic detritus. In contrast, the solutional power of a surface stream running on to limestone is usually lower, although such water contains some carbon dioxide from the air and from biological processes, and perhaps other acids too. The amount of limestone in solution depends not only on the solutional power but also on the rate of solution and the time available. The rate of solution is highest when most of the water is close to the rock surface. Thus water in narrow fissures more rapidly approaches saturation than does water in deep, wide channels. The time available for solution can be expected to follow the general trend of increasing with depth below the limestone surface. Flow through narrow fissures is usually slow, giving long contact times, while flow through large passages can be rapid.

By combining these factors a general idea of limestone solution in a cave is obtained. Percolation water usually has high solutional power, and its slow journey through narrow fissures ensures that most of the dissolving action occurs in the upper part of the limestone. In consequence, the water is nearly saturated with calcium carbonate when it enters the cave as roof drips and rivulets. A surface stream, on the other hand, although usually aggressive when it enters a cave, has low solutional power. Its relatively fast passage through the vadose part of the cave spreads the solutional activity deep into the limestone, and this effect is augmented wherever percolation water enters to boost the solutional power and aggressiveness of the stream. There is usually a progressive increase in both the calcium content and the solutional power of the stream water between the sink and the phreatic region of the cave (Stenner, 1970; Drew, 1970).

DEPOSITION OF CALCIUM CARBONATE

The chemistry of the deposition of calcium carbonate follows readily from what has been described earlier. First, the solution has to become supersaturated with calcium carbonate. The loss to cave air of carbon dioxide from solution is the principal process, and this occurs whenever the water is exposed to air of carbon dioxide content less than the equilibrium value. Figure 7.1 illustrates what happens to an initially aggressive water sample represented by the point A. As carbon dioxide is removed from solution, the point representing the water moves horizontally to the left, and when the point B on the solubility curve is reached the solution is saturated. Further loss of carbon dioxide results in supersaturation as at C.

A supersaturated solution tends to deposit calcium carbonate by the reverse process of Reaction 1. If solid calcium carbonate is already in contact with the solution, deposition will occur until the solution becomes just saturated, the fresh deposit building on the crystal structure of the solid. But if no solid calcium carbonate is present to trigger the process, the tendency for deposition will remain latent and the supersaturated solution will stay unchanged for a long time (Stumper, 1935). Even if solid calcium carbonate is present, coatings of materials like clay can retard deposition. Supersaturated solutions are of widespread occurrence in caves.

The calcium carbonate deposited from solution usually occurs as calcite, aragonite or vaterite, the form depending on a number of factors. First, there is the degree of supersaturation of the solution. Aragonite is stable enough in water to possess a measurable "solubility" which is higher than that of calcite. A solution just supersaturated with respect to calcite will therefore be aggressive towards aragonite, and under such conditions only calcite can be deposited. Vaterite is insufficiently stable in water to have a measurable solubility, but it is considered that similar reasoning can be applied, so that vaterite can only be deposited from solutions highly supersaturated with calcite. In Fig. 7.5, the area above the solubility curve for calcite has been divided into three precipitation zones. Supersaturated water lying in a particular zone can only deposit the crystal forms indicated for that zone.

Another factor governing the variety of calcium carbonate deposited is the presence of a suitable crystalline solid to trigger the process. If calcite crystals are present, then there is a tendency for the deposit to build on these to form more calcite, and the corresponding behaviour occurs if aragonite or vaterite crystals are present. The process is not always straightforward, nor well understood. Magnesium and sulphate at concentrations greater than 10^{-5} molar both tend to inhibit the growth of calcite crystals so that aragonite can develop even in the presence of calcite (Bischoff and Fyfe, 1968). Strontium or lead can initiate aragonite crystallization with the same effect (Zeller and Wray, 1956; Wray and Daniels, 1957). A further factor is the rate of deposition, which is tied to the degree of supersaturation and to the degree of agitation of the solution. Thus at high rates of deposition vaterite is formed in preference to calcite

(Roques, 1964). A final factor is organic matter. Kitano and Hood (1965) reported that some organic substances in solution caused aragonite or even vaterite to form instead of calcite. Aragonite in some shells is stabilized by organic material so that it does not transform to calcite in the presence of water (Fyfe and Bischoff, 1965).

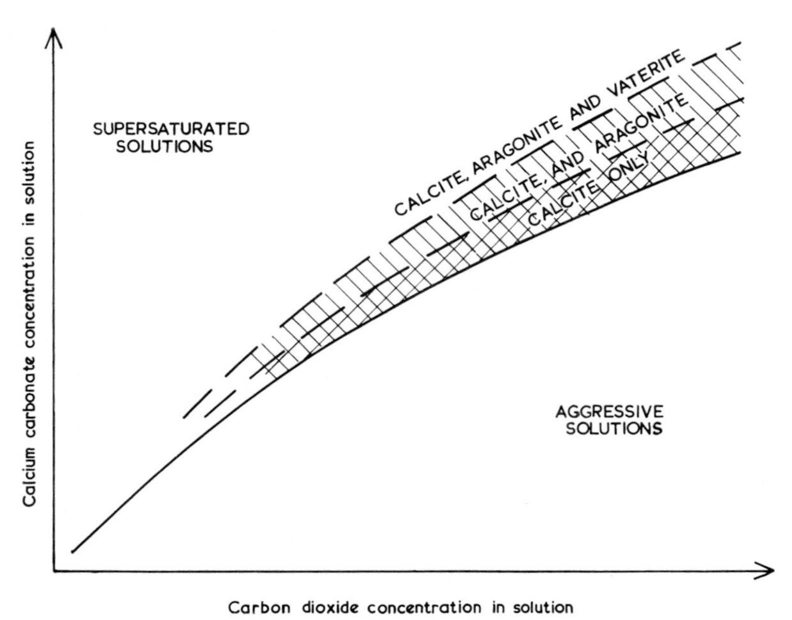

FIG. 7.5. Zones of supersaturation where crystallization of the various solid varieties of calcium carbonate can occur. The solid line gives the solubility curve of calcite: below this no crystallization can occur. The broken lines represent the solubility curves for aragonite and vaterite, respectively.

DEPOSITIONAL EFFECTS IN CAVES

The deposition of calcium carbonate from solution can result from water evaporation near cave entrances, giving rise to deposits which are characteristically impure because all of the dissolved material tends to come out of solution. It can also result from biological processes which remove carbon dioxide from solution, thereby causing supersaturation (Stenner, 1971; Pitty, 1971). (Moonmilk deposits are thought to be often of this kind.) However, these processes are very local: the dominant cause of calcium carbonate deposition is undoubtedly the transfer of carbon dioxide from solution to air, as discussed in Chapter 8. Water of high carbon dioxide content has to be exposed to air of low carbon dioxide content, and in caves this occurs when percolation water first enters a ventilated cave system. The rivulets and roof drips of high calcium carbonate content rapidly lose carbon dioxide to become supersaturated, and then begins the slow process of deposition. Films of water on the rock deposit flowstone; stalactites and draperies form where the water is channelled into drips; stalagmites form where the falling drops land, and still pools become lined with crystals. Since percolation water may penetrate deep below the limestone surface before encountering a ventilated cavity, such speleothems can be found in all parts of the cave system. Even cave streams may become supersaturated by loss of carbon dioxide, and when this happens the resulting deposit coats the stream bed with flowstone and raises rimstone barriers. The physical variety of calcium carbonate deposited is usually calcite, but aragonite is not uncommon and can be found adjacent to calcite even in growing speleothems. Murray (1954) has described aragonite and calcite straw stalactites growing less than 3 cm apart, and has also reported that some large stalagmites have aragonite interiors and calcite exteriors. Vaterite is rare and has not been identified in caves, although it has been found where spring waters emerge to the open air (Rowlands and Webster, 1971).

The attractive colours often found in speleothems are worthy of mention. Both calcite and aragonite are colourless and transparent when pure, although they usually appear white through light scattered by defects and crystal boundaries in the solid. Many impurities cause no colouration, and magnesium, silica and sulphate can be found in pure white speleothems. Other impurities cause colouration, and it is common for speleothems to appear an opaque pale or yellowish brown from material concentrated at the crystal boundaries. Richer colours of red, brown or orange are not infrequent and show themselves to best advantage in the more translucent speleothems. All these colours have been attributed to iron, probably in the form of goethite (FeO . OH), although Moore and Nicholas (1964) have reported brownish staining by carotene, an organic material formed by certain micro-organisms. The black soot-like layers sometimes found coating speleothems (or even whole passages) has been identified as due to manganese compounds, often birnessite ((Ca, Mg, K) $Mn_7O_{14} . 2H_2O$), which may also be responsible for the black bands seen in flowstone curtains. The impurities causing the green and blue-black colourations sometimes found in speleothems have not yet been identified. Colour in transparent speleothems is much less common, but pale yellow to rich red varieties have been found. Once again the agency causing the colour is unknown. It is probably dispersed throughout the solid as in gemstones like the ruby, instead of being concentrated at the grain boundaries.

GYPSUM AND ANHYDRITE

Although limestone is by far the most important of the cave-bearing rocks, solutional caves also occur in anhydrite and gypsum, both of which are forms of calcium sulphate. Anhydrite has the chemical formula $CaSO_4$ and reacts with water to form gypsum, $CaSO_4 . 2H_2O$. There is a difference in density between the two substances, and the transformation effected by water is accompanied by an increase in volume of some 30%. As a result, fissures in anhydrite tend to seal themselves when exposed to water, while gypsum behaves more like limestone.

Gypsum has a solubility of about 2 g per litre, which is far greater than that of limestone. The ions involved in the solution process do not react significantly with hydrogen ions, and so the solubility is almost completely unaffected by carbon dioxide or other acids. Water in contact with gypsum dissolves the rock until saturation is reached, but all sources of water have the same solutional power and there is no possibility of "mischungskorrosion" such as occurs with calcium carbonate. In gypsum or anhydrite rock the deposition of calcium sulphate from solution can occur only as a result of evaporation or cooling; evaporation is very slow in caves except near entrances, and cooling only changes the solubility by about 0·3% per degree C. Deposition, when it occurs, is therefore usually from solutions which are but slightly supersaturated, and large crystals are a typical consequence. Another deposition mechanism is possible when water saturated with gypsum enters limestone. If sufficient carbon dioxide is present, calcium carbonate will dissolve and cause the solution to become supersaturated with gypsum. The effect is only slight, the solution of 300 mg of calcium carbonate per litre causing a supersaturation of 1%. All these depositional mechanisms can produce the large crystals of calcium sulphate which are to be found in cave muds and on cave walls.

PART II: ADVANCED DISCUSSION

R. G. Picknett

In this section the solution chemistry of pure calcium carbonate is studied first, and then the modifying effects of natural impurities in the solid and in solution are considered. The discussion of saturation with the pure solid has been restricted to calcite because this is the only stable form at normal temperatures (Curl, 1962). Aragonite cannot achieve true saturation except at high temperature or in the presence of impurity.

SATURATED CALCIUM CARBONATE SOLUTIONS

Theory

For the pure system the ionic and molecular species present in solution are:* H^+, OH^-, CO_2^0, H_2CO_3, HCO_3^-, CO_3^{2-}, Ca^{2+}, $CaHCO_3^+$,$CaCO_3^0$. At very high pH some $CaOH^+$ is also present, but this is rare in natural waters. These ions and molecules react with one another, the probable reactions being shown in Fig. 7.6. The equilibria resulting from these reactions can be fully described by any one of several sets of equations, but the following has been chosen because the values of the equilibrium constants K_0, K_1, etc., which are used are better known than for other sets.

Equilibrium equations:

$$K_0 = (CO_2^0)/(H_2CO_3) \quad 1$$
$$K_1 = (H^+)(HCO_3^-)/(H_2CO_3) \quad 2$$
$$K_2 = (H^+)(CO_3^{2-})/(HCO_3^-) \quad 3$$
$$K_3 = (Ca^{2+})(HCO_3^-)/(CaHCO_3^+) \quad 4$$
$$K_4 = (Ca^{2+})(CO_3^{2-})/(CaCO_3^0) \quad 5$$
$$K_w = (H^+)(OH^-) \quad 6$$
$$K_s = (Ca^{2+})(CO_3^{2-}) \quad 7$$

Equation of electroneutrality:

$$2[Ca^{2+}]+[CaHCO_3^+]+[H^+] = 2[CO_3^{2-}]+[HCO_3^-]+[OH^-] \quad 8$$

Here the square brackets represent the molar concentration of the ion or molecule, while the round brackets represent the activity with respect to a standard state at infinite dilution (see Garrels and Christ, 1965). For non-ionized molecules in dilute solution, activity and concentration are equal. For ions the following relationship holds:

$$(J) = [J]\gamma_j, \quad 9$$

where γ_j is the individual ion activity coefficient of the ion J. This is determined by the Debye-Hückel equation:

$$-\log_{10}\gamma_j = A z_j^2 I^{\frac{1}{2}} / \{1 + d_j B I^{\frac{1}{2}}\}. \quad 10$$

A and B are constants for water at a given temperature, z_j is the number of electronic charges on the ion J (i.e. 1 for H^+; 2 for CO_3^{2-}), d_j is the effective diameter of the ion in solution and I is the ionic strength. I is defined by:

* For simplicity, the species H_3O^4 has been included with H^+. This does not affect the validity of what follows.

$$I = \tfrac{1}{2}\Sigma\{C_a z_a^2 + C_b z_b^2 + C_c z_c^2 + \ldots\}, \qquad 11$$

where C_a, C_b, C_c . . . represents the molar concentrations of every ion in solution. For amplification of these concepts see, e.g., Garrels and Christ (1965). It is important that ion activities, not concentrations, should be used in the equilibrium equations because the difference is marked, even in dilute solution. For a saturated calcite solution with $[Ca^{2+}]$ equal to 1 mmole/l, (Ca^{2+}) is only 80% of this.

It is customary to combine Eqns 1 and 2 to form a new constant K_1' which is more readily evaluated:

$$K_1' = K_1/(1+K_0) = (H^+)(HCO_3^-)/\{(H_2CO_3)+(CO_2^0)\}. \qquad 12$$

Finally, an equation is needed to relate the carbon dioxide of the air to that in solution:

$$K_c = [CO_2^0]/P. \qquad 13$$

P represents the partial pressure of carbon dioxide in the air.

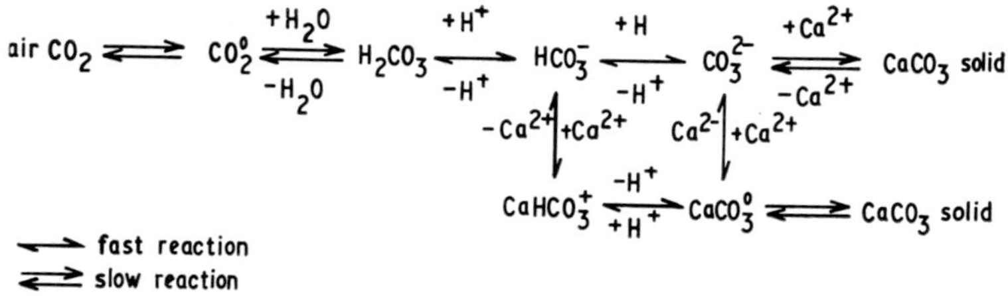

FIG. 7.6. The reactions that can occur in the $CaCO_3$—CO_2—H_2O—air system.

Equations 1 to 13 can be used to calculate the concentrations of all ions and molecules in saturated solution, given only one parameter, e.g. the partial pressure of carbon dioxide. For such calculations to be meaningful, however, values for the constants used must be known with accuracy.

Constants for the Theory

Values for the equilibrium constants and other parameters defined above have been extensively reviewed (Garrels and Christ, 1965; Langmuir, 1971; Wigley, 1971; Jacobson and Langmuir, 1972). Those listed in Table 7.2 have been taken from Picknett (1973). There is general agreement for K_1', K_2, K_W and for the constants of Eqn 10, but for the other constants there is some controversy or lack of precision. Consider K_3 for example. Roques (1964) and Langmuir (1968) claim that the ion $CaHCO_3^+$ is of negligible concentration in natural waters, which implies that K_3 must be greater than about 0·5. However, Garrels and Thompson (1962) give K_3 a value of 0·055 at 25°C while Nakayama (1968) gives a value of 0·056 at the same temperature, both in fair agreement with the tabulated value. The listed values of this constant are considered to be very inaccurate at 10°C, but to be within a factor or two at 25°C. K_4 is also poorly known, the tabulated values from Picknett (1973) being correct

Table 7.2

VALUES OF CONSTANTS FOR THE THEORY

Units: concentration, mole/l; partial pressure, atmospheres; ion size, cm

Constant	Temperature °C				Reference
	10	16	20	25	
Constants for Equation 10					
A	0·4961	0·5011	0·5046	0·5092	1
B	$3{\cdot}258 \times 10^7$	$3{\cdot}269 \times 10^7$	$3{\cdot}276 \times 10^7$	$3{\cdot}286 \times 10^7$	1
Ion sizes H^+	Assumed invariant with temperature			9×10^{-8}	2
Ion sizes OH^-				$3{\cdot}5 \times 10^{-8}$	2
Ion sizes HCO_3^-				$4{\cdot}5 \times 10^{-8}$	2
Ion sizes CO_3^{2-}				$4{\cdot}5 \times 10^{-8}$	2
Ion sizes Ca^{2+}				6×10^{-8}	2
Ion sizes $CaHCO_3^+$				$4{\cdot}5 \times 10^{-8}$	3
Equilibrium constants					
K_1'	$3{\cdot}44 \times 10^{-7}$	$3{\cdot}88 \times 10^{-7}$	$4{\cdot}16 \times 10^{-7}$	$4{\cdot}45 \times 10^{-7}$	4
K_2	$3{\cdot}24 \times 10^{-11}$	$3{\cdot}81 \times 10^{-11}$	$4{\cdot}20 \times 10^{-11}$	$4{\cdot}69 \times 10^{-11}$	5
K_W	$2{\cdot}92 \times 10^{-15}$	$4{\cdot}91 \times 10^{-15}$	$6{\cdot}81 \times 10^{-15}$	$1{\cdot}01 \times 10^{-14}$	6
K_3	0·13	0·09	0·06	0·04	7
K_4	1×10^{-4}	3×10^{-4}	3×10^{-4}	4×10^{-4}	7
K_s (calcite)	$4{\cdot}0 \times 10^{-9}$	$3{\cdot}9 \times 10^{-9}$	$3{\cdot}8 \times 10^{-9}$	$3{\cdot}7 \times 10^{-9}$	7
K_c	0·0526	0·0435	0·0383	0·0329	8
K_0	493	450	420	381	4, 9
K_1	$1{\cdot}7 \times 10^{-4}$	$1{\cdot}75 \times 10^{-4}$	$1{\cdot}75 \times 10^{-4}$	$1{\cdot}7 \times 10^{-4}$	9

1. Manov *et al.* (1943)
2. Klotz (1964)
3. Picknett (1964)
4. Harned and Owen (1958)
5. Harned and Scholes (1941)
6. Harned and Robinson (1941)
7. Picknett (1973)
8. Siedell (1958)
9. Wissbrun *et al.* (1954)

only to within a factor of two. However, there is some confirmation for them in that Garrels and Christ (1965) give a value of 6×10^{-4} for K_4 at 25°C, and Lafon (1970) quotes 2×10^{-4} at the same temperature. The calcite solubility product, K_s, causes some disagreement. In the past considerable use has been made of values published by Larsen and

Buswell (1942), but these are now considered to be erroneously high, particularly at low temperatures, through their method of derivation. Other values have been reviewed by Langmuir (1968). Those listed in the Table are claimed to be accurate to within 5%: they are lower than recent values by Jacobson and Langmuir (1972) by only $0{\cdot}2$ to $0{\cdot}4 \times 10^{-9}$, but are much lower than the value for 25°C quoted by Latimer (1959). Values for K_c from Harned and Davis (1943) have been accepted in the past, but Siedell (1958) has concluded that these are too high. Siedell's analysis seems conclusive and his values are given in Table 7.2. Experimental determinations of K_0 have been inconsistent, and therefore the listed values have been calculated from K_1 and K_1' using Eqn 12.

Ion and Molecule Concentrations

Several methods of obtaining equilibrium ion concentrations from Eqns 1 to 11 have been established (Roques, 1964; Picknett, 1964, 1973; Garrels and Christ, 1965) which involve considerable calculation and will not be considered here. The calculated concentrations depend critically on the values of the equilibrium constants used and it is necessary to study the errors involved. Results for the equilibrium constants of Table 7.2 at temperatures of 10°, 16°, 20° and 25°C are shown in Table 7.3 together with the expected percentage errors, which arise almost entirely from the uncertainty in K_s and K_3 values. Some ions are determinable to within a few per cent, but for $CaHCO_3^+$ and $CaCO_3^0$ the expected errors are large. The calculations have not been extended to more concentrated solutions because of the error which is then introduced in Eqn 10.

These calculated concentrations have been compared with available experimental measurements on saturated calcite solutions by Picknett (1973). For calcite solubility at various partial pressures of carbon dioxide in air (*vide* Fig. 7.2) the data used were those of Frear and Johnston (1929), Shternina and Frolova (1952) and Yanat'yeva (1954) at 25°C, Schloesing (1872) at 16°C and Picknett (1964) at 10°C. The calculated partial pressures for calcite concentrations up to 5·5 mmole/l agreed reasonably well with these experiments, the average (root mean square) error being 7%; but, as expected, there were indications of increased error for more concentrated solutions. For calcite solubility in carbon dioxide solutions without air present (*vide* Fig. 7.3), the experimental data of Weyl (1959) at 10° and 25°C were taken. Again the calculation, this time of the initial carbon dioxide concentration $[CO_2]_{init.}$, agreed reasonably well with the experiments for calcite concentrations up to 5·5 mmole/l, the average (root mean square) error being 6%. For calcite solubility in

Table 7.3

CALCULATED CONCENTRATIONS FOR SATURATED CALCITE SOLUTIONS

°C	$[Ca]_T$	$[CO_2]_{init.}$	$[Ca^{2+}]$	$[HCO_3^-]$	$[CO_2^0]$	$[H_2CO_3]$	$[CO_3^{2-}]$	$[CaHCO_3^+]$	$[CaCO_3^0]$	$[H^+]$	$[OH^-]$	P	pH*
	mmole/l					µmole/l						mAt	
10	5	10·3	4·73	9·69	5·30	10·7	2·10	226	40·0	0·235	0·0155	101	6·66
	3	4·28	2·87	5·82	1·32	2·68	2·91	89·7	40·0	0·0935	0·0374	25·1	7·06
	1	1·01	0·949	1·90	0·0573	0·116	6·60	11·1	40·0	0·0116	0·280	1·09	7·95
	0·6	0·561	0·556	1·09	0·0121	0·0244	10·2	3·94	40·0	0·00414	0·770	0·229	8·40
16	5	10·5	4·67	9·65	5·52	12·3	2·09	320	13·0	0·277	0·0222	127	6·60
	3	4·40	2·86	5·84	1·41	3·13	2·87	129	13·0	0·112	0·0525	32·4	6·99
	1	1·05	0·970	1·94	0·0655	0·145	6·35	16·7	13·0	0·0146	0·375	1·50	7·86
	0·6	0·591	0·581	1·15	0·0147	0·0326	9·64	6·19	13·0	0·00543	0·988	0·337	8·29
20	5	10·5	4·53	9·51	5·49	13·1	2·10	459	12·7	0·300	0·0284	143	6·57
	3	4·41	2·80	5·78	1·43	3·40	2·85	187	12·7	0·123	0·0664	37·2	6·95
	1	1·05	0·963	1·94	0·0679	0·162	6·25	24·8	12·7	0·0164	0·466	1·77	7·82
	0·6	0·592	0·578	1·15	0·0153	0·0365	9·45	9·22	12·7	0·00610	1·22	0·400	8·24
25	5	10·4	4·34	9·33	5·44	14·3	2·12	649	9·25	0·324	0·0390	165	6·54
	3	4·44	2·72	5·71	1·45	3·80	2·86	270	9·25	0·135	0·0896	44·0	6·92
	1	1·06	0·954	1·93	0·0716	0·188	6·16	36·7	9·25	0·0185	0·611	2·18	7·78
	0·6	0·596	0·577	1·15	0·0164	0·0431	9·25	13·8	9·25	0·00698	1·58	0·499	8·17
Average accuracy (%)	—	7	3	2	5 (combined)		4	75	40	5	5	5	0·2

Notes 1. pH* is the calculated pH as measured with a glass electrode. It is obtained by adding a liquid junction potential term to the calculated value of $-\log(H^+)$. See Picknett (1973).

2. $[CO_2]_{init.}$ is the concentration of carbon dioxide which will dissolve the stated amount of calcium carbonate in the absence of air.

$$[CO_2]_{init.} = [HCO_3^-]+[CO_2^0]+[H_2CO_3]+[CO_3^{2-}]-[Ca^{2+}]$$

3. The accuracy figures quoted are deduced from the estimated errors in the constants K_S, K_3 and K_4.

relation to total carbon dioxide in solution (*vide* Fig. 7.1), the experimental data of Schloesing (1872) at 16°C were taken. The calculation of total carbon dioxide in solution agreed very well with the experiments, the average (root mean square) error being only 4%. The final experimental relationship available for comparison is that between calcite solubility and pH. After allowing for the liquid junction potential associated with the calomel reference electrode (Picknett, 1968), the agreement was excellent between the calculated pH and the experiments of Picknett (1972) at 10°, 16°, 20° and 25°C. (This is to be expected since the tabled values of K_s, K_3 and K_4 were determined from these pH data.) The average (root mean square) error in calculated pH was less than 0·01 pH units.

Other experiments have been reported, giving data which could be used to test the theoretical calculations, but most of these are unreliable. However, the careful experiments of Roques (1964) should be noted. He determined both calcite solubility at various partial pressures of carbon dioxide in air and also the pH of saturated calcite solutions, the temperatures used being 10° and 15°C. His results are inconsistent with the experiments described above, but have been confirmed by Stchouzkoy-Muxart (1972) in experiments at 20° and 30°C, leaving an anomaly which merits further investigation. The work of Roques and of Stchouzkoy-Muxart apart, the four sets of experiments described provide the only reliable data applicable to cave conditions which can be used for testing the theoretical predictions, and the success of the tests gives confidence both in the theory and in the values of the constants used. The agreement is by no means perfect and no doubt refinements will be made to the theory and to the equilibrium constants of Table 7.2. Nevertheless, the role played by ion-pairs is made clear, $CaHCO_3^+$ being significant at low pH and $CaCO_3^\circ$ at high pH (Table 7.3). Also revealed is the variation in the individual ion concentrations with both carbon dioxide content and temperature. From more detailed calculations, phenomena such as "mischungskorrosion" can be evaluated, and indeed the data for Figs. 7.1 to 7.5 of this chapter were obtained by this means. Such quantitative calculations, however, cannot always be applied to natural waters because impurities may exert a considerable influence on the chemical equilibria (*vide infra*).

UNSATURATED CALCIUM CARBONATE SOLUTIONS

Even when air and solid calcium carbonate are absent, the chemical reactions of Fig. 7.6 can all proceed except for those on the far left and right sides of the diagram. The reactions in solution are sufficiently rapid for the establishment of equilibrium in only a few minutes. Equilibrium can also occur with air present, provided sufficient time is available for the exchange of carbon dioxide between solution and air, and is little disturbed by the presence of solid carbonate when the rate of solution or deposition is slow enough. For such conditions of equilibrium the same theoretical equations apply as were presented for saturated solutions, with Eqn 7 excepted. The remaining equations can be used to calculate the concentrations of all ions and molecules in solution, given only two parameters, e.g. the calcium concentration and the partial pressure of carbon dioxide.

Table 7.4 gives calculated concentrations for a solution containing 2 mmole/l of calcium carbonate exposed to air of various carbon dioxide contents, as indicated by the column for the partial pressure P. The first row corresponds to aggressive percolation water of high carbon dioxide content. As the table proceeds, carbon dioxide is lost from solution, representing the exposure of the percolation water to ventilated caves without calcite solution or deposition. (This is of practical consequence because both solution and deposition are slow.) Eventually the water is transformed from aggressive to supersaturated, as shown by the degree of saturation in the last column which is the ratio of calcium concentration in solution to that at saturation. The table shows that both the pH and the degree of saturation can vary markedly with P, and so in practical cave research care is needed whenever water samples may be exposed to the outside atmosphere before analysis. Also the ion-pair $CaCO_3^0$ is very important in unsaturated solutions at high pH, and should not be neglected in calculations.

Another calculation, representing the slow solution and deposition of calcite in the absence of air, is shown in Table 7.5. Here different amounts of calcite have been dissolved in a solution containing 2·38 mmole/l of carbon dioxide, the ion concentrations for each amount dissolved being shown in a separate row in the table. The total calcium concentration increases as the table proceeds, reaching calcite saturation at the sixth row. The solution of calcite in a waterfilled passage is therefore modelled by moving down the table to the sixth row, whilst the slow deposition of calcite is modelled by moving up the table from the last row.

These theoretical techniques can be applied to the solution of calcium carbonate in the presence of air, to the mixing of waters of different composition, and to many other practical conditions. They provide a powerful tool for studying the behaviour of calcite solutions and are

Table 7.4

ION CONCENTRATIONS FOR UNSATURATED SOLUTIONS AT 10°C IN EQUILIBRIUM WITH VARIOUS PARTIAL PRESSURES OF CARBON DIOXIDE.
$[Ca]_T = 2{\cdot}0$ mmole/l
(Transformation from aggressive to supersaturated solutions by loss of CO_2 to air)

P	$[Ca^{2+}]$	$[HCO_3^-]$	$[CO_2^0]$	$[H_2CO_3]$	$[CO_3^{2-}]$	$[CaHCO_3^+]$	$[CaCO_3^0]$	pH	Degree of Saturation (calcite)
mAt	mmole/l			μmole/l					
100	1·95	3·95	5·26	10·7	0·326	43·7	3·43	6·30	0·40
50	1·95	3·94	2·63	5·33	0·650	43·6	6·82	6·60	0·52
10	1·92	3·89	0·526	1·07	3·16	42·5	32·8	7·30	0·93
8·11	1·92	3·87	0·426	0·865	3·86	42·2	40·0	7·39	1·00
6·50	1·91	3·85	0·342	0·693	4·76	41·8	49·2	7·48	1·08
1·00	1·73	3·44	0·0526	0·107	24·6	34·3	236	8·25	2·06
0·10	1·22	2·25	0·00526	0·0108	103	16·5	760	9·07	4·26

Table 7.5

ION CONCENTRATIONS FOR UNSATURATED SOLUTIONS OF INITIAL CARBON DIOXIDE CONCENTRATION 2·38 MMOLE/L AT 10°C

(Solution and deposition of calcite in the absence of air)

$[Ca]_T$	$[Ca^{2+}]$	$[HCO_3^-]$	$[CO_2^0]$	$[H_2CO_3]$	$[CO_3^{2-}]$	$[CaHCO_3^+]$	$[CaCO_3]$	pH	Degree of Saturation (calcite)
	mmole/l			μmole/l					
0	0	0·0286	2·35	4·76	0	0	0	4·55	0
0·500	0·497	0·997	1·88	3·81	0·0540	3·23	0·193	6·17	0·25
1·00	0·986	1·98	1·38	2·80	0·300	12·0	1·88	6·59	0·50
1·50	1·47	2·95	0·891	1·81	1·06	25·5	8·98	6·95	0·75
1·90	1·83	3·70	0·514	1·04	2·91	38·7	29·1	7·29	0·95
2·00	1·92	3·87	0·426	0·865	3·86	42·2	40·0	7·39	1·00
2·10	2·00	4·03	0·343	0·697	5·22	45·5	56·9	7·50	1·05
2·50	2·23	4·47	0·119	0·240	18·8	55·5	216	8·00	1·25
3·00	2·33	4·62	0·0462	0·0938	51·4	59·4	611	8·42	1·50

especially important in the analysis of rates of reaction, but caution must be observed in applying the results to natural waters.

The computer is indispensible in calculations for both saturated and unsaturated solutions. The techniques are described in Chapter 14 of this volume.

DEGREE OF SATURATION

The *degree of saturation* of a solution is customarily taken to be the ratio of the calcium concentration to the saturation calcium concentration, and methods of quantifying it have recently raised much interest. What is required is a measure of the weight of calcium carbonate that can dissolve in a given volume of solution in the absence of air (or, for supersaturated solutions, the weight that can be precipitated). For example, one litre of calcite solution containing 1·90 mmole of calcium and 4·29 mmole of all forms of carbon dioxide (Table 7.5 fifth row) is capable of dissolving a further 0·1 mmole of calcium, i.e., a further 10 mg of calcite, which is the information necessary for the evaluation of the degree of saturation. Stenner (1969) has in fact defined *aggressiveness* as the weight of calcium carbonate that can dissolve in one litre of solution, negative values being used for supersaturated solutions. In the above example the aggressiveness is 10 mg of calcium carbonate per litre, and the degree of saturation is 1·90/(0·1 + 1·90), or 0·95.

Less direct methods of quantifying the degree of saturation have often been employed in comparing waters, although interpretation becomes more difficult. The *pH saturation index*, which has been favoured in the field of water quality control, is defined as the difference between the pH values of the water before and after saturation with calcium carbonate (Larson and Buswell, 1942), and can be related to the degree of saturation either theoretically or experimentally. Other measures of saturation all involve the *ion activity product*, IAP, defined as the product of (Ca^{2+}) and (CO_3^{2-}) in solution, which is compared in various ways with the solubility product, K_s. The *calcite saturation index*, or SI_c, has been defined (Jacobson and Langmuir, 1970) by the formula IAP/K_s. (The same name has been used in Europe (Roques, 1964) to represent the degree of saturation.) Shuster and White (1971) have employed the *saturation ratio*, defined as $\log_{10}$ (IAP/K_s), and J. J. Drake has employed the *Satcal index* (Ford, 1971), defined as $\log_{10} K_s/\log_{10}$ IAP. (Wigley (1971) has used the same name for the saturation ratio.) Because the carbonate ions cannot be measured directly, these have to be evaluated from measured data by means of theoretical considerations.

The relationship between all these measures of saturation depends somewhat on the carbon dioxide content of the solution, and is complex. One reason for such a diversity of measures has been the need to evaluate solutions from the readily available parameters of hardness, alkalinity and pH. However, Stenner (1969, 1971) has now established that the experimental "marble method" is capable of measuring the degree of saturation directly, even with natural waters (see Part III of this chapter). It is to be hoped that the use of this simple technique will become widespread, and that the proliferation of measures will be reduced.

RATES OF DEPOSITION AND SOLUTION

During solution, calcium and carbonate ions from the breakdown of the calcite enter the water as they are released from the solid and, were they not carried away, their increased concentration would eventually stop solution before the bulk of the water attained saturation. Normally, however, the ions diffuse from the zone of high concentration to the bulk of the water, where naturally occurring currents complete the mixing process and the solution of calcite proceeds to saturation. The system is more complex if air is in contact with the water, but this will be ignored for now. Analogous effects apply to the reverse process of calcite deposition from supersaturated solution, when calcium and carbonate ions from the solution react to join the calcite crystal, thereby depleting the zone near the surface. Ions have to diffuse from the more concentrated solution in the bulk of the liquid towards this depleted zone so that deposition may continue.

The rate of solution or deposition depends on two main processes:

Surface reaction. The release and recombination of calcium and carbonate ions at the solid surface.

Diffusion. The migration of ions and molecules through the solution towards or away from the solid surface.

If the rate of one process is much slower than the other, then that process will limit the overall rate of solution. The deposition of calcite at the bottom of a cave pool is an example. Close to the calcite floor of the pool the solution is depleted of calcium and carbonate ions by surface reaction and, as a result, ions diffuse from the more concentrated bulk of the pool to this depleted zone. If the pool is deep enough, the ion diffusion is very slow, and calcium and carbonate ions react to deposit calcite as soon as they arrive at the crystal surface. In this case the diffusion process is rate controlling and the rate of deposition is governed by the rate of diffusion.

If the pool is shallow, diffusion is so rapid that the ion concentration of the depleted zone is scarcely less than that in the bulk of the pool. In this case the surface reaction is rate controlling, and the rate of deposition equals the rate of surface reaction.

Surface Reaction Rate

Although the particular surface reactions which are important with calcite have yet to be established, both Roques (1964) and Reddy and Nancollas (1971) have derived similar expressions for calcite deposition on crystals under conditions where surface reaction is rate controlling. For unit volume of solution:

$$R = S\{k_1 - k_2[Ca^{2+}]_s[CO_3^{2-}]_s\}. \qquad 14$$

Here the left-hand side represents the rate of deposition or solution of calcium carbonate (mole s^{-1}), S is the surface area of crystal, k_1 and k_2 are rate constants and the ion concentrations are for the solution near to the surface (suffix s). According to fundamental principles, activities, not concentrations, should be used in this equation and in the diffusion equation which follows, but it is still customary in rate studies to employ concentrations. Equation 14 satisfactorily accounted for the results of the above authors in experiments where surface reaction was made rate controlling by violent stirring. A similar but more detailed equation has been derived by Nancollas and Purdie (1964). Values of k_1 and k_2 at 25°C from Reddy and Nancollas are about $1{\cdot}5 \times 10^{-11}$ mole s^{-1} cm^{-2} and 2200 $mole^{-1}$ s^{-1} cm^4, respectively, and are expected to decrease with decreasing temperature. As a rough estimate, k_1 at 10°C would be 4×10^{-12} mole s^{-1} cm^{-2}, and k_2 would be 600 $mole^{-1}$ s^{-1} cm^4.

Diffusion rate

Diffusional theory has been reviewed by Bircumshaw and Riddiford (1952). For each type of molecule, X, the rate of migration towards or away from the crystal surface is given by:

$$dM_x/dt = k_t S\{[X]_s - [X]_b\}. \qquad 15$$

Here M_x is the number of moles of X diffusing towards or away from the surface, $[X]_s$ and $[X]_b$ are the concentrations near the surface and in the bulk of solution respectively, and k_t is a constant (the transfer coefficient). When calcite deposition or solution is under diffusional control, $dM_x/dt = R$. For ions, Eqn 15 has to be modified because of the electrical forces involved, but the effect is usually significant only when the diffusional path is small and will be ignored here. The transfer coefficient k_t varies with the degree of mixing in the water. For water which is still or slowly moving Bircumshaw and Riddiford (1952) have given the formula:

$$k_t = D/L, \qquad 16$$

where D is the molecular diffusion coefficient (about 10^{-5} cm^2 s^{-1} at 25°C, varying little with temperature) and L is the distance over which diffusion occurs (normally 10^{-2} cm or less). For water moving in streamline flow across the solid surface Allen (1971) has given the formula:

$$k_t = 8\nu/L'. \qquad 17$$

Here ν is the kinematic viscosity (0·0090 cm^2 s^{-1} at 25°C, 0·013 cm^2 s^{-1} at 10°C) and L' is the characteristic length of the system, i.e. the stream depth or the stream width. For turbulent flow over rough surfaces Moody (1944) has given a complicated expression which can be simplified to:

$$k_t \simeq 0{\cdot}01U, \qquad 18$$

where U is the stream velocity in cm s^{-1}.

Deposition of Calcite

Equations 14 and 15 together describe the rate of deposition of calcite from supersaturated solution for the simple case with no air present, e.g. in a waterfilled passage. Since ion concentrations can be calculated as in Tables 7.4 and 7.5, and approximate values of the constants in the equations are available, it is possible to estimate and compare deposition rates by diffusion and by surface reaction. If one rate is much slower than the other, then the calcite deposition rate is established. For this comparison Eqn 14 (surface reaction) is used with bulk values for the ion concentrations. This overestimates the deposition rate (except where surface reaction is rate determining, when concentrations in bulk and near the surface are equal), but is satisfactory for deciding which mechanism is rate controlling.

Equation 15 (diffusion) presents more problems since allowance must be made for the relative ease of interchangeability of the ionic and molecular species. It is not just the diffusion of Ca^{2+} which must be considered, but that of $CaHCO_3^+$ and $CaCO_3^0$ also; similarly the other carbonate species must be included with CO_3^{2-}. The ions and molecules are usually in equilibrium (*vide infra*), and concentrations can be obtained by calculation as in Tables 7.4 and 7.5. The concentrations to be used in the term $\{[X]_s - [X]_b\}$ are therefore $[Ca]_T$ for the calcium ion, and the sum of $[CO_2^0]$, $[H_2CO_3]$, $[HCO_3^-]$, $[CO_3^{2-}]$, $[CaHCO_3^+]$ and $[CaCO_3^0]$ for the carbonate ion. For the comparison, Eqn 15 is used with surface concentrations, $[X]_s$, equal to the values in saturated solution. This again overestimates the deposition rate (except where diffusion is rate controlling, when $[X]_s$ is the saturation value), but is satisfactory for establishing which mechanism is rate controlling.

Consider deposition in the absence of air from the solution represented by the seventh row of Table 7.5 (degree of saturation 1·05). First, Eqns 14 and 15 are compared. In Eqn 14 the bulk concentrations in mole cm^{-3} are used with the rate constants for 10°C:

Surface reaction equation

$$R/S = 4 \times 10^{-12} - 600 \times 2{\cdot}00 \times 10^{-6} \times 5{\cdot}22 \times 10^{-9}$$
$$= -2 \times 10^{-12} \text{ mole s}^{-1} \text{ cm}^{-2}. \qquad 14a$$

(The negative sign denotes deposition.) In Eqn 15 the group of ions giving the smallest value of $\{[X]_s - [X]_b\}$ must be found since this will limit deposition by diffusion. As usual, the calcium group gives the smallest value, i.e. $(2{\cdot}00 \times 10^{-6} - 2{\cdot}10 \times 10^{-6})$, or $-1{\cdot}0 \times 10^{-7}$ mole cm^{-3}. Using 10°C constants, Eqn 15 becomes:

Diffusion equation (still water)

$$R/S = \frac{10^{-5}(-1{\cdot}0 \times 10^{-7})}{10^{-2}}$$
$$= -1 \times 10^{-10} \text{ mole s}^{-1} \text{ cm}^{-2}. \qquad 15a$$

Diffusion equation (streamline flow)

$$R/S = \frac{8 \times 0{\cdot}013\,(-1{\cdot}0 \times 10^{-7})}{L'}$$
$$= -1 \times 10^{-8}/L' \text{ mole s}^{-1} \text{ cm}^{-2}. \qquad 15b$$

Diffusion equation (turbulent flow)

$$R/S = 0{\cdot}01U\,(-1{\cdot}0 \times 10^{-7})$$
$$= -1 \times 10^{-9}U \text{ mole s}^{-1} \text{ cm}^{-2}. \qquad 15c$$

Turbulent flow always involves high velocities, U, and in streamline flow L' cannot be extremely large; it follows that the surface reaction is slowest and is rate controlling for all types of flow. In such a case, ion concentrations in the bulk of solution and near the surface are equal, so that the rate of calcite deposition is given by Eqn 14a. Similar investigations of other supersaturated solutions show that calcite deposition is always controlled by Eqn 14, provided that impurities affecting the surface are absent.

Solution of Calcite

The solution of calcite in the absence of air may be studied in the same manner as deposition. Consider the dissolving action of the solution represented by the fourth row in Table 7.5 (degree of saturation 0·75). Equation 14 becomes:

$$R/S = 4 \times 10^{-12} - 600 \times 1{\cdot}47 \times 10^{-6} \times 1{\cdot}06 \times 10^{-9} = 3 \times 10^{-12} \text{ mole s}^{-1} \text{ cm}^{-2}. \quad 14a'$$

Equation 15 becomes:

Still water

$$R/S = \frac{10^{-5}(2{\cdot}00 \times 10^{-6} - 1{\cdot}50 \times 10^{-6})}{10^{-2}} = 5 \times 10^{-10} \text{ mole s}^{-1} \text{ cm}^{-2}. \quad 15a'$$

Streamline flow

$$R/S = \frac{8 \times 0{\cdot}013(2{\cdot}00 \times 10^{-6} - 1{\cdot}50 \times 10^{-6})}{L'} = 5 \times 10^{-8}/L' \text{ mole s}^{-1} \text{ cm}^{-2}. \quad 15b'$$

Turbulent flow

$$R/S = 0{\cdot}01U(2{\cdot}00 \times 10^{-6} - 1{\cdot}50 \times 10^{-6}) = 5 \times 10^{-9}U \text{ mole s}^{-1} \text{ cm}^{-2} \quad 15c'$$

As with deposition, Eqn 14 always gives the slowest rate: solution is controlled by surface reaction, the rate of solution being given by Eqn 14a′.

This conclusion is in conflict with laboratory experiments by Weyl (1958), who found diffusional control of solution with carbonated water in streamline flow through packed calcite crystals. However, similar experiments by Howard and Howard (1967) with carbonated water in streamline flow through a limestone slit showed that solution was not controlled by diffusion. More experiments are necessary to resolve this matter.

Reactions of Carbon Dioxide in Solution

Thus far it has been assumed that the ion and molecule reactions in water are fast enough to have no effect on calcite deposition or solution. Kern

(1960) has reviewed the literature and has shown that, although most reactions are fast, two involving CO_2^0 are slow, one being dominant at high pH and the other at a lower pH. Only the latter is important in natural waters because the equilibrium concentration of CO_2^0 is relatively very small at high pH. The important reaction is:

$$CO_2^0 + H_2O \rightleftharpoons H_2CO_3.$$

The rate of reaction is given by:

$$d[CO_2^0]/dt = k_3[H_2CO_3] - k_4[CO_2^0]. \qquad 19$$

At 25°C, k_3 is roughly 10 s^{-1} and k_4 0·01 s^{-1}. This slow reaction has little effect on solution or deposition under diffusional control because the important factor is then the concentration gradient of the calcium group of ions and molecules, and this is unaffected by the $CO_2^0 - H_2CO_3$ reaction. Under surface reaction control, however, the reaction slows the rate of solution or deposition. For deposition the effect is minor because only a small fraction of the total carbon dioxide in solution is in the form of CO_2^0 (see Table 7.4); for aggressive waters it is expected that the rate of solution will be reduced to some extent, but a quantitative estimate of the reduction has yet to be made.

Rate of Carbon Dioxide Exchange with Air

When air is in contact with the calcite solution, transfer of carbon dioxide has a profound effect on the degree of calcite saturation, and therefore on the rate of solution or deposition. For example, aggressive water entering a cave where the air has a low P value will emit carbon dioxide to the air and, instead of dissolving calcite, it may lose its aggressiveness or even deposit calcite. The transfer of carbon dioxide between water and air has three components: diffusion of the gas through the air, diffusion through the water in the various ionic and molecular forms, and transfer across the air-water interface. The rate of diffusion through the air is described by the same type of equation as was given for ion diffusion (Eqn 15), except that S is now defined as the area of the air-water interface. The transfer coefficient is normally of the order of 1 cm^2 s^{-1}, which is much greater than the values discussed for water. Diffusion through the solution is also defined by an equation like Eqn 15, but the key molecule, CO_2^0, can be converted to H_2CO_3 and the other ionic forms only by the slow reaction described by Eqn 19, which complicates the rate mechanism. Transfer across the air-water interface is thought to be fast.

The slow step in carbon dioxide exchange is usually diffusion through the liquid, associated with the slow chemical reaction described by Eqn 19. It has not been possible to calculate the overall effect of gas exchange rate on the rate of deposition or solution of calcite, but Roques (1964) has established experimentally that, in the case of a pendant drop, the gas exchange takes hours to come near to completion, which is far shorter than the time taken for surface reaction. Surface reaction (Eqn 14) is probably rate determining for calcite solution and deposition in most cave situations where impurities and foreign acids are absent.

Application to Cave Waters

Under some conditions it is possible to evaluate rates of calcite deposition or solution in caves on theoretical grounds, although the results must be regarded as approximations. Consider the growth of a stalagmite wetted by a feeder drip from a source which is saturated with calcite at 2 mmole/l, and let the air in the cave contain carbon dioxide of *P* equal to 1 mAtm. These values are not atypical of temperate caves (10°C). The dripping water rapidly loses carbon dioxide (Dixon and Hands, 1957), and by the time it spreads over the stalagmite it has reached equilibrium with the cave air, so that the degree of saturation is 2·06 (sixth row of Table 7.4). During deposition the slow interconversion of CO_2^0 and H_2CO_3 is unimportant, and so the rate of calcite crystallization is controlled only by the rate of surface reaction. Use of Eqn 14 with rate constants for 10°C and with ion concentrations from the sixth row of Table 7.4 gives the rate of deposition as 2×10^{-11} mole^{-1} cm^{-2}, equivalent to a growth rate of 0·02 cm per year for the wetted part of the stalagmite, which is not unreasonable (Moore, 1962). The rate of water supply must be sufficient to maintain growth, but 0·1 litre per year for each square centimetre would probably be enough.

The rate of solution of calcite is harder to evaluate because of the effect of the slow interconversion of CO_2^0 and H_2CO_3 in aggressive water, but an upper limit can be obtained by assuming that surface reaction is rate controlling. Consider the simple case where aggressive percolation water enters calcite rock in the absence of air. If the water is at 10°C and has a typical degree of saturation of 0·75 (fourth row, Table 7.5), the rate of surface reaction will be 3×10^{-12} mole s^{-1} cm^{-2}, as shown previously, and this is an upper limit to the rate of calcite solution. The calcite surface exposed to the water will therefore recede at less than 0·003 cm per year, which is again not unreasonable. Acids other than carbon dioxide in the water may give much greater rates of calcite solution, but no studies of this have been made.

Impurities in the water can markedly alter the rate of calcite solution and deposition when surface reactions are rate controlling. Small concentrations of magnesium or sulphate diminish the rate of deposition (Bischoff and Fyfe, 1968; Bischoff, 1968a and b), but the mechanism is not clear. Terjesen *et al.* (1961) showed that traces of some metal ions in solution (*vide infra*) can reduce the rate of solution by altering the effective value of k_2. If the rate of surface reaction is diminished sufficiently, control of the rate of solution will transfer from surface reaction to diffusion. That this does occur in nature is demonstrated by the occurrence of regular scallops and flutes in caves, which Allen (1971) has shown to be caused by periodic patterns in the water flow. Water flow can only affect rates of diffusion, not rates of surface reaction.

THE EFFECT OF FOREIGN SUBSTANCES

Foreign Ions in Solution

For the purpose of comparison, calcite solubility in this section will refer to solutions exposed to air of constant carbon dioxide content. Substances in solution other than carbon dioxide can have an important influence on the solubility of calcite. Dissolved materials such as sodium chloride, which have no ion in common with the carbonate system, enhance calcite solubility by the *ionic strength effect*. The foreign electrolyte raises the ionic strength of the solution, and this in turn reduces all ionic activity coefficients (see Eqns 10 and 11). As a result, higher concentrations of calcium and carbonate ions can be supported in solution without the product of the calcium and carbonate activities exceeding the solubility product. The ionic strength effect can be accurately evaluated by including terms for the foreign ions in Eqns 8 and 11 (Akin and Lagerwerff, 1965a). The effect is slight at low concentrations of foreign electrolyte, as is demonstrated by the NaCl curve of Fig. 7.7, but at high concentrations such as occur in sea water the calcite solubility can be nearly doubled.

Solutions of soluble substances which have an ion in common with the carbonate system reduce calcite solubility by the *common ion effect*. With calcium chloride, for example, the presence of calcium ions already in solution means that little calcite can dissolve before the product of calcium and carbonate activities exceeds the solubility product. Like the ionic strength effect, this can be accurately evaluated by including terms for the foreign ions in Eqns 8 and 11. The magnitude of the change is shown by the $CaCl_2$ curve of Fig. 7.7: a significant reduction in calcite solubility is caused by relatively small concentrations of added electrolyte.

These two general principles, the ionic strength effect and the common ion effect, act whenever foreign electrolytes are present in solution.

Certain substances which often occur in nature have additional properties, and the action of acids on calcite solubility has already been discussed superficially. Theoretically, the influence is determined by including a term for the acid anion in Eqns 8 and 11. All strong acids of equal normality have the same effect, and the curve for hydrochloric acid is

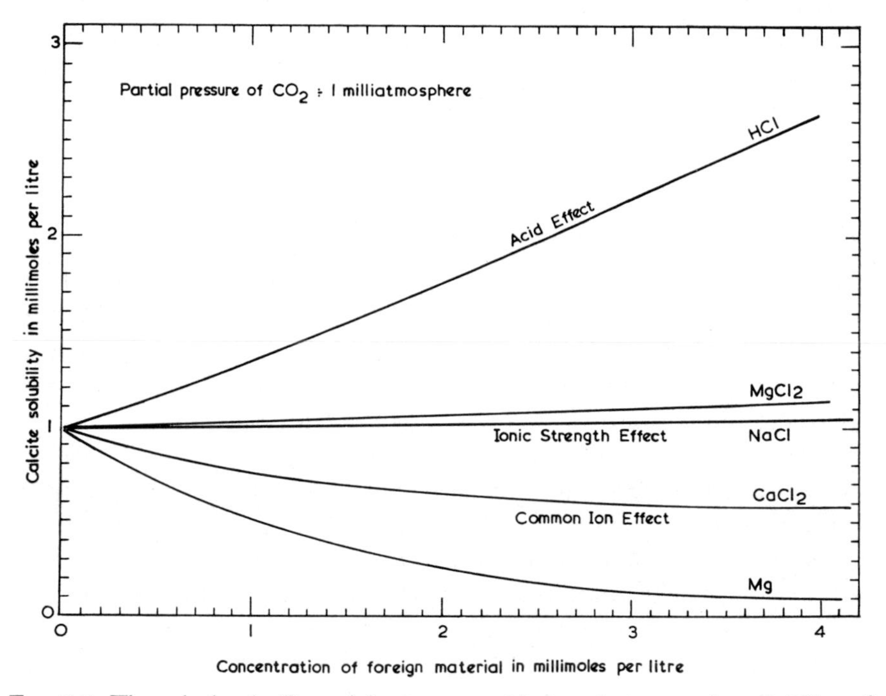

FIG. 7.7. The calculated effect of foreign materials in solution on the solubility of calcite at 10°C.

shown in Fig. 7.7. The increase in calcite solubility is pronounced, even at low concentration, but the extra calcite dissolved is only equivalent to a fraction of the amount of acid added. Weak acids have a lesser effect, the criterion being that the ionization constant of the acid should be much greater than (H^+) for full effect. However, most organic acids of importance in limestone solution act as strong acids.

When sulphate ion is present in solution it tends to react with calcium to form the ion-pair $CaSO_4^0$, the equilibrium constant being about 0·005 at 25°C (Garrels and Christ, 1965), and this ion-pair affects calcite solubility. With sulphuric acid in solution, for example, the $CaSO_4^0$ ion-pairs which are formed increase calcite solubility above the level expected from the acid concentration. Similarly, with calcium sulphate in solution, calcite solubility is somewhat greater than expected from the common ion effect. The presence of ion-pairs can be accommodated in theoretical calculations of ion concentration and degree of saturation (Wigley, 1972).

Magnesium is important because it is a constituent of many limestones and natural waters. In solution the ion reacts with HCO_3^- and CO_3^{2-} to form the ion-pairs $MgHCO_3^+$ and $MgCO_3^0$, the equilibrium constants at 25°C being about 0·07 and 0·0004, respectively (Garrels and Christ, 1965). When magnesium is the only foreign ion in solution, as for example when both magnesium carbonate and calcite have dissolved in the water, the calcite solubility is slightly greater than expected from the common ion effect because of these ion-pairs. A typical calculation is given by the curve labelled Mg in Fig. 7.7. (It differs from the common ion effect curve of $CaCl_2$, not only because of the magnesium ion-pairs, but also because of the different ion valencies involved.) When foreign anions are present as well as magnesium, calcite solubility can be very different from the case just considered. In magnesium chloride solution, where both Mg^{2+} and Cl^- are present, calcite solubility is increased, not decreased, as shown by the $MgCl_2$ curve of Fig. 7.7. In waters where the ratio of Cl^- to Mg^{2+} is less than in $MgCl_2$ solution, the calcite solubility follows a curve intermediate between those for $MgCl_2$ and Mg. Experimental verification of the $MgCl_2$ curve has been obtained by Roques (1964). The solubility of calcite in water containing mixtures of ions such as Mg^{2+}, K^+, Na^+,Cl^-, SO_4^{2-} and NO_3^- can be determined by calculation, thus providing a powerful tool for the study of natural waters (Wigley, 1972). However, there is some evidence of a discrepancy with magnesium carbonate solutions, where the Mg curve of Fig. 7.7 is not consistent with the observations of Picknett (1972) on the pH of saturated calcite solutions with magnesium ion present. The reason for this may lie in unknown ion-pair effects, or in changes at the surface of solid calcite when placed in magnesium solution.

Experiments by Terjesen *et al.* (1961) have shown that a number of metal ions reduce the rate of solution of calcium carbonate in carbonated waters. Only trace concentrations are needed (μmole/l or less) and a true equilibrium is not reached, but the final rate of solution is so slow that the phenomenon can be described in terms of an apparent calcite solubility which is reduced as the metal concentration is increased. Roques (1964) disagreed with these results, his experiments indicating that metal ions increased calcite solubility, but the work of Picknett (1964) and Nestaas and Terjesen (1969) support Terjesen's conclusions. The metal ions investigated by Terjesen were, in order of decreasing effectiveness:

Pb(II), La(III), Y(III), Sc(III), Cd(II), Cu(II), Au(III), Zn(II), Ge(IV), Mn(II), Ni(II), Ba(II), Mg(II), Co(II).

Of these, lead, copper, manganese and magnesium are often present in natural limestones and their effect on the apparent solubility is considerable. Copper, in the middle of the range of metals, halves the apparent solubility of calcite at a concentration of 10 μmole/l: scandium, one of the more effective inhibitors, has a significant effect on the apparent solubility at a concentration of only 50 nmole/l. Recent work (Nestaas and Terjesen, 1969) has shown that the effect is due to absorption of the metal ions on the calcite surface. Clearly, all the properties of saturated calcium carbonate solutions, including pH and calcium content, can be profoundly altered by traces of such metals in solution.

Foreign ions also affect the deposition of calcium carbonate from solution. Akin and Lagerwerff (1965b) found less calcite deposited than expected from supersaturated solutions containing Mg^{2+} and SO_4^{2-}, the solubility being enhanced by factors of between 1·3 and 2 for solutions containing these ions in concentrations between 0·3 and 10 mmole/l. A formula was given relating solubility enhancement to the ratios $[Mg^{2+}]:[Ca^{2+}]$ and $[SO_4^{2-}]:[Ca^{2+}]$. The calcite deposited always contained magnesium and sulphate, and it was considered that surface absorption caused the enhancement of solubility. A similar process probably accounts for the effect of certain phosphates in inhibiting the deposition of calcium carbonate from solution. Reitemeier and Buehrer (1940) found that sodium hexametaphosphate was particularly effective in this respect, even in concentrations as low as one μmole/l. Whether the phenomenon is important in nature remains to be discovered, but it has been much used in water softening. Raistrick (1949) has discussed the absorption mechanism.

Foreign Materials in the Solid

Besides calcium carbonate, other carbonates such as those of magnesium, manganese and strontium are of common occurrence in limestones, and, provided they are of sufficiently low concentration, can exist in solid solution in the calcium carbonate. With magnesian calcite, for example, Graf and Lamar (1955) report that the magnesium carbonate can be in solid solution up to a mole fraction of 0·06 (5% by weight). The significance of solid solutions is that the activities of the components of the solution are altered. So far it has been assumed that the activity of the solid is unity, which is correct for pure solids, but for solid solutions Raoult's law states that the activity of each component is proportional to its mole fraction. Thus, for the magnesian calcite mentioned above where the calcite mole fraction is 0·94, the calcite activity would be 0·94 instead of 1. Equation 7 has to be replaced by a more complete version:

$$K_S = (Ca^{2+})\,(CO_3^{2-})/(CaCO_3,\ \text{calcite}). \qquad 7a$$

For the present example the denominator has the value of 0·94. The effect of the magnesium carbonate component on calcite solubility can be determined by calculation, using Eqn 7a instead of Eqn 7, the result always being a reduction in solubility. With the present example, the calcite component of the magnesian calcite is 3% less soluble than pure calcite. Calcite solubility must not be confused with the overall solubility of the magnesian calcite; both magnesium carbonate and calcite dissolve, the total amount of material entering solution being greater than for pure calcite. Thus the overall solubility is greater than the solubility of pure calcite, and in fact increases with the magnesium carbonate content (Chave *et al.*, 1962).

When the magnesium carbonate component of a magnesian calcite has a molefraction greater than 0·06 but less than about 0·25, the excess material exists as a separate phase mixed with the calcite. Berner (1971) stated that such magnesian calcites dissolve with preferential solution of the magnesium carbonate, but Chave and Schmaltz (1966) reported that calcium and magnesium entered solution in the same proportions as in the solid. Over several days an apparent saturation equilibrium can be reached, but over several months a purer calcite is redeposited, magnesium being preferentially retained in solution.

Magnesium carbonate and calcium carbonate can also exist in combination as the mineral dolomite, Ca, $Mg(CO_3)_2$. This is a chemical compound and has a characteristic crystal structure. It dissolves so that the ratio of calcium to magnesium in solution equals that of the solid, a true saturation equilibrium being attainable (Berner, 1967) in which the solubility depends on the carbon dioxide content of the solution, as with calcite. In the past there has been doubt about the value of the dolomite solubility product, defined as $(Ca^{2+})(Mg^{2-})(CO_3^{2-})^2$, but Langmuir (1971) has now established that it is 10^{-17} at 25°C, and calculations of dolomite solubility can be made using this figure in the same way that calcite solubility can be evaluated using K_S. Dolomite is less soluble than calcite when in contact with solutions where the molar ratio of magnesium to calcium is greater than 0·6 (Bricker and Garrels, 1967), but despite this it is extremely difficult to crystallize from solution at normal temperatures, magnesian aragonite usually being deposited instead (Bathurst, 1971). It is the rate of deposition of dolomite which is extremely slow (Fyfe and Bischoff, 1965) so that nuclei of other crystals grow preferentially to dolomite nuclei. Dolomites with a slight excess of calcium carbonate or magnesium carbonate can be much more soluble than pure dolomite, the apparent solubility product being as large as 10^{-15} (Bricker and Garrels, 1967), but even with this material the rate of deposition is extremely slow.

CONCLUSIONS

This part of the chapter has shown that in some ways our understanding of the basic chemistry of calcite solutions is fairly satisfactory. The properties of dilute saturated solutions can be predicted to an accuracy of about 7%; those of unsaturated solutions can also be forecast, although they remain as yet untested. More accurate laboratory experiments are needed before improvement in this field is to be expected. What is less satisfactory is our understanding of rates of solution and deposition. A start has been made, but more information is required concerning surface reactions and rate constants, with the ultimate aim of establishing accurate rates of recession of limestone surfaces and rates of growth of speleothems.

When we turn from calcite solutions to natural waters, the situation becomes even less satisfactory. With the ubiquitous magnesium ion present, allowance must be made for $MgHCO_3^+$ and $MgCO_3^0$, but values for the equilibrium constants are only available at 25°C. It is probable that unknown effects, perhaps due to magnesium absorption on the crystal surface, can cause deviations from present theory. Absorption effects are also known to influence calcite solution in the presence of foreign metals, and it remains to devise means of predicting the behaviour of natural waters under these circumstances.

It is in this context that chemical field work becomes important. Detailed analyses of natural waters are needed in order to test modifications to theory, and in these analyses all the ions present in appreciable concentration must be determined, plus those ions which are found to have a significant effect in trace quantities. A start on this work of comparing the results of detailed analyses with theory has already been made (Hall, 1964; Roberson, 1964; van Everdingen, 1969; Jacobson and Langmuir, 1970; Langmuir, 1971). In order to foster this approach some aspects of field sampling and analysis will now be considered.

Part III: PRACTICAL TECHNIQUES

L. G. Bray and *R. D. Stenner*

GENERAL CONSIDERATIONS

For studies in limestone areas much useful work can be done using relatively inexpensive equipment and techniques which can be mastered quite readily. Many of the textbooks and references likely to be consulted by cavers give details of analytical procedures which can be misleading for the individual operating under less than ideal conditions in cave studies. The analysis of rock, of sediment and of air needs complex equipment and considerable experience, so it is recommended that the inexperienced worker restricts himself to the analysis of water samples.

PREMISES

Good results are possible without a fully equipped laboratory. The essentials are a working surface, supplies of distilled water, reagents and equipment, and lighting of daylight quality. Chemical work has been performed in a cave using a discarded ironing board as the working surface (Bray, 1971) and in tents far from civilization. The luggage compartment of a Mini estate car offers sufficient working space for simple investigations. The South Wales Caving Club has established a very efficient field laboratory at its Penwyllt headquarters which has accommodated a wide range of investigations (Bray, 1972).

SUPPLIES

Equipment and reagents are needed, and most suppliers require an individual customer to offer some indication of status to prevent inherently dangerous materials falling into the hands of children. Catalogues are too expensive to be available freely, but most suppliers will offer a quotation against a definite enquiry, although they may stipulate a minimum value of £10 per order. Possible suppliers are listed in Appendix I.

Only a modest amount of glassware is needed to cater for a range of estimations (Appendix II). For those without access to an accurate balance use may be made of the standard solutions offered by some manufacturers. If enough money is available to allow the purchase of a simple instrument the Lovibond Comparator could be considered. With accessories it covers a wide range of trace constituent analyses as well as the estimation of pH. If a portable pH meter is already available a simple colorimeter could be useful to cater for trace constituent estimations (Vogel, 1968, p. 738). The WPA CO65 would be a suitable instrument at a realistic cost.

PRECISION

The newcomer to analytical work may be assured that good results may be obtained with only a little practice; 16-year-old pupils embarking on an "A" level chemistry course are expected to work to a precision of better than 0·5%.

PLANNING

In planning a study, textbooks and published articles may be consulted. The cost in terms of equipment, skill, effort, time and money should be assessed, also the help available from other cavers; it is best to avoid asking others to commit themselves to a disproportionate contribution of time and effort. Choices must be made between apparently similar analytical methods, between different routes for sample collection and between the complementary approaches of very detailed analysis of a few samples or less detailed examinations of many samples. Only the purpose of the study and individual situations can suggest which choices will be right. The planning should include discussions with those taking part, so that everyone knows how his contribution fits, sampling procedures are standardized (Stenner, 1969) and careful notes are made at the sites chosen for sample collection. The need for sample collection to proceed upstream must be emphasized; results may be of value in more than one field, and work can become of much wider significance if other measurements, such as water temperature (measured to 0·1°C) and discharge values, are made at the time of sample collection.

WATER ANALYSIS TECHNIQUES

Through lack of space here the authors have prepared a monograph in which working details are given for a wide range of techniques for hydrological and geomorphological work in caves (Bray and Stenner, in press). Readers are referred to the texts and original papers cited for details of the methods of calculation.

The general field of chemical analysis is covered by many textbooks (e.g. Vogel, 1968; Belcher and Nutten, 1970). A useful outline of water analysis has been published (BDH, 1969) and includes useful references. A survey of water examination from the viewpoint of water supply includes outlines of routine analytical procedures (Holden, 1970). More complete information on routine analytical methods is available (SAC, 1958; IWE, 1960; APHA, 1965).

The newcomer should first acquire the techniques for determining calcium, magnesium and alkaline hardness (hence non-alkaline hardness, an approximation to sulphate), aggressiveness and chloride (an approximation to sodium), and should establish them, using the local tap-water. When consistent results are obtained with ease, work on cave waters can proceed with some confidence that the results will be of value. Detailed procedures might well depend upon the nature of the waters to be examined. Clearly, the hard waters from Mendip allow the use of smaller volumes in titrations than the relatively soft waters from South Wales. To indicate possible variations some cave water values are quoted (Appendix III) and analyses of other waters have been published elsewhere (Holden, 1970, p. 33).

Calcium and Magnesium

These estimations are fundamental to cave study and they rely usually on titration with complexing agents such as EDTA, EGTA and DCTA (West, 1969, pp. 7, 143, 145). In one method widely used in industry the total hardness (i.e. calcium + magnesium) is estimated by titration with EDTA at pH 10 using Solochrome Black T as indicator. The calcium hardness is estimated separately by titration with EDTA at pH 12, using murexide as indicator. The magnesium content may be found by subtracting the calcium concentration from the total hardness, expressed in suitable units (West, 1969, p. 159). This method suffers from deficiencies in that certain indicators are liable to interference from trace metals, especially iron III; others work only in the presence of quite high magnesium concentrations; and others do not keep well. Solutions to these problems have been suggested (West, 1969, p. 156; Stenner, 1969), but the method involving the use of highly toxic cyanides cannot be recommended to non-chemists. A modified method (Stenner, 1971, p. 284) uses the good end-points available for total hardness estimations and an instrumental technique for magnesium estimation. The difference between the results gives the calcium content.

An alternative approach (Pribil and Veseley, 1966) has been modified for use with cave waters (Bray, 1969). A portion of the water is used to estimate calcium by titration with EGTA at pH 12 using calcein as indicator. A second portion of the water is treated with a slight excess of EGTA to bind the calcium as a tight complex, leaving the magnesium, bound as a loose complex, to be estimated by titration with DCTA at pH

10 using methylthymol blue as indicator. As with Stenner's method, this procedure gives good precision.

Whichever methods and indicators are used, care must be taken to check on "blanks" (titrations with distilled water instead of sample) and on the standardization of the complexing agents employed by the use of solutions containing known concentrations of calcium and/or magnesium.

pH

The very simple portable pH meters in current use can give rise to serious errors from electrode deterioration and, in the view of the present authors,

Table 7.6

pH RANGES OF VARIOUS INDICATORS

Dye	pH range	Colour Change
Bromocresol green	3·6-5·2	Yellow→blue
Bromocresol purple	5·2-6·8	Yellow→violet
Bromothymol blue	6·0-7·6	Yellow→blue
Phenol Red	6·8-8·4	Yellow→red

cannot be recommended to non-chemists. The more expensive pH meters, such as the Pye Unicam Model 293, have compensation for electrode conditions and can be used reliably once the calibration procedure has been mastered. The user of a pH meter relies absolutely on the condition of the buffer solutions (solutions having known pH) used for calibration, which may be purchased or prepared from powder or from tablets.

For simple work the use of pH-sensitive dyes can give results of reasonable reliability, especially if the colours developed are examined down the column of liquid rather than across it (Chamberlin, 1967, p. 194; BDH, 1970, p. 54). The Lovibond Comparator with Nessler attachment may be used. For cave waters the range of dyes needed is not large, and these are available ready-prepared. Some examples are given (Table 7.6) and the methods are given in the literature (Chamberlin, 1967, pp. 191, 197; BDH, 1970).

Alkaline Hardness

This result gives an approximation to the bicarbonate concentration of the sample. It is estimated by titrating a portion of the sample with 0·01M or 0·02M hydrochloric acid to pH 4·5, using either the mixed BDH 4·5 indicator or a pH meter to detect the end-point (Stenner, 1969).

Aggressiveness to Calcium Carbonate

The ability of a water to attack limestone (or conversely to deposit calcite) may be measured directly, quantitatively and relatively easily (Stenner, 1969). In outline the method requires two samples to be taken simultaneously at a site. One sample is treated with AnalaR calcium carbonate at the site. In the laboratory the untreated sample is examined for total hardness. The treated sample is filtered and the filtrate analysed for total hardness. The change in hardness brought about by the addition of the calcium carbonate is known as the aggressiveness of the water. An increased hardness (a positive value) indicates a water which is capable of dissolving more limestone; and a decreased hardness (a negative value) indicates a water tending to deposit calcite. Certain precautions are necessary and the conditions in which the samples need to be filtered in the cave are important. This measurement is of great importance in the study of the chemical processes taking place within caves.

Electrical Conductivity

At low ionic concentrations the ability of waters to conduct an alternating electric current is almost proportional to the total ionic concentration. The measurement of conductivity has considerable value in indicating that waters from certain sites have special features (Bray, 1971). Measurement of electrical conductivity requires the use of alternating current, and modified Wheatstone bridges were used almost exclusively until recently (Vogel, 1968, p. 972; Bray, 1969, 1971). The development of integrated circuit devices has produced direct-reading conductivity meters at realistic prices. The WPA CM25 generates a square wave signal, applies this across the sample between the electrodes of a dip-type conductivity cell (Vogel, 1968, p. 973) and measures the current flowing. Internal switching allows a wide range of conductivity to be covered on a linear-reading meter. As the electrical conductivity of solutions is dependent upon temperature, comparison work requires either all of the measurements to be made at the same temperature or a temperature correction to be applied (Vogel, 1968, p. 970; Bray, 1969, 1971).

Oxygen Demand

Attention has been drawn to the importance of organic matter in cave development and to this estimation as a means of gaining a measure of the easily oxidized organic matter in cave waters (Bray, 1972). Of the several methods available (Holden, 1970, p. 177), it is unfortunate that the dichromate value test, which gives most complete oxidation, cannot be recommended to newcomers as it requires the use of boiling sulphuric acid and a highly toxic catalyst. A method which gives an arbitrary measure of oxidizable organic matter relies on incubating a portion of the sample with a controlled amount of potassium permanganate in acid conditions for 4 hours at 27°C. At the end of the test the permanganate remaining is estimated and the value obtained subtracted from that given by a similar test on distilled water. The difference can be used as a basis for comparing waters; the greater the difference the more organic matter present.

Dissolved Oxygen

This estimation is of value in assessing the oxidation/reduction system of a water. The sample is collected in a glass bottle, manganese II sulphate and a sodium hydroxide/potassium iodide mixture are added at once, and the bottle is capped. The oxygen dissolved in the water reacts with the manganese II hydroxide first precipitated and gives a higher valency manganese compound which, on acidification, oxidizes iodide to iodine. The iodine is titrated with sodium thiosulphate, using starch as indicator (Bray, 1972). The temperature of the water must be measured at the site. Oxygen is more soluble in cold water than in warm water and tables of solubility are required so that the percentage saturation can be calculated (Chamberlin, 1967, p. 40). Present indications are that percolation waters tend to be low in dissolved oxygen.

Trace Constituents

Discussion of these has been limited to cover different analytical techniques as well as those constituents of importance in cave work.

Sulphate

Sulphate may be estimated by measurement of the colour shown when water is treated with barium chloranilate (Vogel, 1968, p. 806) or by comparison of the cloudiness developed on treatment of the water with barium chloride against that developed by standard sulphate solutions (Bray, 1972). The sulphate concentration is often very close to the non-alkaline hardness when compatible units are used.

Chloride

Chloride may be titrated with silver nitrate using potassium chromate as indicator (Vogel, 1968, p. 259) or estimated colorimetrically using mercury II thiocyanate and iron III sulphate (Chamberlin, 1967, p. 86).

Sodium and Potassium

These may be estimated by using a flame photometer (Vogel, 1968, p. 882). If many are to be examined, separate samples may be collected in small plastic vessels which fit under the sampling jet of the instrument (Bray, 1972). The sodium concentration is often very close to the chloride concentration, since both have wind-borne sea water as their normal source.

The list of trace constituents could be extended, but how important these additions might be for speleology remains to be seen. If standard solutions can be prepared, the simple colorimeter or even visual comparison allows colorimetric estimation of many trace constituents. Where standard solutions cannot be prepared, the Lovibond Comparator discs allow estimations to be made. In many cases the methods cited, although intended for use with the Lovibond Comparator, can be modified for colorimeter or visual use. Additional estimations include dissolved silica (Chamberlin, 1967, p. 227), iron (Chamberlin, 1967, p. 137), manganese (Chamberlin, 1967, p. 153) and phosphate (Chamberlin, 1967, p. 230).

FOR THE FUTURE

New chemistry courses in schools and a growing interest in the environment have led to a demand for low-cost instruments for monitoring chemical changes: fortunately there has been very rapid progress in instrument technology, and the cave scientist is able to take advantage of a new generation of chemical equipment. It may, however, be many years before electrical measurements of hardness, based upon the use of specific-ion electrodes, can become as precise as those measurements obtained far more cheaply using a burette. The development of an inexpensive electrical method for measuring organic matter is still remote, but a low-cost dissolved oxygen electrode and associated circuitry is on the market (part of Environmental Unit from Walden Precision Apparatus, Ltd.).

Activity seems likely in an aspect of cave science neglected for far too long, the long-term recording of various features of the waters entering a cave system at the sink and leaving from the resurgence. The prime need is for a cheaper chart recorder capable of operation at a very slow chart speed. New components and attitudes have led to the production of a very flexible recorder, the WPA CQ75, at a more realistic price, able to work with other WPA equipment and with home-made equipment. In a field as complex as cave chemistry the continuous monitoring of the chemical changes taking place under flood conditions could be of immense value.

THEORY AND PRACTICE

The techniques described are very important in cave studies, but there is an application of analytical chemistry which deserves emphasis. When the theory of calcite solution is applied to the solution of natural limestone samples by natural waters, the experimentally observed facts do not fit the theory. A great deal of painstaking work will need to be done before it becomes possible to predict exactly what will happen when a given natural water meets a given limestone, and the work falls into two complementary categories.

The major constituents of the water before and after contact with the limestone must be known accurately. The work of Thrailkill (1971) includes the measurement in the field of conductivity and of pH, the latter to a genuine 0·05 pH unit. After analysis the cation concentration is compared with the anion concentration. If the values match, the expected conductivity is computed and compared with the experimental result. If there is agreement the whole experiment is accepted as valid and the results used.

In parallel with this work trace element analysis must proceed, together with examination for organic matter and for dissolved gases, especially oxygen. Limestone analysis must be conducted equally rigorously. With enough information it should be possible to compute the changes which will take place with various combinations of water and limestone, though there will still be the problem of establishing the kinetics of the processes taking place. Only then will it be possible to link practical results with theoretical predictions.

Appendix I

SUPPLIERS

General Apparatus

Baird and Tatlock Ltd., Freshwater Road, Chadwell Heath, Essex.
Griffin and George Ltd., Ealing Road, Alperton, Wembley, Middlesex.
Scientific Supplies Co. Ltd., Scientific House, Vine Hill, London, EC4.

Special Items

Pye Unicam Ltd., York Street, Cambridge.
(Portable pH meter, Model 293)
The Tintometer Ltd., Waterloo Road, Salisbury, Wiltshire.
(Lovibond Comparator, accessories, reagents)

Walden Precision Apparatus Ltd., Shire Hill, Saffron Walden, Essex.
(Conductivity meter CM25, colorimeter CO65, chart recorder CQ75)

Chemicals

BDH Chemicals Ltd., Poole, Dorset.
Hopkin and Williams Ltd., Freshwater Road, Chadwell Heath, Essex.
May and Baker Ltd., Dagenham, Essex.

Appendix II

REQUIREMENTS

General Apparatus

Conical flasks, 100 cm^3 or 250 cm^3	3
Pipettes, 25 cm^3 and 50 cm^3	1 each
Burettes, 10 $cm^3 \times 0{\cdot}02$ cm^3 or 50 $cm^3 \times 0{\cdot}1$ cm^3	3 each
Graduated flask, 250 cm^3, 100 cm^3, 500 cm^3	1 each
Beakers, glass and plastic, 100 cm^3	1 each
Stirring rods	
Graduated pipettes, 2 cm^3 and 10 cm^3 (with pipette controller for safety)	1 each
Droppers for indicators	
Plastic wash bottle for distilled water	1
Plastic bottles for reagent storage	
Plastic bottles for sample collection	as required

Reagents

EDTA 0·01M or 0·02M
Buffer pH 10 (ammonia/ammonium chloride)
Sodium hydroxide for pH 12 buffer
Solochrome Black T indicator or Total Hardness tablets
Murexide indicator or calcium hardness tablets
(*The materials listed above are available in a pack from BDH.*)

BDH 4·5 indicator	100 cm^3
Hydrochloric acid 0·1M for dilution	500 cm^3
Dyes for pH tests (Table 1)	25 cm^3

Buffer, pH 7, solution or (better) tablets or sachets
Buffer, pH 4, solution or (better) tablets or sachets
Buffer, pH 7, solution or (better) tablets or sachets
(The buffer solutions will be needed only if a pH meter is to be used.)

Provision of these materials will allow basic hardness estimations to be performed and pH to be measured.

Appendix III

EXPECTED VALUES OF SOLUTES IN UNPOLLUTED WATERS IN LIMESTONE AREAS

Solute	*Average values*	*Extreme Values*	*Units*
Total hardness	15 to 300	10 to 400	ppm $CaCO_3$
Alkaline hardness	5 to 250	0 to 350	ppm $CaCO_3$
Aggressiveness	−15 to 15	−40 to 60	ppm $CaCO_3$
Magnesium hardness	10 to 30	2 to 60	10^{-5}M Mg^{2+}
Sulphate	10 to 30	2 to 80	10^{-5}M SO_4^{2-}
Chloride, sodium	15 to 40	0 to 130	10^{-5}M NaCl
Silica	5	2 to 15	ppm SiO_2
Potassium	2	0·1 to 15	ppm K
Iron	0·05	0·01 to 0·4	ppm Fe
Oxygen demand 4 hr	0·5 to 4	0 to 10	ppm O_2

High values of phosphate, nitrate, ammonia and oxygen demand indicate contamination by sewage, silo drainage or excessive use of fertilizers. Higher values of heavy metals might originate from mining wastes and can have great ecological importance.

Electrical conductivity varies markedly with temperature change and values at 25°C from 34 to 600 μmho.cm^{-1} have been recorded in South Wales.

BIBLIOGRAPHY

Akin, G. W. and Lagerwerff, J. V. (1965a). Calcium carbonate equilibria in aqueous solutions open to the air. I. The solubility of calcite in relation to ionic strength. *Geochimica et Cosmochimica Acta 29*, 343-352.

Akin, G. W. and Lagerwerff, J. V. (1965b). Calcium carbonate equilibria in solutions open to the air. II. Enhanced solubility of $CaCO_3$ in the presence of Mg^{2+} and SO_4^{2-}. *Geochimica et Cosmochimica Acta 29*, 353-360.

Allen, J. R. L. (1971). Transverse erosional marks of mud and rock: their physical basis and geological significance. *Sedimentary Geology 5*, 167-385.

APHA (1965). *Standard Methods for the Examination of Water and Waste Water*, 12th ed. American Public Health Association, New York.

Bathurst, R. G. C. (1971). *Carbonate Sediments and their Diagenesis*. Elsevier, Amsterdam, 250.

BDH (1969). *Chemical Methods of Water Testing*. 3rd ed. BDH Chemicals Ltd., Poole.

BDH (1970). *pH Values and their Determination*. 8th edn. BDH Chemicals Ltd., Poole.

Belcher, R. and Nutten, A. J. (1970). *Quantitative Inorganic Analysis*. 3rd edn. Butterworths, London.

Berner, R. A. (1967). Comparative dissolution characteristics of carbonate minerals in the presence and absence of aqueous magnesium ion. *Amer. J. Sci. 265*, 45-70.

Berner, R. A. (1971). *Principles of Chemical Sedimentology*. McGraw-Hill, New York.
Bircumshaw, L. L. and Riddiford, A. C. (1952). Transport control in heterogeneous reactions. *Quart. Rev. Chem. Soc.* *6*, 157-185.
Bischoff, J. L. (1968a). Kinetics of calcite nucleation: magnesium ion inhibition and ionic strength catalysis. *J. Geophys. Res.* *73* (10), 3315-3322.
Bischoff, J. L. (1968b). Catalysis, inhibition and the calcite-aragonite problem. II. The vaterite-aragonite transformation. *Amer. J. Sci.* *266*, 80-90.
Bischoff, J. L. and Fyfe, W. S. (1968). Catalysis inhibition and the calcite-aragonite problem. I. The aragonite-calcite transformation. *Amer. J. Sci.* *266*, 65-79.
Bögli, A. (1964). Corrosion par mélange des eaux. *Internat. J. of Speleo.* *1* (1-2), 61-70.
Bögli, A. (1971). Corrosion by mixing of karst water. *Trans. Cave Res. Grp. G.B.* *13* (2), 109-114.
Bray, L. G. (1969). Some notes on the chemical investigation of cave waters. *Trans. Cave Res. Grp. G.B.* *11* (3), 165-174.
Bray, L. G. (1971). Some problems encountered in a study of the chemistry of solution. *Trans. Cave Res. Grp. G.B.* *13* (2), 115-122.
Bray, L. G. (1972). Preliminary oxidation studies on some cave waters from South Wales. *Trans. Cave Res. Grp. G.B.* *14* (2), 59-66.
Bray, L. G. and Stenner, R. D. In preparation.
Bricker, O. P. and Garrels, R. M. (1967). Mineralological factors in natural water equilibria. In *Principles and Applications of Water Chemistry*, 449-468. Faust, S. D. and Hunter, J. V. (eds). Wiley, New York.
Brooks, R., Clark, L. M. and Thurston, E. F. (1950). Calcium carbonate and its hydrates. *Phil. Trans. Roy. Soc.* *A243*, 145-167.
Caro, P. (1965). La chimie du gaz carbonique et des carbonates, et les phénomènes hydrogéologiques karstiques. *Chronique d'Hydrogéologie* 7, 51-77.
Chamberlain, G. J. (1967). *Colour Measurement and Public Health*. 2nd edn. The Tintometer Ltd., Salisbury.
Chave, K. E., Deffeyes, K. S., Weyl, P. K., Garrels, R. M. and Thompson, M. E. (1962). Observations on the solubility of skeletal carbonates in aqueous solutions. *Science* *137*, 33-34.
Chave, K. E. and Schmaltz, R. F. (1966). Carbonate-seawater interactions. *Geochimica et Cosmochimica Acta* *30*, 1037-1048.
Curl, R. L. (1962). The aragonite-calcite problem. *Nat. Speleo. Soc. Bull.* (*U.S.A.*) *24*, 57-71.
Dixon, B. E. and Hands, G. C. (1957). Desorption and absorption of gases by drops during impact. *J. Applied Chemistry* 7, 342-348.
Drew, D. (1970). Limestone solution within the East Mendip area, Somerset. *Trans. Cave Res. Grp. G.B.* *12* (4), 259-270.
Fetzer, W. G. (1946). Humic acids and true organic acids as solvents of minerals. *Economic Geol.* *41* (i), 47-56.
Fischbeck, R. and Müller, G. (1971). Monohydrocalcite, hydromagnesite, nesquehonite, dolomite, aragonite and calcite in speleothems of the Fränkische Schweiz, Western Germany. *Contrib. Mineralogy Petrology* *33*, 87-92.
Ford, D. C. (1971). Characteristics of limestone solution in the Southern Rocky Mountains and Selkirk Mountains, Alberta and British Columbia. *Canadian J. Earth Sci.* *8* (6), 585-609.

Frear, G. L. and Johnston, J. (1929). The solubility of calcium carbonate (calcite) in certain aqueous solutions at 25°C. *J. Amer. Chem. Soc. 51*, 2082-2093.

Fyfe, W. S. and Bischoff, J. L. (1965). *The Calcite-Aragonite Problem*. Society of Economic Paleontologists and Mineralogists Special Publication No. 13 "*Dolomitization and Limestone Diagenesis—a Symposium*", 3-13.

Garrels, R. M. and Christ, C. L. (1965). *Solutions, Minerals and Equilibria*. Harper and Row, New York.

Garrels, R. M. and Thompson, M. E. (1962). A chemical model for seawater at 25°C and one atmosphere total pressure. *Amer. J. Sci. 260*, 57-66.

Gorham, E. (1958). Free acid in British soils. *Nature 181*, 106.

Graf, D. L. and Lamar, J. E. (1955). *Economic Geology*. 50th Anniversary Volume, 639-713.

Grèzes, G. and Basset, M. (1965). Contribution à l'étude de la solubilité du carbonate de calcium. *Comptes Rendus de l'Académie des Sciences 260*, 869-872.

Hall, F. R. (1964). *Calculated Chemical Composition of some Sulphate-bearing Waters*. International Association of Scientific Hydrology, Commission on Subterranean waters, Publication 64 (General Assembly of Berkeley), 7-15.

Harned, H. S. and Davis, R. (1943). The ionization constant of carbonic acid in water and the solubility of carbon dioxide in water and aqueous salt solutions from 0 to 50°C. *J. Amer. Chem. Soc. 65*, 2030-2037.

Harned, H. S. and Owen, B. B. (1958). *The Physical Chemistry of Electrolytic Solutions*. 3rd edn. Reinhold Publishing Corporation, New York.

Harned, H. S. and Robinson, R. A. (1940). A note on the temperature variation of the ionization constants of weak electrolytes. *Trans. Faraday Soc. 36*, 973-978.

Harned, H. S. and Scholes, S. R. (1941). The ionization constant of HCO_3^- from 0 to 50°C. *J. Amer. Chem. Soc. 63*, 1706-1709.

Holden, W. S. (1970). *Water Treatment and Examination*. J. and A. Churchill, London.

Howard, A. D. (1964). Processes of limestone cave development. *Internat. J. Speleo. 1* (1-2), 47-60.

Howard, A. D. and Howard, B. Y. (1967). Solution of limestone under laminar flow between parallel boundaries. *Caves and Karst*, *9* (4), 25-38.

IWE (1960). *Approved Methods for the Physical and Chemical Examination of Water*. 3rd edn. Institution of Water Engineers, London.

Jacobson, R. L. and Langmuir, D. (1970). The chemical history of some spring waters in carbonate rocks. *Ground water 8* (3), 5-9.

Jacobson, R. L. and Langmuir, D. (1972). An accurate method for calculating saturation levels of ground waters with respect to calcite and dolomite. *Trans. Cave Res. Grp. G.B. 14*, 104-108.

Kern, D. M. (1960). The hydration of carbon dioxide. *J. Chem. Educ. 37* (1), 14-23.

Kitano, Y. and Hood, D. W. (1965). The influence of organic material on the polymorphic crystallization of calcium carbonate. *Geochimica et Cosmochimica Acta 29*, 29-41.

Klotz, I. M. (1964). *Chemical Thermodynamics*. Rev. edn. W. A. Benjamin Inc., New York.

Lafon, G. M. (1970). Calcium complexing with carbonate ion in aqueous solutions at 25°C and 1 atmosphere. *Geochimica et Cosmochimica Acta 34*, 935-940.

Langmuir, D. (1968). Stability of calcite based on aqueous solubility measurements. *Geochimica et Cosmochimica Acta 32*, 835-851.

Langmuir, D. (1971). The geochemistry of some carbonate ground waters in central Pennsylvania. *Geochimica et Cosmochimica Acta 35*, 1023-1045.

Larsen, T. E. and Buswell, A. M. (1942). Calcium carbonate saturation index and alkalinity interpretations. *J. Amer. Water Works Assoc. 34* (11), 1667-1684.

Latimer, W. M. (1959). *Oxidation Potentials*. Prentice Hall Inc., New York.

Manov, G. G., Bates, R. G., Hamer, W. J. and Acree, S. F. (1943). Values of the constants in the Debye-Hückel equation for activity coefficients. *J. Amer. Chem. Soc. 65*, 1765-1767.

Moody, L. F. (1944). Friction factors in pipe flow. *Trans. Amer. Soc. Mech. Eng. 6*, 673-684.

Moore, G. W. (1962). The growth of stalactites. *Nat. Speleo. Soc. Bull. 24* (2), 95-106.

Moore, G. W. and Nicholas, G. (1964). *Speleology: The Study of Caves*. D. C. Heath & Co., Boston.

Murray, A. N. and Love, W. W. (1929). Action of organic acids upon limestone. *Bull. Amer. Assoc. Petrol. Geol. 13* (11), 1467-1475.

Murray, J. W. (1954). The deposition of calcite and aragonite in caves. *J. Geol. 62* (5), 481-492.

Nakayama, F. S. (1968). Calcium activity, complex and ion-pair in saturated $CaCO_3$ solutions. *Soil Science 106* (6), 429-434.

Nancollas, G. H. and Purdie, N. (1964). The kinetics of crystal growth. *Quart. Rev. Chem. Soc. 18*, 1-20.

Nestaas, I. and Terjesen, S. G. (1969). The inhibiting effect of scandium ions upon the dissolution of calcium carbonate. *Acta Chemica Scandinavica 23*, 2519-2531.

Picknett, R. G. (1964). A study of calcite solutions at 10°C. *Trans. Cave Res. Grp. G.B.* 7 (1), 41-62.

Picknett, R. G. (1968). Liquid junction potential between dilute electrolytes and saturated potassium chloride. *Trans. Faraday Soc. 64*, 1059-1069.

Picknett, R. G. (1972). The pH of calcite solutions with and without magnesium carbonate present, and the implications with regard to rejuvenated aggressiveness. *Trans. Cave Res. Grp. G.B. 14*, 141-149.

Picknett, R. G. (1973). Saturated calcite solutions from 10 to 40°C: a theoretical study evaluating the solubility product and other constants. *Trans. Cave Res. Grp. G.B. 15* (2), 67-80.

Pitty, A. F. (1966). *An Approach to the Study of Karst Water*. Occ. Papers in Geography 5, University of Hull, pp. 32-46.

Pitty, A. F. (1971). Biological activity and the uptake and redeposition of calcium carbonate in natural water. *Environmental Letters 1* (2), 103-109.

Poulson, T. L. and White, W. B. (1969). The cave environment. *Science 165*, 971-981.

Pribil, R. and Veseley, V. (1966). Contributions to the basic problems of complexometry. XX. Determination of calcium and magnesium. *Talanta 13*, 223.

Raistrick, B. (1949). The stabilization of the supersaturation of calcium carbonate solutions by anions possessing O—P—O—P—O chains. *Discussions Faraday Soc. 5*, pp. 234-236.

Reddy, M. and Nancollas, G. H. (1971). The crystallization of calcium carbonate I. Isotopic exchange and kinetics. *J. Colloid Interface Science 36* (2), 166-172.

Reitemeier, R. F. and Buehrer, T. F. (1940). The inhibiting action of minute amounts of sodium hexametaphosphate on the precipitation of calcium carbonate from ammoniacal solutions. *J. Phys. Chem.* *44*, 535-551, 552-573.

Roberson, C. E. (1964). Carbonate equilibria in selected natural waters. *Amer. J. Sci.* *262*, 56-65.

Rodhe, H. (1971). *A Study of the Sulphur Budget for the Atmosphere over Northern Europe*. Report AC-17, International Meteorological Institute in Stockholm.

Roques, H. (1963). Sur la répartition du CO_2 dans les karsts. *Ann. Spéléologie* *18* (2), 11-184.

Roques, H. (1964). Contribution à l'étude statique et cinétique des systèmes gas carbonique-eau-carbonate. *Ann. Spéléologie* *19* (2), 258-484.

Rowlands, D. L. G. and Webster, R. K. (1971). Precipitation of vaterite in lake water. *Nature Physical Science* *229*, 158.

Runnells, D. D. (1969). Diagenesis due to the mixing of natural waters: a hypothesis. *Nature* *224*, 361-363.

SAC (1968). *Recommended Methods for the Analysis of Trade Effluents*. Soc. for Analytical Chemistry, London.

Schloesing, T. (1872). Sur la dissolution du carbonate de chaux par l'acide carbonique. *Comptes Rendus de l'Académie des Sciences* *74*, 1552-1556 and *75*, 70-73.

Schoeller, H. (1962). *Traité des eaux souterraines*. Masson, Paris.

Shternina, E. B. and Frolova, E. V. (1952). The solubility of calcite in the presence of CO_2 and NaCl. *Izv. Sektor. Fiz.-Khim. Anal., Inst. Obschei. Kim., Akad. Nauk. SSSR* *21*, 271-287. (A.T.S. Translations RJ-379.)

Shuster, E. T. and White, W. B. (1971). Seasonal fluctuations in the chemistry of limestone springs: a possible means for characterising carbonate aquifers. *J. Hydrology* *14*, 93-128.

Siedell, A. (1958). *Solubilities of Inorganic and Metal-Organic Compounds*, Volume 1. 4th edn (revised by Linke, W. F.). Nostrand Co. Inc., New York.

Stchouzkoy-Muxart, T. (1972). Contribution à l'étude des courbes de solubilité de la calcite dans l'eau en présence d'anhydride carbonique. *Annales de Spéléologie* *27* (3), 465-478.

Stenner, R. D. (1969). The measurement of the aggressiveness of water towards calcium carbonate. *Trans. Cave Res. Grp. G.B.* *11* (3), 175-200.

Stenner, R. D. (1970). Preliminary results of an application of the procedure for the measurement of aggressiveness of water to calcium carbonate. *Trans. Cave Res. Grp. G.B.* *12* (4), 283-289.

Stenner, R. D. (1971). The measurement of the aggressiveness of water to calcium carbonate. *Trans. Cave Res. Grp. G.B.* *13* (4), 283-295.

Stumper, R. (1935). La cinétique et la catalyse de la décomposition du bicarbonate de calcium en solution aqueuse. *Bull. Soc. Chimique de Belgique* *44*, 176-209.

Terjesen, S. G., Erga, O., Thorsen, G. and Ve, A. (1961). The inhibitory action of metal ions on the formation of calcium bicarbonate by the reaction of calcite with aqueous carbon dioxide. *Chem. Eng. Sci.* *44*, 277-288.

Thrailkill, J. (1971). Carbonate deposition in Carlsbad Cavern. *J. Geology* *79* (6), 683-695.

Van Everdingen, R. O. (1969). Degree of saturation with respect to $CaCO_3$, $CaMg(CO_3)_2$ and $CaSO_4$ for some thermal and mineral springs in the Southern

Rocky Mountains, Alberta and British Columbia. *Canadian J. Earth Sci. 6*, 1421-1430.

Vogel, A. I. (1968). *A Text-book of Quantitative Inorganic Analysis*. 3rd edn. Longmans, London.

Warwick, G. T. (1962). Cave formations and deposits. In *British Caving*. 2nd edn. Cullingford, C. H. D. (ed.). Routledge and Kegan Paul Ltd., London.

West, T. S. (1969). *Complexometry with EDTA and Related Reagents*. 3rd edn. BDH Chemicals Ltd., Poole.

Weyl, P. K. (1958). The solution kinetics of calcite. *J. Geol. 66*, 163-176.

Weyl, P. K. (1959). The change in solubility of calcium carbonate with temperature and carbon dioxide content. *Geochimica et Cosmochimica Acta* 17, 214-225.

Wigley, T. M. L. (1971). Ion-pairing and water quality measurements. *Canadian J. Earth Sci. 8* (4), 468-476.

Wigley, T. M. L. (1972). *A Computer Programme for Water Quality Analysis*. Technical Note No. 15, Dept. Mechanical Engineering, University of Waterloo, Ontario.

Wissbrun, K. F., French, D. M. and Patterson, A. (1954). The true ionisation constant of carbonic acid in aqueous solution from 5 to 45°C. *J. Phys. Chem. 58*, 693-695.

Wray, J. L. and Daniels, F. (1957). Precipitation of calcite and aragonite. *J. Amer. Chem. Soc. 79*, 2031-2034.

Yanat'yeva, O. K. (1954). Solubility in the system $CaCO_3$—$MgCO_3$—H_2O at different temperatures and pressures of CO_2. *Doklady̆ Akademii Nauk SSSR, 96*, 777-779.

Zeller, E. J. and Wray, J. L. (1956). Factors influencing precipitation of calcium carbonate. *Bull. Amer. Assoc. Petrol. Geol. 40* (1), 140-152.

GLOSSARY OF SCIENTIFIC TERMS AND SYMBOLS

The definitions given here refer only to the usage in this chapter and are not intended to be universally applicable.

Activity: A measure of the effective concentration of a substance. For an ion in dilute solution, activity is related to concentration by Eqn 9; for other substances activity is related to molefraction.

Aggressiveness: The ability of a solution to dissolve further calcium carbonate or limestone; quantitatively the mass of calcium carbonate that can dissolve in 1 litre of solution, negative values being used for supersaturated solutions.

Alkaline: The opposite of acidic. A solution is alkaline if its pH is greater than 7.

Alkalinity: Alkaline hardness, the concentration in solution of calcium, magnesium and other metal ions *less* the concentration of acids other than carbonic.

Anion: A negatively charged ion.

Calomel reference electrode: An electrode customarily used with a glass electrode to measure the pH of water by electrical means.

Common ion effect: The reduction in solubility of a substance caused by the presence in solution of another substance, the two substances having an ion in common.

Concentration: The amount of a substance present in a given volume. Usually measured in units of milligrams per litre (mg/l), parts per million (ppm) or millimoles per litre (mmole/l). For calcium carbonate,

$$100 \text{ mg/l} = 100 \text{ ppm} = 1 \text{ mmole/l}.$$

Degree of saturation: The ratio of the concentration of calcium carbonate in a solution to the concentration when that solution is saturated with the solid.

Diffusion coefficient (molecular): A measure of the rate with which a substance diffuses from high to low concentration.

Electrolyte: A substance which exists in the form of ions, especially in solution.

Equilibrium: The state of balance that exists in some reactions when the concentrations of reactants and products of reaction remain constant.

Equilibrium constant: A number which defines the relationship between the activities of the ions and molecules in a reaction at equilibrium. Constant for a given temperature and pressure.

Glass electrode: An electrode sensitive to hydrogen ions which is commonly used to measure pH by electrical means.

Hardness: In water analysis, the concentration of calcium (calcium hardness), magnesium (magnesium hardness), and calcium plus magnesium (total hardness). For alkaline hardness, see Alkalinity.

Ion: An electrically charged atom or molecule, important in solution.

Ion activity product: The product of the activities of two or more ions, especially $(Ca^{2+})(CO_3^{2-})$ for calcite solutions and $(Ca^{2+})(Mg^{2+})(CO_3^{2-})^2$ for dolomite solutions. Related to the solubility product.

Ion pair: A pseudo-molecule formed by the association of ions in solution, e.g. $CaHCO_3^+$ formed from Ca^{2+} and HCO_3^-.

Ionic strength: A fundamental measure of the total concentration of dissolved electrolytes, as defined by Eqn 11.

Ionic strength effect: The increase in solubility caused by the addition of foreign electrolyte to increase the ionic strength.

Liquid junction potential: A voltage that causes error in the measurement of pH in dilute solution when a calomel electrode is used.

Mischungskorrosion: The rejuvenation of aggressiveness towards calcium carbonate that occurs when different saturated solutions of calcium carbonate are mixed.

Molar: Concentration measured in moles per litre.

Mole: A unit of mass: the molecular weight of a substance expressed in grams, e.g. 1 mole of $CaCO_3$ = 100 g of $CaCO_3$.

Molefraction: A unit of concentration for one substance mixed with others. For a given amount of the mixture, the number of moles of the substance divided by the total number of moles of all substances present.

Normality: A measure of concentration dependent on valency. For HCl, normality equals mole/l; for H_2SO_4, normality is 2 times mole/l.

Oxidation: Chemical reaction with oxygen, e.g. acetic acid ultimately oxidizes to carbon dioxide and water: $C_2H_4O_2 + O_2 \rightarrow 2O_2 + 2H_2O$.

Oxygen demand: The amount of oxygen needed to oxidize materials such as organic matter in a sample of solution.

Partial pressure: A measure of carbon dioxide concentration in air. The ratio of partial pressure to total air pressure is equal to the volume of carbon dioxide in unit volume of air. At 1 atmosphere total pressure, partial pressure × 100 = $\%CO_2$ in air.

Precipitation: The deposition of solid from solution.

Rate constant: A constant relating the rate of reaction to the concentrations of reactants.

Saturated: A solution of a solid or gas that is in equilibrium with excess of the solid or gas.

Solid solution: A solution of a solid in a solid.

Solubility: The concentration of a substance in saturated solution.

Solubility curve: The curve produced by graphically plotting solubility against some property of the solution such as carbon dioxide content.

Solubility product: The product of the activities in saturated solution of (*a*) (Ca^{2+}) and (CO_3^{2-}) for calcite, (*b*) (Ca^{2+}), (Mg^{2+}) and $(CO_3^{2-})^2$ for dolomite.

Solutional power: The ability of water to dissolve a solid, e.g. limestone.

Standard solution: A solution of definite concentration, used in chemical analyses.

Supersaturated: A solution which contains more dissolved solid than the corresponding saturated solution.

Transfer coefficient: The constant of proportionality relating rate of diffusion and concentration difference, as in Eqn 15.

Unsaturated: Not saturated, i.e. either aggressive or supersaturated.

Viscosity (kinematic): The ratio of viscosity to density.

X-ray diffraction: A method of determining crystal structure from the diffraction pattern produced by X-rays.

A	A constant in Eqn 10 which depends on the properties of water.
B	A constant in Eqn 10 which depends on the properties of water.
b	A suffix relating to the bulk of solution where the concentration is uniform.
C	The concentration of an ion in mole/l.
D	Diffusion coefficient (molecular).
d	The effective diameter of an ion in solution, as used in Eqn 10.
g	Gram.
I	Ionic strength of the solution.
IAP	Ion activity product.
J	Symbol for a given ion.
K	Equilibrium constant. K_0, K_1, K_2, K_3, K_4, K_W, K_1' and K_C are defined by Eqns 1-6 and 12-13; K_S is the calcite solubility product.
k	Reaction rate constant. k_1 and k_2 refer to Eqn 14, k_3 and k_4 to Eqn 19.
k_t	Transfer coefficient.
L	The effective distance over which diffusion occurs in still water (Eqn 16).
L'	The characteristic distance over which diffusion occurs in streamline flow (Eqn 17).
mg	Milligram (0·001 g).
P	The partial pressure of carbon dioxide in air.
pH	A measure of the acidity of water. Low pH is acidic, pH 7 is neutral and high pH is alkaline. pH $= -\log_{10}(H^+)$.
R	The rate of solution or deposition of calcium carbonate in mole s^{-1}.
S	The surface area of calcite crystals (cm^2).

s	Seconds. Also a suffix relating to the solution near the solid surface.
U	The flow velocity of water across the solid surface (cm s^{-1}).
z	The number of electron charges per ion, all numbers being taken as positive.
γ	The ion activity coefficient.
ν	The kinematic viscosity.
μ mole/l	Micromole per litre (10^{-6} molar).
$[Ca]_T$	The sum of the molar concentrations of all forms of calcium in solution.
$[CO_2]_{init}$	The initial carbon dioxide in solution when water is exposed to air with carbon dioxide before, but not during, the solution of calcium carbonate.
mole/l	Mole per litre, or molar concentration.
mmole/l	Millimole per litre (10^{-3} molar).
nmole/l	Nanomole per litre (10^{-9} molar).
ppm	Parts (mass) per million parts (volume). 1 ppm = 1 mmole/l for $CaCO_3$.
X	Molar concentration of a molecule or ion X.
[ ]	The concentration of the ion or molecule inside the brackets.
( )	The activity of the ion or molecule inside the brackets.

8. Cave Minerals and Speleothems

W. B. White

INTRODUCTION

A cave, at constant temperature and invaded by percolating solutions carrying various substances, forms an excellent environment for the slow deposition of minerals. Because of the slow growth and nearly constant conditions, minerals in caves often show spectacular crystal development and indeed much of the aesthetic appeal of caves, especially those commercially developed, is derived from these deposits. The number of minerals commonly occurring in caves is small, but because of the variety of depositional mechanisms (by evaporation, from dripping water, from flowing water, in pools of standing water, etc.) there exist a large number of characteristic shapes. These shapes are known as *speleothems* (Moore, 1952) or *cave formations*.

There are three main categories of materials found in caves: (1) fragments of bedrock and debris left behind from the dissolution of the bedrock, (2) material transported into the cave by mechanical action of water, wind, or gravity and (3) material formed in the cavern by chemical deposition. Materials considered to be cave minerals are restricted to the final category. The other two are considered as clastic sediments and are discussed elsewhere in this volume (Chapter 2). The chemical deposits could be considered as chemical sediments by analogy with the chemical sediments formed in the lakes and oceans of the earth's surface. These may be classified in terms of their dominant mineralogy as follows:

I. Carbonate minerals.
II. Evaporite minerals.
III. Phosphate and nitrate minerals.
IV. Oxide and other minerals.

Table 8.1

CHECKLIST OF CAVE MINERALS

Aragonite, $CaCO_3$
Skyline Cavern, Virginia
Ardealite, $Ca_2HPO_4SO_4.4H_2O$
Csoklovina Cave, Romania
Azurite, $Cu_3(CO_3)_2(OH)_2$
Copper Queen Cave, Arizona
Barite, $BaSO_4$
Madoc Cave, Ontario
Beudantite, $PbFe_3AsO_4SO_4(OH)_6$
Island Ford Cave, Virginia
Birnessite $(Mg,Ca,K)Mn_7O_{14}.2H_2O$
Webers Cave, Iowa
Blödite*, $Na_2Mg(SO_4)_2.4H_2O$
Lee Cave, Kentucky
Brushite, $CaHPO_4.2H_2O$
Pig Hole Cave, Virginia
Calcite, $CaCO_3$
Grand Caverns, Virginia
Carbonate-apatite, $Ca_{10}(PO_4)_6CO_3.H_2O$
El Chapote Cave, Nuevo Léon
Celestite, $SrSO_4$
Miller Cave, Texas
Cerussite, $PbCO_3$
Herman Smith Cave, Illinois
Cimolite, $Al_4(SiO_2)_9(OH)_{12}$
Lookout Cave, Washington
Crandallite, $CaAl_3(PO_4)_2(OH)_5.H_2O$
Pajaros Cave, Puerto Rico
Cristobalite, SiO_2
Wind Cave, South Dakota
Dolomite, $CaMg(CO_3)_2$
Lehman Caves, Nevada
Endellite, $Al_2Si_2O_5(OH)_4.2H_2O$
Carlsbad Caverns, New Mexico
Epsomite, $MgSO_4.7H_2O$
Wyandotte Cave, Indiana
Ferghanite, $U_3(VO_4)_2.6H_2O$
Tyuya-muyun Cave, USSR
Fluorapatite, $Ca_5(PO_4)_3F$
Poorfarm Cave, West Virginia
Fluorite, CaF_2
Kootenay Florence Cave, British Columbia
Galena, PbS
Herman Smith Cave, Illinois
Goethite, FeO(OH)
Herman Smith Cave, Illinois
Gypsum, $CaSO_4.2H_2O$
Mammoth Cave, Kentucky
Hematite, Fe_2O_3
Wind Cave, South Dakota
Hemimorphite, $Zn_4Si_2O_7(OH)_2.H_2O$
Broken Hill Cave, Zambia
Hexahydrite*, $MgSO_4.6H_2O$
Lee Cave, Kentucky
Hibbenite, $Zn_7(PO_4)_4(OH)_2.7H_2O$
Hudson Bay Cave, British Columbia
Hopeite, $Zn_3(PO_4)_2.4H_2O$
Broken Hill Cave, Zambia
Huntite, $Mg_3Ca(CO_3)_4$
Titus Canyon Cave, California
Hydromagnesite, $Mg_5(CO_3)_4(OH)_2.4H_2O$
Carlsbad Caverns, New Mexico
Hydroxyapatite, $Ca_5(PO_4)_3(OH)$
Negra Cave, Puerto Rico
Hydrozincite, $Zn_5(CO_3)_2(OH)_6$
Island Ford Cave, Virginia
Ice, H_2O
Fossil Mountain Cave, Wyoming
Jarosite, $KFe_3(SO_4)_2(OH)_6$
Tintic Cave, Utah
Lecontite, $NaNH_4SO_4.2H_2O$
Piedras Cave, Honduras
Limonite*, $Fe_2O_3.nH_2O$
Porters Cave, Pennsylvania
Magnesite, $MgCO_3$
Titus Canyon Cave, California
Malachite, $Cu_2CO_3(OH)_2$
Copper Queen Cave, Arizona
Martinite, $Ca_3(PO_4,CO_3,OH)_2$
Negra Cave, Puerto Rico
Melanterite, $FeSO_4.7H_2O$
Wilson Cave, Nevada

Table 8.1 (*continued*)

Mineral	Locality
Mirabilite, $Na_2SO_4.10H_2O$	Flint-Mammoth Cave System, Kentucky
Monetite, $CaHPO_4$	Aggteloker Cave, Hungary
Monohydrocalcite*, $CaCO_3.H_2O$	Eibengrotte, Germany
Nesquehonite, $MgCO_3.3H_2O$	Moulis Cave, France
Newberyite, $MgHPO_4.3H_2O$	Skipton Cave, Australia
Nitre, KNO_3	Goyder Cave, Australia
Nitrammite, NH_4NO_3	Nicajack Cave, Tennessee
Nitrocalcite, $Ca(NO_3)_2.4H_2O$	Mammoth Cave, Kentucky
Nitromagnesite, $Mg(NO_3)_2.6H_2O$	Great Cave, Kentucky
Ondrejite	[*See* sepiolite]
Opal	[*See* cristobalite]
Palygorskite, $Mg_3(Al_2)Si_8O_{20}(OH)_2.8H_2O$	Broken Hill Cave, New Zealand
Parahopeite, $Zn_3(PO_4)_2.4H_2O$	Hudson Bay Cave, British Columbia
Psilomelane, $BaMn_9O_{16}(OH)_4$	Jasper Cave, Wyoming
Pyrite, FeS_2	Herman Smith Cave, Illinois
Pyrrhotite, FeS	Playa Pajaro Cave, Puerto Rico
Quartz, SiO_2	Wind Cave, South Dakota
Salmonite, Basic zinc phosphate?	Hudson Bay Cave, British Columbia
Scholzite, $CaZn_2(PO_4)_2.2H_2O$	Island Ford Cave, Virginia
Sepiolite, $Mg_2Si_3O_8.2H_2O$	Zbraov Cave, Czechoslovakia
Smithsonite, $ZnCO_3$	Herman Smith Cave, Illinois
Spencerite, $Zn_4(PO_4)_2(OH)_2.3H_2O$	Hudson Bay Cave, British Columbia
Sphalerite, ZnS	Herman Smith Cave, Illinois
Struvite*, $NH_4MgPO_4.6H_2O$	Skipton Cave, Australia
Taranakite, $H_6K_3Al_5(PO_4)_8.18H_2O$	Pig Hole Cave, Virginia
Tarbuttite, Zn_2PO_4OH	Broken Hill Cave, Zambia
Tenorite, CuO	Calumet Cave, Arizona
Thaumasite	[*See* sepiolite]
Thenardite, Na_2SO_4	Wind Cave, South Dakota
Tinticite, $Fe_3(PO_4)_2(OH)_3.3H_2O$	Tintic Cave, Utah
Turanite, $Cu_5(VO_4)_2(OH)_4$	Tyuya-muyun Cave, USSR
Tyuyamunite, $Ca(UO_2)_2(VO_4)_2.nH_2O$	Tyuya-muyun Cave, USSR
Vanadinite, $Pb_5(VO_4)_3Cl$	Havasu Canyon Cave, Arizona
Variscite, $AlPO_4.2H_2O$	Drachen Cave, Hungary
Whitlockite, $Ca_3(PO_4)_2$	El Chapote Cave, Nuevo Léon

* All entries are from Moore (1970) except for addenda marked (*)

The carbonate minerals are essentially fresh-water limestones, although usually very coarsely crystalline; they are often known under the general term of *travertine*. The cave evaporites, by analogy with lacustrine and marine evaporites, consist of sulphate minerals, mainly gypsum and, rarely, halides. Phosphate minerals occur mainly in tropical caves where there has been a reaction between organic debris and the limestone wall rock. Nitrates occur in the soils of many temperate climate caves. Thin coatings of black manganese oxides or brown iron oxide hydrates appear in many caves. Perennial ice occurs in caves of mountainous areas, and it deserves a separate entry as a cave mineral.

Nearly 80 minerals have been recognized from solution caves according to the compilations of Moore (1970) and Broughton (1972). The list in Table 8.1 is taken directly from Moore's (1970) compilation with a few additions. However, only about 20 of these are found in the "normal" cave environment. The rest are mainly the result of some very special conditions of mineralization such as the invasion of cavernous openings in limestone by ore-forming solutions, such as in the lead-zinc district of Iowa and Illinois in the United States. Large euhedral quartz linings in several western U.S. caves are apparently the result of hydrothermal water, and an occurrence of fluorite in Spirit Mountain Cave, Wyoming, was formed by gases escaping from a lava flow that overrode the cavernous limestone. Although interesting in themselves, these exotic minerals have little to do with the main processes of cave formation and filling. Each must be analysed within the framework of its peculiar geology. For this reason this chapter will be concerned only with the "normal" cave minerals, those deposited from cold groundwater percolating through a section of sedimentary rocks and into the cavernous opening where the minerals occur.

Systematic treatments of cave mineralogy have been published. The most comprehensive descriptive work is a recent book published by the National Speleological Society (Hill, 1976). Extensive chapters on cave minerals appear in the books by Gèze (1965) and Trombe (1952). Many of the other recent monographs on caves contain at least a short section on chemical sedimentation in caves.

CLASSIFICATION OF SPELEOTHEMS

Mineral-depositing solutions move vertically through the vadose zone under the influence of gravity. The mineral deposits left behind by the dripping or flowing water, therefore, tend to take on shapes in which the gravitational control of the solution is much in evidence. The crystals themselves, however, develop under the influence of other forces. Particular growth directions of crystals are much faster than other directions; modifications in the chemistry of the solution or the rate of growth can change the crystal habit and thus change the relative rates of growth in different directions. There is a competition between shapes guided by the flow path of the solution and shapes guided by the particular mineral and its crystal habit. This gives rise to two broad classes of speleothems, dripstone and flowstone forms and erratic forms. These terms are equivalent to the words gravitomorphic and non-gravitomorphic used by Halliday (1962).

Within the context of these basic mechanisms there is an immense and indeed continuous variety of shapes for the travertine deposits depending on the vagaries of exact flow path, flow rate, chemical characteristics of the water, and relative humidity and CO_2 pressure of the cave atmosphere. Because caves derive much of their charm from speleothems, the deposits have collected many fanciful names and indeed constitute the bulk of the guides' lore in commercial caves. It is thus possible to over-classify speleothem shapes and to arrive at an elaborate nomenclature that adds little to understanding. As a compromise, this review utilizes a two-level hierarchy of classification: *forms* of speleothems are the basic shapes that can be distinguished because of growth habit of minerals or basic depositional mechanisms. Thus stalactites, helictites, and rimstone pools (i.e. those shapes of cave mineral deposits that have distinctive names) are *forms*. Any particular form can be further classified into *styles*, which describe the different shapes that result from different flow rates, rates of deposition, or other factors. Classification of speleothem shapes according to style could be made as complex as needed for any particular purpose. In general, forms are described by proper nouns, styles by adjectives. Since there seems little value in inventing a new scientific-sounding jargon, the colloquial (or even slang) adjective used to describe the styles have been retained as in straw stalactites or broomstick stalagmites.

The compromise listing of speleothem forms is

A. Dripstone and Flowstone Forms:
1. Stalactites
2. Stalagmites
3. Draperies
4. Flowstone sheets

B. Erratic Forms:
1. Shields
2. Helictites
3. Botryoidal forms
4. Anthodites
5. Moonmilk

C. Sub-aqueous Forms:
1. Rimstone Pools
2. Concretions of various kinds
3. Pool deposits
4. Crystal linings

A classification such as that above is of limited application because it was devised with respect to a very restricted group of minerals, mainly carbonates. It appears to have fairly wide geographic application, because speleothems from limestone caves in most regions of the world seem to have very similar shapes.

THE CARBONATE MINERALS

MINERALOGY OF THE CARBONATES

Carbonate-bearing groundwaters contain mainly the cations Ca^{++} and Mg^{++} with smaller concentrations of Sr^{++} and Na^{+}. Other cationic species are rare. Some typical analyses are shown in Table 8.2. When these solutions percolate into the cave and deposit their mineral load, the dominant mineral is calcite, the stable form of calcium carbonate under the temperature, total pressure, and carbon dioxide partial pressure of the cave environment. Other anhydrous carbonates that occasionally occur in caves are listed with their crystallographic properties in Table 8.3. Of these, magnesite and huntite have been found only in association with moonmilk and will be discussed later. Vaterite has not been reported from a cave locality.

Most cave carbonates are low magnesian calcites. A summary of the available analyses is given in Table 8.4. The magnesium concentration is typically of the order of one or two per cent, very much in the range of calcites from other environments and typical of fresh-water carbonates in general (Goldsmith *et al.*, 1955; Friedman, 1964). Co-existing aragonite has a composition similar to that of the calcite, usually with a somewhat higher strontium content. Calcite frequently appears as single crystals of which the scalenohedral (dog-tooth spar) and rhombohedral (nail-head spar) habits are most common (Fig. 8.1).

Aragonite is the high-pressure polymorph of $CaCO_3$ and is, in theory, thermodynamically stable at cave temperatures only above pressures of 3000 atm (Simmons and Bell, 1963; Boettcher and Wyllie, 1968). Its occurrence in the cave environment is metastable, but in spite of that aragonite is very common in caves and is perhaps the third most common

Table 8.2

SELECTED CHEMICAL ANALYSES FOR SPELEOTHEM-DEPOSITING CAVE WATERS*†

Sample location	pH	Ca^{++}	Mg^{++}	HCO_3^-	SO_4^-	Reference
Calcite-depositing stalactite Canoe Cave, Virginia	—	46	35	282	8	Murray (1954)
Calcite-depositing stalactite New River Cave, Virginia	—	64	33	437	15	Murray (1954)
Stalactite drip Luray Caverns, Virginia	7·85	84	55	522	13	Holland *et al.* (1964)
Aragonite-depositing stalactite New River Cave, Virginia	—	27	27	164	38	Murray (1954)
Calcite-encrusted pool New River Cave, Virginia	—	23	13	134	—	Murray (1954)
Rimstone Pool Luray Caverns, Virginia	8·00	52	56	420	—	Holland *et al.* (1964)

* Complete chemical analyses of travertine-depositing waters are rare compared with published analyses of carbonate-dissolving waters. Trace element analyses have not been made.

† All concentrations are given in parts per million.

Table 8.3

CRYSTALLOGRAPHIC DATA FOR THE CARBONATE MINERALS

Mineral	Composition	System and Space Group		Unit Cell Parameters	Reference*
Calcite	$CaCO_3$	Trigonal	$R\bar{3}c$	$a_0 = 4{\cdot}990$ $c_0 = 17{\cdot}061$	Graf (1961)
Magnesite	$MgCO_3$	Trigonal	$R\bar{3}c$	$a_0 = 4{\cdot}633$ $c_0 = 15{\cdot}664$	Graf (1961)
Dolomite	$CaMg(CO_3)_2$	Trigonal	$R\bar{3}$	$a_0 = 4{\cdot}808$ $c_0 = 16{\cdot}010$	Graf (1961)
Huntite	$CaMg_3(CO_3)_4$	Trigonal	R32	$a_0 = 9{\cdot}505$ $c_0 = 7{\cdot}821$	Graf and Bradley (1962)
Aragonite	$CaCO_3$	Orthorhombic	Pmcn	$a_0 = 4{\cdot}960$ $b_0 = 7{\cdot}964$ $c_0 = 5{\cdot}738$	Dickens and Bowen (1971)
Vaterite	$CaCO_3$	Hexagonal	$P6_3/mmc$	$a_0 = 7{\cdot}15$ $c_0 = 16{\cdot}96$	Meyer (1965)
Nesquehonite	$MgCO_3 \cdot 3H_2O$	Monoclinic	$P2_1/n$	$a_0 = 7{\cdot}705$ $b_0 = 5{\cdot}367$ $c_0 = 12{\cdot}121$ $\beta = 90{\cdot}45°$	Stephen and MacGillavry (1972)
Lansfordite	$MgCO_3 \cdot 5H_2O$	Monoclinic		$a_0 = 12{\cdot}369$ $b_0 = 7{\cdot}529$ $c_0 = 7{\cdot}315$ $\beta = 99°36'$	XRPD File Card No. 18-769
Artinite	$MgCO_3 \cdot Mg(OH)_2 \cdot 3H_2O$	Monoclinic	C2/m	$a_0 = 16{\cdot}561$ $b_0 = 6{\cdot}298$ $c_0 = 6{\cdot}220$ $\beta = 99°9'$	Jagodzinski (1965)
Hydromagnesite	$4MgCO_3 \cdot Mg(OH)_2 \cdot 4H_2O$	Orthorhombic	$C222_1$	$a_0 = 18{\cdot}58$ $b_0 = 9{\cdot}06$ $c_0 = 8{\cdot}42$	Murdoch (1954)
Dypingite	$4MgCO_3 \cdot Mg(OH)_2 \cdot 5H_2O$	Unknown			Raade (1970)

* References are given for most recent or most complete crystal structure determination. X-ray powder data for most compounds may be found in the X-ray Powder Data File (XRPD).

Table 8.4

SELECTED CHEMICAL ANALYSES OF SPELEOTHEMS

Specimen	SiO_2	R_2O_3*	MgO	SrO	Reference
Calcite flowstone Stydle Vody Cave, Czechoslovakia	7·90†	4·84	0·60	0·43	Skřivánek and Hodac (1958)
Calcite Helictite New River Cave, Virginia	0·01	0·03	0·90	—	Murray (1954)
Calcite stalactite New River Cave, Virginia	0·57	0·36	1·68	—	Murray (1954)
Calcite stalactite Great Onyx Cave, Kentucky	—	—	0·27	0·14	Siegel (1965)
Calcite stalactite Wind Cave, South Dakota	—	0·23 (Fe_2O_3) 0·13 (Al_2O_3)	0·38	—	White and Deike (1962)
Sclalenohedral calcite Wind Cave, South Dakota	—	0·054 (Fe_2O_3) 0·063 (Al_2O_3)	0·25	—	White and Deike (1962)
Aragonite druses Wind Cave, South Dakota	—	0·068 (Fe_2O_3) 0·20 (Al_2O_3)	0·29	—	White and Deike (1962)
Anthodite New River Cave, Virginia	0·03	0·01	—	—	Murray (1954)
Aragonite crystals Stydle Vody Cave, Czechoslovakia	2·20	1·44	0·77	0·68	Skřivánek and Hodac (1958)

* $R_2O_3 = Fe_2O_3 + Al_2O_3$.
† All analyses in weight per cent.

cave mineral (the second being gypsum). The solubility of aragonite in water at cave temperatures is about 11% greater than that of calcite. The precipitation of aragonite, therefore, requires some mechanism that will inhibit the nucleation and precipitation of calcite until this level of supersaturation can be built up. Mechanisms commonly involved include

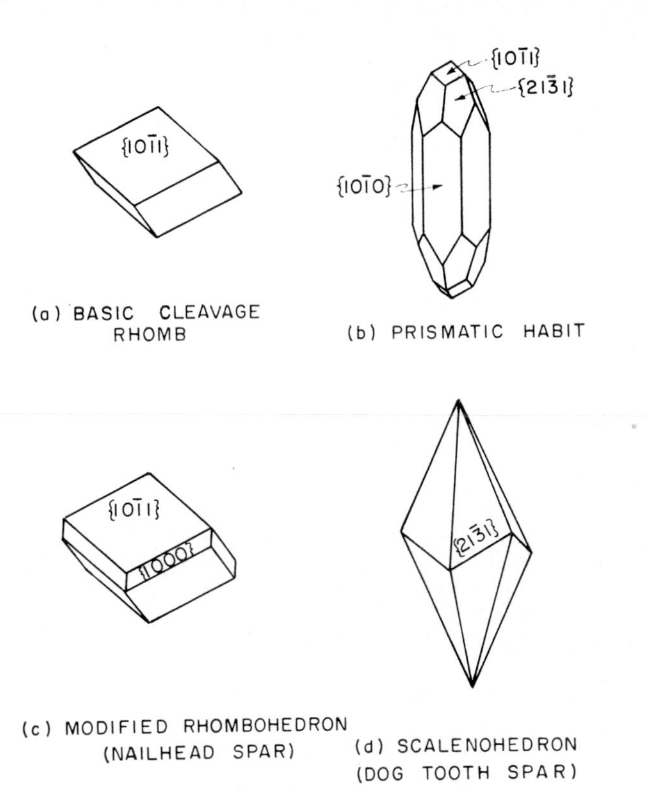

FIG. 8.1. Crystal habits of calcite commonly occurring in cave deposits.

poisoning of calcite growth by adsorption of foreign ions such as Sr^{++}, Mg^{++} or SO_4^{-2} on screw dislocations, or enhanced epitaxial growth of aragonite on other precursor nuclei such as nesquehonite, hydrocalcite or strontianite. Organic constituents, particularly amino acids, have also been shown to play an important role in enhancing aragonite through adsorption on growth steps and dislocations, thus inhibiting the growth of calcite. These mechanisms were reviewed up to 1962 by Curl. An immense literature has appeared since then, mainly on the kinetics of the calcite-aragonite transformation and on controlling mechanisms (Roques, 1964, 1968, 1969; Davis and Adams, 1965; Metzger and Barnard, 1968;

Bischoff and Fyfe, 1968; Bischoff, 1968a and b, 1969; Kinsman and Holland, 1969; Allen *et al.*, 1970; Jackson and Bischoff, 1971). A more detailed account of the physical chemistry of aragonite precipitation is presented in Chapter 7.

Dolomite is an ordered structure closely related to calcite but with alternating layers of magnesium and calcium that lower the crystal symmetry. Although dolomite is a common constituent of carbonate rocks, it is difficult to form in the laboratory at low temperatures and is rarely observed actively forming in any sedimentary environment. The several reports of a primary dolomite in speleothem deposits are therefore of interest. Thrailkill (1968) identified dolomite in Carlsbad Caverns as a secondary modification of moonmilk minerals. However, other occurrences in Lehman Caves, Nevada (Moore, 1961), and as crystallized overgrowths on calcite spar crystals in Jewel and Bethlehem Caves, South Dakota (Deal, 1962), appear to be genuine primary dolomite.

Magnesium generally occurs at lower concentrations in groundwater than does Ca^{++}, and since pure magnesium carbonates are more soluble than calcite, magnesium compounds become a residue deposited only when extensive evaporation or CO_2 degassing has taken place. At the low temperatures of the cave environment, magnesium carbonate is deposited in the form of hydrated basic carbonates, of which five distinct minerals are now known (Table 8.3). These usually take the form of pasty masses, or coatings known as moonmilk.

Calcium carbonate hydrates also exist, of which the monohydrate $CaCO_3 . H_2O$ has been identified in low-temperature deposits (Marschner, 1969) and one cave (Fischback and Müller, 1971). The hexahydrate is also known and its structure has been determined (Dickens and Brown, 1970). Both hydrated calcites may in fact occur in caves at low temperatures, and it has been argued that the calcite moonmilks found in some alpine caves precipitated as the hydrated carbonates. The necessary *in situ* analysis or sample collection under refrigeration has not been done.

Models for Carbonate Deposition

Carbonate minerals are deposited in caves by a transport process in which material is carried into the cave by moving groundwater and left behind as a carbonate speleothem. The chemical and physical mechanisms for calcite deposition are outlined in Figs. 8.2 and 8.3 following the suggestion of Holland *et al.* (1964), Roques (1964, 1968, 1969) and Thrailkill (1971).

Rain water reaches the earth in equilibrium with the nominal CO_2 pressure of the atmosphere, $10^{-3.5}$ atm. The infiltration fraction of the rainfall seeps through the A-horizon of the soil where it absorbs carbon dioxide generated by biological processes and becomes soil water with a CO_2 pressure as high as 0·1 atm as calculated from measurements on soil waters collected in lysometers. Direct measurement of CO_2 gas in the soil

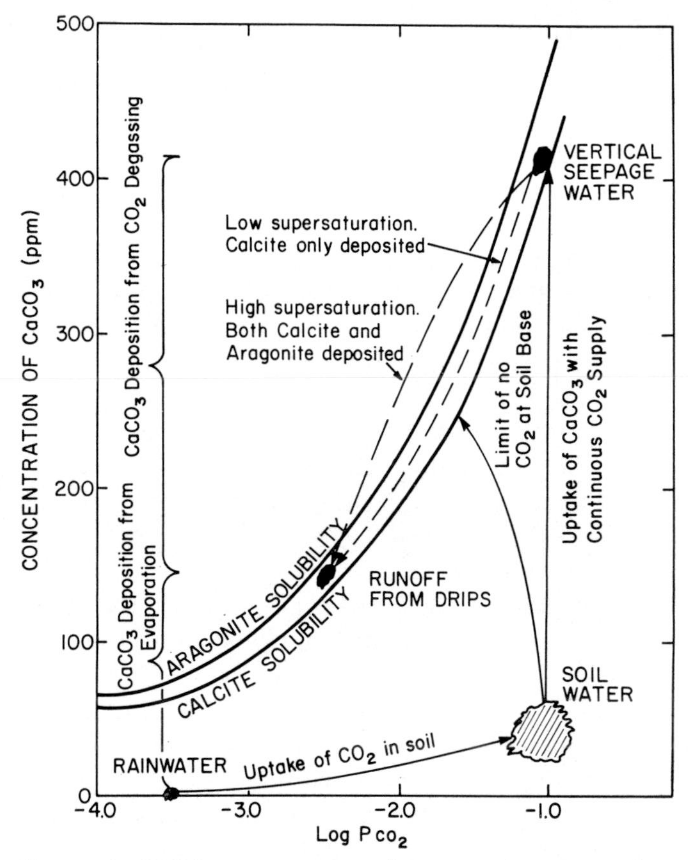

FIG. 8.2. Changes in $CaCO_3$ concentration of downward-seeping infiltration water from its source as rain water to the deposition of carbonates in dripstone deposits. Solubility curves for calcite and aragonite are based on results of Roques (1968). CO_2 pressures of water at different stages are typical experimental values.

atmosphere gives generally comparable values (Miotke, 1972). Soil water has a low pH and is highly aggressive toward carbonate rocks. As the water seeps to the lower C-horizon of the soil near the soil-bedrock interface there is vigorous solution of the limestone, and at the limit the water becomes saturated with calcite in equilibrium with the high CO_2

pressure of the soil. Exactly what happens in individual localities is complicated by the relative thickness and permeability of the different soil horizons. If the soil water has continued access to the CO_2 from the upper biologically active zone, the CO_2 pressure remains essentially fixed, and the amount of $CaCO_3$ taken into solution is large (as shown by the vertical arrow in Fig. 8.2). If the soil water reacting with limestone at the soil base is completely isolated from a fresh CO_2 supply, for example by the

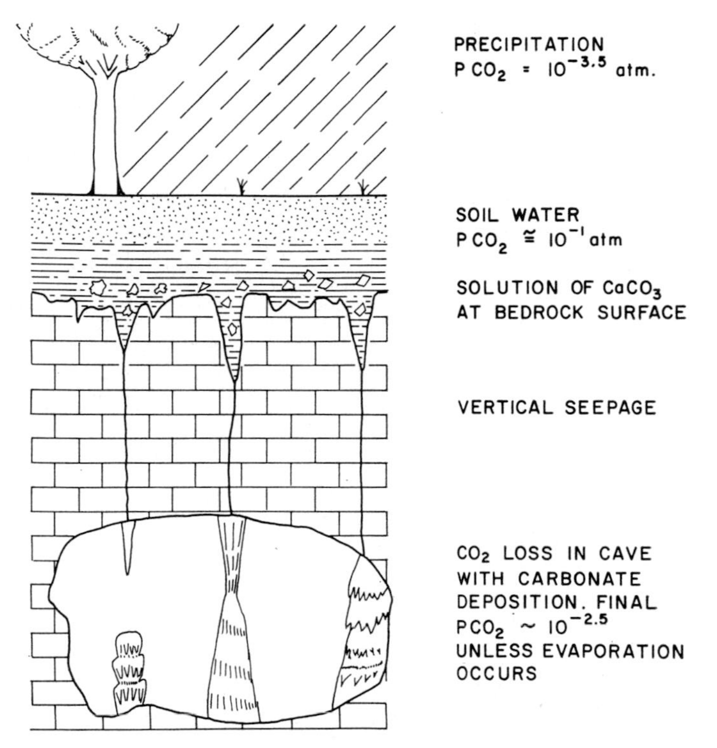

FIG. 8.3. Pictorial model for calcite deposition in caves according to the suggestions of Holland *et al.* (1964).

existence of a thick and clay-rich B-horizon, the CO_2 pressures will fall as the available dissolved CO_2 is used up in reacting with the limestone. The limiting case is indicated by the curved arrow of Fig. 8.2. Most real situations are probably somewhere in between and, as can be seen from the diagram, the amount of $CaCO_3$ in solution when the soil water begins its downward course into the limestone can vary greatly, depending on the exact details of the soil zone reactions.

If it is assumed that the small vertical joints and fissures carrying the soil water, now a vadose ground water, into the depths of the carbonate aquifer are so tight that there is no gain or loss of CO_2 during the downward course, the chemistry remains fixed during this part of the transport process. The walls of the joint are not attacked and no travertine is deposited. When the water emerges from the cave roof as a hanging drop of water, it is far out of chemical equilibrium with the new environment. The CO_2 partial pressure in the cave air is higher than that of the outside atmosphere, but actual measurements show that it is rarely more than a factor of ten higher (Ek *et al.*, 1968, 1969). If the drop hangs from the ceiling long enough to regain equilibrium, it will degas CO_2 with concurrent deposition of calcite or aragonite until the remaining dissolved CO_2 comes into equilibrium with the cave atmosphere. The quantity of calcite deposited in the stalactite is therefore given by the difference in $CaCO_3$ concentration shown in Fig. 8.2. Usually, however, the drop falls before the process is complete and the $CaCO_3$ deposited is distributed in proportion between the stalactite and the stalagmite, the ratio being determined primarily by the drip rate. The process so far assumes that liquid water is conserved. If evaporation occurs, there will be some additional deposition of calcite and if the limit of complete evaporation of the dripping water is reached, all calcite will be deposited. Figure 8.2 indicates, however, that degassing of CO_2 is a quite adequate mechanism for calcite deposition.

The looped pathway illustrated in Fig. 8.2 is effectively a $CaCO_3$ pump that carries calcite from the upper weathering zone at the top of the limestone down into the cave and redeposits it as travertine. The effectiveness of this mechanism is clearly related to the amount of water available and the amount of CO_2 in the overlying soils. Both of these parameters and also the rate factors involved are related to climate. It was observed as long ago as 1952 by Corbel that the travertine deposits in tropical caves tend to be massive and extensively developed, while travertine is sparse in northern caves. A similar situation prevails in North America. Travertine in caves of the temperate northern United States is not nearly so prolific or massive (on the average) as that of the caves of the Caribbean Islands or of northern South America. Climatic factors which can account for this difference within the context of the model are:

1. Higher CO_2 pressures in tropical soils owing to more intense biological activity.
2. Longer growing seasons in the tropics, thereby increasing the time scale of the process.
3. Thicker soils in the tropics that allow longer contact with the percolating soil water and thus a closer approach to equilibrium.

4. Higher temperatures in the tropics (although they decrease CO_2 solubility somewhat) increase reaction rates and thus promote a closer approach to equilibrium.
5. Precipitation rates are higher in the tropics, thereby making more infiltration water available for the transport process.

Quantitative evaluation of any of the above factors is sparse. Data are becoming available on the amount of CO_2 in the soil atmosphere through the work of Miotke (1972) and others who have adopted his techniques, and first results indicate that the differences in CO_2 content between tropical and northern soils is not as great as one might have expected.

The massive travertine deposits of Carlsbad Caverns and other caves of the American Southwest present an enigma within the context of the model presented above. In Texas and New Mexico, soils are thin and the climate is semi-arid. Many of these massive speleothems must be relic features and record warmer or at least wetter climates of the Pleistocene period or earlier.

Flowstone and Dripstone Speleothems

Water emerging from joints in the cave ceiling hangs in drops for a short time before the drops fall to the floor below. During the time in which the drop hangs, CO_2 is lost to the cave atmosphere, the solution becomes supersaturated, and a small amount of mineral matter is deposited in a ring with a diameter similar to that of the drop. This ring grows downward at constant diameter as more material is deposited until a slender tube of calcite known as a straw stalactite is formed (Fig. 8.4). The tube is somewhat porous, and water can seep between grain boundaries and along cleavage cracks to deposit mineral matter on the outside. Likewise, the tube may become blocked because of the introduction of foreign matter or through the growth of crystals in the orifice, thus forcing all water to percolate to the outside of the stalactite (Andrieux, 1962, 1965). Fungi may play a role in tube growth (Went, 1969). Additional seepage may flow down the outside of the stalactite and deposit mineral matter there. Thus there may be a gradual transformation of the stalactite from a straw style to a more or less conical or icicle-like shape. The resulting stalactite is hollow, or there is at least the remnant of the central canal. A section cut perpendicular to the growth axis usually shows a layered structure of alternating rings reminiscent of the growth rings on trees. This at least suggests that most of the nutrient for thicker stalactite styles comes from additional seepage down the outside, rather than seepage outward from the central canal.

The channels which feed stalactites are sometimes large enough to allow sand grains and other clastic material to be transported, and these may also be incorporated in the growing speleothem. Many of the more complicated styles of stalactites arise because of partial blockage of the channel and because of large seasonal variations in growth rate (Plates 8.1 and 8.2). More than one drip point may occur near by, and stalactites grow together to form more complicated speleothems. Their ultimate size is limited by the weight that can be supported, and stalactites that have broken and fallen under their own weight are common.

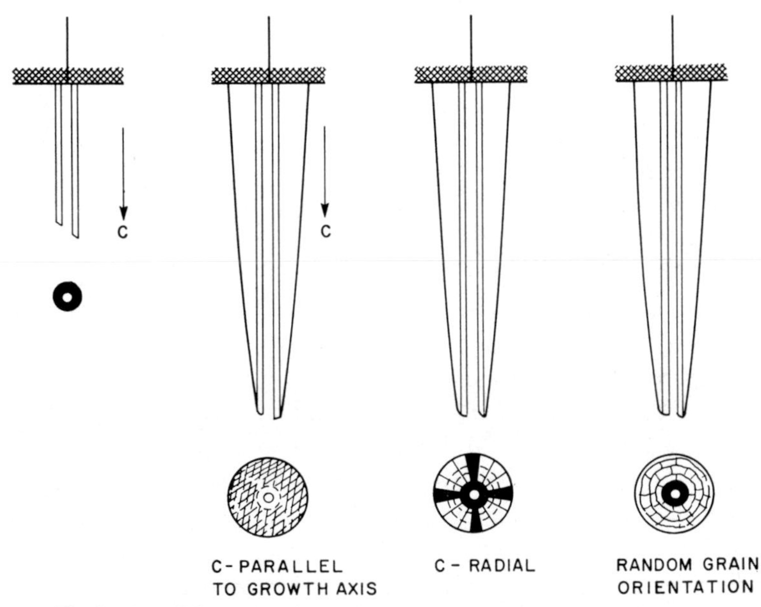

FIG. 8.4. Evolution of the straw depending on rate and constancy of deposition. Slow constant growth yields internally monocrystalline stalactite with *c*-axis of calcite grains parallel to the axis orientation of the parent straw. Still faster and/or fluctuating growth conditions lead to randomly oriented grains. The external morphology of the stalactite need not change.

Drops that fall to the cave floor continue to deposit materials, and these build up into mounds known as *stalagmites*. There is a limiting radius to the stalagmite determined by the drip rate because of exhaustion of super-saturation or by complete evaporation of the moisture film as it spreads outward from the drip point. Stalagmites, therefore, often appear with tall cylindrical shapes. Uniform diameter of a tall stalagmite can be taken as evidence that both the drip rate and the chemical composition of the solution have remained constant for a considerable period of time.

PLATE 8.1. Various styles of stalactites and stalagmites. (a) Straws, Sites Cave, West Virginia. (b) Straw stalactites and columnar stalagmites, The Colonnade, Lancaster Hole (*Photo*: T. D. Ford). (c) Stalactitic canopies and flowstone, Shenandoah Caverns, Virginia.

PLATE 8.2. Various styles of stalactites and stalagmites. (a) Straw grotto, Cnoc nan Uamh caves, Sutherland, Scotland (*Photo*: T. D. Ford). (b) Conical style, Cripps Mill Cave, Tennessee. (c) Terraced or Pagoda style, Luray Caverns, Virginia. (d) Massive cylindrical stalagmite (height about 4 m), Cumberland Caverns, Tennessee.

Stalagmites are not limited by weight considerations, and some grow to heights of many metres. It is apparent that most of the calcite deposition takes place in the stalagmite and some of the most massive have only tiny stalactites above them. Many stalagmites are fed by more than one stalactite, with the result that shapes become very complex (Plate 8.2).

The growth processes of dripstone can be quantified to produce mathematical relationships between size and shape parameters and the characteristics of the solution. Curl's (1973a) analysis of straw stalactites shows that the diameter is related to the gravitational force on the pendant drop and the surface tension of the liquid through a dimensionless Bond number by the relationship:

$$B_0 = \frac{\rho g d^2}{\sigma},$$

where B_0 is the Bond number, ρ is density of the solution, g is the acceleration due to gravity, d is the diameter of the straw, and σ is the surface tension of the solution. Experimental determination shows that $B_0 = 3{\cdot}50$ for minimum diameter straws, yielding a lower limit to the diameter of straw stalactites of 5·1 mm, in good agreement with field observations. The analysis could be done of this stalactite style because of the relatively simple physics of pendant drops. More complicated stalactite shapes have not been analysed.

Stalagmites have an internal structure consisting of cuspate layers or caps (Fig. 8.5). Those stalagmites which maintain a uniform cross-section can be described by an equilibrium diameter, d, which measures the lateral spread of solution before deposition is complete. Franke (1961, 1963, 1965) made use of the equilibrium diameter to evaluate growth rate and therefore conditions of the feeding solution. Uniform diameters indicate constant conditions for a long period of time. Terraced styles imply periodic variations in growth rate, and conical styles imply decreases in growth rate. From simple mass balance considerations the equilibrium diameter is given by

$$d = 2\sqrt{\frac{c_0 q}{\pi \dot{Z}}},$$

where c_0 is the calcite in solution available for deposition (i.e. the supersaturation if only CO_2 degassing is considered, or the total concentration of the solution if the liquid is assumed to evaporate to dryness on the stalagmite), q is the flow rate of the solution, and $\dot{Z} = \partial/\dot{Z}\partial t$ is the vertical rate of growth of the stalagmite. Curl (1973b) has shown that there are really two flow regimes for stalagmites: a rapid flow regime in which the equilibrium diameter varies with the square root of the flow rate as in Franke's theory, and a slow flow regime which is independent of the

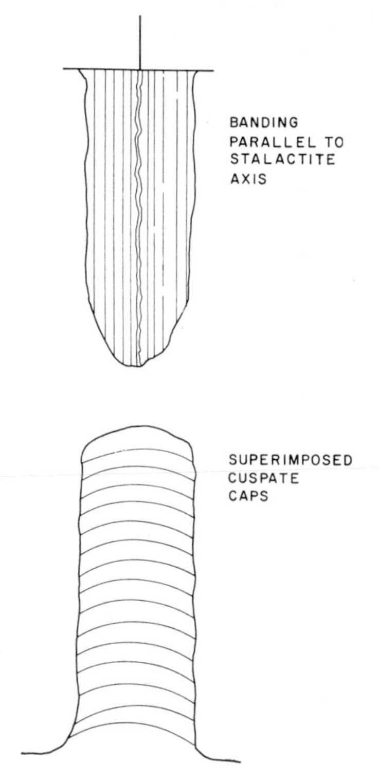

FIG. 8.5. Longitudinal sections of stalactite and stalagmite: comparison of growth patterns.

flow rate. The slow flow regime leads to a minimum diameter for stalagmites given by

$$d_{\min} = 2\sqrt{\frac{\nu}{\pi\delta}},$$

where ν is the volume of the water drop and δ is the thickness of the moisture film left on the surface of the stalagmite by the falling drop. The minimum diameter is of the order of 3 cm.

There is a discrete minimum size to both stalactites and stalagmites that arises because the water drops which feed the speleothems themselves have a discrete minimum size, for drop characteristics are determined by the gravitational field, and the surface tension of the particular liquid, water, both of which are constants in the temperature regime of limestone caves on our planet. The minimum sizes for both dripstone forms are thus essentially constant for all caves.

Flowstone is deposited where supersaturated solutions flow over cave walls. Precipitation can take place in very thin moisture films, and there is no minimum thickness. Flowstone deposits can become very massive and indeed can completely choke cave passageways. If walls are irregular with ledges, intermediate forms between flowstone cascades and dripstone forms appear. Small trickles of water run along on the under sides of ledges before dripping off, so that thin curved plates of travertine result. These may take the appearance of "bacon strips" where the growth bands are parallel to the slope of the ledge, or more stalactite-like curving folds known as draperies. Descriptions of individual dripstone and flowstone deposits include Allison (1923), Bernasconi (1967), Hicks (1950), and Nemeč and Panoš (1960). Horizontally ribbed flowstone is of frequent occurrence but as yet is unexplained; perpetuation of irregularities in the original rock surface may be a factor which causes increased degassing of CO_2 by turbulence, but other factors such as a critical limit to surface adhesion of the water film may also be important.

The individual grains which make up dripstone and flowstone deposits often show a strong preferred orientation. The growth rate in calcite is fastest along the direction of the *c*-axis (0001). Assuming that the first small crystals deposited from the drop in the proto-stalactite have a random orientation, those whose *c*-axes are vertical will grow faster than the others and will ultimately outrun them (Moore, 1962). If growth is slow and uniform, the initial "seeds" will be propagated through the entire length of the stalactite, and it will have the appearance of a calcite single crystal. Single-crystal straws are not perfect; they contain grain boundaries and in reality consist of a small number of large calcite grains, all with nearly the same orientation. These straws break along the calcite cleavage direction $(10\bar{1}1)$, so that the fracture plane makes an angle of $44° 36{\cdot}5'$ with the straw axis (Andrieux, 1962).

If growth rate remains slow, an entire stalactite can be built up as a single calcite crystal (Fig. 8.6). Since the very restricted conditions of slow and uniform growth are seldom found, such stalactites are rare. Occasionally, stalactites are found with external crystal faces (Halliday, 1959; Bassett and Bassett, 1962). These speleothems are termed internally monocrystalline and externally monocrystalline, respectively.* At

* Halliday used the term "holocrystalline" but this term has been pre-empted in petrography to mean merely "composed entirely of crystals".

somewhat higher supersaturations (and hence rates of growth) new nuclei can form on the external surface of the straw and permit growth with a different orientation. The most competitive grains in this case are those whose *c*-axes are oriented perpendicular to the axis to the straw, and thus stalactites are found with the core calcite grains oriented parallel to the

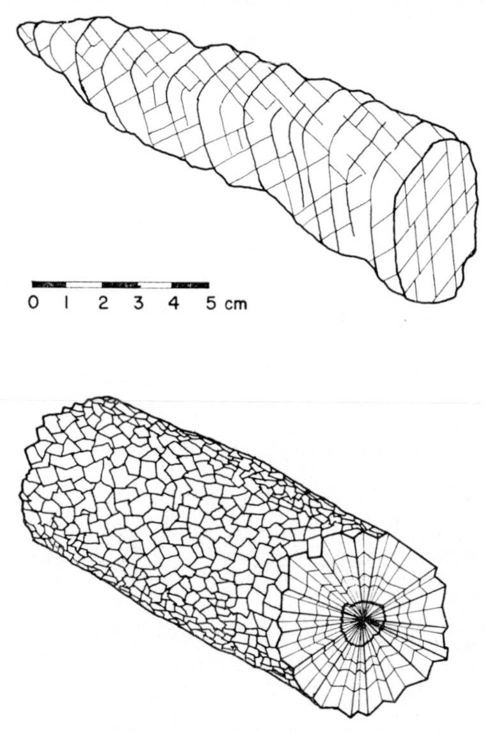

FIG. 8.6. Oriented stalactites. Top: Internally monocrystalline style with *c*-axis parallel to stalactite axis showing pattern of cleavage cracks. Bottom: Stalactite with wedge-shaped grains oriented with *c*-axes perpendicular to growth axis. After White, Jefferson and Haman (1965).

stalactite axis surrounded by an overgrowth of wedge-shaped crystals with their *c*-axes radial to the stalactite axis (Fig. 8.6).

Oriented growth takes place because the supersaturation (a measure of the excess free energy) required for propagation of a pre-existing nucleus or seed crystal by means of growth on screw dislocations or by means of two-dimensional nucleation is less than the supersaturation required to form a new stable three-dimensional nucleus. There is, therefore, a competition between the growth of existing grains and the formation of new grains. As supersaturation increases, three-dimensional nucleation becomes energetically more favourable, more grains are formed, the grain size becomes smaller, and the orientational effects disappear. At high supersaturations, aragonite may be precipitated and is found either intergrown or interlayered with calcite in some dripstone and flowstone deposits (Siegel, 1965). Aragonite, belonging to the orthorhombic system and with a different growth habit, appears in dripstone as small radiating needles. The polished cross-section has a satiny appearance, and the external surface has a plaster-like or matt surface compared to the shiny smooth surface of calcite. Small stalactites composed entirely of aragonite sometimes take on a spreading petal-like form (Fig. 8.7) termed spathites by White and Stellmack (1959).

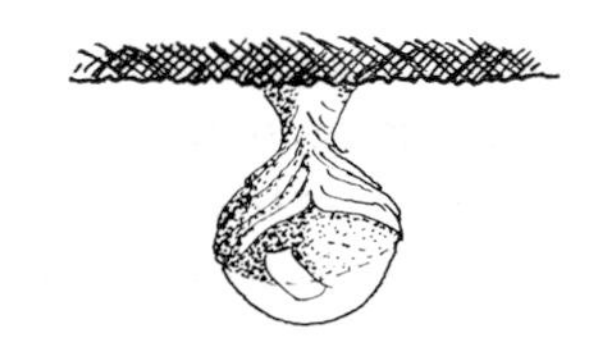

FIG. 8.7. Crystallization pattern for aragonite stalactites. These speleothems were described *spathites* by White and Stellmack (1959).

Grain size and mineral composition of dripstone and flowstone are rather delicately determined by rate of growth and supersaturation which is in turn controlled by rate of CO_2 degassing and solution evaporation. Growth is also controlled by foreign ions in the solution. Mg^{++} and Sr^{++} have been frequently assumed to inhibit calcite nucleation, and foreign solid impurities may act as centres for heterogeneous nucleation. The distribution and influence of solid impurities has been little studied except for the work of Mills (1965), who found clay, sand particles and opaline silica in addition to calcite and aragonite in many stalactites. Crystal growth is continuous across the impurity zones that give rise to colour banding in many stalactites, but higher concentrations of impurities occur along circular bands in which grain size becomes very small and growth discontinuous.

The finest development of dripstone and flowstone occurs in caves

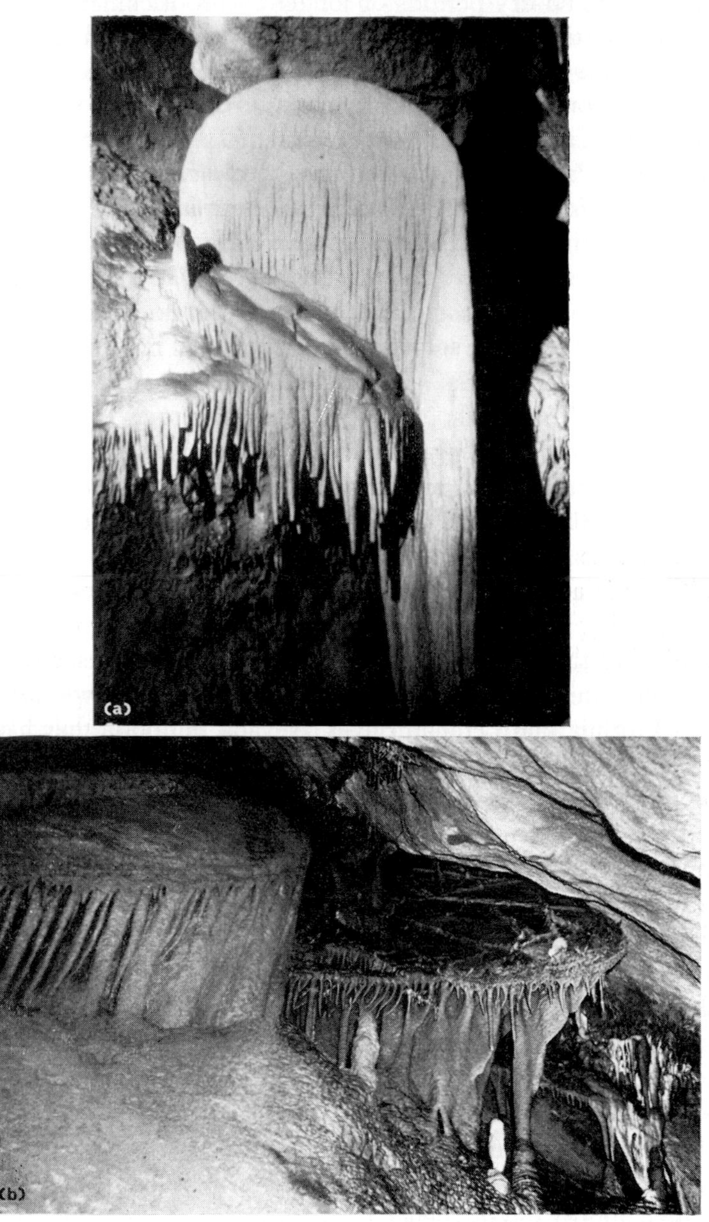

PLATE 8.3. Shields. (a) Vertical shield with stalactite canopy, Grand Caverns, Virginia. (b) Sloping shields about 2 m in diameter, Sites Cave, West Virginia.

The helictite form develops when crystal growth forces are dominant over the hydraulic forces of vertically moving water. Specifically, this implies that the flow to the helictite must be sufficiently slow to prevent

PLATE 8.4. Helictites. (a) Vermiform style of curving calcite crystals from Sites Cave, West Virginia. (b) Helictites among straw, rod and curtain stalactites, Traligill Swallet, Sutherland (*Photo*: T. D. Ford).

drops forming. Crystallization from the moisture film at the tip then follows the fast growth direction of calcite. Indeed, monocrystalline helictites are found in which the *c*-axis of the calcite follows the intricate winding path of the helictite. The grains themselves are not curved but are stacked in a twisting pattern, an observation made long ago by Prinz (1908) and confirmed by the petrographic microscope observations of Moore (1954). The origin of the spiral or twisting habit is not known, although Andrieux (1965b) proposed that impurities from overlying soils modified the growth habit. However, Huff's (1940) laboratory-grown helictites had similar curved habits when only pure sodium thiosulphate solutions were used.

The spicular or angulate style of helictite is more obviously controlled by stray clastic impurities which clog the canal, causing growth in a new direction. Likewise, some helictites are known with cores of aragonite which appear to be calcite overgrowths on anthodites.

Similar to helictites but larger are a variety of eccentric stalactites found in many caves. These have the appearance of normal stalactites except that they curve, twist or occasionally contain angular bends. Viehmann (1955) tabulated many styles of eccentric stalactites. These arise because of blockages in the feeder canal, strong air currents, and quite possibly other unknown effects.

Common in many caves are globular protuberances from the cave walls. These *botryoidal* forms range in size from tiny beads a few millimetres in diameter atop a slender stalk, to gargoyle-like masses nearly 1 m in diameter (Plate 8.5). The smaller varieties are known as cave coral, cave grapes or cave popcorn. The smaller styles have been called *globulites* and the larger ones, more obviously related to splash phenomena, *spattermites*. Globulites have a smooth exterior. In cross-section they are layered like stalactites but with the origin or centre of the layering at the point of contact of the globe with the wall. Many are seen to grow on small projections on rough limestone walls. There is little evidence for a central canal, and field observations indicate that the nutrient is derived from moisture films bathing the outside of the speleothem. Most consist of calcite, but interlaying of calcite and aragonite has been observed. Some are associated with moonmilk (Gradzinski and Unrug, 1960). Botryoidal forms may be produced by several different mechanisms. Some, the small stalky beads or globulites, are associated with vertical sheet flow of water. Some appear on etched and corroded bare limestone walls. Others appear as overgrowths on scalenohedral calcite or on aragonite crystals, and some appear to be transitional with pool deposits. The largest, the spattermite style, are clearly of splash origin. They have been found in vertical shafts where water falling from the top of the shafts splashes against the walls. The largest speleothems are near the shaft bottoms and they get smaller and gradually disappear at greater heights. The texture is also different; spattermites are porous and contain a great deal of included sand and clay. Growth banding is not nearly so much in evidence.

The term *anthodite* (from the Greek anthos, "flower") has been applied

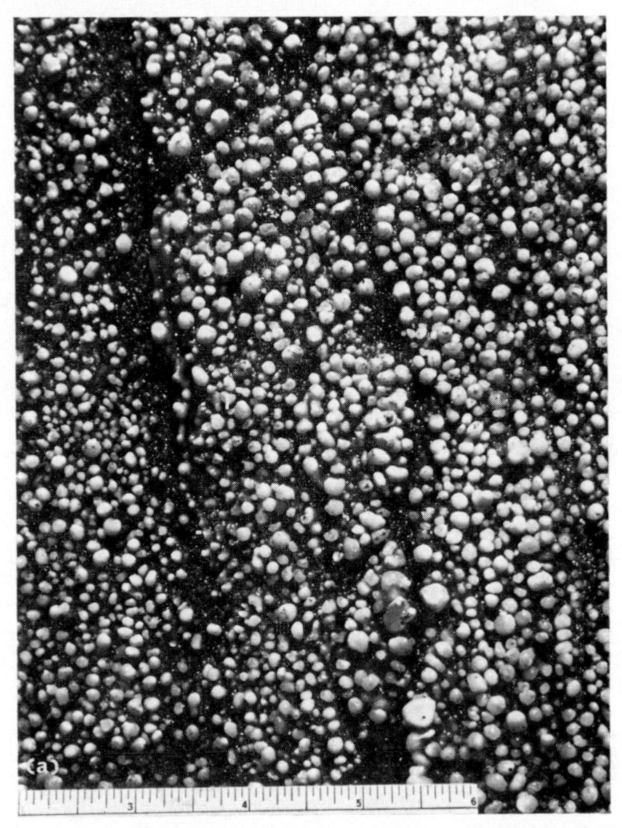

PLATE 8.5. Botryoidal and encrusting speleothems. (a) Small bead-like globulites from Harlansburg Cave, Pennsylvania.

to radiating clumps of crystalline aragonite. Typical anthodites grow in tufts of elongate acicular crystals radiating from a common centre (Plate 8.6). Dendritic growth of the individual crystals is common, resulting in a spiky appearance. The length of the crystals ranges from a few millimetres to many centimetres. A well-known locality for this type of speleothem is Skyline Caverns, Virginia, where exceptionally fine development occurs (Henderson, 1949). These speleothems occur sparsely in many caves throughout the United States such as Appalachian Caves (Murray 1951, 1954a and b), the frostwork of the Black Hills Caves

(White and Deike, 1962), in Timpanogos Cave, Utah (White and Van Gundy, 1974). Similar speleothems occur in the Moulis Cave, France

PLATE 8.5. (b) Larger nodular globulites growing on sides of stalactites, Swago Pit, West Virginia. (c) Encrustations on strands of "boot-lace" fungus, Golconda Cavern, Derbyshire (*Photo*: T. D. Ford).

(Gèze, 1957), in a number of caves in Czechoslovakia (Skřivánek, 1958; Kralik and Skřivánek, 1963) and in Cuba (Nemec *et al.*, 1967).

Anthodites grow under sub-aerial conditions presumably by accretion of new material at the base. They are often associated with helictites. The presence of aragonite suggests solutions highly supersaturated with respect to $CaCO_3$ and, as in the case of helictites, the rate of supply of the nutrient must be so slow that drops cannot form. Aragonite crystals overgrown by calcite are common, and transitional forms between anthodites and helictites occur. There is some evidence that the best

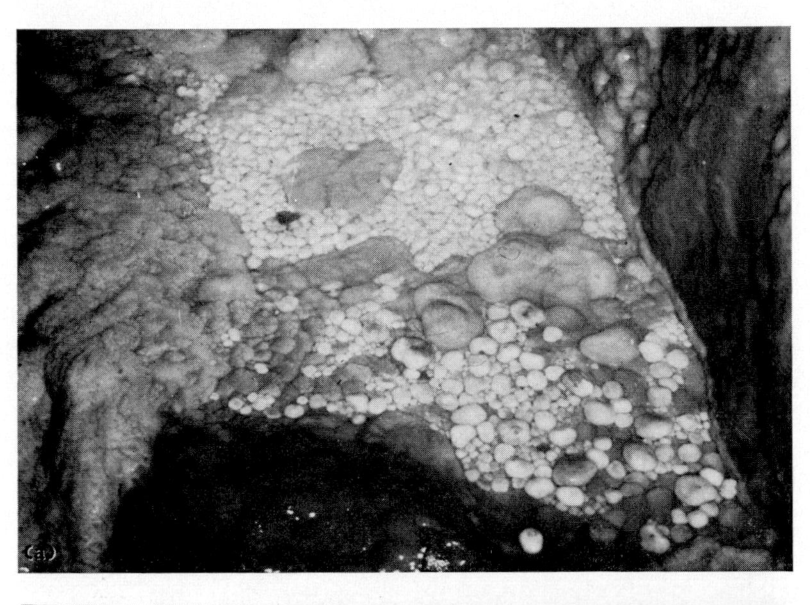

PLATE 8.6. Pisoliths and anthodites. (a) Cave pearls, up to 2 cm diameter, encrusting sand grains, Golconda Cavern, Derbyshire (*Photo*: T. D. Ford). (b) Anthoditic forms on sides of stalactites, Carlsbad Caverns, New Mexico.

anthodite growth takes place in highly humid caves, where there is sufficient water to maintain a moisture film over the entire speleothem. Loose masses of hydrated magnesium carbonates are often found on the tips of anthodites, suggesting that deposition of the final mineral matter took place there. Likewise, small calcite crystals forming on aragonite needles suggest that solutions moving along the aragonite crystals have lost supersaturation and begun to precipitate calcite. Conditions for anthodite growth might therefore be summarized as slow percolation of highly mineralized and highly supersaturated solutions into a sealed humid chamber under constant conditions.

Anthodites have much in common with oulopholites (gypsum flowers), to be described later. The chemical processes are different but the growth mechanisms are similar. It is only the differing habits of aragonite and gypsum that result in straight or dendritic crystals in the anthodites and curved crystals in the gypsum flowers. Strict classification would probably insist that these are two varieties of the same speleothem.

Concretions are roughly spherical deposits of calcium carbonate that are not connected to the cave walls or floor. They occur as several styles, of which the *cave pearls* have attracted most attention (Stone, 1932; Mackin and Coombs, 1945; Baker and Frostick, 1947, 1951; Coleman, 1949; Viehmann, 1954; Perna, 1959; Thrailkill, 1963; Gradzinski and Radomski, 1967). Cave pearls usually occur in "nests" in shallow depressions in cave floors. Typically, they range from perfect spheres to irregular equant shapes. The exterior surface of the pearls is often white and highly polished. In cross-section they are seen to be built up of concentric layers of carbonate coating a nucleus of some foreign material. Nuclei of bits of gravel, sand grains and even bat bones have been observed. They occur in places where there is intensive water dripping so that enough agitation of the pearls is maintained to prevent cementation to the floors of the depressions. Although no measurements seem to be available, there must also be a somewhat different balance between drip rate and water chemistry than occurs elsewhere, because otherwise a stalagmite would form instead. Indeed, pearls may eventually become cemented to the sides of the nests, and these may in turn be overgrown with dripstone. Kunsky (1954) has an excellent series of photographs showing the transition from splash holes in the mud to a nest of cave pearls to a small stalagmite, the root of which extends into the soil, filling the pre-existing splash hole. How much motion is required to produce pearls that are round and highly polished has been subject of some argument, but it seems clear that no great amount of motion is needed.

Other concretions form in small pools or at least where they are covered with water for long periods of time. These are also nearly spherical but with a porous texture. They consist of aggregate crystals growing radially from a common centre. The grain is much larger than the grain size of cave pearls, and the surface has usually a spongy rather than a polished appearance.

Sub-aqueous Carbonate Deposits

Water in cave pools and lakes can also become supersaturated with respect to calcite, through the slow loss of CO_2 from the pool surface, allowing calcite to deposit on the walls and floors of the pool. The smooth surfaces characteristic of sheet flow of moisture films on sub-aerial speleothems are absent. Pool deposits are typically rough and irregular with many projecting crystal faces. The calcite crystals exhibit a well-developed external morphology, of which the scalenohedron is the most common habit. Growth tends to be most rapid at the water line, because this is closest to the site of CO_2 loss. As a result, crusts of calcite tend to grow outward from the borders of pools, partially roofing them over. Thin crusts of calcite crystal nucleate and grow in free-floating form on the water surface. Piles of sunken "rafts" are found on the bottoms of some pools. Nucleation of calcite on bubbles has been observed (Warwick, 1950). Since the fast growing direction of calcite (the *c*-axis) is also the principal axis of the scalenohedron, most pool linings are composed of scalenohedral crystals with the long axis perpendicular to the walls.

Flowing water in caves has a quite remarkable property to create its own dams through the deposition of calcite. *Rimstone dams* (French: *Gours*) are usually vertical to slightly overhung on the pool side, sometimes vertical but often tapering to a flowstone slope on the overflow side. They are very sinuous in plan and often highly convoluted into complex folds. They vary in size from a few centimetres in height and width to many metres in width and several metres in height (Plate 8.7). Some form on flat floors and some on steep flowstone slopes. Rimstone dams tend to be rough and porous in texture and much less dense than flowstone, except for those dams that are part of the flowstone slopes themselves.

Rimstone dams can form in deep cave streams, those in Padirac being so large as to be a nuisance to navigation of the rubber rafts of the explorers. Rimstone dams also form on surface streams fed by supersaturated spring waters. Gurnee (1962) reports an undermined rimstone dam complex that has formed a natural bridge across the Cahabon River in Guatemala. Dams become very large in tropical caves, and flowstone slopes several hundred metres in height may be a giant's stairsteps of rimstone dams.

The mechanisms responsible for rimstone dams have not been elucidated. Warwick (1952) reviewed the occurrence of these features and summarized a number of mainly speculative hypotheses. Since then the main contribution has been Varnedoe's (1965) comment that rimstone dams

PLATE 8.7. Pool deposits. (a) Small rimstone pools (lily pads) from Poorfarm Cave, West Virginia.

seem to require a critical balance between degree of supersaturation of the water and shifts in flow regime of the water. Highly supersaturated water would deposit flowstone continuously. At lower supersaturations, something is needed to trigger the deposition. Varnedoe argued that shifts in flow regime from laminar to turbulent in the sheet flow down flowstone cascades, or shifts from subcritical to critical flow, provide additional mixing that enhances the degassing of CO_2 and thus triggers calcite deposition. This hypothesis at least accounts for the horizontal lip of the dams, because the height of the dam would be self-adjusting.

Very small dams or microgours often develop on ribbed flowstone slopes, probably by the mechanism suggested above by Varnedoe (1965).

On certain rare occasions, caves may be mineralized while still water-filled. These are the so-called *crystal caves*. Faceted crystals often completely line the cave walls and ceiling. Most spectacular in North America

Plate 8.7. (b) Rimstone pools on flowstone wall, Luray Caverns, Virginia.

PLATE 8.7. (c) Tufa encrusted stalactite columns which later submerged in a pool. The last water-level is shown by the flange on each pillar. The "Inverted Forest", Ingleborough Cavern, Yorkshire (*Photo*: T. D. Ford).

are the caves of the Black Hills of South Dakota (Tullis and Gries, 1938; Deal, 1962; White and Deike, 1962). The history of solution, deposition, re-solution, and redeposition in the Black Hills caves is quite complicated, but the end product has been to line many of the cave passages with dog-tooth or nail-head spar. In Sitting Bull Crystal Caverns, essentially a gigantic geode, the spar crystals protrude from every surface and reach lengths of 20 cm. It is difficult to achieve supersaturated solutions under phreatic conditions, and most crystal caves contain evidence of either higher than normal temperatures or of unusual chemical compositions. A number of similar calcite-lined caverns occur around Matlock in Derbyshire.

It happens not infrequently that caves previously airfilled and containing speleothems are again flooded with water. This can happen by rising water-tables associated with climatic change, with sediment-filling of nearby river valleys, or by shifts in the regimen of underground watercourses. If the reflooding solutions are unsaturated, the existing speleothems will be etched or partially redissolved. Sometimes the surface etching is only visible by microscopic examination (De Saussure, 1955; Bourguignon and Melon, 1963a and b). At other times there is a massive removal of material. Later deposition may cover the redissolved layers, forming a break in the microstratigraphy of the speleothem analogous to the unconformities in sedimentary rocks and equally useful in deciphering stages in the cave's history. Stalagmite-cemented gravels in Gaping Gill have been undercut by subsequent stream erosion, and now have stalactites pendent from their bases.

If the reflooding solutions are supersaturated, sub-aqueous deposition of calcite on the pre-existing speleothems may occur as in the Inverted Forest of Ingleborough Cave. If the speleothems at the time of flooding are clean, or if there is some initial solution to remove dirt and damage layers, new crystals may be deposited on the old without a break. These may grow into faceted crystals. One of the most spectacular examples of calcite overgrowth from later solutions is Caverns of Sonora (formerly Mayfield Cave), Sutton County, Texas (Tandy, 1962). Here an entire cavern has been filled with monocrystalline stalactites, stalagmites, helictites of unusual size and perfection as well as many unclassified forms.

Moonmilk

Moonmilk is a term applied to white formless pasty or powdery masses found in caves. Some moonmilk is wet and has a cheese-like consistency. Other occurrences are dry masses of somewhat sticky powder. Moonmilk may be composed of several carbonate minerals, including calcite, aragonite, monohydrocalcite, magnesite, hydromagnesite, nesquehonite, huntite. It has been known since the Middle Ages (Gèze, 1955; Bernasconi, 1959) and has attracted an immense amount of attention and descriptions of its occurrence, and a variety of explanations for its origin have appeared (Gèze *et al.*, 1956; Baron *et al.*, 1957; Davies and Moore, 1957; Gradzinski and Radomski, 1957; Pobequin, 1959; Bernasconi, 1961; Gèze, 1961; Melon and Bourguignon, 1962; Fischbeck and Müller, 1971; Thrailkill, 1971).

Moonmilk appears to occur frequently in alpine caves, often as semi-liquid pasty masses of calcite or aragonite. It occurs widely but in much less massive deposits in a variety of temperate or low-altitude caves, including those of the arid American Southwest. In these caves the moonmilk is more likely to be the drier, more powdery variety, and the mineralogy is more likely to consist of hydrated magnesium carbonates, although calcite has been observed. Moonmilk appears to be rare in tropical caves, although observations are very incomplete.

The wet pasty forms of moonmilk are so striking that some special explanation for their origin seems to be necessary, since calcite usually has a completely different habit in the cave environment. A bacterial origin seems probable (Williams, 1959). Moonmilk specimens have been found to contain a number of viable bacterial organisms which could be cultured. The mechanism by which the organisms break down the calcite and redeposit it as moonmilk has not been determined.

Although microbiological mechanisms cannot be excluded, they are not necessary to explain the hydrated magnesium carbonate minerals. These often occur as clumps and tufts on the ends of other speleothems and appear to be a final depositional product of the evaporating water. Magnesium is excluded from the calcite and aragonite precipitation and concentrated in the residual liquid. The final product therefore is a magnesium carbonate. Figure 8.8 shows the magnesium carbonate phases stable in the presence of liquid water at various temperatures and carbon dioxide pressures. At nominal cave temperatures near 10°C hydromagnesite is the stable phase, as observed, unless the CO_2 pressure is exceptionally high, when nesquehonite may be formed. The pentahydrate, lansfordite, has not yet been reported from a cave but would be stable at low temperatures. It is possible that the moonmilks of cold alpine caves indeed contain lansfordite but that the mineral has decomposed during transport to the laboratory. Collection of samples under refrigeration would be useful in this case. The mineral artinite has likewise not been reported from a cave and has been shown by Langmuir to be everywhere metastable. The stability of dypingite is not known, since the mineral was only recently discovered and is very similar to hydromagnesite in most of its properties.

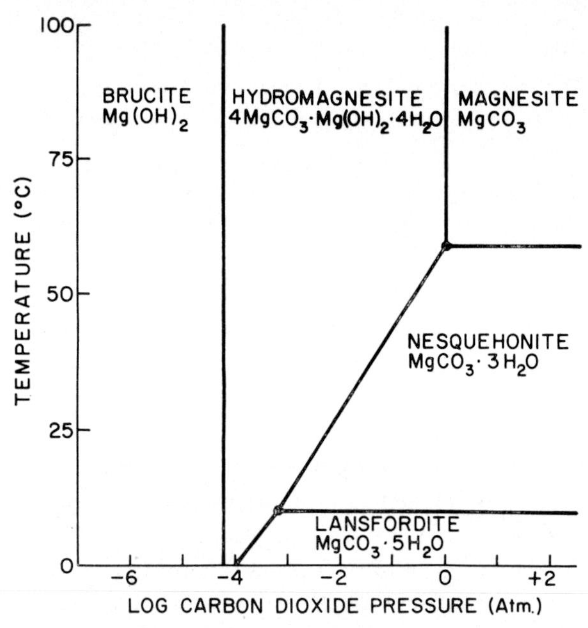

FIG. 8.8. Phase diagram showing regions of stability of the magnesium carbonate minerals as a function of temperature and CO_2 pressure. All phases assumed to be in equilibrium with liquid water. Adapted from Langmuir (1965).

THE EVAPORITE MINERALS

By "evaporite minerals" we mean the spelean analogues of ocean floor and playa lake sediments. These consist mainly of sulphate and a few halide minerals. Borates, a common constituent of playa lake sediments, do not occur in caves. Similarly the potassium ion, a common constituent in most evaporite deposits, is of such low concentration in cave waters that potash minerals do not occur.

MINERALOGY OF SULPHATE AND HALIDE MINERALS

The sulphate and halide minerals so far identified in caves are listed in Table 8.5. The cations most common in cave waters are Ca^{++}, Mg^{++} and Na^{+} in order of concentration. Sulphate salts of all of these occur as cave minerals, and some as the double salts. Gypsum is perhaps the second most common cave mineral and is found in a variety of forms and environments in many caves of the United States and Europe (see Perna and Pozzi, 1959). Gypsum is moderately soluble in water, and transport and recrystallization can take place by simple solution and redeposition.

Table 8.5

CRYSTALLOGRAPHIC DATA FOR EVAPORATE MINERALS

Mineral	Composition	System and Space Group		Unit Cell Parameters			Reference
Gypsum	$CaSO_4 \cdot 2H_2O$	Monoclinic	A2/a	$a_0 = 5{\cdot}68$ $\beta = 113°50'$	$b_0 = 15{\cdot}18$	$c_0 = 6{\cdot}29$	Deer *et al.* (1962)
Epsomite	$MgSO_4 \cdot 7H_2O$	Orthorhombic	$P2_12_12_1$	$a_0 = 11{\cdot}86$	$b_0 = 11{\cdot}99$	$c_0 = 6{\cdot}858$	XRPDF 8-467
Hexahydrite	$MgSO_4 \cdot 6H_2O$	Monoclinic	C2/c	$a_0 = 10{\cdot}06$	$b_0 = 7{\cdot}16$ $\beta = 98°34'$	$c_0 = 24{\cdot}39$	XRPDF 1-354
Mirabilite	$Na_2SO_4 \cdot 10H_2O$	Monoclinic	$P2_1/a$	$a_0 = 12{\cdot}84$	$b_0 = 10{\cdot}37$ $\beta = 107{\cdot}77°$	$c_0 = 11{\cdot}52$	XRPDF 11-647
Blödite	$MgSO_4 \cdot Na_2SO_4 \cdot 4H_2O$	Monoclinic	$P2_1/a$	$a_0 = 11{\cdot}128$	$b_0 = 8{\cdot}246$ $\beta = 100°51{\cdot}9'$	$c_0 = 5{\cdot}543$	XRPDF 19-1215
"Labile salt"	$CaSO_4 \cdot Na_2SO_4 \cdot 2H_2O$	Unknown					
Anhydrite	$CaSO_4$	Orthorhombic	Amma	$a_0 = 6{\cdot}991$	$b_0 = 6{\cdot}996$	$c_0 = 6{\cdot}238$	Deer *et al.* (1962)
Celestite	$SrSO_4$	Orthorhombic	Pnma	$a_0 = 8{\cdot}359$	$b_0 = 5{\cdot}352$	$c_0 = 6{\cdot}866$	Deer *et al.* (1962)
Barite	$BaSO_4$	Orthorhombic	Pnma	$a_0 = 8{\cdot}878$	$b_0 = 5{\cdot}450$	$c_0 = 7{\cdot}152$	Deer *et al.* (1962)
Thenardite	Na_2SO_4	Orthorhombic	Fddd	$a_0 = 5{\cdot}863$	$b_0 = 12{\cdot}304$	$c_0 = 9{\cdot}821$	Deer *et al.* (1962)
Halite	NaCl	Cubic	Fm3m	$a_0 = 5{\cdot}639$			Deer *et al.* (1962)

At temperatures of the order of 98°C, gypsum loses part of its water of crystallization to form the hemihydrate salt bassanite, $2CaSO_4 . H_2O$, the main constituent of common plaster of Paris. This reaction is reversible and bassanite easily takes up water to re-form gypsum. The anhydrous sulphate, anhydrite, is more difficult to form from gypsum, and the reaction is less readily reversible. Anhydrite, however, precipitates directly from brines where the presence of a high salt concentration has lowered the activity of water at low temperatures, and it is actually a common mineral in certain evaporites. In fresh-water solutions, gypsum is the stable solid phase, and anhydrite slowly hydrolyzes to gypsum. It has not been reported from caves. The solubilities of gypsum and anhydrite are given in Fig. 8.9.

The other sulphate minerals listed in Table 8.5 are less common. Epsomite occurs in Wyandotte Cave, Indiana, in many of the caves in South Central Kentucky, and in caves of the Guadalupe Mountains in the American Southwest. It has been reported in Italian caves (Bernasconi, 1962). At nominal cave temperatures of the order of 10°C and at high relative humidities, epsomite is the stable hydrate of magnesium sulphate. Under extremely dry conditions, one water of hydration may be lost, and the mineral hexahydrite, $MgSO_4 . 6H_2O$, will be formed. Hexahydrite occurs in Lee Cave, southern Kentucky, as a cave mineral (Freeman *et al.*, 1973) but has not been identified elsewhere. The lower hydrates of magnesium sulphate require either very low water partial pressures or higher temperatures and are not stable in the cavern environment.

Mirabilite is stable in a dry cavern environment but is very unstable in the normal atmosphere outside the cave, where it quickly decomposes to thenardite, Na_2SO_4. Although it occurs as a minor mineral in a number of caves, it is a dominant mineral only in the Flint-Mammoth System, Kentucky. Massive crystals up to a metre in length are found in some localities in the cave. It is also well known in East African lava caves where it is eaten by elephants! The double salts blödite and an unnamed "labile salt" have only been identified in the South Central Kentucky Karst. Blödite is a major constituent of the thick floor crust that covers much of a 2 km long avenue in Lee Cave (Freeman *et al.*, 1973). The labile salt occurs as a trace component in mirabilite stalactites from Turner Avenue in the Flint-Mammoth system (Benington, 1959).

The anhydrous sulphates occur mainly as authigenic constituents of cave soils and only rarely occur as speleothems. Anhydrite has not been identified as a cave mineral at all, although it may be a source material for gypsum. Celestite occurs as authigenic crystals in the soils of central

Kentucky caves (Davies and Chao, 1958) and as wall crusts in a few caves including Flint-Mammoth and Cumberland Caverns (Tennessee). Baryte occurs in the soils of the Flint-Mammoth System (Davies and Chao, 1958). Both occur in thermal spring caves of Hungary as fairly large crystals (Ozaray, 1960). Pseudo-stalactitic brown baryte has been used as an ornamental stone from a fissure-filling in Derbyshire (Ford and Sarjeant, 1964). Thenardite has been identified in sulphate deposits from

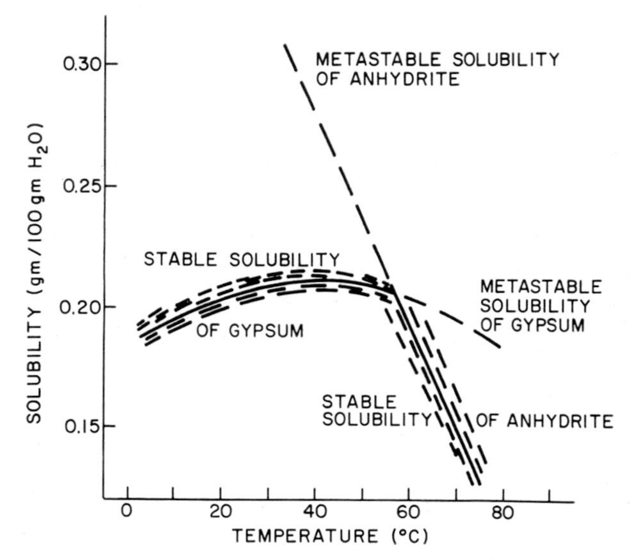

FIG. 8.9. Solubility of gypsum and anhydrite. Transition temperature accepted as 56-58°C. Diagram derived from Hardie (1967) and Blount and Dickson (1973).

a number of dry caves, but since it forms so easily by decomposition of mirabilite, its identity as a separate cave mineral remains in doubt.

Among the halide minerals, only halite, NaCl, has been identified with certainty. It is found as crusts and crystals in caves in the American Southwest and as very spectacular flower-like speleothems in the Mullamullang Cave, Australia (Dunkley and Wigley, 1967). It is most likely that other halide-bearing minerals occur in caves of the world's desert areas, but little systematic mineralogy of these areas has been published.

THE DEPOSITION OF EVAPORITE MINERALS

The transport of the evaporite minerals within the cave environment is apparently a simple matter of percolating aqueous solutions followed by deposition when the solutions evaporate. Gypsum has quite a high solubility, and the other minerals are so highly soluble, that they only occur in dry portions of caves.

The ultimate source of the sulphates is a more difficult matter. No exhaustive investigations have been made, but the following mechanisms can be demonstrated in certain specific localities:

1. Solution and transport by percolating ground water of evaporites that occur as part of the sedimentary sequence.
2. Hydration of anhydrite occurring with the limestone.
3. Oxidation of sulphides locally within the cavernous limestone.
4. Oxidation of sulphides from elsewhere in the stratigraphic section and transport into the cavern by percolating water.

The first mechanism is responsible for many of the gypsum and other sulphate speleothems in the caves of the American Southwest. Above the Capitan reef limestone in which Carlsbad Caverns and many of the caves of the Guadalupe Mountains are located is the Permian Rustler formation, primarily gypsum with some salt, which may well be the source of sulphates in the caves. It is unnecessary for transport of gypsum into the cave to be from above its roof, for movement of water under artesian conditions such as occur along the Pecos River Valley in Eastern New Mexico, is known to transport gypsum dissolved from evaporite rocks at depth.

Formation of gypsum, indeed the formation of entire caves by the hydration of anhydrite, has occurred in the small caves of the Lake Erie Islands north of Ohio (Verber and Stansbury, 1955). Small pockets in recently shattered breakdown and irregular nodules of gypsum provide evidence that anhydrite occurs in small pockets in the Greenbrier limestone of West Virginia and may be the source for the relatively minor gypsum mineralization of the caves in this area.

A variety of sulphide minerals occur in argillaceous sediments; of these pyrite and marcasite, both FeS_2, are the most common. In the presence of water and sometimes catalyzed by the sulphur- and iron-metabolizing bacteria *Thiobacillus thiooxidans* and *Thiobacillus ferroxidans*, the sulphide is oxidized to sulphate (Barnes and Clarke, 1964). Iron is relatively insoluble and is left behind as iron oxide hydrates, while the sulphate goes into solution as sulphuric acid. The reaction of the sulphuric acid with limestone is a source of gypsum in caves. Indeed, it has been claimed that the solution of limestone by sulphuric acid attack is occasionally sufficiently intense to be regarded as a cave-forming process (Morehouse, 1968).

If the sulphide minerals occur in shaly horizons within the cavernous limestone, the reactions merely provide a local source of gypsum and other sulphates to be redistributed by percolating water. Many caves of the folded Appalachians have minor gypsum mineralization that may derive from this source. The situation is more complicated if the sulphides occur elsewhere in the stratigraphic column. The source of sulphate minerals in the South Central Kentucky caves has been argued (Pohl and White, 1965) to be a pyrite-rich zone at the top of the Big Clifty sandstone overlying the cavernous Mississippian limestone. Seeping groundwater can oxidize the pyrite, producing a sulphuric acid solution which moves slowly through the permeable sandstone to the limestone below. Sulphuric acid attack on the limestone at the contact is inhibited because of the build up of carbon dioxide pressure which stops the reaction. This occurs because the reaction

$$2H_2O + H^+ + SO_4^{-2} + CaCO_3 \rightleftharpoons WCaSO_4 . 2H_2O + HCO_3^-$$

is in fact rather delicately balanced with an equilibrium constant

$$\frac{[HCO_3^-]}{[H^+]\,[SO_4^{-2}]} = 10^{6 \cdot 6}.$$

When the solutions move into the vicinity of a cave passage, however, it acts as a sink for release of CO_2, and the reaction proceeds to the right. The attack of sulphuric acid on the limestone, therefore, is in the walls of the cave passage. It results in replacement of calcite by gypsum in the limestone, of formation of gypsum speleothems on the cave walls directly without any intermediate transport and redeposition, and it results in considerable shattering and breakdown of the bedrock. The characteristic mineral-activated breakdown is perhaps the best field evidence for the occurrence of this process. The same reaction may occur in calcareous cave sediments, resulting in the growth of large selenite crystals as in Agen Allwedd, South Wales.

The list of processes is not exhaustive. Caves are known, such as Cumberland Caverns in Tennessee and other caves of the Highland Rim Province, where extensive sulphate mineralization occurs and where none of the criteria for one of the listed processes has been identified. In these caves, the most extensive gypsum deposits are in the cave soils, which suggests that some lateral transport by water has occurred. The ultimate sources have still not been identified. Similarly, gypsum apparently derived from bat guano occurs as very fine white powder consisting of flat platelets (Dietrich and Murray, 1958).

EVAPORITE SPELEOTHEMS

Partly because of higher solubilities, but more because of the different crystal habit of the sulphate minerals, evaporite speleothem forms appear that have few analogues among the carbonate minerals. Rather broadly these may be classified as either dripstone, crusts, oulopholites or monocrystalline forms. Evaporite minerals tend to form only in dry portions of caves, and liquid water is rarely in evidence, so that stalactites and stalagmites of evaporite minerals are rare. Gypsum stalactites usually appear as gypsum overgrowths or intergrowths with calcite stalactites. These speleothems have a very rough knobby appearance. Stalactites of epsomite or mirabilite are often water-clear and transparent and indeed have much the appearance of icicles.

Because the evaporite minerals form from seeping waters or from reactions with sulphides, the minerals appear uniformly on cave walls rather than only from drip points in the ceiling. Many caves of the Appalachian Plateaus and the Interior Lowlands and some in South Wales have walls coated with thin crusts of gypsum and, more rarely, other sulphates. Most of these crusts are composed of roughly equant grains of gypsum a few millimetres in diameter, giving the crust a granular appearance. The minerals are usually a translucent white colour, although occasionally the minerals are brown from included clays. There occurs more rarely a fibrous crust in which individual mineral grains grow perpendicular to the wall. Fibrous crusts break parallel to the fibre axes and have a satiny appearance.

Oulopholites are more commonly known as "gypsum flowers" in both English and French writing. They consist of branching and curving bundles of gypsum crystals. The individual gypsum crystals are not bent. They consist of long fibres, and the curvature of the gypsum flower petal comes from rotating the fibres in the bundle. The structure is open and porous with the fibres only loosely bound together. They apparently grow from the base, and the curvature results from more rapid growth at the centre of the cluster. Gypsum wall crusts often show a blister which bulges out, breaks, and develops into a flower, presumably because of a localized point of higher nutrient supply. Since the other sulphate minerals have crystal habits similar to gypsum, oulophites of epsomite and mirabilite also occur (Plate 8.8). The largest such may well be a single mirabilite "petal" in the Flint-Mammoth system that forms a semicircle a metre in

PLATE 8.8. Sulphate mineral speleothems. (a) Gypsum flowers, Cumberland Caverns, Tennessee. (b) Gypsum needles, Cumberland Caverns, Tennessee. (c) Dentate gypsum crystals growing from gypsum crust on limestone pendant, Cumberland Caverns, Tennessee. (d) Massive crystals of mirabilite pushing through loose breakdown, Flint-Mammoth System, Kentucky.

diameter. In addition to many localities in the United States, oulopholites are well developed in the Cigalère Cavern in France and have been described in many of Casteret's writings.

Most remarkable of the evaporite speleothems are the monocrystalline forms. Gypsum tends to grow most rapidly in the direction of the *c*-axis and therefore easily forms long fibres or blades with the habit of a (001) prism. Three styles can be recognized. *Gypsum needles* take the habit of straight acicular crystals ranging in length from a few centimetres to several metres, and from fractions of a millimetre to a few millimetres in diameter. They are massive crystals, water clear, and often fish-tail twins with {100} as the twin plane. Among the best displays known are those in Agen Allwedd, South Wales, and Cumberland Caverns, Tennessee. The needles here grow from loose clay floor and are loosely attached to it. There is very little "root" embedded in the floor sediment (Plate 8.8). Growth appears to be from the base with the needle gradually extruded as more material is added. Very rarely, bits of clay will be entrapped in the growing crystals and carried out with it as the crystal grows. *Fibrous gypsum* is also known as "angel hair" and occurs as tufts of long (few centimetres to a metre), very thin (fractions of a millimetre) fibres of gypsum which occur on roofs and walls of cave passages with only a loose attachment. Some are very tiny fibres closely spaced with the appearance of tufts of cotton. Others are longer fibres, very loosely packed. They occur in many gypsum-containing caves, but little is known of their crystallography or mechanism of growth. *Gypsum blades* and other more irregular gypsum crystals grow within the cave soils of many gypsum-containing caves. These range from a few to 20-30 cm in length and are a centimetre or more in width. Some are massive crystals with a single {010} cleavage; others are bundles of parallel [001] fibres. The exterior surfaces of the blades tend to be irregular, and crystal faces are not usually developed. Very complex gypsum crystals from Bolognese cave soils have been described by Casali and Forti (1969). Perhaps because of their higher solubility, monocrystalline forms of the magnesium and sodium sulphates have not been observed.

A final style of monocrystalline gypsum is rather different from the fibrous or bladed varieties. Gypsum recrystallizes easily in the presence of a small amount of liquid water into an externally monocrystalline shape. These occur as chunky masses of dentate crystals (Plate 8.8).

PHOSPHATE AND NITRATE MINERALS

Caves are habitats for many organisms whose urine, excrement and decomposing remains interact with the limestone wall rock and with inorganic cave sediments. The reaction products contain authigenic chemical species which must be regarded as cave minerals. The phosphate minerals found in caves are always associated with animal populations; the nitrate minerals are more equivocal. Crystallographic data for the better established minerals are given in Table 8.6. A few others are listed in Table 8.1.

MINERALOGY OF THE CAVE PHOSPHATE MINERALS

Since the cave phosphates are formed by reaction of phosphate-bearing organic material with the carbonate wall rock, one would expect mainly calcium and magnesium compounds by analogy with the carbonate and sulphate minerals previously discussed. Recent literature, however, is very sparse and only three localities are known to the writer to have been described in detail: the caves of Mona Island off the coast of Puerto Rico (Kaye, 1959), Pig Hole Cave, Giles County, Virginia (Murray and Dietrich, 1956), and the Domica Caves, Czechoslovakia (Kettner, 1948). The phosphate minerals occur as white, yellow or brown, loose or pasty masses. They are usually microcrystalline, and mineral identification by X-ray methods is necessary.

Calcium phosphate itself, whitlockite, occurs sparsely in the Mona Island caves as a hydrated and carbonated modification known as martinite. The dominant phosphate mineral appears to be hydroxyapatite. Hydroxyapatite makes up the bulk of the phosphate minerals in the Mona Island caves and also has been observed in other tropical caves. Many caves on Jamaica contain stalagmites that have been corroded and degraded from bat urine, and analysis of the corrosion layer shows it to be a mixture of calcite and hydroxyapatite (White and Dunn, 1962). A similar corrosion occurs in the Domica Caves. A carbonate-apatite, the mineral francolite, appears as shiny brown crusts in some Appalachian caves. The acid phosphate, monetite, and its hydrate brushite also occur commonly. The hydration-dehydration reaction takes place at 25°C in equilibrium with liquid water but at lower temperatures when the water vapour partial pressure is reduced. In the damp sediments of Pig Hole Cave, only brushite was found, but in the hotter and drier Mona Island caves both minerals occur. The magnesium phosphates struvite,

Table 8.6

CRYSTALLOGRAPHIC DATA FOR PHOSPHATE AND NITRATE MINERALS

Mineral	Composition	System and Space Group		Unit Cell Parameters	Reference
Whitlockite	$\beta Ca_3(PO_4)_2$	Trigonal	R3c	$a_0 = 10{\cdot}32$ $c_0 = 37{\cdot}0$	Dickens (1972)
Hydroxyapatite	$Ca_5(PO_4)_3(OH)$	Hexagonal	$C6_3/m$	$a = 9{\cdot}432$ $c = 6{\cdot}881$	Dickens (1972)
Monetite	$CaHPO_4$	Triclinic	$P\bar{1}$	$a = 6{\cdot}910$ $b = 6{\cdot}627$ $c = 6{\cdot}998$ $\alpha = 93{\cdot}34°$ $\beta = 103{\cdot}82°$ $\gamma = 88{\cdot}83°$	Dickens *et al.* (1972)
Brushite	$CaHPO_4 \cdot 2H_2O$	Monoclinic	I2/a	$a = 5{\cdot}812$ $b = 15{\cdot}180$ $c = 6{\cdot}239$ $\beta = 115{\cdot}42°$	Dickens (1972)
Crandallite	$CaAl_3(PO_4)_2(OH)_5H_2O$	Hexagonal		Unknown	
Taranakite	$3(K,Na,NH_4,Ca)_2O \cdot 5(Al,Fe)_2O_3 \cdot 7P_2O_5 \cdot 43H_2O$	Hexagonal	(?)	Unknown	Murray and Dietrich (1956)
Soda Nitre (Nitronatrite)	$NaNO_3$	Trigonal	$R\bar{3}c$	$a_0 = 5{\cdot}070$ $c_0 = 16{\cdot}829$	XRPDF 7-271
Nitre (Nitrokalite)	KNO_3	Orthorhombic	Pcmn	$a_0 = 5{\cdot}414$ $b_0 = 9{\cdot}114$ $c_0 = 6{\cdot}431$	XRPDF 5-377
Nitrocalcite	$Ca(NO_3)_2 \cdot 4H_2O$	Monoclinic	$P2_1/c$	Unknown	
Nitromagnesite	$Mg(NO_3)_2 \cdot 6H_2O$	Monoclinic	$P2_1/c$	$a = 6{\cdot}194$ $b = 12{\cdot}707$ $c = 6{\cdot}600$ $\beta = 92{\cdot}99°$	Mozzi and Berkebrede (1961)
Nitrammite	NH_4NO_3	Orthorhombic	Pnmm	$a_0 = 4{\cdot}942$ $b_0 = 5{\cdot}438$ $c_0 = 5{\cdot}745$	XRPDF 8-452

$NH_4MgPO_4 \cdot 6H_2O$, and newberyite, $MgHPO_4 \cdot 3H_2O$, occur in Skipton Cave, Australia (Bridge, 1971).

The solutions leached through the organic debris may attack and decompose clay minerals to form the complex calcium aluminium phosphates crandallite and taranakite. Crandallite was found in the Mona Island caves admixed with montmorillonite. Taranakite from Pig Hole Cave occurred where the bat guano was in direct contact with the clayey clastic cave sediments. Leucophosphite, $K_2(Fe,Al)_7(PO_4)_4(OH)_{11} \cdot 6H_2O$ has been found in caves in iron-rich sediments in Brazil (Simmons, 1963).

All phosphates so far investigated have been associated with thick deposits of bat guano (and possibly bird guano in some tropical caves). The phosphate is obtained from the guano through leaching by downward percolating water. The leachates react with the limestone to form first the acid phosphate minerals. As the interaction layer thickens the more soluble, acid phosphates are leached and redeposited as the less soluble hydroxyapatite. There is observed a stratification in the phosphate deposits with the acid phosphate minerals in contact with the bedrock, overlain by layers in which basic or normal phosphates are dominant, and above which lies the organic material.

MINERALOGY OF CAVE NITRATE MINERALS

The nitrate minerals that are reputed to occur in caves are listed in Table 8.6. This group is rather an enigma. Nitrate minerals occur in soils of moderately dry, temperate-climate caves such as those of the Appalachian Mountains and Interior Lowlands of the United States. They have received considerable notoriety, because the deposits were mined for saltpetre for use in gunpowder manufacture in the early years of the nineteenth century and at least as recently as the American Civil War. A water-soluble nitrate, reputedly nitrocalcite, was extracted from the cave soils in leaching vats, and the solutions so obtained passed through other vats of wood ashes to convert the material to potash nitre. The process ended with a series of recrystallization steps to purify the KNO_3 and to separate it from the other salts, primarily sodium and magnesium sulphates that were carried along in the leaching process. The activity was extensive and many caves were mined for saltpetre at one time or another (Faust, 1949, 1955, 1964, 1967).

The difficulty for the mineralogist is this: none of the nitrate minerals reported to occur in cave soils has been verified by modern methods. They occur in finely divided form greatly diluted with other clastic material and also with other water-soluble minerals, particularly sulphates.

Occasionally druses of fibrous crystals (nitre moss) are found, but these are rare.

The soils in which saltpetre is found are not associated with guano or obvious organic debris. The ultimate source of the material has been subject to considerable argument but few experiments. There is some evidence that nitrogen-fixing bacteria (*Nitrobacter*) are responsible (Hess, 1900). The environment in which nitrates are found, conditions of temperature, and soil moisture are quite restricted. Previously leached saltpetre-bearing soils are said to be renewed after a time and may be leached again. By and large, the question remains open.

OXIDE AND HYDRATE MINERALS

ICE

The crystalline forms of solid water are among the most complicated substances known: At least nine polymorphs of ice have been described. The crystal structure of ice under earth surface conditions is hexagonal (known as ice I_h). In the cave environment there is both ephemeral ice forming in the entrance areas of caves during the cold seasons of the year and perennial ice occurring in caves of regions where the mean annual temperature is below 0°C or where special airflow conditions maintain low temperatures in the cave during the warm season. Caves containing perennial ice are known as *glacières*. These occur in the mountains of Europe (Balch, 1900), as for example the Eisriesenwelt and the Riesen-eishöhle of the Austrian Alps (Kyrle, 1925) and the Grotte Casteret of the Pyrenees. Many ice caves are also known in North America (Halliday, 1954).

Ice is deposited in caves by two mechanisms: freezing of seeping or flowing water, and freezing of water vapour. Ice speleothems formed from freezing water take on forms similar to calcite speleothems. Both stalactites and stalagmites occur as do massive flowstone icefalls. However, the different mechanism of deposition of ice manifests itself in the stalactites. Ice stalactites (icicles) are formed by water films moving down the outside of the deposit; they do not usually contain a central canal. Ice stalagmites appear to be very similar in texture to the calcite stalagmites. Tall broom-stick styles are most common, although massive domed or parabolic styles occur in perennial ice caves. Speleothems formed from freezing of dripping or flowing water have smooth exterior surfaces; crystal faces are rarely in evidence. Speleothems formed from freezing water vapour have no counterparts among the other cave minerals. Spectacular masses of loosely connected crystals (frost crystals) can fill entire cave chambers. In Fossil Mountain Ice Cave, Wyoming, and in caves in the Canadian Rockies chambers are found to contain transparent hexagonal platelets of ice 10 to 30 cm across during certain seasons. Much of these caves are filled with large snow crystals of great complexity.

IRON AND MANGANESE OXIDES AND HYDRATES

The cavern environment can be characterized as wet, mildly alkaline, and oxidizing. The activity of water is near unity, pH is typically in the range of 7 to 8, and the redox potential (eH) is in the range of +0·4 to +0·6 volts. Under these conditions, the stable form of iron is a ferric hydrate and of manganese a mixed Mn^{3+}—Mn^{4+} oxide. Both the iron hydrates and the manganese oxide compounds are highly insoluble, so that these materials tend to form crusts and coatings in caves. Little definitive mineralogy has been done, but the most likely minerals are tabulated for reference in Table 8.7.

Most iron oxides are transported into caves as suspended particulate material and therefore do not qualify as chemical sediments under the definition given earlier. However, some are formed *in situ* by oxidation of iron sulphides and other iron-bearing compounds. Most appear as brown coatings of uncertain structure and carry the name of limonite. Some thicker coatings have been analysed by X-ray diffraction and shown to consist of goethite. The other oxyhydroxide polymorph, lepidocrocite, has not been identified, although its characteristic deep red colour appears on some iron oxide speleothems. The iron oxide hydrates dissociate to form hematite and water at temperatures of the order of 130°C (Schmalz, 1959) or under extremely dry conditions. Well-crystallized particles of hematite occur encapsulated in crystals of calcite in Wind Cave, South Dakota (White and Deike, 1962).

Speleothems of iron oxide hydrates are rare; the materials usually take the form of thin wall coatings. Iron hydrate stalactites have been observed in a few caves. Thick limonite layers appear in the sediments of some Czech caves (Kukla and Skrivanek, 1955).

The oxides of manganese appear as coatings on pebbles and stream bed material and more rarely as thick masses of poorly crystalline wad. Deal (1962) observed black manganese oxides in many localities in Jewel Cave, South Dakota, including one passage where the floor was covered to a depth of nearly a metre. Crabtree (1962) described masses of bog ore (50·7% MnO_2, 21·4% Fe_2O_3+25%H_2O) occurring as vertical barriers across the surface of the water in Black Reef Cave, Ribblesdale. The

Table 8.7

CRYSTALLOGRAPHIC DATA FOR OXIDE AND HYDRATE MINERALS

Mineral	Composition	System and Space Group		Unit Cell Parameters	Reference
Ice	H_2O	Hexagonal	P6mc	$a_0 = 7{\cdot}82$ $c_0 = 7{\cdot}36$	
Hematite	Fe_2O_3	Trigonal	$R\bar{3}c$	$a_0 = 5{\cdot}0345$ $c_0 = 13{\cdot}749$	Deer *et al.* (1962)
Goethite	FeOOH	Orthorhombic	Pbnm	$a_0 = 4{\cdot}59$ $b_0 = 9{\cdot}94$ $c_0 = 3{\cdot}02$	Deer *et al.* (1962)
Lepidocrocite	FeOOH	Orthorhombic	Amam	$a_0 = 3{\cdot}87$ $b_0 = 12{\cdot}53$ $c_0 = 3{\cdot}06$	Deer *et al.* (1962)
Limonite	$Fe_2O_3 \cdot nH_2O$			Amorphous or cryptocrystalline	
Birnessite	δMnO_2	Hexagonal	(?)	$a_0 = 5{\cdot}82$ $c_0 = 14{\cdot}62$	Jones and Milne (1956)
Cryptomelane	KMn_8O_{16}	Tetragonal	I4/m	$a_0 = 9{\cdot}8$ $c_0 = 2{\cdot}86$	Byström and Byström (1950)
Hollandite	$BaMn_8O_{16}$	Tetragonal	I4/m	$a_0 = 9{\cdot}8$ $c_0 = 2{\cdot}86$	Byström and Byström (1950)
Psilomelane	$(Ba,H_2O)_2Mn_5O_{10}$	Monoclinic	A2/m	$a_0 = 9{\cdot}56$ $b_0 = 2{\cdot}88$ $c_0 = 13{\cdot}85$ $\beta = 92°30'$	Wadsley (1953) Fleischer (1960)
Todorokite	$(Na,Ca,K,Mn^{2+})(Mn^{4+},Mn^{2+},Mg)_6O_{12} \cdot 3H_2O$	Monoclinic (Pseudoorthorhombic	(?)	$a_0 = 9{\cdot}75$ $b_0 = 2{\cdot}85$ $c_0 = 9{\cdot}59$ $\beta = 90°$	Straczek *et al.* (1960)
Quartz	SiO_2	Trigonal	$P3_121$	$a_0 = 4{\cdot}913$ $c_0 = 5{\cdot}405$	Deer *et al.* (1962)
Cristobalite	SiO_2	Tetragonal	$P4_12_1$	$a_0 = 4{\cdot}97$ $c_0 = 6{\cdot}92$	Deer *et al.* (1962)

black manganese coatings in most caves are very thin, fractions of a millimetre, and are concentrated in the beds of streams that flow into the cave from the surface. The probable source for the material is traces of divalent manganese in solution, leached from other rocks in the source area of the stream. The pH-eH diagram for the Mn-O-H_2O system is such (Bricker, 1965) that small increases in pH when the stream flows on to the limestone floor of the cave would be sufficient to allow oxidation of Mn^{2+} with concurrent precipitation of the higher-valent insoluble oxides. The localization in fresh surface streams may be due to a higher oxidation potential from the higher concentration of dissolved oxygen in the stream. Bacteria may also play a part.

Mineralogical identification of the manganese oxides is very incomplete. Although quite a large number of samples have been examined, most exhibit very diffuse X-ray diffraction patterns, indicating that the particle size is below the coherent scattering length of X-rays, about 100 Å. Chemical analyses of a collection of specimens from Appalachian and Interior Lowland province caves (White and Fisher, unpublished work) show calcium as the main cation in addition to manganese, implying a todorokite-type mineral. Analyses of several Jewel Cave specimens (Deal, 1962) revealed barium concentrations of several percent with low calcium, suggesting psilomelane or hollandite as the principal mineral.

As noted, manganese oxides form a black stain on some speleothems but do not form speleothems themselves.

SILICA

Silica, SiO_2, occurs in caves in two principal crystalline forms, quartz and cristobalite (Table 8.7). The solubility of SiO_2 in cold water is of the order of 10 ppm as crystalline quartz and more than 100 ppm as amorphous silica (Siever, 1962). The solubility rises rapidly with temperature reaching 80 ppm at 100°C for crystalline quartz. Similarly, the recrystallization kinetics of quartz are very sluggish at low temperatures and increase rapidly with temperature. It seems possible, therefore, that the presence euhedral quartz crystals in cave minerals is an indication of higher than normal groundwater temperatures. Low temperature silica deposition usually takes the form of poorly crystallized cristobalite or an amorphous form of silica known as opal.

Quartz occurs as large clear euhedral crystals in the Black Hills Caves (Deal, 1962; White and Deike, 1962), including spectacular red helictite-like quartz speleothems referred to by Deal (1964) as *scintillites*. Quartz also occurs in a number of other Western caves. Cristobalite is widely distributed in speleothems and has been reported by Mills (1965), from a number of Eastern U.S. localities, in Black Hills Caves (Tullis and Gries, 1938), and in speleothems from an Argentine cave (Siegel *et al.*, 1968). Silica stalactites of fine-grained chalcedony have been found in limestone caves of Platte County, Wyoming (Broughton, 1973). The source of silica was interpreted to be overlying Oligocene ash beds which, consisting of easily weathered volcanic glass, would be more soluble than most silica sources.

OTHER SILICATES

Silicate minerals in caves, mainly clays, are usually transported as clastic sediments. An exception may be the endellite, $Al_2Si_2O_5(OH)_4 \cdot 2H_2O$, reported from the soils of Carlsbad Caverns (Davies and Moore, 1957). Sepiolite (ondrejite), $Mg_3Si_4O_{10}(OH)_2 \cdot 4H_2O$, occurs as tufts on the ends of speleothems in the Zbrazov Caves of Czechoslovakia.

ACKNOWLEDGMENTS

This Chapter draws heavily on unpublished analyses of minerals in North American Caves carried out in cooperation with G. W. Moore over the past 15 years. The detailed examination of sulphate mineralization is part of the programme of the Cave Research Foundation in Mammoth Cave National Park and its environs.

REFERENCES

Allen, R. J., Martin, D. F. and Taft, W. H. (1970). A study of the effects of selected amino acids on the recrystallization of aragonite to calcite. *J. Inorg. Nucl. Chem. 32*, 2963-2970.

Allison, V. C. (1923). The growth of stalagmites and stalactites. *J. Geol. 31*, 106-125.

Andrieux, C. (1962). Étude cristallographique des édifices stalactitiques. *Bull. Soc. Franc. Mineral. Crist. 85*, 67-76.

Andrieux, C. (1965a). Étude des stalactites tubiformes monocristallines. Mécanisme de leur formation et conditionnement de leurs dimensions transversales. *Bull. Soc. Franc. Mineral. Crist. 88*, 53-58.

Andrieux, C. (1965b). Morphogenèse des helictites monocristallines. *Bull. Soc. Franc. Mineral. Crist. 88*, 163-171.

Baker, J. and Frostick, A. C. (1947). Pisoliths and öoliths from some Australian caves and mines, *J. Sed. Petrol. 17*, 39-67.

Baker, J. and Frostick, A. C. (1951). Pisoliths, öoliths and calcareous growths in limestone caves at Port Campbell, Victoria. *J. Sed. Petrol. 21*, 85-104.

Balch, E. S. (1900). *Glacières or Freezing Caverns*. Allen, Lane and Scott, Philadelphia. Reprinted, 1970, by Johnson Reprints, New York.

Barnes, I. and Clarke, F. E. (1964). Geochemistry of groundwater in mine drainage problems. *U.S. Geol. Survey Prof. Paper 473A*, A1-A6.

Baron, C., Caillere, S., Lagrange, R. and Pobequin, T. (1957). Sur la presence de huntite dans une grotte de l'Herault (la Clamouse). *Compt. Rend. Acad. Sci., Paris 245*, 92-94.

Bassett, W. A. and Bassett, A. M. (1962). Hexagonal stalactite from Rushmore Cave, South Dakota. *Nat. Speleo. Soc. Bull. 24*, 88-94.

Benington, F. (1959). Preliminary identification of crystalline phases in a transparent stalactite. *Science 12* , 1227.

Bernasconi, R. (1959). Contributio allo studio del mondmilch studio storico. *Rass. Speleol. Ital. 11*, 39-56.

Bernasconi, R. (1961). L'évolution physico-chimique du mondmilch. *Rass. Speleol. Ital., Memoir* 5 (2), 75-100.

Bernasconi, R. (1962). Nota sul giacimento di epsomite di tana di v. serrata (Ticino). *Rass. Speleol. Ital. 14*, 3-11.

Bernasconi, R. (1967). Il deposito chimico del carbonato di calcio in relazione con il fenomeno dello stillicidio. *Rass. Speleol. Ital. 19*, 1-41.

Bischoff, J. L. (1968a). Catalysis, inhibition, and the calcite aragonite problem. II. The vaterite-aragonite transformation. *Amer. J. Sci. 266*, 80-90.

Bischoff, J. L. (1968b). Kinetics of calcite nucleation: magnesium ion inhibition and ionic strength catalysis. *J. Geophys. Res. 73*, 3315-3322.

Bischoff, J. L. and Fyfe, W. S. (1968). Catalysis, inhibition and the calcite-aragonite problem. I. The aragonite-calcite transformation. *Amer. J. Sci. 266*, 65-79.

Bischoff, J. L. (1969). Temperature controls on the aragonite-calcite transformation in aqueous solution. *Amer. Mineral. 54*, 149-155.

Blount, C. W. and Dickson, F. W. (1973). Gypsum-anhydrite equilibria in the systems $CaSO_4$—H_2O and $CaCO_3$—NaCl—H_2O. *Amer. Mineral. 58*, 323-331.

Boettcher, A. L. and Wyllie, P. J. (1968). The calcite-aragonite transition measured in the system CaO—CO_2—H_2O. *J. Geol. 76*, 314-330.

Bourguignon, P. and Melon, J. (1963a). Cristallisation et corrosion de calcites flottantes en grotte. *Ann. Soc. Géol. Belgique 86*, 351-358.

Bourguignon, P. and Melon, J. (1963b). Étude cristallographique d'un plancher stalagmitique friable de la Grotte de Remouchamps. *Ann. Soc. Géol. Belgique 86*, 345-350.

Bricker, O. P. (1965). Some stability relations in the system Mn—O_2—H_2O at 25° and one atmosphere total pressure. *Amer. Mineral. 50*, 1296-1354.

Bridge, P. J. (1971). Analyses of altered struvite from Skipton, Victoria. *Miner. Mag. 38*, 381-382.

Broughton, P. L. (1972). Secondary mineralization in the cave environment. *Studies in Speleology 2* (5), 191-207.

Broughton, P. L. (1974). Silica deposits from Eastern Wyoming caves. *Nat. Speleo. Soc. Bull. 36*, 9-11.

Byström, A. and Byström, A. M. (1950). The crystal structure of hollandite, the related manganese oxide minerals, and αMnO_2. *Acta Cryst. 3*, 146-154.

Casali, R. and Forti, P. (1969). I cristalli di gesso del Bolognese. *Speleol. Emiliana*, ser. ii, *1* (7), 25-64.

Coleman, J. C. (1949). Irish cave pearls. *Proc. Univ. Bristol Speleol. Soc. 6*, 68-71.

Corbel, J. (1952). A comparison between the karst of the Mediterranean Region and of North Western Europe. *Trans. Cave Res. Grp. G.B. 2*, 1-25.

Crabtree, P. W. (1962). Bog ore from Black Reef Cave. *Cave Science 4* (32), 360-361.

Curl, R. L. (1962). The aragonite-calcite problem. *Nat. Speleo. Soc. Bull. 24*, 57-73.

Curl, R. L. (1972). Minimum diameter stalactites. *Nat. Speleol. Soc. Bull. 34*, 129-136.

Curl, R. L. (1973). Minimum diameter stalagmites. *Nat. Speleol. Soc. Bull. 35*, 1-9.

Davies, W. E. and Moore, G. W. (1957). Endellite and hydromagnesite from Carlsbad Caverns. *Nat. Speleol. Soc. Bull. 19*, 24-27.

Davies, W. E. and Chao, E. C. T. (1959). *Report on Sediments in Mammoth Cave, Kentucky*. Administrative Rpt., U.S. Geol. Surv. to Nat. Park Service.

Davis, B. L. and Adams, L. H. (1965). Kinetics of the calcite-aragonite transformation. *J. Geophys. Res. 70*, 433-411.

Deal, D. E. (1962). *Geology of Jewel Cave National Monument, Custer County, South Dakota, with special reference to cavern formation in the Black Hills*. M.S. Thesis in Geology, University of Wyoming, 183.

Deal, D. E. (1964). Scintillites: A variety of quartz speleothems. *Nat. Speleo. Soc. Bull. 26*, 29-31.

Deer, W. A., Howie, R. A. and Zussman, J. (1962). *Rock-Forming Minerals*, Vol. 5, *Non-Silicates*. John Wiley and Sons, New York.

De Saussure, R. (1955). The solution of speleothems. *Cave Studies*, No. 8.

Dickens, B. and Brown, W. E. (1970). The crystal structure of calcium carbonate hexahydrate at —120°. *Inorg. Chem. 9*, 480-486.

Dickens, B. (1972). *The crystallography of other biologically significant inorganic compounds*. In *Proc. Internatl. Symp. on Structural Properties of Hydroxypatite and Related Compounds*. W. F. Brown and R. A. Young (eds) (in press).

Dickens, B. and Bowen J. S. (1971). Refinement of the crystal structure of the aragonite phase of $CaCO_3$. *J. Res. Nat. Bur. Std.* (USA), *75A*, 27-32.

Dickens, B., Bowen, J. S. and Brown, W. E. (1972). A refinement of the crystal structure of $CaHPO_4$ (synthetic monetite). *Acta Cryst.* B28, 797-806.

Dietrich, R. V. and Murray, J. W. (1958). A peculiar type of cave gypsum. *Nat. Speleo. Soc. Bull. 20*, 25-30.

Dunkley, J. R. and Wigley, T. M. L. (1967). *Caves of the Nullarbor*. Speleological Res. Council, Sydney.

Ek, C., Delecour, F. and Weissen, F. (1968). Teneur en CO_2 de l'air de quelques grottes Belges. *Ann. Spéléol. 23*, 243-257.

Ek, C., Gilewska, S., Kaszowski, L., Kobylecki, A., Oleksynowa, K. and Oleksynowa, B. Some analyses of the CO_2 content of the air in five Polish caves. *Zeit. für Geom. 13*, 267-286.

Faust, B. (1949). The formation of saltpetre in caves. *Nat. Speleo. Soc. Bull. 11*, 17-23.

Faust, B. (1955). Saltpetre mining tools used in caves. *Nat. Speleo. Soc. Bull. 17* 8-18.

Faust, B. (1964). *Saltpetre caves and Virginia history*. In Douglas, H., *Caves of Virginia*. Privately published, 31-35.

Faust, B. (1967). Saltpetre mining in Mammoth Cave, Kentucky. *Filson Club History Quarterly*.

Fischbeck, R. and Müller, G. (1971). Monohydrocalcite, hydromagnesite, nesquehonite, dolomite, aragonite, and calcite in speleothems of the Fränkische Schweiz, Western Germany. *Contr. Mineral. Petrol. 33*, 87-92.

Fleischer, M. (1960). Studies of manganese oxides. III. Psilomelane. *Amer. Mineral. 45*, 176-187.

Ford, T. D. and Sarjeant, W. A. S. (1964). The stalactitic barytes of Arborlow, Derbyshire. *Proc. Yorkshire Geol. Soc. 34*, 371-386.

Franke, H. W. (1961). Formgesetze des Höhlensinters. *Rass. Speleol. Italiana*, Mem. 5, 185-202.

Franke, H. W. (1963). Formprinzipien des Tropfsteins. *Proc. III Internat. Congr. Speleol. 2*, 63-71.

Franke, H. W. (1965). The theory behind stalagmite shapes. *Studies in Speleo. 1*, 89-95.

Freeman, J. P., Smith, G. L., Poulson, T. L., Watson, P. J. and White, W. B. (1973). Lee Cave, Mammoth Cave National Park, Kentucky. *Nat. Speleo. Soc. Bull. 35*, 109-125.

Friedman, G. M. (1964). Early diagenesis and lithification in carbonate sediments. *J. Sed. Petrol. 34*, 777-813.

Gèze, B. (1955). A propos du montmilch ou mondmilch. *Bull. Comm. Nat. Speleo.* No. 3, 1-4.

Gèze, B. and Renault, P. (1955). Morphologie des concrétions de la Grotte de Moulis. *Bull. Soc. Franc. Mineral. Crist. 78*, 400-409.

Gèze, B., Lagrange, R. and Pobequin, T. (1956). Sur la nature du revetement occasionnel des parois ou du sol des grottes ("Montmilch"). *Compt. Rend. Acad. Sci., Paris 242*, 144-145.

Gèze, B. (1957). *Les cristallisations excentriques de la Grotte de Moulis*. Laboratoire Souterrain du CNRS.

Gèze, B. (1961). Etat actuel de la question du "mondmilch". *Spelunca*, Mem. 1, 25-30.

Gèze, B. (1965). *La Spéléologie Scientifique*. Editions du Seuil, Paris.

Goldsmith, J. R., Graf, D. L. and Joensuu, O. I. (1955). The occurrence of magnesium calcites in nature. *Geochim. Cosmochin. Acta* 7, 212-230.

Gradzinski, R. and Radomski, A. (1956). Utwory naciekowe z "mleka wapiennego" w jaskini Szczelinie Chocholowskiej (Cavern deposits of "rock milk" in the Szczelina Chocholowska Cave). *Ann. Geol. Soc. Poland 26*, 63-90.

Gradzinski, R. and Unrug, R. (1960). Uwagi o powstawaniu nacieku grzybkowego w jaskiniach (Remarks on the formation of fungoidal concretions in limestone caves). *Ann. Geol. Soc. Poland 30*, 273-287.

Gradzinski, R. and Radomski, A. (1967). Pizolity z jaskiń kubańskich (Pisoliths from Cuban Caves). *Ann. Geol. Soc. Poland 37*, 243-265.

Graf, D. L. (1961). Crystallographic tables for the rhombohedral carbonates. *Amer. Mineral. 46*, 1283-1316.

Graf, D. L. and Bradley, W. F. (1962). The crystal structure of huntite, $Mg_3Ca(CO_3)_4$. *Acta Cryst. 15*, 238-242.

Gurnee, R. H. (1962). The caves of Guatemala. *Nat. Speleo. Soc. Bull. 24*, 25-30.

Halliday, W. R. (1954). Ice Caves of the United States. *Nat. Speleo. Soc. Bull. 16*, 3-28.

Halliday, W. R. (1959). Holocrystalline speleothems. *Nat. Speleo. Soc. Bull. 21*, 15-20.

Halliday, W. R. (1962). *Caves of California*. Privately published.

Hardie, L. W. (1967). The gypsum-anhydrite equilibrium at one atmosphere pressure. *Amer. Mineral. 52*, 171-200.

Harmon, R. S. (1970). The chemical evolution of cave waters, Inner Space Cavern, Texas. *Caves and Karst 12*, 1-8.

Henderson, E. P. (1949). Some unusual formations in Skyline Caverns, Virginia. *Nat. Speleo. Soc. Bull. 11*, 31-34.

Hendy, C. H. (1971). The isotopic geochemistry of speleothems—I. The calculation of the effects of different modes of formation on the isotopic composition of speleothems and their applicability as paleoclimatic indicators. *Geochim. Cosmochim. Acta 35*, 801-824.

Hess, W. H. (1900). The origin of nitrates in cavern earths. *J. Geol. 8*, 129-134.

Hicks, F. L. (1950). Formation and mineralogy of stalactites and stalagmites. *Nat. Speleo. Soc. Bull. 12*, 63-72.

Hill, C. (1974). *Field Guide to Cave Minerals*, Nat. Speleo. Soc. Spec. Publication No. 2.

Holland, H. D. Kirsipu, T. V., Huebner, J. S. and Oxburgh, U. M. (1964). On some aspects of the chemical evolution of cave waters. *J. Geol. 72*, 36-67.

Huff, L. C. (1940). Artificial helictites and gypsum flowers. *J. Geol. 48*, 641-659.

Jackson, T. A. and Bischoff, J. L. (1971). The influence of amino acids on the kinetics of the recrystallization of aragonite to calcite. *J. Geol. 79*, 493-497.

Jagodzinski, H. (1965). Kristallstruktur und Fehlordnung des Artinits $Mg_2[CO_3(OH)_2]\cdot 3H_2O$. *Tschermaks mineral. petrograph. Mitteil. 10*, 297-330.

Jones, L. H. P. and Milne, A. A. (1956). Birnessite, a new manganese mineral from Aberdeenshire, Scotland. *Miner. Mag. 31*, 283-288.

Kaye, C. A. (1959). *Geology of Isla Mona, Puerto Rico, and notes on the age of Mona Passage*. U.S. Geol. Surv. Prof. Paper 317C, 141-178.

Kettner, R. (1948). *O netopýřím guanu a guanových korosích v jeskyni Domici* (*Bat guano and corrosion phenomena in the Domica Caves*) Zvlaštní Otisk ze Sborníku Státního Geologického Ústavu Československé Republicky *15*, 41-64.

Kinsman, D. J. J. and Holland, H. D. (1969). The co-precipitation of cations with $CaCO_3$—IV. The co-precipitation of Sr^{2+} with aragonite between 16° and 96°C. *Geochim. Cosmochim. Acta 33*, 1-17.

Kral, Z. (1971) Studie vzniku a barevnosti krapnikovych utvaru (study of the origin and colouring of dripstone). *Ceskosl. kras 23*, 7-15.

Kralik, F. and Skrivanek, F. (1963). Aragonit v Ceskoslovenskych jeskynich (aragonite in Czechoslovak caves). *Ceskosl. kras 15*, 11-35.

Kukla, J. and Skrivanek, F. (1955). Limoniticka vypln jeskyne u Strasina na Susicku (The limonitic filling of the cavern near Strasin in the District of Susice). *Vestnik UUG 30*, 113-126.

Kundert, C. J. (1952). The origin of the palettes, Lehman Caves National Monument, Baker, Nevada. *Nat. Speleo. Soc. Bull. 14*, 30-33.

Kunsky, J. (1950). *Kras a Jaskyne*. Priradovedecke Nakladatelstis Praze, Prague.

Kunsky, J. (1954). *Homes of Primeval Man*. Artia, Prague.

Kyrle, G. (1925). *Theoretische Speläologiee*. Austrian State Press, Vienna, 110-163.

Langmuir, D. (1965). Stability of carbonates in the system $MgO—CO_2—H_2O$. *J. Geol. 73*, 730-754.

Mackin, J. H. and Coombs, H. A. (1945). An occurrence of "cave pearls" in a mine in Idaho. *J. Geol. 53*, 58-65.

Marschner, H. (1969). Hydrocalcite ($CaCO_3 \cdot H_2O$) and nesquehonite ($MgCO_3 \cdot 3H_2O$) in carbonate scales. *Science, 165*, 1119-1121.

Melon, J. and Bourguignon, P. (1962). Étude du mondmilch de quelques grottes de Belgique. *Bull. Soc. Franc. Mineral. Crist. 85*, 234-241.

Merrill, G. P. (1945). On the formation of stalactites and gypsum incrustations in caves. *Nat. Speleo. Soc. Bull.* 7, 45-48.

Metzger, W. J. and Barnard, W. M. (1968). Transformation of aragonite to calcite under hydrothermal conditions. *Amer. Mineral. 53*, 295-300.

Meyer, H. J. (1965). Bildung und Morphologie des Vaterits. *Zeits. Krist. 121*, 220-242.

Mills, J. P. (1965). *Petrography of selected speleothems of carbonate caverns*. M.S. Thesis, Department of Geology, University of Kansas.

Miotke, F-D. (1972). Die Messung des CO_2-Gehaltes der Bodenluft mit dem Dräger-Gerät und die beschleunigte Kalklösung durch höhere Flieszgeschwindigkeiten. *Zeits. Geomorph. 16*, 93-102.

Moore, G. W. (1952). Speleothem—a new cave term. *Nat. Speleo. Soc. News 10* (6), 2.

Moore, G. W. (1954). The origin of helictites. *Nat. Speleo. Soc. Occasional Pap.* 1.

Moore, G. W. (1961). Dolomite speleothems. *Nat. Speleo. Soc. News 19*, 82.

Moore, G. W. (1962). The growth of stalactites. *Nat. Speleo. Soc. Bull. 24*, 95-106.

Moore, G. W. (1962). Role of earth tides in the formation of disc-shaped cave deposits. *Proc. Second Internat. Congr. Speleol. 1*, 500-506.

Morehouse, D. F. (1968). Cave development via the sulfuric acid reaction. *Nat. Speleo. Soc. Bull. 30*, 1-10.

Mozzi, R. L. and Bekebrede, W. R. (1961). Cell dimensions and space group of magnesium nitrate hexahydrate. *Acta Cryst. 14*, 1296-1297.

Murdoch, J. (1954). Unit cell of hydromagnesite. *Amer. Mineral. 39*, 24-29.

Murray, J. W. (1951). Report on the mineralogy of New River Cave. *Nat. Speleo. Soc. Bull. 13*, 50-54.

Murray, J. W. (1954a). Supplemental report on mineralogy of New River Cave. *Nat. Speleo. Soc. Bull. 16*, 77-82.

Murray, J. W. (1954b). The deposition of calcite and aragonite in caves. *J. Geol. 62*, 481-492.

Murray, J. W. and Dietrich, R. V. (1956). Brushite and taranakite from Pig Hole Cave, Giles County, Virginia. *Amer. Mineral. 41*, 616-626.

Němec, F. and Panoš, V. (1960). Stagmalitove formy jeskyni vapencoveho bradla spranku v severomoravskem krasu (Stalagmite forms of caves of the limestone block Spranek, in the northern Moravian Karst). *Acta Universitatis Palackinanae Olomucensis, Facultas Rerum Naturalium 4*, 63-96.

Nemeč, F., Panoš, V. and Štelcl, O. (1967). Duté aragonitové stalagmity z jeskyně 'La Gran Caverna de Santo Tomás' na západní Kubě (Hollow aragonite stalagmites in the cave "Gran Caverna de Santo Tomas", western Cuba.) *Ceskosl. Kras. 19*, 101-105.

Ozoray, G. (1961). The mineral filling of the thermal spring caves of Budapest. *Rass. Spel. Ital. Mem. 3*, 152-170.

Perna, G. and Pozzi, R. (1959). Osservazioni au alcuni fenomeni concrezionari della Grotta del Fiume (Ancona). *Rass. Speleol. Ital. 11*, 3-17.

Perna, G. (1959). Perle di grotta Poliedriche della Galleria Ferroviaria de Bergeggi (Savona). *Rass. Speleol. Ital. 11*, 18-20.

Pobeguin, T. (1959). Étude, au moyen des rayons infrarouges, de quelques concrétions et specimens d'argiles rencontrés dans les grottes. *Compt. Rend. Acad. Sci., Paris 248*, 2220-2222.

Pohn, E. R. and White, W. B. (1965). Sulfate minerals: their origin in the Central Kentucky Karst. *Amer. Mineral. 50*, 1461-1465.

Prinz, W. (1908). The crystalline structures of the caves of Belgium. (Translation by S. Melmore, 1949). *Cave Science 2* (9), 1-14; (10), 63-78; (11), 102-117; (13), 194-205.

Roques, H. (1964). Contribution à l'étude statique et cinétique des systèmes gaz carbonique-eau-carbonate. *Ann. Speleol. 19*, 255-484.

Roques, H. (1968). Chimie des carbonates et hydrogéologie karstique. *Mém. et Documents 4*, 113-141.

Roques, H. (1969). A review of present-day problems in the physical chemistry of carbonates in solution. *Trans. Cave Res. Grp. G.B. 11*, 139-163.

Raade, G. (1970). Dypingite, a new hydrous basic carbonate of magnesium from Norway. *Amer. Mineral. 55*, 1457-1465.

Schmalz, R. F. (1959). A note on the system Fe_2O_3—H_2O. *J. Geophys. Res. 64*, 575-579.

Siegel, F. R. (1965). Aspects of calcium carbonate deposition in Great Onyx Cave, Kentucky. *Sedimentology 4*, 285-299.

Siegel, F. R. and Reams, M. W. (1966). Temperature effect of precipitation of calcium carbonate from calcium bicarbonate solutions and its application to cavern environments. *Sedimentology 7*, 241-248.

Siegel, F. R., Mills, J. P. and Pierce, J. W. (1968). Aspectos petrograficos y geoquimicos de espeleotemas de opalo y calcita de la Cueva de la Bruja, Mendoza, Republica Argentina. *Revista Assoc. Geol. Argentina 23*, 5-19.

Siever, R. (1962). Silica solubility, 0-200°C, and the diagenesis of siliceous sediments. *J. Geol. 70*, 127-150.

Simmons, G. C. (1963). Canga Caves in the Quadrilatero Ferrifero, Minas Gerais, Brazil. *Nat. Speleo. Soc. Bull. 25*, 66-72.

Simmons, G. and Bell, P. (1963). Calcite-aragonite equilibrium. *Science 139*, 1197-1198.

Skřivánek, F. (1958). Výskyt aragonitu v česckslovenských jeskyních (the presence of aragonite in Czechoslovak caves). *Ochrana přírody 13*, 177-182.

Skřivánek, F. and Hodac, V. (1958). Stroncium v aragonitech a vapencich ceskeho krasu. (Strontium in aragonite and limestone in the Bohemian Karst.) *Ceskosl. kras 11*, 177-180.

Stephan, G. W. and MacGillavry, C. H. (1972). The crystal structure of nesquehonite, $MgCO_3 \cdot 33H_2O$. *Acta Cryst. 28*,, 1031-1033.

Stone, R. W. (1932). Cave Concretions. *Proc. Pennsylvania Acad. Sci. 6*, 1-4.

Straczck, J. D., Horen, A., Ross, M. and Warshaw, C. M. (1960). Studies of manganese oxides. IV. Todorokite. *Amer. Mineral. 45*, 1174-1184.

Tandy, M. (1962). Mayfield Cave, Texas. *Nat. Speleo. Soc. Bull. 24*, 31-39.

Thompson, P. (1973). *Speleochronology and late Pleistocene climates inferred from O, C, H, U and Th isotopic abundances in speleothems.* Ph.D. Dissertation, McMaster University.

Thrailkill, J. (1963). Moonmilk, cave pearls, and pool accretions from Fulford Cave, Colorado. *Nat. Speleo. Soc. Bull. 25*, 88-90.

Thrailkill, J. (1968). Dolomite cave deposits from Carlsbad Caverns. *J. Sed. Pet. 38*, 141-145.

Thrailkill, J. (1971). Carbonate deposition in Carlsbad Caverns. *J. Geol. 79*, 683-695.

Trombe, F. (1952). *Traité de Spéléologie*. Payot, Paris.

Tullis, E. L. and Gries, J. P. (1938). Black Hills Caves. *Black Hills Engineer 24*, 233-271.

Verber, J. L. and Stansbury, D. H. (1955). Caves in Lake Erie Islands. *Ohio J. Sci. 53*, 358-362.

Viehmann, I. (1954). Contributiuni la cunoasterea formatiunilor stalagmitice din pesteri (Contribution to the understanding of stalagmite formations in caves). *Dari de Seana ale Sedintelor Comitetului Geologic 42*, 579-610.

Viehmann, I. (1959). Příspěvsky k vývoji jeskynních perel (A new process for the origin of cave pearls). *Ceskosl. kras. 12*, 177-185.

Varnedoe, W. W., Jr. (1965). A hypothesis for the formation of rimstone dams and gours. *Nat. Speleo. Soc. Bull. 27*, 151-152.

Wadsley, A. D. (1953). The crystal structure of psilomelane $(Ba,H_2O)_2Mn_5O_{12}$. *Acta Cryst. 6*, 433-438.

Warwick, G. T. (1950). Calcite bubbles—a new cave formation. *Nat. Speleo. Soc. Bull. 12*, 38-42.

Warwick, G. T. (1952). Rimstone pools and associated phenomena. *Trans. Cave Res. Grp. G.B. 2*, 153-165.

Went, F. W. (1969). Fungi associated with stalactite growth. *Science, 166*, 385-386.

White, W. B. and Stellmack, J. A. (1959). Introductory report on the mineralogy of Carroll Cave. *Missouri Speleol. 1* (3), 3-8.

White, W. B. and Deike, G. H. (1962). Secondary mineralization in Wind Cave, South Dakota. *Nat. Speleo. Soc. Bull. 24*, 74-87.

White, W. B. and Dunn, J. R. (1962). Notes on the caves of Jamaica. *Nat. Speleo. Soc. Bull. 24*, 9-24.

White, W. B., Haman, J. F. and Jefferson, G. L. (1963). Note on the mineralogy of Cueva del Guacharo. *Bol. Soc. Venezolana Cien. Nat. 25*, 155-162.

White, W. B. and Van Gundy, J. J. (1974). Reconnaissance of Timpanogos Cave, Wasatch County, Utah. *Nat. Speleo. Soc. Bull. 36*, 5-17.

Williams, A. M. (1959). The formation and deposition of moonmilk. *Trans. Cave Res. Grp. G.B. 5*, 133-138.

9. The Physics of Caves

T. M. L. Wigley and *M. C. Brown*

INTRODUCTION

There are many speleological questions which can be reduced to problems in physics. For example, how do water temperatures vary in karst aquifers and what determines the temperature of spring waters? How do micro-erosion features (scallops, flutes, etc.) form on cave walls and what determines their size; what determines the size and growth rate of stalactites and stalagmites; what controls the temperature and humidity distributions in caves? How can caves be located from the surface? These are all essentially problems of physics. The scientific discipline "physics" covers such a wide range of topics that there are few aspects of speleology which do not require some knowledge of the laws of physics for their proper understanding. The purpose of this chapter is to illustrate some of the many applications of physics in speleology.

Unfortunately, it is not possible to discuss all things which might strictly be called cave physics. To narrow the field, we will fall back on convention and discuss in detail two topics which are not traditionally within the realm of other sciences such as geomorphology, geology and chemistry. These topics are *cave meteorology* and *geophysical detection* of caves. Of necessity, the discussion will be mainly descriptive since some of the mathematics involved is beyond the scope of this book. It is not meant to be definitive, nor to provide a complete guide to the literature in these topics. Rather it is an attempt to give a logical and organized presentation, outlining the physical principles and describing how they act, thus providing a basis for both understanding and further work.

CAVE METEOROLOGY

Cave meteorology is the study of the distribution and variation of the atmospheric variables, temperature, moisture content (humidity) and wind (or, more precisely, air movement), within caves. Here, the term meteorology is used in its broader sense, embracing climatology. Strictly speaking, the term applies to the dynamic and rapidly changing aspects of the atmospheric variables, while climatology is concerned with their average values and long-term changes in these averages. Inside caves, where the variables remain more nearly constant than in the outside atmosphere, this distinction is less valuable. However, as will be seen below, the cave is a dynamic environment, and temperature, humidity and wind show considerable spatial and temporal variability. Air movement will be discussed first, since this is the main communication mechanism between the cave environment and the continually changing outside world and so is largely responsible for any changes in conditions within the cave.

AIR MOVEMENT

Air movements are detectable in nearly all caves. In some instances movement is slow and may only be noticeable in restrictions where the cross-section of a passage narrows and conservation of mass ensures an increase in speed, while in other instances air movement may be quite rapid and observable as a "wind". In discussing cave winds it is common to divide caves into two types, single- and multiple-entrance caves. This division was popularized by Geiger (1966), who used the terms static cave and dynamic cave, terms which were introduced many years ago. We do not advocate the use of this classification, partly because it says nothing about the cause of any air movement, and partly because most single-entrance caves, which Geiger would classify as static, contain quite active circulations of air.

The single- or multiple-entrance distinction is valuable because it separates caves in which there is a thermally-induced air current produced by the so-called *chimney effect* from those in which there is not. The chimney effect is the most important type of cave air movement. Cave winds may also be caused by changes in barometric pressure outside a cave, entrainment of air by flowing water, gravitational drainage of air in single-entrance caves, resonance of air in large chambers, the influence of external weather patterns, or by changes in the volume of air in a cave when it is being flooded.

The chimney effect is in response to temperature differences inside and outside a multiple-entrance cave. Such temperature differences produce a pressure difference at the lower entrance (or entrances). This is illustrated in Fig. 9.1. At the lower entrance (L) the external pressure is

$$p_{ext} = p_0 + \bar{\rho}_{ext} gh,$$

where p_0 is the pressure at the level of the upper entrance (U), g is the gravitational acceleration, h is the elevation difference between entrances, and $\bar{\rho}_{ext}$ is the mean density of the air column of height h above L. The pressure exerted by the cave air at the lower entrance is

$$p_{int} = p_0 + \bar{\rho}_{int} gh,$$

where $\bar{\rho}_{int}$ is the mean air density over the vertical column within the cave. Note that density is averaged over a vertical column, so that

$$\bar{\rho} = \frac{1}{h}\int_0^h \rho dz,$$

where z is the vertical coordinate measured from L. This means that differences in density along the horizontal passage in Fig. 9.1 are not important. These results, and some given below, follow from the hydrostatic equation used in meteorology (see, for example, Haltiner and Martin, 1959).

If $\bar{\rho}_{ext}$ is not equal to $\bar{\rho}_{int}$, then there will be a pressure imbalance at L and air will flow either into or out of the cave in an attempt to restore the balance. Since warm air is less dense than cold air, movement will be *into* the lower entrance if the cave air is warmer than the air outside the cave, and *out of* the lower entrance if the cave air is colder than the external air. To be precise, because air density depends on moisture content as well as temperature, "colder" and "warmer" should relate to differences in *virtual* temperature. The virtual temperature is the temperature that a sample of perfectly dry air must have in order to have the same density as a given sample of moist air. Virtual temperature is therefore a function of temperature and moisture content. The difference between virtual temperature and temperature is small. The cave air is, nevertheless, usually much more moist than air outside the cave so that this difference may be significant in determining the direction of chimney effect air flow.

The speed of air movement is determined by the magnitude of the pressure difference at L (i.e. by the size of $(\bar{\rho}_{ext} - \bar{\rho}_{int})\, gh$), a driving force in one direction or the other, and the resistance to flow within the cave. The resistance increases as flow rate increases and so, by a type of negative feedback, keeps chimney effect winds from becoming embarrassingly strong. Even so in cave systems where h is large (the pressure difference is proportional to h) quite strong winds do occur, especially at restrictions in passages. For instance, in the Dachstein Rieseneishöhle, in Austria, winds of more than 5 m/sec are quite common (von Saar, 1956).

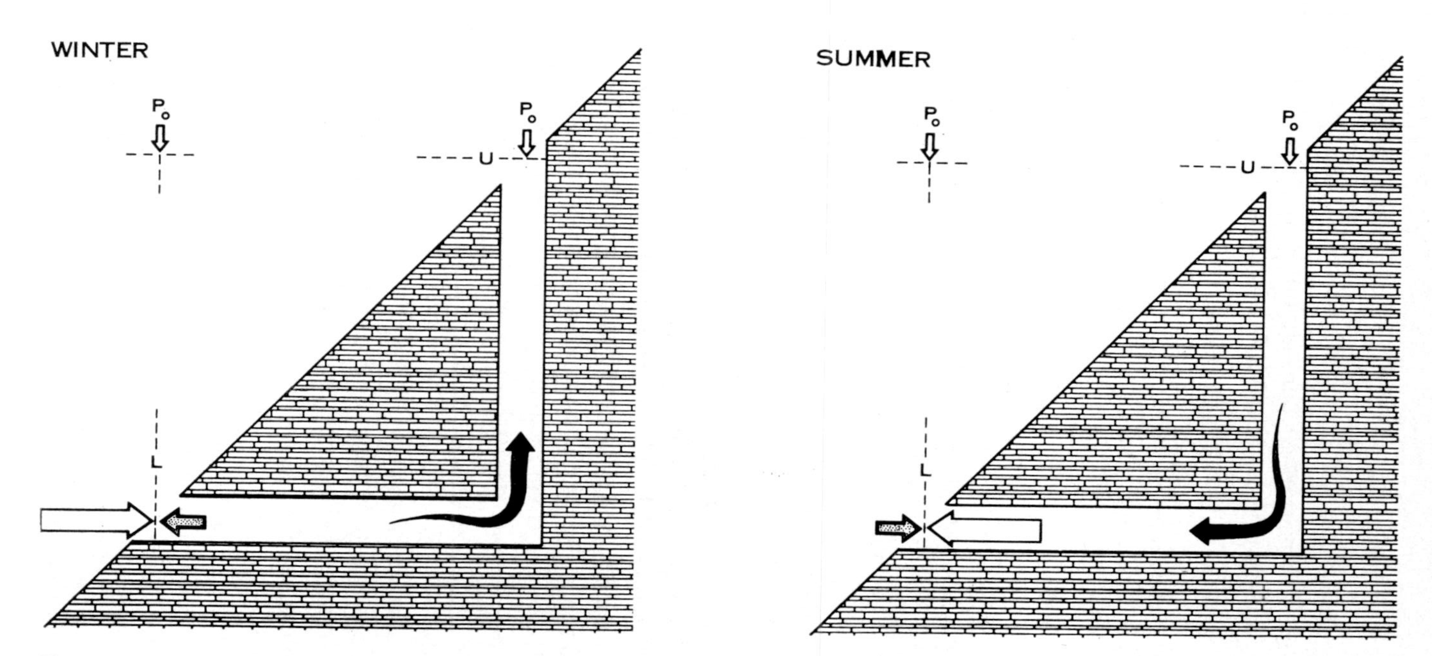

FIG. 9.1. Cave winds caused by the "chimney effect". In winter the cold outside air column above L creates a pressure (left-hand arrow) which is greater than that due to the air column inside the cave (smaller right-hand arrow at L). The pressure imbalance causes air to move within the cave from L to U. In summer the pressure imbalance is in the opposite direction and air moves from U to L.

Winds produced in this way necessarily change direction seasonally, blowing into lower entrances during the winter and out of lower entrances in the summer. During these seasons the air movement is persistently in the one direction. In autumn and spring, when $\bar{\rho}_{ext}$ and $\bar{\rho}_{int}$ have close to the same value, diurnal temperature variations outside the cave may produce daily direction changes. We have observed this phenomenon in Windsor Great Cave, Jamaica. The persistence of direction in winter or summer is often (though not always) a good guide to the existence and relative positions of other entrances in caves with only one known entrance. However, strong persistent winds may be caused by other effects (for example, entrainment by flowing water) and, even when the chimney effect is responsible, the undiscovered entrance may be smaller than man-size. This is the case, for example, in the Eisriesenwelt, Austria, where the upper "entrance" is thought to be a complex of cracks and fissures which permit movement of air but not people.

There is a need for further studies on the chimney effect; no one has presented an adequate theory to predict the strength of these thermally-induced cave winds from temperature and cave geometry data. Such a theoretical development would not be difficult. Representative resistence values could be obtained from mine ventilation texts (for example, McElroy, 1966). However, a well-planned and lengthy observational programme would be required either to support or to reject such a theory and the site would have to be chosen to minimize other wind-producing mechanisms.

External pressure fluctuations may be responsible for air movement in caves. Whenever the pressure is changed outside the entrance to a closed space, air will flow into or out of the space to maintain a pressure balance. Since the atmospheric pressure is continually changing, cave air is always responding to these changes, but the speed of air movement so produced is generally small. The speed is larger for caves with larger volume and/or smaller entrances, and for more rapid external pressure changes. Normally the cave volume must be extremely large to give significant air movement by this mechanism: for a cave with volume 3×10^5 m^3 a rapid drop in external pressure at the rate of, say, 5 mb per hour would induce an outward airflow at the entrance of less than 0·5 m^3 per second. Nevertheless, in some cases cave volume may be much larger than the part which can be penetrated by man. In the Nullarbor Plain in southern Australia many of the caves are developed in extremely porous limestones and the volume which can respond to external pressure changes is much larger than the explorable cave volume. The changes in direction of "breathing" of these caves lags behind the changes in the rate of change of pressure and the caves can breath out (or in) for some hours after the pressure changes from falling to rising (or rising to falling). Strong winds occur in a number of caves in the Nullarbor Plain, and speeds in excess of 11 m/sec have been reported from Mullamullang Cave (Hill, 1966). Further details and a theoretical explanation of this phenomenon have been given by Wigley (1967). It has also been observed in the United States in Wupatki National Monument, Arizona (Sartor and Lamar, 1962), and in Jewel Cave, South Dakota (Conn, 1966).

Usually only weak air flow occurs in caves through changing pressure. "Cave breathing" (above) is an exception. As another exception, strong air movements in single-entrance caves occur where cave streams pull or drag an air current along with them (the word "entrain", though not strictly correct, is often used in this context). This *entrainment effect* is illustrated in Fig. 9.2. Winds of a few metres per second or more may be caused in this way. It is always necessary that there be a return air flow. This may be near the passage roof and/or through other passages, as shown in Fig. 9.2, or possibly through smaller impenetrable passages. Stream-induced winds with no return flow may occur when a cave is flooding. Rapidly rising water may displace air sufficiently fast to produce a noticeable outward flowing air current (Myers, 1962). For example, a 3 m^3/sec flood in a system which can only take 0·3 m^3/sec will produce a flow of displaced air at 2·7 m^3/sec and an appreciable air current where passage cross-section is small.

In caves of small volume and with a single entrance which lack vadose streams, air flow may be caused by gravitational drainage. As with the entrainment mechanism, a return flow is also a necessary feature in this case. In a simple downward-trending cave (such as that shown in Fig. 9.2, but without the stream) cold external air can drain into the cave in winter and produce a slow flow of air at and close to floor level. This is accompanied by a higher-level return flow of displaced warmer air. Roughly speaking, such a flow may be initiated whenever the external air temperature falls below the near-entrance cave air temperature, although usually a few degrees overshoot is necessary to produce a noticeable air movement, especially when the cave floor slopes down only gently. Flow rates produced by this mechanism are invariably low, generally a few to a few tens of centimetres per second, and it is often masked by other effects. The reverse phenomenon may occur in upward-trending caves with cold air draining out of the cave at floor level in summer.

Cave winds are occasionally observed to change direction regularly

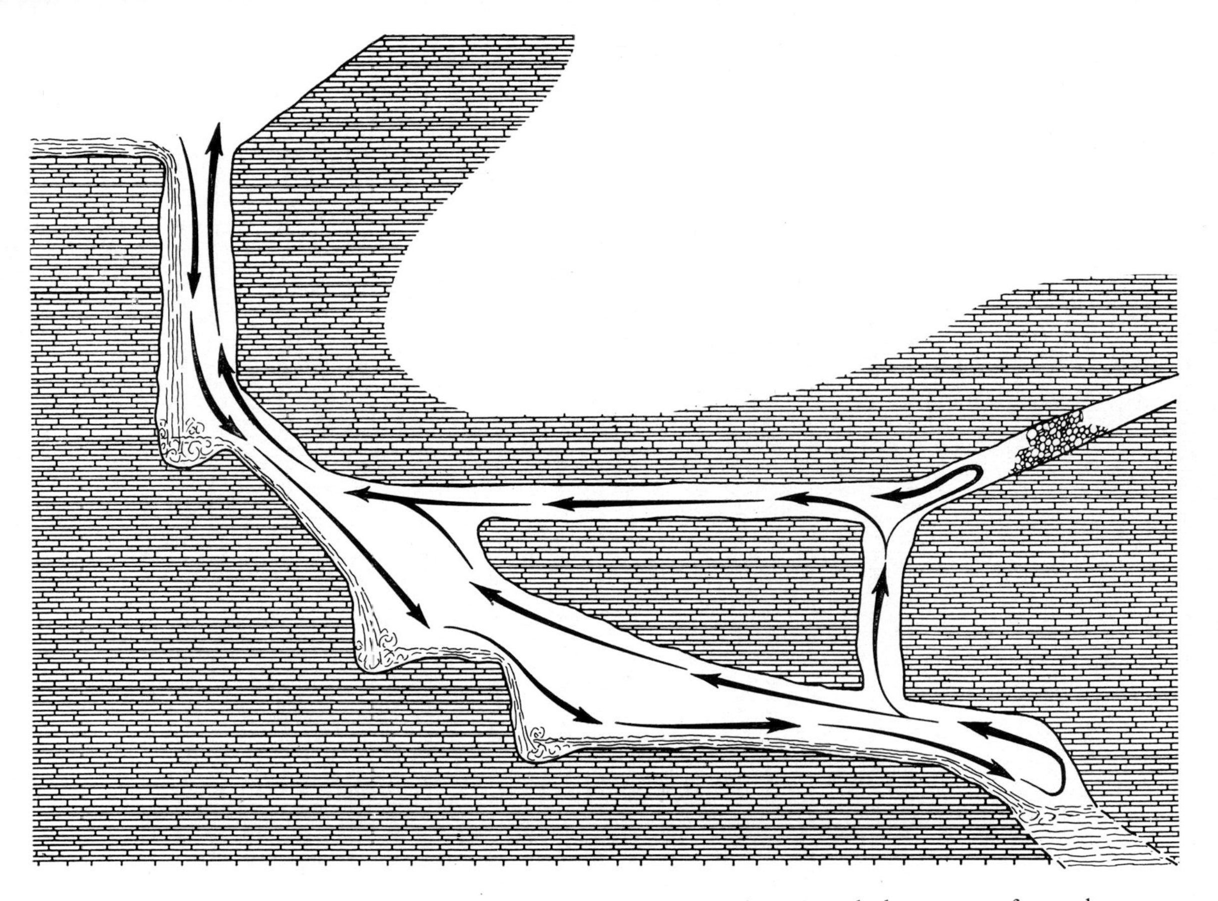

FIG. 9.2. Air flow induced by flowing water. Air is dragged along by friction with the water surface and causes a low-level flow of air. A high-level return current is necessary for mass balance: this may be through branch passages.

with a period which may range from a few seconds to a few minutes. In the United States this is called "breathing", a term which we have associated with longer period pressure-change induced air movements above. Since both forms of "breathing" involve periodic, but not seasonal, changes in direction of air flow there is no real conflict here. Short-period breathing was first studied by Faust (1947) in Breathing Cave, Virginia. He, and others following him, explained these oscillations as being a resonance phenomenon, similar to that produced when one blows over the neck of a bottle. With a bottle the resonant frequency is high and the oscillations are observed as sound waves; in a cave, with a much larger resonating volume the resonant frequency is much lower and is observed as periodic reversals in the direction cf air flow at the mouth of the resonating chamber. This form of air movement may be quite common (it has been observed in the United States, Canada and Australia), but it is often masked by other effects and may, in some cases, only be detectable using spectral analysis and/or quite sensitive instrumentation. A comprehensive account and explanation of these resonance winds has been published by Plummer (1969).

Air movements may occur in caves in direct response to external atmospheric conditions. For instance, if a cave entrance is properly exposed, external winds may funnel into the entrance and in some cases produce a strong and persistent air current right through a multiple-entrance cave. Winds caused solely by this effect and exceeding 5 m/sec have been observed in Cleft Cave in the Canadian Rocky Mountains (D. C. Ford, personal communication).

It has also been speculated that, when external surface pressure gradients are high, two widely separated cave entrances may experience quite different pressures which could set up a compensating air current within the cave. This might occur if an intense low pressure system, or a strong thunderstorm cell, were moving in the vicinity of the cave.

In any particular cave a number of the air movement mechanisms discussed above may act simultaneously. In many instances, however, it is possible to isolate one mechanism as being the major cause of air movement. In most caves of even moderate dimensions there is invariably some air movement. It may be detectable only by the motion of a condensed breath, by the flickering of an acetylene flame, or by some other more sophisticated method (see, for example, Went, 1970, and Halbert and Michie, 1971); or it may be so strong that an exposed flame is extinguished or that small pebbles are rolled along a cave floor. No matter what its magnitude, air movement is the most important factor in cave meteorology because it is the mechanism responsible for exchange of air between the cave and the external environment and so for bringing external conditions, and their changes, into the cave.

TEMPERATURE AND HUMIDITY

The fundamental measure of humidity is called *specific humidity*. This is a direct measure of the concentration (and hence amount) of water vapour in the air. The ratio of this amount to the amount needed to saturate the air, expressed as a percentage, is the *relative humidity* (strictly speaking, relative humidity is defined as a ratio of vapour pressures, but the difference is negligible). The word humidity may apply to either of these terms.

The cave environment is commonly conceived as one in which temperature and humidity are effectively constant. In reality, caves can often experience appreciable changes in these variables, though not so large nor so rapid as the changes which occur outside the cave. The greater is the air exchange between the cave and the outside, the greater will be the temperature and humidity changes within the cave. Cave temperature and humidity variations are thus inexorably linked to cave winds. The other important determining factors are external conditions, passage configurations and distance from the cave entrance (or entrances), the heat transfer processes of conduction through passage walls and liberation or absorption of latent heat by condensation (or sublimation) and evaporation.

All of these factors are considered in a theory developed by Wigley and Brown (1971) to calculate temperature and humidity distributions in caves. A basic assumption in their theory is that cave air temperatures can change much more rapidly than cave wall temperatures. They then use established principles of heat and mass transfer to determine how external conditions change as air flows into a simple model cave, a long wet-walled cylindrical passage.

As air moves into a cave it is either cooled or warmed by contact with the cave walls. At the same time moisture may either evaporate from or condense on to the walls, depending on whether the walls are warmer or cooler than the *dewpoint* temperature of the air. Thus both direct heat transfer between wall and air and latent heat absorption or release are important in determining cave air temperature distributions. Most investigators have ignored one or other of these processes. The problem becomes one of coupled heat and mass transfer which can be solved quite simply.

External temperature fluctuations decay as one moves deeper into a cave; the same effect occurs in soils where surface temperature fluctuations are increasingly damped as one moves deeper into the ground. In the cave this decay, or damping, is characterized by a relaxation length, x_0, which is determined by the Prandtl and Reynolds numbers of the flow. (These numbers are non-dimensional groups which frequently occur in heat transfer and fluid mechanics. The Reynolds number, for instance, determines whether flow is laminar or turbulent, and is a measure of the relative importance of viscous and inertial forces. The reader is referred to standard texts such as Kays (1966) for further details.)

If external conditions are constant at temperature T, the same relaxation length also characterizes the decay of temperature from T to its asymptotic value T_a. If, for example, latent heat effects were neglected, cave air temperature would decay exponentially towards T_a, with the difference reducing to $1/e$, where $e \approx 2{\cdot}718$ of its initial value after one relaxation length (this statement defines the term *relaxation length*). For fluctuating external conditions (external conditions will generally fluctuate diurnally) it is the *amplitude* of the fluctuations which would decay exponentially to zero with relation length x_0. If latent heat is included in the theory, the decays are no longer precisely exponential, but they are still characterized by the same length scale, x_0.

Wigley and Brown (1971) distinguish three typical longitudinal temperature and moisture profiles. If we let T and T_d be the external temperature and dewpoint temperature, and let T_a be the asymptotic cave air temperature (the temperature deep within the cave), then the *summer profile* occurs when both T and T_d are greater than T_a, the *transition profile* occurs when T_a is between T and T_d, and the *winter profile* occurs when both T and T_d are less than T_a (see Fig. 9.3).

In the summer case, close to the cave entrance, latent heat released to the air by condensation on the cave walls may exceed the direct heat loss to the walls and cause a slight increase in temperature. The decay of temperature from T to T_a will always be somewhat slower than exponential. This contrasts with the transition case where both evaporative and direct cooling of the air reinforce each other and produce a decay towards T_a which is faster than exponential. In the winter case air is simultaneously warmed by direct contact with the walls and cooled by evaporation of wall moisture and the resultant temperature profile shows a slower-than-exponential decay. In both the transition and winter profiles, the theoretical development predicts the possible existence of a "cold zone" close to the cave entrance where the temperature is lower than both T and T_a.

Humidity profiles corresponding to the summer, transition and winter cases are shown in Fig. 9.4. The summer profile indicates that the cave air never reaches saturation, although realistic deviations from the model may easily override this expectation. In the transition and winter cases relative humidity rises rapidly as one moves into the cave, reaching

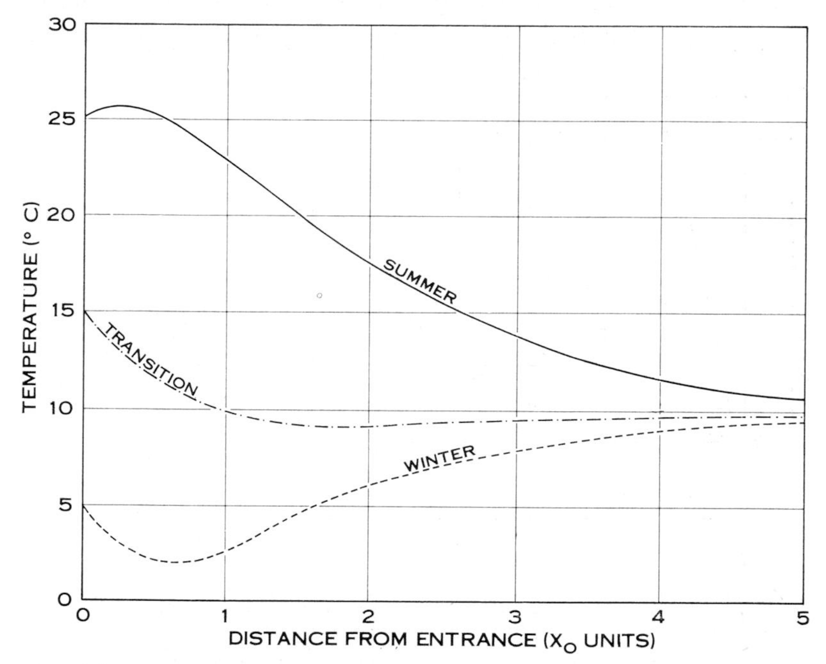

FIG. 9.3. Temperature profiles in a well-ventilated cave (from Wigley and Brown, 1971). The summer profile occurs when external temperature (T) and dew-point temperature (T_d) both exceed the asymptotic temperature (T_a). The case illustrated is for a relatively high value of T_d, 21°C, chosen to show the possible increase in temperature close to the entrance. This need not always be a feature of the summer profile. In the transition case, defined by $T_d < T_a < T$, T_d was chosen as 5°C. A cold zone always occurs in the transitional case, but is slight and may be masked by other effects. The winter case is characterized by having both T and T_d below T. The example shown ($T = 5$°C, $T_d = -5$°C) has a marked cold zone close to the cave entrance. The cold zone is less pronounced for larger values of $T_a - T$ and may not appear at all under very cold external conditions. Distance is measured in multiples of the relaxation length which is a function of rate of air movement and passage size (see text).

saturation in one or two relaxation lengths. The distance to saturation is less in the winter profile.

The relaxation length is a very important parameter. Its value depends on whether the air flow is laminar or turbulent, and in both cases it is determined by the rate of air movement and the passage dimensions. For turbulent flow, as is generally the case for cave air movement, Wigley and Brown derive the following relation (stated here in S.I. units):

$$x_0 = 100\ D^{1 \cdot 2}\ V^{0 \cdot 2},$$

where D is the passage diameter (m), V is the cross-sectional mean flow speed (m/sec) and x_0 is the relaxation length (m). Thus, larger passages

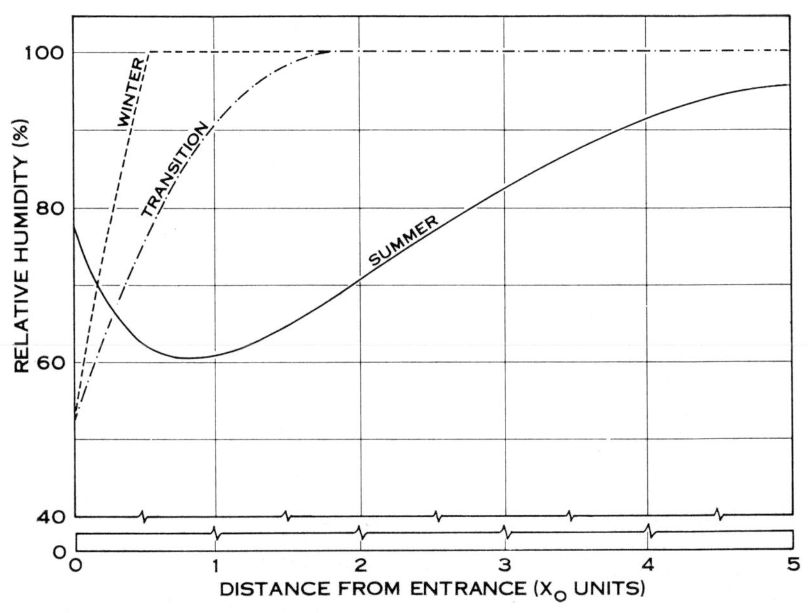

FIG. 9.4. Profiles of relative humidity in a well-ventilated cave for the summer, transition and winter season cases (from Wigley and Brown, 1971). In the summer case a pronounced humidity minimum may occur after which relative humidity rises asymptotically towards saturation. In the transition and winter season cases, relative humidity always rises rapidly to reach saturation quite close to the cave entrance, reaching saturation sooner in the winter case. Distance is measured in multiples of the relaxation length, and external and deep-cave conditions are the same as used in Fig. 9.3.

and/or greater wind speeds give larger values for x_0 and indicate that external conditions and/or variations will be felt further inside the cave than for small passages and/or slower cave winds. The relaxation length is comparatively insensitive to V, which appears only to the power 0·2. Typical values of D and V yield relaxation lengths ranging from 10 to 500 m.

These theoretical results are explained in more detail in the original paper by Wigley and Brown (1971). Not all aspects of their work have been tested, particularly the predictions regarding relative humidity profiles. This is mainly because of the incompleteness of published cave meteorological data. Ideally observers should record dry- and wet-bulb temperatures (the latter with a strongly ventilated device such as a whirling or Assman psychrometer), mean wind speeds and passage dimensions; most investigators omit one or more of these variables. Data quoted by Wigley and Brown agree well both qualitatively and quantitatively with their theoretical results. Naturally many caves depart considerably from the simple model used by these authors, and only qualitative agreement can be expected in such cases. However, the theory does serve to isolate the important processes which operate and to evaluate their relative importance.

One of the theoretical predictions, the near-entrance cold zone, is a commonly observed phenomenon in caves. The cold zone is important in explaining the formation of large ice deposits found in some caves, since it allows the existence of a zone of below freezing temperatures even when both the external air temperature and the deep cave temperature are above freezing. Under these conditions seepage water is still available as an ice "source" which will freeze if it enters the below-zero zone. If external temperatures are much above zero, the cold zone, although it may still exist, will not be below the freezing point. The cold zone therefore allows a relatively brief period of ice accumulation during early winter (and possibly during late winter) in caves which are situated in certain climatic regions (Köppen D types and high-altitude variants of other types).

Plate 9.1 shows a cold zone ice deposit in Castleguard Cave in the Canadian Rocky Mountains. The ice here extends from close to the entrance to about 1000 m inside the cave, a distance of approximately three relaxation lengths. The cave is located in a region where external summer temperatures are well above zero and winter temperatures are well below zero; deep cave temperature is about 3°C and conditions are ideal for cold zone ice formation.

Cave ice is not always associated with a cold zone, although it is obviously essential to have a water supply and a part of the cave below freezing. The water supply may occasionally be in the vapour form and produce extremely beautiful sublimation deposits. An example from another cave in the Canadian Rockies is shown in Plate 9.2. These unusually large crystals can only grow under fairly static conditions where temperatures stay within the quite narrow range which favours growth of the particular crystal form (approximately −3°C).

Plate 9.1. Ice near entrance of Castleguard Cave, Banff National Park, British Columbia. Deep cave temperature is $+3°C$, and ice persists near this lower entrance all year if not destroyed by summer flood water. (*Photo*: D. C. Ford.)

Plate 9.2. Hexagonal sublimation crystals from a small cave in the Front Ranges of the Canadian Rockies. Up to 45 cm in diameter and 0·5 cm thick, these crystals have annual growth lines similar to tree rings. (*Photo*: D. C. Ford.)

One of the parameters in the above theoretical discussion of cave temperature distribution is the asymptotic or deep cave temperature. This is often assumed to be equal to the mean annual temperature outside the cave, and this is probably quite a good approximation. Regional surveys of cave temperatures have shown that deep cave temperatures tend to be lower the higher the elevation or latitude of the cave (see, for example, Moore and Nicholas, 1964, p. 22, and Renault, 1961). There are naturally exceptions to this rule, and in some caves there will be no part which is not noticeably affected by external climatic fluctuations.

Myers (1962, p. 142) states that deep cave temperatures in England are generally below the mean annual external temperature, but it is doubtful whether a careful enough survey has ever been made to verify this point. (In conducting such a survey one would first have to confirm that temperatures were measured at points within the caves where diurnal and seasonal changes were negligible.) Nevertheless, there are theoretical grounds for expecting Myers' statement to be correct. Because the latent heat effect causes cooling for a longer period of the year than it causes warming of cave air, deep cave temperatures should lie below the mean annual temperature—at least for caves which fit the model of Wigley and Brown. In single-entrance caves where air movement is a consequence of either gravitational drainage or entrainment by flowing water the existence of a return air flow and of an additional heat source (the stream) necessitates some qualitative modification of the model's predictions. In these situations one can argue for deep cave temperatures being either above or below the mean annual external temperature depending on local conditions.

The principles of heat and mass transfer used by Wigley and Brown are also applicable to such speleological problems as solution rates and temperature distributions in karst conduits. Some details have been given by Wigley and Brown (1971) and by Wigley (1972).

ELECTRICAL PROPERTIES OF CAVE AIR

Myers (1962), in discussing lightning near caves, points out Casteret's observation that trees near cave entrances are often struck by lightning, and speculates that this may be caused by venting of ionized cave air. Trombe (1947) has measured the degree of ionization of cave air. We have observed that cavers sometimes experience a "speleotherapeutic" effect which may be caused by this ionization; heavy colds or sinus conditions become alleviated. It is known that negatively ionized air causes increased cilia movement in the nasal passages, and in fact negatively ionized air is used as medication for persons suffering from severe asthma.

A physical property of cave air which is of some interest is that it often contains liquid water droplets which may have high concentrations of solutes. Cser and Maucha (1968) have demonstrated that such droplets may be attracted to the walls of the cave at electrically conductive points, and this could be responsible for helictite growth. Their hypothesis is supported by theoretical refutations of most of the other possible theories, and by actual growth of helictites from metal needles connected to a battery and placed in a cave.

GEOPHYSICAL DETECTION OF CAVES

All methods of geophysical prospecting depend on the fact that the object for which a search is made (such as a cave or ore body, etc.) exhibits physical differences from its host material, which can be rock, soil, or even water. These differences may be detected by a variety of sensors which respond to physical forces (such as gravity, electric or magnetic fields, or vibrations) in the host material. It is possible to divide all the different ways of detecting caves geophysically into two classes: *passive* and *active*. Passive methods are those which do not change the medium being investigated, while active methods deliberately cause changes in the medium. However, both active and passive methods still depend on contrasts between the cave and its host rock.

The purpose of this section is, once again, not to provide a definitive discussion about geophysical prospecting for caves (which would require a book in itself) but rather to give the reader a general idea of the possibilities and limitations of the various methods. The bibliography is not meant to be complete and is limited to English language publications, despite the fact that much relevant work has been reported in other languages.

PASSIVE METHODS

1. *Gravimetry*

One of the most obvious methods which has been used, with mixed success, in hunting for caves, is gravimetry. An extremely sensitive (and delicate and expensive) device known as a gravimeter is placed on the surface and measures the force of the attraction of the earth's mass at that point (i.e. the "force of gravity"). The force of gravity varies in a well-defined way with distance from the earth's centre. Superimposed on these variations are variations due to inhomogeneities in the earth's crust. A cave is such an inhomogeneity and one would expect less gravitational attraction over a void in bedrock than if there were no void. Thus by moving a gravimeter from point to point it is possible to locate a cave if it is large enough and close enough to the surface.

Portable gravimeters, such as one made by Worden, are sensitive to variations of about 0·03 milligal (one milligal is 10^{-3} cm/sec^2 or approximately 1/980 000 of the earth's gravitation). The anomaly produced by

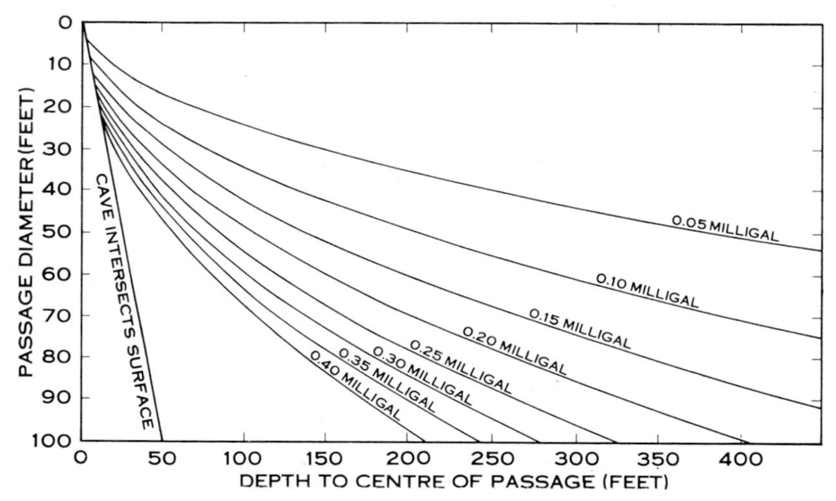

FIG. 9.5. Gravity anomaly produced by air-filled cylinder of infinite length (after A. N. Palmer).

air-filled or water-filled caves (assuming these to be buried horizontal cylinders) has been calculated by many writers (e.g. Colley, 1963; Palmer, 1969):

$$\Delta g_{\max} = 12{\cdot}77(\Delta\rho)\frac{r^2}{z},$$

where $\Delta g_{\max}$ is the maximum gravitational anomaly, $\Delta\rho$ is the density contrast between the host rock and air or water (most limestone has a density of about 2·6 gm/cc, so $\Delta\rho \simeq 2{\cdot}6$ for an air-filled cave and $\Delta\rho \simeq 1{\cdot}6$ for a water-filled cave), r is the passage radius, and z is the depth to the centre of the passage. Figures 9.5 and 9.6 show the maximum anomaly one would expect for cylinders of various sizes at various depths, and thus indicate the quantitative limits of this method. Some of the corrections necessary for ordinary gravity surveying may not be necessary for cave detection. This is because the cave anomaly will generally be well defined spatially, and roughly symmetrical. Thus, although elevation corrections for a station are necessary, corrections for distant mountains may not be.

For a simple model cave, such as the buried horizontal cylinder above, it should be possible to obtain some information about cave depth from a gravity traverse, despite the intuitive conclusion that a small shallow cave would produce an anomaly similar to a large deep cave. This is because

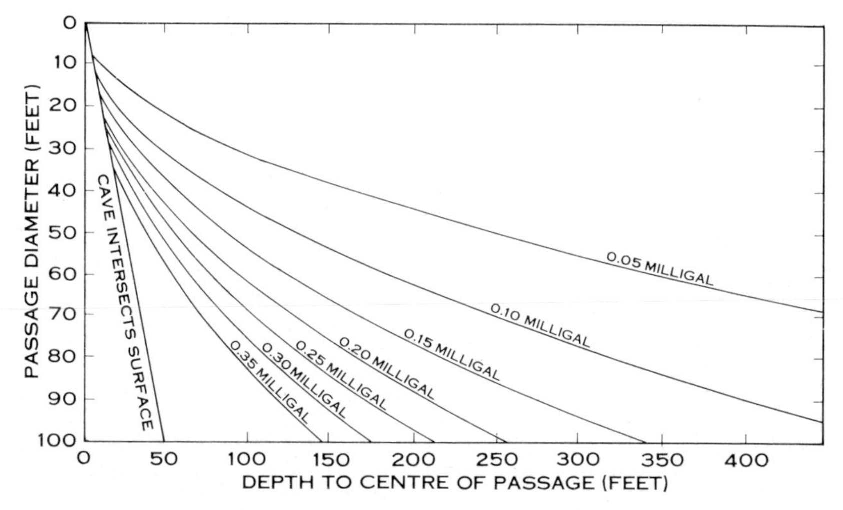

Fig. 9.6. Gravity anomaly produced by water-filled cylinder of infinite length (after A. N. Palmer).

the spread of the anomaly varies with the depth of the feature, as shown in Fig. 9.7. Other cave shapes, however, can also cause this spread to vary, and other geological structures (different rock types, etc.) may obscure the interpretation. It is also often advantageous to use filtering methods on gravity data, such as Fourier filtering, or trend surface analysis, to derive more information about the anomaly. This analysis is easily done by computer, and the techniques are described in Chapter 14.

2. *Passive Seismic* (*Acoustic*)

If a cave has a river flowing through it, then the river will generate noise across a fairly broad spectrum. Lange (1971, 1972) has discussed the detection of this sound, which is transmitted through rock with little attenuation, using a geophone placed on the surface and coupled to a high gain amplifier and meter or tape recorder. Lange's preliminary results show good potential for this method: running water and even cavers talking could be heard through up to 50 feet of marble!

Problems which this method entail include the initial cost of equipment, and background noise from surface streams, wind-induced tree movement, and other sources. It may be possible to minimize the background noise problem by using shallow holes for the geophones, or by

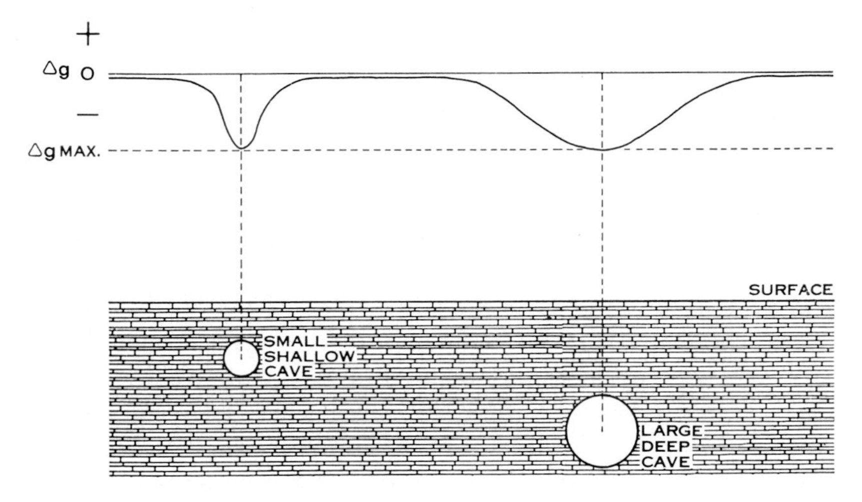

FIG. 9.7. Gravity anomalies produced by a small shallow cave and a large deep cave, with Δg_{max} equal.

filtering the recorded data. If the size of the underground river could be estimated independently, then bandpass filters admitting its "characteristic" frequencies could be used to eliminate unwanted noise.

Alternatively, Morrison (personal communication) has suggested that cross-spectral analysis of multiple-site geophone data, which would require the use of a multi-channel recorder and computing facilities, might eliminate background noise problems. This is a relatively new idea, but its potential for cave detection appears promising.

3. *Magnetic and Telluric*

Rocks may have magnetic properties of two distinct types. Magnetization may have been induced in the recent past by the earth's magnetic field if the rock were heated to above its Curie point temperature. The characteristics of such magnetization depend on the magnetic susceptibility of the rock and the strength and direction of the inducing magnetic field. In addition, certain rocks may have a permanent or remanent magnetization, the characteristics of which depend more on the geological history of the rock (Parasnis, 1971). Lange (1965) has investigated the possibility of detecting a cave in bedrock by measuring either the induced or remanent magnetism at the surface, using a medium sensitivity magnetometer. He found that the method could be used for rocks with high magnetic susceptibility and small variations in remanent magnetism. North/south trending caves will be more easily detected. Although Lange has used this method successfully, he states that it is *not* useful for detection of caves in soluble rocks (i.e. for almost all non-lava-tube caves), a rather severe restriction.

Astronomic phenomena, e.g. solar electron streams, the earth's rotation, etc., cause electric currents to flow in the earth. These cover extremely large areas, and are called telluric currents (Parasnis, 1971). Knowing this, Day (1965) has suggested an interesting possibility for cave detection which is as yet untried. Graphical representation of measurements of the telluric field vector of the earth at a point over a 24-hour period yields an envelope, the shape of which varies with the subsurface electrical resistance. By comparing such measurements of the telluric field at different points one might detect resistive anomalies which would indicate the presence of a cave. In a sense this method might be called a "passive resistance method". However, the method is more appropriate to oilfield work. A magneto-telluric method appears to give similar results to resistivity in other situations, as over a railway tunnel (Guineau, 1975).

ACTIVE METHODS

1. *Resistivity*

Resistivity methods are the most frequently used in searching for caves. While some workers (e.g. Palmer, 1954, 1963; Wilcock, 1965; Bristow, 1966) claim to have located (and later dug into) previously unknown caves, we are aware of at least one large-scale failure of the technique to locate a cave (Brown, 1972). We will deal with resistivity in some detail, because the methods are not over-expensive and the technique yields more information than any other (except seismic, which is both costly and difficult to interpret).

Essentially, resistivity methods involve passing an electrical current between points on the earth's surface, and, by means of two measuring electrodes at two other points, measuring the resistance of the earth to that current. Because current flows along the path of least resistance, and in a uniform medium flows in well-defined lines (somewhat akin to a series of hemispheres), resistance on the surface varies with the conductivity of objects below the surface. For example, air has an extremely high resistance to electric current (as does an air-filled cave) relative to limestone, while water (and a water-filled cave) has a very low resistance. The type of current used is important. Day (1964, 1965) discussed this: direct current (D.C.) is not used because the ground near the electrodes becomes polarized (although this can be overcome by seating the electrodes in a porous pot in the ground), and usually either a normal alternating current (A.C.) or a commutated D.C. is used. In the latter case, the polarity of the potential or measuring electrodes is reversed in synchronization with the current so that a D.C. voltmeter can be used. If A.C. is used, the choice of frequency is important; very low frequencies penetrate the earth quite deeply allowing D.C. theory to be used, but high frequencies cause a "skin effect" with a concentration of current near the surface. Operating frequencies of less than 10 Hz are usual.

A considerable amount of work has been done on earth resistivity anomalies, concerning limitations of the method for finding a buried sphere (Van Nostrand, 1953), dealing with layered media (Mooney and Wetzel, 1956), treating vertical discontinuities (Logn, 1954), and even discussing filled sinks (Cook and Van Nostrand, 1954). There is, however, scope for further valuable experimental model work, using, for example, resistivity paper for two-dimensional experiments. In three dimensions, tanks filled with a conducting liquid and high- or low-resistance objects could simulate caves of various shapes and positions. The insulating sphere problem has been discussed by Habberjam (1969).

The problem of exactly how to conduct a resistivity survey is a difficult one. One obviously wishes to maximize the amount and quality of data obtained, with as few people and in as short a time as possible. There are several different electrode configurations or "arrays" from which to choose. Figure 9.8 illustrates these: the Wenner, Schlumberger, and dipole-dipole arrays are most commonly used for general geophysical work. For the Wenner array all electrodes are equally spaced ($= a$), for the Schlumberger spacing P_1P_2 ($= 2Z$) $\ll$ C_1C_2 ($= 2L$), while for the dipole-dipole $P_1P_2 = C_1C_2$ ($= a$) but P_1P_2 are outside C_1C_2.

Following Parasnis (1971), the apparent resistivity, P_a, is given in general by the relation:

$$P_a = 2\pi \frac{\Delta V}{IG},$$

where ΔV is the potential difference between P_1 and P_2, I is the current passing into the ground, and

$$G = \frac{1}{C_1P_1} - \frac{1}{C_2P_1} - \frac{1}{C_1P_2} + \frac{1}{C_2P_2}.$$

The apparent resistivity therefore varies with electrode configuration as well as with location on the surface. It is easy to show that for the Wenner array

$$P_a = 2\pi aR,$$

while for the Schlumberger array with $C_1P_1 = C_2P_2$

$$P_a = \frac{\pi(L^2 - Z^2)\Delta V}{2ZI}$$

and for the dipole-dipole

$$P_a = \pi n(n+1)\ (n+2)a\frac{\Delta V}{I},$$

where $C_2P_1 = na$ defines n.

If depth information is desired, for the Wenner array a is increased keeping the midpoint constant, while for the Schlumberger C_1 and C_2 are moved outward symmetrically. To complement this information, lateral anomalies can be discovered by (for the Wenner) moving the whole array along a fixed a profile. For the Schlumberger, this is done by moving P_1 and P_2 between C_1 and C_2 although such repositioning will also cause G to vary.

Bristow (1966) has provided a summary of the way in which resistivity analysis is done, as well as an example of its apparent potential for finding caves. Figure 9.8 shows his array: here a single current probe (equivalent to a Schlumberger array with C_2 at infinity) has been used to derive a series of high anomaly "shells" the intersections of which gave three-dimensional cave locations. Known caves were located, and at least one unknown cave was identified and subsequently dug into and explored. Bristow was apparently unaware of Searle's (1910) work which throws doubt upon the efficacy of the method.

Other valuable work using resistivity apparatus includes the successful pioneering work of L. S. Palmer (1954, 1963), who re-located Pen Park Hole near Bristol; and A. N. Palmer (1969) who used gravimetry together with resistivity to indicate the location of a known cave, and Dutta *et al.* (1969) among others.

There are as well other types of electrode arrays, such as Wilcock's (1965) which was suggested by L. S. Palmer, and that referenced in Brown

(1972) which was designed by J. Bonniwell. Although in general these decrease the number of different stations required and the total number of electrode movements, they also increase the ambiguity of the results.

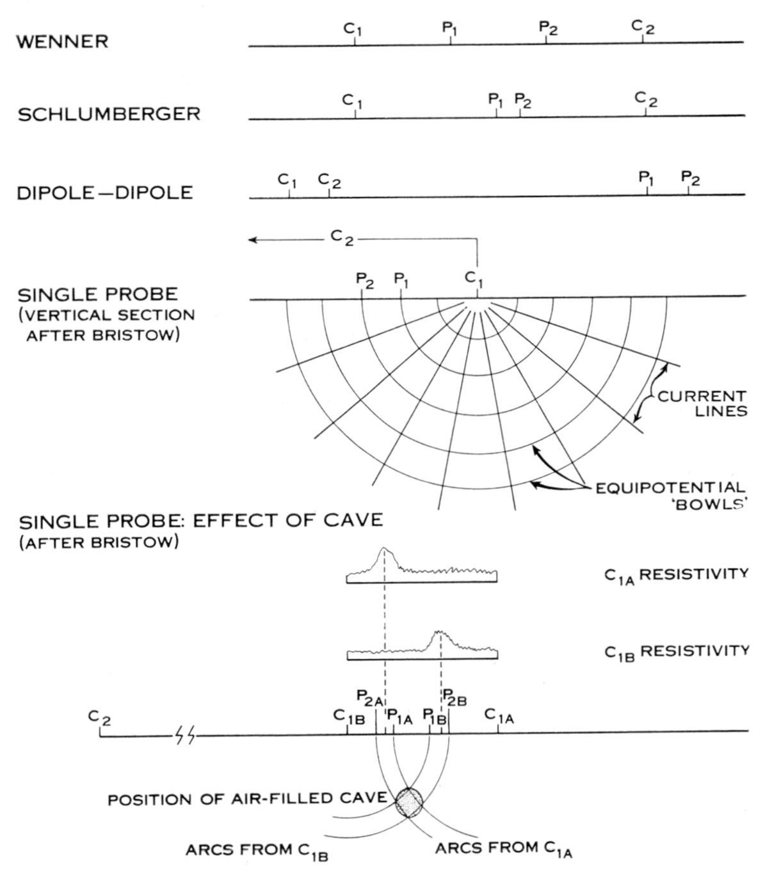

FIG. 9.8. Various electrode configurations for resistivity surveys, with Bristow's single probe method shown in detail.

Chapter 14 provides further references to resistivity analysis, in particular with respect to ways in which computers can help in analysing the data once it is obtained.

In summary, resistivity analysis, especially when combined with gravity or seismic studies, offers an extremely useful tool to the speleologist. Apparatus need not be inordinately expensive, and frequencies can be chosen to define a cave at a suspected depth. Various probe configurations may be used, although the single current-probe array method seems most successful up to now. However, interpretation of resistivity graphs is still very difficult (but see Chapter 14).

2. *Seismic*

A technique somewhat more active than those discussed so far is seismic surveying. Any physical shock (such as an explosion or hammer blow) applied to the earth's surface causes vibrations to travel through it. These vibrations move through rock as compressional waves of two types: longitudinal or "P" waves when the vibration direction is parallel to the direction of wave movement, and transverse or "S" waves when the vibration direction is at right angles to the direction of wave propagation.

Both P and S waves travel at different speeds through different media and so both may be *refracted* (curved) by passing from one medium to another. They may also be *reflected* from the interface between two different media. For most geophysical seismic work, P waves, which travel faster and attenuate more slowly than S waves, are used. P waves are sound waves, and travel through air at about 330 m/sec, through water about 1450 m/sec, and through limestone from 3500 to 6500 m/sec. Thus an air-filled cave causes a major discontinuity in the wave, and reflected or refracted signals, detected on the surface by *geophones*, can indicate its presence by a "shadow" (see Fig. 9.9).

Watkins (1967) discusses three effects which caves have on seismically induced vibrations: amplitude attenuation, arrival time delay, and free oscillation. The last of these is the most interesting, since the former two may be produced by other geological structures and may therefore be difficult to interpret. Free oscillation, or amplitude enhancement, occurs when Rayleigh waves (a type of free surface S wave) cause the walls of an air-filled cave to resonate. For a cylindrical cavity the resonant wavelength is about 1·5 times its diameter, so the frequency of the induced wave (detected as an S wave at the surface), may yield information about cave size. Unfortunately S waves require greater generating energy than P waves, and are not commonly used in conventional geophysical searches. Further information regarding this phenomenon was given by Cook (1965) and Rechtien and Stewart (1975).

There are two other detection methods which are closely related to usual seismic methods, both of which have yet to be tested. Firstly, Arandjelovic (1969) has suggested that seismic exploration for caves could be improved if the energy source were *inside* the cave, and therefore proposed floating a small time-bomb, with a fail-safe device, into the cave. A precise timer would explode the bomb and the seismic wave so generated would be detected on the surface by a geophone array. Lange (1970) has pointed out several problems with this concept, aside from the obvious environmental difficulties inherent in setting off explosives in areas in which the geological and biological resources have not been assessed.

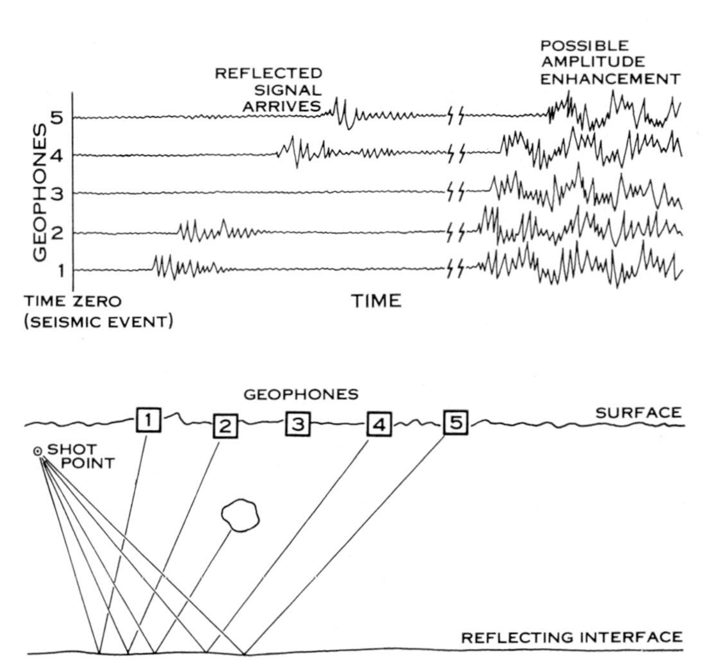

FIG. 9.9. Seismic "shadow" effect of cave, and possible later seismic wave amplitude enhancement.

Secondly, a more esoteric method of finding caves is possible. It may be possible, using a *continuous* hammer seismic source to measure continuously the reflection and refraction patterns of the induced waves by using a grid of surface geophones. This is the VIBROSEIS (a trademark of the Continental Oil Company) method, first discussed by Edelmann (1966), then Krey (1969), and lately Barbier and Pouleau (1970). Advantages of this method for cave detection would include not using explosives, lower momentary seismic energy requirement, shorter field time (relative to "stacked" explosive shots), and ability to change frequencies, filters, and geophone positions in the field during the test. Barbier and Pouleau suggest optical filtering of the data by coherent light, to produce a hologram of the subsurface.

In summary, while it is possible to use seismic methods to find caves (see for example Cook, 1965; Watkins, 1965), it seems that this method is best employed in conjunction with other methods such as gravimetry or resistivity. Seismic equipment is also very expensive, and interpretation of seismic records, even with computers, is an art not easily mastered.

CONCLUSION

This chapter has been limited to a review of only two aspects of cave physics. We conclude with a brief description of how cave physics relates to the problem of speleogenesis, the formation of caves.

Speleogenesis, usually considered as a part of geomorphology, involves both physical and physico-chemical processes. Investigation of the processes alone, and in detail, requires a knowledge of the physics of phenomena such as mass transfer, soil and rock collapse, etc. It is quite possible for a physicist or engineer to study these processes in their own right. The karst geomorphologist becomes involved when the results of these studies are applied to karst or cave problems. The geomorphologist should therefore have some knowledge of the basic principles of physics as he is often called on to adapt and modify results from outside geomorphology to suit his own particular problems.

One aspect of speleogenesis concerns the determination of the rate of growth of a karst conduit in the phreatic zone, or a cave passage in the vadose zone. These growth rates depend on how rapidly material can be removed from a wall by fluid flowing past it, which in turn depends on two factors: the chemical "driving force" which pushes material from the wall through the boundary layer into the body of the fluid, and the hydrodynamic properties of the fluid itself. The driving force is proportional to the difference between the concentration of dissolved substance in the fluid and the saturation concentration appropriate to the prevailing conditions of temperature and carbon dioxide content. The physical process of mass transfer in response to this driving force occurs at a rate determined by other parameters such as the Reynolds number. This application of physics to speleogenesis has been considered by White and Longyear (1962), Palmer (1969), Curl (1971) and Wigley (1971). Some results point to the possible significance of the transition from laminar to turbulent flow. But all models so far have been oversimplified and have neglected factors such as homogeneous and heterogeneous chemical reaction rates (see Curl (1967) and Plummer and Wigley (1976)).

Some of the principles of mass transfer theory which are directly applicable to this aspect of speleogenesis have been given in this chapter in the discussion of cave climate. The transfer of moisture from cave walls into the cave atmosphere is essentially the same process as the transfer of solid wall material into a water-filled conduit (although the former problem is complicated by latent heat effects, while the latter may be complicated by chemical reactions). The same principles are also relevant to the study of solutional features such as scallops: the reader is referred to the work of Curl (1966), Blumberg (1970) and Goodchild and Ford (1971) for a more completed discussion of such features. Curl (1973) has also published a fascinating investigation into the minimum diameter of straw stalactites which he related to a non-dimensional parameter called the Bond number: other non-dimensional parameters have been used in the cave climate section above. Soil-collapse and cavern-roof breakdown are other speleogenetic processes which can only be properly interpreted by geomorphologists with an understanding of soil and rock mechanics or by civil engineers with considerable familiarity with geomorphology. This latter approach was used by J. E. Jennings (1966) in an analysis of the origin of some spectacular and catastrophic collapses in South Africa which were triggered by the lowering of the water-table in a region of soil-covered karst. Reviews of the origin of diverse types of dolines have been published by J. N. Jennings (1971) and Quinlan (1972).

REFERENCES

Arandjelović, D. (1969). A possible way of tracing groundwater flows in karst. *Geophysical Prospecting* 17, 404-418.

Barbier, M. and Pouleau, J. (1970). A pulse compression technique using coherent light. *Geophysical Prospecting 18*, 571-580.

Blumberg, P. N. and Curl, R. L. (1974). Experimental and theoretical studies of dissolution roughness. *J. Fluid Mech. 65*, 735-751.

Bristow, C. (1966). A new graphical resistivity technique for detecting air-filled cavities. *Studies in Speleology*, *4* (1), 204-227.

Brown, M. C. (1972). *Karst Hydrology of the Lower Maligne Basin.* Cave Studies 13. Cave Research Associates, California.

Colley, G. C. (1963). The detection of caves by gravity measurement. *Geophysical Prospecting 11*, 1-9.

Conn, H. W. (1966). Barometric wind in Wind and Jewel Caves, South Dakota. *Nat. Speleo. Soc. Bull. 28* (2), 55.

Cook, J. C. (1965). Seismic mapping of underground cavities using reflection amplitudes. *Geophysics 30* (4), 527-538.

Cook, K. L. and Van Nostrand, R. G. (1954). Interpretation of resistivity data over filled sinks. *Geophysics 19*, 761-790.

Cser, F. and Maucha, L. (1968). Contribution to the origin of "excentric" concretions. *Karszt-és Barlangkutatás* (Hungarian Speleological Society), 83-98.

Curl, R. L. (1968). Solution kinetics of calcite. *Proc. 4th Int. Cong. Speleol., Ljubljana*, 61-66.

Curl, R. L. (1971). Cave conduit competition—1; power law models for short tubes (abstract only). *Caves and Karst 13*, (5) 39.

Curl, R. L. Minimum diameter stalactites. *Nat. Speleo. Soc. Bull.* (in press).

Day, J. G. (1964). Cave detection by geoelectrical methods, Part 1, Resistivity. *Cave Notes 6* (6), 41-45.

Day, J. G. (1965). Cave detection by geoelectrical methods, Part 2, Transient and inductive methods. *Cave Notes* 7 (3), 17-22.

Dutta, N. P., Bose, R. N. and Saika, B. C. (1969). Detection of solution channels in limestone by electrical resistivity method. *Geophysical Prospecting 18*, 405-414.

Edelman, H. (1966). New filtering method with "VIBROSEIS". *Geophysical Prospecting 14*, 455-469.

Faust, B. S. (1947). *Nat. Speleo. Soc. Bull. 9*, 52-54.

Geiger, R. (1966). *The Climate near the Ground* (rev. edn). Harvard Univ. Press, Cambridge, Mass.

Goodchild, M. F. and Ford, D. C. (1971). Analysis of scallop patterns under controlled conditions. *Journal of Geology 79*, Discussion by Wigley, T. M. L.; *Journal of Geology 80* (1972), 121-122.

Guineau, B. (1975). La méthode magnéto-tellurique de prospection . . . *Geophysical Prospecting 23*, 104-124.

Habberjam, G. M. (1969). The location of spherical cavities using a tri-potential resistivity technique. *Geophysics 34*, 780-784.

Halbert, E. J. and Michie, N. A. (1971). The use of titanium tetrachloride in the visualization of air movement in caves. *Helictite 9*, 85-88.

Haltiner, G. J. and Martin, F. L. (1959). *Dynamic Meteorology*. McGraw-Hill, New York.

Hill, A. L. (1966). *Nullarbor Expeditions*. Occasional Paper No. 4 of Cave Exploration Group of South Australia.

Jennings, J. N. (1972). *Karst*. Australian National Univ. Press.

Jennings, J. E. (1966). Building on dolomites in the Transvaal. *The Civil Engineer in South Africa* 41-62.

Kays, W. M. (1966). *Convective Heat and Mass Transfer*. McGraw-Hill, New York.

Krey, Th. (1969). Remarks on the signal to noise ratio in the VIBROSEIS system. *Geophysical Prospecting* 17, 206-218.

Lange, A. L. (1972). Mapping underground streams using discrete natural noise signals: a proposed method. *Caves and Karst*, *14* (6), 41-44.

Lange, A. L. (1971). Acoustic tracing of karst streams. *Caves and Karst 13* (2), 9-14.

Lange, A. L. (1970). Review of Arandjelović (1969). *Caves and Karst 12* (3), 20-24.

Lange, A. L. (1965). Cave detection by magnetic surveys. *Cave Notes* 7 (6), 41-54.

Logn, O. (1954). Mapping nearly vertical discontinuities by earth resistivities. *Geophysics 19*, 739-760.

McElroy, G. E. (1966). Mine ventilation, in *Mining Engineers Handbook* I, 3rd edn (Ed. R. Peele). John Wiley and Sons, New York.

Mooney, H. M. and Wetzel, W. W. (1956). *The Potentials About a Point Electrode*. Univ. of Minnesota Press, Minneapolis.

Moore, G. W. and Nicholas, G. (1964). *Speleology*. D. C. Heath and Co., Boston.

Myers, J. O. (1962). Cave physics, Chap. 7 in *British Caving* (ed. C. H. D. Cullingford), 226-250.

Palmer, A. N. (1969). *A Hydrologic Study of the Indiana Karst*. Cave Research Foundation Publication (Columbus, Ohio).

Palmer, L. S. (1954). Location of subterranean cavities by geoelectrical methods. *Mining Mag*. London *91*, 137-141.

Palmer, L. S. and Glennie, E. A. (1963). The geoelectric survey and excavation, in Tratman, E. K. (ed.), *Reports on the Investigation of Pen Park Hole, Bristol*. Cave Res. Grp. G.B. publication 13, 15-24.

Parasnis, D. S. (1971). *Principles of Applied Geophysics*. Chapman and Hall, London.

Plummer, L. N. and Wigley, T. M. L. (1976). The dissolution of calcite in CO_2-saturated solutions at 25°C and 1 atmosphere total pressure. *Geochim. et Cosmochim. Acta 40*, 191-202.

Plummer, W. T. (1969). Infrasonic resonances in natural underground cavities. *J. Acoustical Soc. Amer. 46*, 1074-1080.

Quinlan, J. F. (1972). Outline for a genetic classification of major types of sinkholes and related karst depressions. *Proceedings, 22nd International Geological Congress*, Montreal, Sect. 6, pp. 156-168.

Rechtien, R. D. and Stewart, D. M. (1975). A Seismic Investigation over a near-surface cavity. *Geoexploration 13*, 235-246.

Renault, Ph. (1961). Première Étude météorologique de la Grotte de Moulis (Ariège). *Ann. Spéléol.* (16).

Saar, R. von (1956). Eishohlen—Ein meteorologische, geophysikalisches Phänomen. *Geografiska Annaler 38*, 1-63.

Sartor, J. D. and Lamar, D. L. (1962). *Meteorological-Geological Investigations of the Wupatki Blowhole System*. Rand Corporation Memorandum RM-3139-RC.

Searle, G. F. C. (1910). On resistances with current and potential terminals. *The Electrician 66*, 999-1002.

Van Nostrand, R. G. (1953). Limitations on resistivity methods as inferred from the buried sphere problem. *Geophysics 18*, 423-433.

Watkins, J. S. *et al.* (1967). *Seismic Detection of Near-Surface Cavities*. U.S. Geological Survey Prof. Paper 599-A.

Watkins, J. S. *et al.* (1965). *Seismic Investigation of Near Surface Cavities—a preliminary report; "Investigation of in situ physical properties of surface and sub-surface materials by engineering geophysical techniques*. Project, Annual Report; U.S. Geol. Survey, Flagstaff, Arizona, B1-B26.

Went, F. W. (1970). Measuring cave air movements with condensation nuclei. *Nat. Speleo. Soc. Bull. 32*, 1-9.

White, W. B. and Longyear, Judith (1962). Some limitations on speleogenetic speculation imposed by the hydraulics of ground-water flow through limestone. *Nat. Speleo. Soc. Nittany Grotto Newsletter*, *1*, 155-167 (also in *Speleo Digest* for 1962)

Wigley, T. M. L. (1967). Non-steady flow through a porous medium and cave breathing. *J. Geophys. Res.* 72, 3199-3205.

Wigley, T. M. L. (1971). Solution of pipes by turbulent fluids (abstract). *Caves and Karst 13*, 50.

Wigley, T. M. L. and Brown, M. C. (1971). Geophysical applications of heat and mass transfer in turbulent pipe flow. *Boundary-Layer Meteorology 1*, 300-320.

Wilcock, J. D. (1965). Geophysical survey and results; Oxford University Expedition to Northern Spain 1961. *Trans. Cave Res. Grp. G.B.*, publication 14, 31-38.

10. Cave Faunas

G. T. Jefferson

INTRODUCTION

It is something of a biological axiom that any place on the earth where conditions are capable of supporting life will come to be occupied by organisms adapted to those conditions. There are a few regions where conditions are too severe, but life exists over most of the earth's land surface and in its waters, including the greatest depths of the oceans; it also exists in caves, where both plants and animals are to be found. The plants, which are dealt with in Chapter 11, are, at least in the dark regions of caves, mainly bacteria and fungi; but the animals are drawn from a wide range of both invertebrate and vertebrate groups.

Animal life, however, does not feature very prominently in British caves; the animals, being few in number and usually small in size, are mostly inconspicuous. Elsewhere cave faunas may be richer and include more striking creatures. This is true of some parts of continental Europe and probably explains why the biological study of caves started there and was slow to develop in Britain. Although scattered reports on individual underground animals such as the cave salamander, *Proteus*, had appeared earlier, real cave biology can be said to date from about the middle of last century when a number of continental authors started to publish accounts of animals from the great Balkan caves and, to a lesser extent, from Mammoth Cave, Kentucky. The most significant step in the development of the subject occurred in 1907 when E. G. Racovitza, a Romanian at that time living in France, joined with the French entomologist R. Jeannel and others, to form an association of cave biologists. This group called itself Biospeologica and, under that general title, started to publish, in the *Archives de Zoologie Expérimentale et Général*, what was to become a long and important series of papers. The work of this group continued in

Romania after 1920 when Racovitza was invited to take charge of the newly formed Institute of Speleology at Cluj. In an appendix to his first report as director of the Institute Racovitza (1926, annex B, p. 2) listed some thirty eminent biologists whose co-operation he had secured; among these there was just one British name, the mammalogist, M. A. C. Hinton, who collaborated in the work on bats. Biospeleology, or Biospeology as many Continental workers prefer to call it, developed apace, particularly in Europe where several research stations and even underground laboratories were established in the period before the Second World War. Some of these were short-lived, but others survived and new ones have been set up since the war; outstanding among the latter is the splendid Laboratoire Souterrain at Moulis in the French Pyrenees. The number of biologists working on cave organisms has increased markedly in the last 20 years, and the volume of work published, although small compared with many branches of biological science, is nevertheless considerable.

In Britain investigation started slowly and developments have been along rather different lines. There was a little sporadic collecting of animals in a few caves around the turn of the century; palaeontological work on a substantial scale had started well before this, but the systematic study of the British cave flora and fauna only started with the pioneering work of E. A. Glennie in the late 1930s. This was interrupted by the war but was resumed in 1946, and soon Brigadier Glennie, his niece, Mary Hazelton, and several other enthusiasts were collecting specimens in all the major cave areas of Britain. The Cave Research Group of Great Britain was involved, right from its inception, in this biological work, and indeed the Group's first publication was on cave fauna (Glennie, 1947). A considerable body of specialists collaborated in this early work by providing an expert identification service, but the collecting and recording was done by amateurs, and it was their enthusiasm which carried the programme forward. This indeed has been the pattern of biospeleological work in Britain ever since; very few professional biologists have done any work on cave organisms except for this most important service of providing authoritative identifications. This is in marked contrast to the situation in many countries, but nevertheless the results have been extremely valuable. The cave fauna of Britain, admittedly not a particularly extensive one, is very well known, and it is this considerable body of faunistic work which makes it possible to put forward some of the generalizations discussed later.

Ecological and physiological work has been much less extensive, and no permanently staffed biospeleological research stations or underground laboratories comparable with those of some other countries have been set up in Britain. This is not altogether surprising, as such research stations naturally tend to be located in regions where the underground fauna is particularly rich. Britain, however, has some facilities. The William Pengelly Cave Studies Trust has established its attractive centre at Buckfastleigh in South Devon (Sutcliffe, 1965), and although the main interest here is palaeontological, a wide range of speleological studies, including biology, can be pursued in the area. The only underground biological laboratory is a small one set up in an old Yorkshire lead-mine adit at Starbotton by Mrs Jean Dixon and her colleagues of the Northern Cavern and Mine Research Society. Observations and experiments are in progress (Dixon, in Waltham, 1974) and this commendable venture deserves to succeed.

The early stages in the development of any branch of biology tend to be largely descriptive, often involving the collection, identification and recording of organisms. Later more integrated studies can be undertaken; form and function, development, the interrelationship of organisms, can all be studied and generalizations can be made and tested. Biospeleology is now in a position to move in this direction; it has already done so in some countries where ecological and physiological studies have been in progress for some little time. These stages in the development of the subject are not sharply separated; the very collecting of material naturally led to some ecological observations and speculations being made at an early stage. It is clearly also true that plenty of faunistic work still needs to be done, both in Britain and elsewhere. Nevertheless, more and more attention will be focussed on ecological, biogeographical, physiological and evolutionary problems, though little will be said here about the last two of these aspects.

A considerable amount of physiological and behavioural work is being done on cave animals, particularly in relation to their metabolism, their sensory equipment and the extent to which any rhythmical behaviour is retained. Thinès and Tercafs (1972) discuss these, among many other topics, and there are extensive earlier reviews by Poulson (1964) and in Parts 4 and 5 of Vandel's *Biospeleology* (1965). The evolution of cave animals has been a fruitful field for conjecture and disputation since the time of Darwin, and even today there are almost as many conflicting views as there are interesting problems. A most useful paper by Barr (1968a), dealing with many aspects of cave biology, can be consulted for a review of the main theories that have been put forward and for a discussion of the speciation of cave animals in the light of current evolutionary theory.

THE HYPOGEAN DOMAIN

The biosphere—the thin shell around our planet in which life occurs—contains a great variety of habitats which can be classified in various ways. On the land areas several horizontally separated major habitat regions or *domains* can be recognized, and this approach is a convenient one for the cave biologist. Glennie's (1965) discussion serves as a basis for the following classification of land-area domains.

Domain	Habitat
Para-epigean	Forest canopy, particularly in the tropics.
Epigean	The land surface, including trees and other plants on it, and bodies of water such as rivers, lakes, ponds and marshes.
Endogean	The soil extending down to the greatest depth to which the roots of epigean vegetation may penetrate; also muds, sands and gravels associated with epigean bodies of water.
Pro-epigean (or Para-hypogean)	The cave threshold from the surface opening to the furthest point to which daylight penetrates.
Hypogean	The dark zone beyond the Pro-epigean and below the Endogean domains. Both airfilled and waterfilled caves, cavities and fissures are included and so also are deposits of clay, earth or sand occurring in this zone.

The Para-epigean domain does not concern us; its inhabitants are epiphytes and animals confined to the forest canopy. Most familiar organisms occur in the Epigean domain; surface plants, animals living on the ground or among the plants (but not *confined* to the canopy), and also plants and animals living in epigean waters. The Endogean flora and fauna, although less familiar, are both extensive and mostly specialized. The soil contains vast numbers of bacteria, fungi and animals either adapted to burrow through it or to lie between its particles. Where tree roots penetrate into the dark zone of caves through rock fissures, even if these appear quite tight, they must be looked upon as representing an incursion of the Endogean domain into the Hypogean. Aquatic organisms also live in the waterfilled spaces between the particles in those endogean beds of sand and gravel associated with bodies of water.

The remaining two domains are the particular interest of biospeleology. Para-hypogean might be a more obvious name for the Pro-epigean domain, but Glennie prefers the latter since it serves to emphasize the relationship of the threshold fauna to that of epigean regions. It is, however, the Hypogean domain with its own specialized fauna which is the main preoccupation of cave biologists. Nevertheless, caves in the ordinarily accepted sense form only part of this domain, which includes any system of crevices and fissures deeper than the soil layer, whether above or below the water-table. It would clearly be quite arbitrary to separate off from the rest of the Hypogean domain those caves large enough for man to enter, and it would be equally artificial for biospeleology to confine its studies to such caves. Strictly speaking, biospeleology and hypogean biology may not be synonymous, but in practice they very nearly are; it is normally only in caves that the Hypogean domain is open to study by man. The domain is penetrated artificially, particularly by mines and wells, both of which often yield hypogean organisms, and floods may bring underground animals to the surface in springs and bourne risings (Glennie, 1960, 1962). The records of the Cave Research Group cover all hypogean fauna and most biospeleologists extend their activities to mines, wells and springs. There is an authoritative precedent for this. Racovitza took the view that speleology was concerned with the whole subterranean region; and when the Institute of Speleology was set up at Cluj in 1920, its programme was laid down as "the general study of the underground domain . . ." (Racovitza, 1926, annex A).

THE INTERSTITIAL MEDIUM

An appreciable part of the total volume of a bed of sand or gravel consists of the interstices between the particles. The size of these interstices depends on the grain size of the material; in fine sand, silt or mud they will be minute, but their total volume will be far from negligible. If the gravel or sand is under water, the interstices will be waterfilled and in communication with the main body of water; this may also be the case in parts of the beds above water-level, since water will rise into the interstices by capillarity. The biological significance of this special aquatic habitat, the *Interstitial Medium*, has only been appreciated fully in the last few decades. There is, in fact, a whole fauna, a vast range of mostly minute or thread-like animals, adapted to life within these waterfilled interstices. The realization of this came as something of a surprise to many biologists with the publication in 1960 of Delamare-Deboutteville's monograph, "Biologie des Eaux Souterraines, Littorales et Continentales".

The first interstitial media to be studied as such were the sandy beaches of seas, estuaries and lakes. These are respectively marine, brackish and fresh-water habitats, but the faunas are not as dissimilar as might be expected. Most interstitial animals are euryhaline—able to withstand a range of salinities—and the faunas intergrade one into another; indeed, this seems to have been the route followed in the past by many lines of originally marine animals that have entered the world of underground fresh waters.

The interstitial medium as at present understood is not confined to what most people would call sand, but is extended to include the rather larger interstices of shingles or gravels and much of the medium constituted by groundwater. The latter occupies cavities ranging from pores, through crevices and anastomosing honeycombs, to large waterfilled cavities. Where this continuum ceases to be interstitial and becomes part of a cave system is an academic point and it is now the generally accepted practice to refer to organisms living in hypogean phreatic waters, whether truly interstitial or not, as "phreatobites". Clearly not all interstitial media are hypogean; in many situations they must be considered endogean and in some places may even contain photosynthetic plants (algae) since light can penetrate a short distance through sand. The interstitial medium is important to cave biologists particularly because it provides an important aquatic link between the Hypogean and Epigean domains.

THE CAVE ENVIRONMENT

It is usual to divide a cave into two zones, the threshold and the dark zone or true cave. The threshold, which constitutes the Pro-epigean domain, extends from any surface opening of the cave to the furthest point to which daylight ever penetrates. It is a zone where the physical environmental factors are relatively variable. Light is always reduced and its intensity falls off rapidly with distance from the entrance; it also obviously varies with time of day and with external conditions. Other factors such as temperature and humidity are similarly variable. The dark zone, on the other hand, is a region of relative constancy. It is, by definition, completely constant in its total absence of daylight at all times, and this is undoubtedly its most important feature ecologically; it also achieves a remarkable level of near constancy in several other physical features. Temperature is substantially constant in most of the deeper parts of caves, both air temperature and the temperature of seepage water and static pools. It rarely varies more than 1°C throughout the year at any one place and approximates to the mean annual surface temperature for the locality, being largely dependent on latitude and altitude. Deep parts of caves remote from surface ventilation may, however, be slightly warmer owing to the effect of the geothermal gradient. In Britain cave temperatures range from about 13°C in the south-west to about 7°C in the north.

Conditions are often less constant in places close to streamways which are fed by surface water or in passages subject to through-draughts in caves having more than one entrance. They are also less constant in those parts of the true cave just beyond the threshold. This has led some workers, particularly Americans, to recognize three zones: a twilight zone identical with the threshold as defined above, a middle zone of permanent darkness but variable temperature, and a zone of constant temperature and complete darkness (see, for example, Mohr and Poulson, 1966, p. 55).

The situation with regard to humidity, which naturally concerns only the terrestrial cave environment, is similar to that for air temperature; where the latter is constant, the humidity is substantially so, but elsewhere it may vary to some extent. It is always high, the relative humidity usually being between 90% and 100%. Evaporation rates are consequently low, although the air is by no means always still and most caves are self-ventilating. The composition of the air is usually similar to that outside, although the concentration of carbon dioxide may sometimes be higher. This is often due to deposits of rotting flood debris, but in parts of caves remote from moving water carbon dioxide is sometimes found associated with beds of clay, and it may also accumulate in choked shafts and passages where there is little movement of air.

The characteristics of underground waters are dealt with elsewhere in this book and only the biologically more important factors need be mentioned here. The water of pools is usually distinctly alkaline (pH 7·2 to 8+) and has a high alkaline hardness; this is to be expected since the caves that concern us are in limestone. Organic matter is sparse in such water and, although this means that food for animals is short, it also means that the biochemical oxygen demand is very low and even in relatively static pools there is little depletion of oxygen. Since evaporative cooling is very slight, the temperature of pools is close to that of the rock and, in the deep cave, is usually within $\frac{1}{2}$°C of the air temperature.

Underground stream waters are much more variable. Where they are derived entirely from seepage through the rock, their characteristics may not differ much from those of pool water; but where any appreciable part of the flow is supplied by the sinking of surface streams both the chemical and physical features will be affected to some extent by surface conditions. Such streams usually contain more organic matter than pools; most of this is carried in from the surface, both as living organisms and as detritus or other dead organic material which itself can provide a source of food for animals. Even in such favoured situations, however, the cave habitat is poorly supplied with food in comparison with most surface situations, and caves as a whole, at least in temperate regions, are marked by the sparsity of food available in them. This is due to the absence of light and the consequential lack of any photosynthetic production of food material by plants.

The cave environment, then, is characterized by a general shortage of food resulting in low population densities and by a remarkable constancy of climatic factors. All the familiar daily and seasonal rhythms of light and temperature are effectively absent and climatic conditions are usually far from severe. Terrestrial animals do not require physiological or behavioural adaptations to protect them from desiccation and some of the complex homeostatic mechanisms employed by surface forms are unnecessary. It is in many ways an undemanding environment, except for the limitations imposed by a sparse food supply and the sensory problems arising from a complete absence of light.

The most obvious modifications shown by true cave animals are the negative ones associated with darkness, a tendency to lose their pigmentation and to be blind. It has often been assumed that other senses are hyper-developed in compensation for the lack of sight, but there is little firm experimental evidence for this (Barr, 1968a, p. 90). Some cave arthropods have, on their bodies and appendages, hair-like setae which are larger than those of their surface relatives, and, in some cases, their antennae are also longer. This seems to be associated with an increase in the tactile sense and perhaps also in the ability to detect vibrations in the air or water, although cave animals in general seem to be rather insensitive to low amplitude vibrations in the form of sound. Some cave fish are extremely sensitive to water-borne vibrations, and this is associated with an increase in the number and size of neuromast organs in their lateral lines (Poulson and White, 1969, p. 975). There is also some evidence of increased olfactory powers in cave fish, and it seems likely, though it is not yet proved, that the chemical senses are well developed in cave animals generally.

Studies on a wide range of cave animals show that many, but not all of them, are adapted to life on a limited food supply. This applies particularly to species which have been long established in the underground habitat, and is achieved by having a low metabolic rate with considerable economy in the use of energy (Poulson and White, 1969, p. 976); rates of growth, development and maturation are often low, and the life-span extended.

When maturity is reached, there is a tendency for few, but very large, eggs to be produced; and in some cases, although development is slow, there is a telescoping of the life cycle in that the number of juvenile or larval stages is reduced.

Many biological rhythms are clearly related to those of the environment. Animals often show peaks of activity, or other physiological functions, with a periodicity of 24 hours, the so-called *circadian* rhythm, while other cycles, particularly those concerned with reproduction, may be geared to the seasons; a cycle which repeats itself yearly is referred to as *circannual*. Many cycles, particularly those showing a circadian rhythm, appear to be controlled from within the organism by some sort of "biological clock"; but this has to be regulated or set, so as to maintain an accurate time base, by environmental fluctuations such as night following day. It might be expected that, in their substantially constant environment, cave animals would not show rhythmical behaviour, and to a considerable extent this is true. Most of them have periods of activity followed by quiescence, but in those animals which are most completely adapted to underground life this behaviour seems to be essentially non-rhythmic. Evidence has been found of circadian metabolic rhythms in some cave animals (Jegla and Poulson, 1968; Poulson and Jegla, 1969); but, as might be expected, these appear to be "free running" and not synchronized with the external day. Circannual rhythms are shown by the seasonal breeding cycles of some underground species (Jegla, 1969, p. 136), and it seems likely that, in regions where the flooding of caves is a seasonal phenomenon, this can act as the trigger which synchronizes the cycles on a calendar basis (Poulson and Smith, 1969; see also Hawes, 1939). Whether there are other underground seasonal variations large enough to trigger annual cycles of growth and reproduction, is a matter for conjecture. Nor is there agreement as to how widespread such cycles are among cave animals; where they do occur, they are certainly not as clear-cut as in most surface forms.

THE ECOLOGICAL CLASSIFICATION OF CAVE ANIMALS

The animals found in caves not only belong to a wide range of different systematic groups, some of which will be discussed shortly, but they also fall into several distinct categories when their status in relation to the habitat is considered; some species are found *only* in underground habitats while others which occur there are also found elsewhere. Controversy about the method of classification is as old as cave biology itself, but there is a broad measure of agreement concerning the major categories to be adopted, in what is usually called the Schiner-Racovitza system, after its originator and subsequent modifier (Schiner, 1854; Racovitza, 1907, p. 437).

If we confine ourselves to the Dark Zone, that is to hypogean cavernicoles, the following scheme, which is essentially similar to that adopted by Hazelton and Glennie in *British Caving* (1962, p. 361), can be used:

Troglobites. Species which are obligatory cavernicoles and, in nature, are unable to survive except in caves or similar hypogean habitats. This is substantially the definition used by Barr (1968a, p. 43).

Troglophiles. Species which are facultative cavernicoles. They are found living permanently, and successfully completing their life cycles, in caves; but they also do this in suitable epigean or endogean habitats.

Trogloxenes. Species which occur in caves but do not complete their whole life cycle there. These can be subdivided into accidental and habitual trogloxenes.

Accidental Trogloxenes. Those which have come into the dark zone accidentally, often by being carried in by streams sinking from the surface. They survive for varying lengths of time, and larval forms may metamorphose to the adult, but further generations are not established within the cave.

Habitual Trogloxenes. Those which frequent the dark zone of caves at certain times during their life cycle, but at other times live elsewhere and normally feed above ground. Some may use caves as roosts or lairs and "commute" regularly to the surface, others may merely enter caves for hibernation or aestivation.

Bats are the most familiar habitual trogloxenes; but there are birds which nest deep in caves, notably several species of Asiatic swiftlets and the South American oilbird. Some invertebrates, including flies and moths, overwinter in caves, but such hibernating forms seldom penetrate far beyond the threshold.

This system of classification was originally devised for cavernicolous animals; but some, at least, of its categories can usefully be applied to other cave organisms (see Chapter 11). It is unfortunate that the terms employed are tied etymologically to caves; a set of names which could be used for any hypogean situation would be convenient. The difficulty of delimiting caves allows some latitude, but there are hypogean habitats which cannot be considered as parts of caves, and in such situations terms like troglophile or troglobite seem hardly appropriate. Nevertheless, the terminology is well established, and there would have to be more cogent reasons than these for introducing completely new sets of terms.

Numerous suggestions have been made, however, for minor modifications or improvements to the basic scheme. One of the most recent discussions along these lines was introduced by Hamilton-Smith in the light of his work on the Australian cave fauna (Hamilton-Smith, 1971, 1972; Richards, 1971b, 1972). Hamilton-Smith proposed several additional categories and would restrict the term troglobite to species "modified to the cave environment". Many workers only use the term in this sense; but this introduces a morphological criterion and it seems preferable, as in the scheme adopted here, to keep the classification purely ecological, relying essentially on where the animal does or does not occur. The status

PLATE 10.1. *Proasellus cavaticus*, a hypogean isopod crustacean (Mendip specimen). Scale mark = 1 mm. (*Photo*: A. E. McR. Pearce.)

accorded to a species can only be provisional, but this is inevitable with any ecological system of classification since our knowledge of distribution must always be incomplete. In spite of its shortcomings the Schiner-Racovitza system is very useful in practice and, with minor variations, is almost universally adopted by cave biologists.

BRITISH CAVE ANIMALS

The part played by the Cave Research Group in the study of the British cave fauna has been mentioned. As early as 1947 a preliminary fauna list was published (Hazelton and Glennie, 1947) and in 1955 the Group started the publication of all its biological records, first as "Biological

PLATE 10.2. *Niphargellus glenniei*, a small hypogean amphipod known only from Devon (see Fig. 10.1 for distribution). Scale mark = 1 mm. (*Photo*: A. E. McR. Pearce.)

Supplements", then "Biological Records" and latterly as part of the "Transactions" series (Hazelton, 1955 *et seq.*). In addition to sequential lists of records, consolidated lists for individual caves and other hypogean locations are published on an area and Vice County basis (Hazelton, 1965 *et seq.*). The first and second editions of *British Caving* both contained

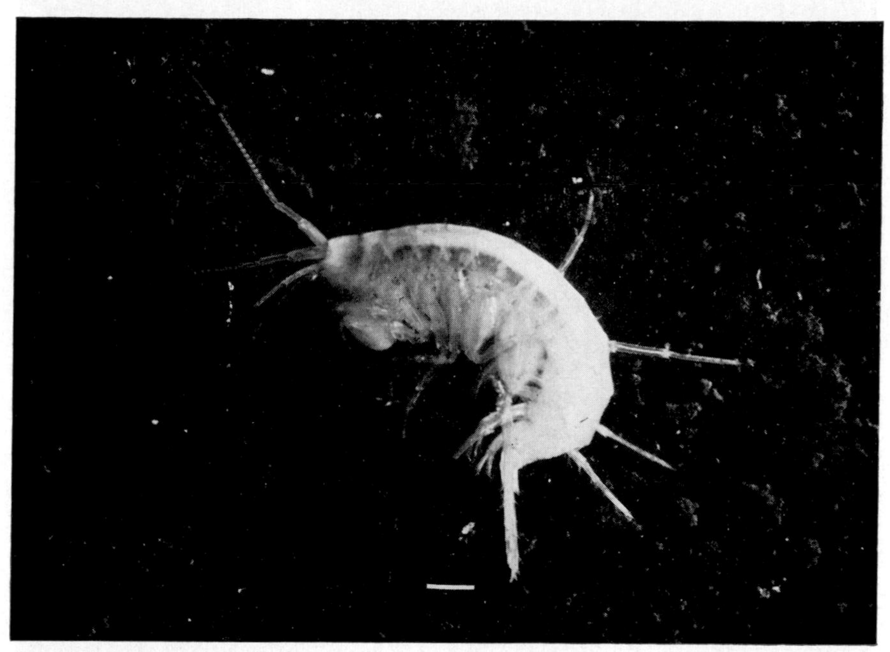

PLATE 10.3. *Niphargus fontanus*, a hypogean amphipod crustacean (see Fig. 10.1 for distribution). Scale mark = 1 mm. (*Photo*: G. T. Jefferson.)

lists of cave fauna (Hazelton and Glennie, 1953, p. 267, 1962, p. 367) and although these are obviously not up to date, they provide a picture which is still generally valid. This all amounts to a very considerable fund of biological information and makes it quite unnecessary to give fauna lists here; all that will be attempted is a review of some of the more interesting cavernicolous animals in their respective systematic groups. Anyone requiring full lists of the animals recorded and details of distribution should consult the relevant issues of the "Transactions", obtainable from the Cave Research Group,* and also the biological sections of *British Caving* and of the Group's Regional Monographs. For information about cave animals throughout the world the extensive review given by Vandel (1965) or Wolf's (1934-1938) monumental *Animalium Cavernarum*

* Now merged with the British Speleological Association as the British Cave Research Association.

PLATE 10.4. Cavernicolous collembolans belonging to the families Neanuridae, Hypogastruridae, and Onychiuridae. Scale mark = 1 mm. (*Photo*: A. E. McR. Pearce.)

Catalogus should be consulted. Leruth's work (1939) is particularly relevant since it deals with an adjacent part of Europe.

The systematic review will be confined to animals of the dark zone, and particular attention will be paid to troglobites. The threshold fauna can be no more than mentioned; animals are relatively numerous in this zone, but the presence of many of them is no doubt fortuitous, and workers have tended to neglect this part of the cave fauna. Nevertheless, there are

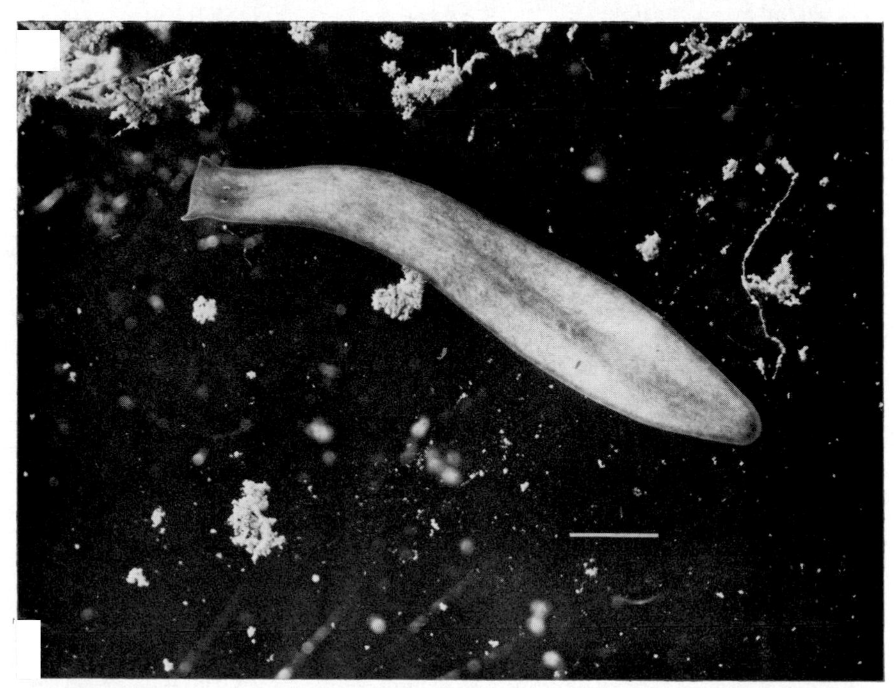

PLATE 10.5. *Crenobia alpina*, a troglophilic flatworm (Turbellaria, Tricladida). Scale mark = 1 mm. (*Photo*: A. E. McR. Pearce.)

some relatively conspicuous animals which are so consistently associated with the threshold that mention must be made of them. Outstanding among these are several spiders, and in particular two species, both belonging to the genus *Meta*, whose rather loose orb-webs are to be seen in many cave thresholds. *Meta merianae* (Scopoli) does not usually penetrate far beyond the entrance, but *Meta menardi* (Latreille) goes much deeper into the threshold and is so common there that, although sometimes found elsewhere, it has come to be known as the Cave Spider.

Many animals, particularly insects, are to be found overwintering in cave thresholds. Most of these are rather inconspicuous creatures such as small flies, and they attract little attention. There are, however, two moths which are quite large and so often to be seen on walls in cave thresholds that many cavers comment on them. These are the Tissue

PLATE 10.6. *Trechoblemus micros*, a carabid beetle (see Fig. 10.3. for its distribution in underground localities). Scale mark = 1 mm. (*Photo*: A. E. McR. Pearce.)

Moth, *Triphosa dubitata* (L.), and the rather handsome Herald Moth *Scoliopteryx libatrix* (L.). Both of these overwinter as adults which remain quiescent in some sheltered place, such as a cave, from late autumn until spring.

PLATE 10.7. *Meta menardi*, a spider which is very common in the deep threshold of caves. Scale mark = 5 mm. (*Photo*: A. E. McR. Pearce.)

The dark zone, although it normally carries a much lower density of animals than the threshold, has been studied more comprehensively, and a systematic treatment can be attempted. Known troglobites from British caves are confined to one phylum, the Arthropoda, but there are other groups which merit mention, either because they include well-established

PLATE 10.8. A short-bodied springtail (Collembola, Symphypleona). Scale mark = 1 mm. (*Photo*: A. E. McR. Pearce.)

troglophiles or because troglobitic forms occur in other parts of the world. This still leaves a number of groups, including all those that are exclusively marine, which do not concern us.

INVERTEBRATES OTHER THAN ARTHROPODS

Free-living Protozoa occur in caves and underground waters, but there are not thought to be any troglobites among them. This is virtually an untouched field as far as British caves are concerned; if a sample of water and silt from the bottom of a cave pool is examined under the microscope, ciliates can often be found and more careful searching sometimes reveals the beautiful quartz-grain case of *Difflugia* (Sarcodina). The Protozoa are

PLATE 10.9. *Scoliopteryx libatrix*, the Herald Moth. A threshold trogloxene. Scale mark = 5 mm. (*Photo*: A. E. McR. Pearce.)

PLATE 10.10. Larva of a fungus-gnat (Diptera, Mycetophilidae). Scale mark = 1 mm. (*Photo*: A. E. McR. Pearce.)

PLATE 10.11. Collembola belonging to the families Neanuridae, Onychiuridae, Entomobryidae, and Sminthuridae. Scale mark = 1 mm. (*Photo*: A. E. McR. Pearce.)

not the easiest animals to study, but anyone with patience and a microscope could make a useful contribution in this neglected field.

Sponges seem never to have been recorded from British caves and only very rarely from any others. This is surprising, since fresh-water sponges such as *Spongilla lacustris* have often been reported encrusting the insides of water pipes, as also have certain ectoproct polyzoans or "moss-animalcules", another group almost unrecorded from caves. There is no

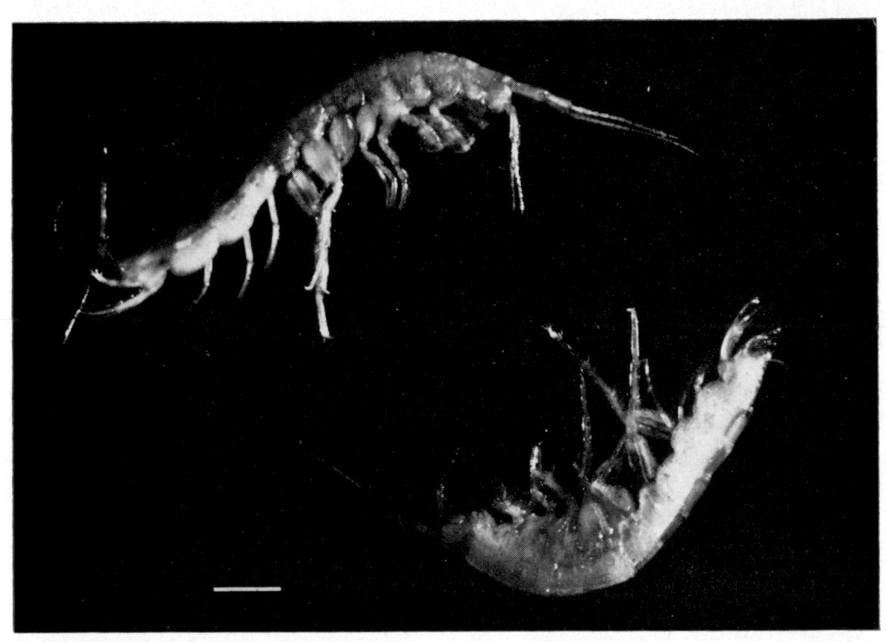

PLATE 10.12. *Niphargus aquilex*, a hypogean amphipod usually found in relatively superficial groundwater. Scale mark = 1 mm. (*Photo*: A. E. McR. Pearce.)

obvious reason why either of these should not occur in gently flowing underground streams and it would be worth while to look out for them.

The phylum Platyhelminthes, or flatworms, includes one class of free-living animals, the Turbellaria, and there are many underground forms among these. Some species belonging to the order Rhabdocoela are interstitial and a few, unrepresented in Britain, are cavernicolous. The order Tricladida includes marine and terrestrial forms, but it is the fresh-water triclads or planarians which are most familiar, and this is the group which is particularly rich in troglobites in continental Europe, North America and probably Asia. Planarians are not common in British caves and most of the species recorded are probably accidentals except for

Phagocata vitta (Dugès) (= *Fonticola vitta* (Dugès)) and *Crenobia alpina* (Dana) which are troglophiles. Both of these species are usually found on the surface, either at rather high altitudes or in spring-heads where the water is always cool, and, no doubt, the temperature factor is conducive to their occurrence in cave waters. Although neither species is found frequently underground, both have been reported from most cave areas in Britain.

The Nemertea are a largely marine phylum and, although two cavernicolous species are known, none has ever been reported from a British cave. Free-living nematodes, on the other hand, are present; if samples of cave water and silt are examined, they can often be seen, writhing in their characteristically sinuous way; but these animals are notoriously difficult to identify, and records are few. It is not certain whether there are any troglobitic nematodes, but there is little doubt that many rank as troglophiles.

Among the annelids, or segmented worms, the polychaetes are almost entirely marine, but three most interesting species are known from underground fresh waters in central and eastern Europe, the Balkans and Japan. They are generally thought to be ancient marine relicts, possibly having left the sea in Tertiary times.

The oligochaetes are mostly terrestrial and fresh-water forms, and this is the class to which the majority of cavernicolous annelids belong. Troglobitic oligochaetes, both aquatic and terrestrial, are known from various parts of the world, but all those recorded so far from British caves are normal surface forms. This is not to say that none of them are troglophiles; several of the lumbricid earthworms which have been recorded from caves seem able to establish themselves in mud banks containing organic detritus, and it is almost certain that some of the British enchytraeid worms are troglophiles. Many of the aquatic oligochaetes in caves are no doubt accidental trogloxenes, but some should be troglophiles. Ladle (1971, p. 313) listed several obtained from the phreatic waters of a gravel bed in Hampshire. Aquatic oligochaetes are mostly very small worms which are easily overlooked, and they have tended to be neglected.

The leeches belong to another class of the Annelida, the Hirudinea. The few which have been recorded from British caves are accidentals, but there are two or three species known from other parts of the world which appear to be troglobites and no doubt others are troglophiles.

The molluscs constitute one of the major phyla of the animal kingdom, but several groups are exclusively marine, and only two classes, the Bivalvia and the Gastropoda, concern us. Bivalves or lamellibranchs, are poorly represented in caves, although a few cavernicolous forms have been reported from various parts of the world. Very few bivalves of any kind have been found in British caves, and any that are present are probably accidental trogloxenes, although Moseley (1970, p. 48) reported finding numerous specimens of the tiny bivalve *Pisidium personatum* in Hazel Grove Main Cave in the Morecambe Bay area. These formed an isolated population in rich peaty mud in a stagnant pool and, if this population was maintaining itself, the species must rank as a troglophile.

The gastropods are much better represented in caves, numerous troglobitic species, both terrestrial and aquatic, being known from many parts of the world. In Britain, however, they do not comprise a very important element in the cave fauna; some slugs and a fair number of snails, including a few water snails, have been recorded, but most of them seem to be accidentals. This is a little surprising, since most of the species recorded belong to a family of small, mainly omnivorous, snails, the Zonitidae, many of which favour dark damp habitats. On a world scale the Zonitidae contain more cavernicolous species than any other family of molluscs and some of the British forms must surely be potential troglophiles. Empty snail shells are often numerous in parts of caves close to the surface, but these are of little significance since most of them have come in accidentally by one route or another.

ARTHROPODA

The majority of cavernicolous animals belong to the phylum Arthropoda, all four major classes of which contain troglobites somewhere in the world and three of them in Britain. The majority of known British troglobites are aquatic and belong to the class Crustacea, but those in the Insecta and Arachnida mostly belong to groups where there are taxonomic problems and when these are resolved more troglobitic species may be recognized.

The Class Crustacea

The Crustacea are largely marine, but there are many fresh-water and some terrestrial groups which could contribute to the cave fauna. Although as a class the Crustacea are very well represented underground, the contributions of its various groups are curiously uneven. In Britain only three subclasses, the Ostracoda, Copepoda and Malacostraca, are of interest to the cave biologist.

The ostracods are small crustaceans, either marine or fresh-water, with a bivalved carapace, easy to recognize as ostracods but difficult to identify.

Some of them swim while others crawl, and, particularly among the latter, there are many interstitial forms including some which occur in the deeper, hypogean, waters. Many ostracods are found in springs (see, for instance, Fox, 1964 and 1967) and for some this may be their normal habitat, but others, such as *Eucypris anglica* which appears in bourne risings, are, no doubt, underground forms which are occasionally carried to the surface. Such species are phreatobites, and some at least must warrant the title of troglobite; unfortunately our information is so sparse that little can be said with assurance on this topic.

Some ostracods have been found in British caves and long ago Lowndes (1932b) reported two species from the Corsham stone mines. *Cypridopsis subterranea* Wolf can provisionally be considered a troglobite; it certainly occurs in caves, being quite numerous in the film of water running over flowstone slopes in parts of Ogof Ffynnon Ddu, Breconshire. It has been recorded from surface situations, but only in springs (Fox, 1967), and seems to be a true hypogean animal. There must be other troglobitic ostracods in Britain, particularly in the genus *Candona*. The systematics of this genus need revision, but it includes many cavernicolous species in Europe and some, which may prove to be troglobites, certainly occur in British caves and underground waters.

A fair range of copepods has been recorded from British caves, and they are often present in wells and other groundwater sources. None of them appears to be troglobitic, but several are well-established troglophiles. Since there are troglobite copepods in other parts of the world, some may be found here.

Among the Malacostraca, the subclass containing most of the larger and more familiar crustaceans, only the Syncarida, Isopoda and Amphipoda are represented in the British underground fauna, but elsewhere there are also cavernicolous mysids and decapods. The syncarid *Bathynella natans* Vejdovski was first discovered in Britain in the Corsham stone-mine (Lowndes, 1932a and b), but in more recent years several have been found in the White Scar and Great Douk caves in Yorkshire (Gledhill and Driver, 1964). It seems, however, that this species is essentially an interstitial form (see Nicholls, 1946) and it has been found quite abundantly in this type of habitat in Devon and the Thames Valley (Efford, 1959; Spooner, 1961) and also in Stirlingshire (Maitland, 1962). Serban and Gledhill (1965) have shown that specimens from Great Douk cave belong to a particular subspecies, widespread in Europe, *Bathynella natans stammeri* Jakobi, which they feel might prove to be a separate species. Subsequent papers by Serban (1966a and b) and Husmann (1968) have produced a considerable amount of evidence in support of this, and it seems likely that the Great Douk specimens and all the British records should be named *Bathynella stammeri* (Jakobi).

Only one troglobitic isopod is known from Britain. This is the small aquatic *Proasellus cavaticus* (Leydig), more familiar under its old name of *Asellus cavaticus* Schiödte. It was first recorded in Britain from a well in Hampshire (Tattersall, 1930), and Lowndes (1932b) found a number in the Corsham stone-mines, but its main areas of distribution are in South Wales and Mendip, where it is essentially cavernicolous, occurring particularly on the underside of stones in underground streams and in the film of water flowing over stalagmite slopes. It has occasionally been found in gravels and in springs, presumably washed out from hypogean waters. Specimens from Mendip caves are consistently smaller than those from South Wales, the largest from Mendip being about 4 mm long, perhaps half the length of fully grown South Wales specimens. Dr E. M. Sheppard has examined many specimens and is inclined to think that the two forms might represent different subspecies.

There are also terrestrial isopods, and many troglobites are numbered among them, but not in Britain. Nevertheless, we have several interesting troglophiles in this group, perhaps the most widely distributed in caves and mines being *Androniscus dentiger* Verhf. This is a pretty little pink woodlouse—although usually darker when in surface situations—which seems almost as much at home in water as on land. Another interesting species is *Trichoniscoides saeroeensis* Lohmander, which was unknown in Britain until found by Moseley (1970) in disused mines in the Morecambe Bay area from 1964 onwards. The species was also found in 1968 in two caves in Co. Clare, Ireland. This minute woodlouse, under 3 mm in length, has so far, in Britain and Ireland, only been found underground, but in Scandinavia and France it occurs in endogean situations close to the coast (see Sheppard, 1968, 1971). Presumably it also occurs in similar situations here and must be considered a troglophile. Some of the other terrestrial isopods reported from British caves may also be troglophiles, but most are probably accidentals.

The amphipods are particularly well represented in the British cavernicolous fauna, there being five species which can be considered troglobites. Detained accounts of the distribution of these species have been given by Glennie (1956a, 1967) and Jefferson (1974) and they need only be commented on briefly here. *Crangonyx subterraneus* Bate has been recorded from two caves (Ogof Pant Canol, South Wales, and Gough's Cave in Mendip), but it has also been found in wells and, in some numbers, in interstitial waters in southern England. It appears to be primarily an inhabitant of interstitial waters.

Niphargus aquilex aquilex Schellenberg is probably the most widely distributed underground amphipod in Britain and right across to Central Europe (Straškraba, 1972a, p. 44). It is found occasionally in caves, but is essentially an animal of rather superficial interstitial waters. It occurs widely in Wales and southern England, but seems to be much less common in the more northerly parts of this range. *Niphargus kochianus kochianus* Schellenberg is a phreatobite; there are numerous records from wells and similar groundwater sources, particularly in chalk and limestone areas, south of a line running from the Bristol Channel to the Wash, but not including the extreme south west (Fig. 10.1). Outside this area there are only isolated and somewhat dubious records of this species. It is surprisingly rare in caves, only having been recorded from St Cuthbert's Swallet in Mendip, Pen Park Hole, Bristol, and Holwell Cave in the Quantock Hills. There is an Irish subspecies, *Ni kochianus irlandicus* Schellenberg, known from caves in Co. Clare and from deep water in Lough Mask.

Niphargus fontanus Bate is the hypogean amphipod most frequently found in British caves, where it has much the same distribution as *Proasellus cavaticus*, namely the cave areas of South Wales and Mendip and a few adjacent localities. It is, however, not confined to caves but seems also to occupy deep phreatic water, and it is frequently reported from wells in many parts of southern England, but again, not in the extreme south-west (Fig. 10.1). This is precisely the area where the remaining British species, *Niphargellus glenniei* (Spooner), does occur (Fig. 10.1). This species was originally described (Spooner, 1962) under the name of *Niphargus glenniei*, but it differs in a number of ways from other members of the genus *Niphargus* and has been transferred (Straškraba, 1972b, p. 85) to a related genus, *Niphargellus*. It has so far been found only in Devon and nowhere else in the world; it seems to be essentially an interstitial form, but also occurs in caves both in the north and south of the county.

Another amphipod frequently found in caves in all parts of the country is the common fresh-water shrimp *Gammarus pulex* (L.) (= *Rivulogammarus pulex*). From the frequency with which this is found underground it might be expected to be a troglophile, but its status is by no means certain. There is some doubt whether it ever establishes permanent populations in caves, and it may be merely an accidental trogloxene.

In the whole of the Malacostraca, which is admittedly mainly a marine group, there are something less than twenty fresh-water species which are truly native to Britain, and that seven of these should be hypogean forms is surprising in view of the paucity of our underground fauna. Details, including identification and distribution, of all British fresh-water malacostracans are given by Hynes *et al.* (1960).

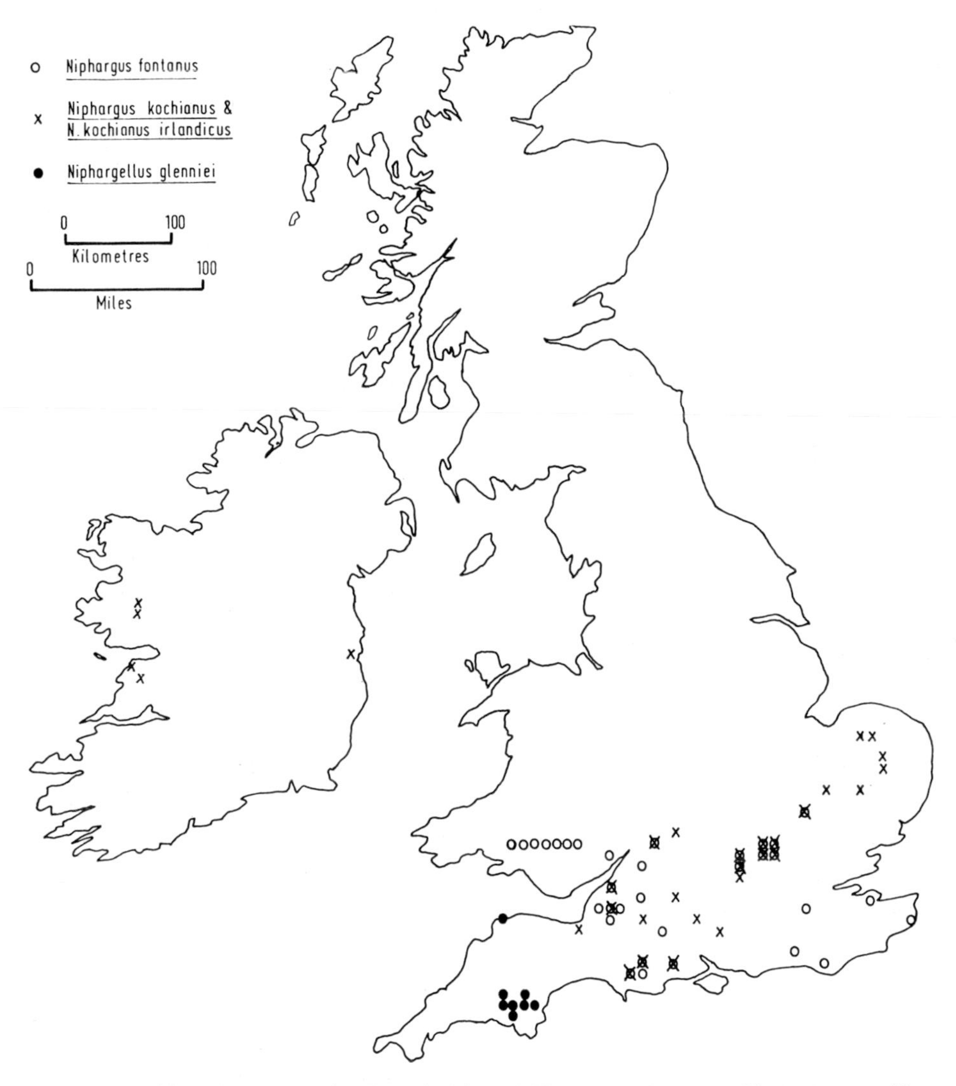

FIG. 10.1. Distribution in the British Isles of *Niphargus fontanus*, *N. kochianus*, *N. kochianus irlandicus* and *Niphargellus glenniei* (maps by courtesy of Nature Conservancy Council, Biological Records Centre).

The Class Myriapoda

It is convenient to treat the whole of the myriapods as one class even while recognizing that this may be an artificial grouping. They are not particularly well represented in British caves and none of them appears to be troglobitic here, although many are elsewhere in the world.

The Pauropoda and Symphyla are essentially endogeans and usually said to be of little interest to the cave biologist. Nevertheless there has been a steady trickle of reports of symphylans from British caves and some five species have been recorded. This is a most unusual situation in that symphylans appear to be much more common in British than in Continental caves (Turk, 1967, p. 157). Some symphylans feed on organic detritus and there is no obvious reason why they should not be troglophiles. At least *Symphylella isabellae* Grassi, the commonest species in British caves, may be troglophilic.

The diplopods or millipedes include many completely cavernicolous species and, although none of these occurs here, some 25 species of diplopods have been recorded from British caves and some of these are troglophiles. This is undoubtedly the case with the commonest of our cave millipedes, *Polymicrodon polydesmoides* (Leach), which is very widely distributed. Turk (1967, p. 146) has found that in underground populations of this species the number of ocelli varies from none to 23 as compared with a maximum of 30 or so in surface individuals, and he suggests (p. 155) that some populations may be in the process of becoming troglobites. Several of the other species recorded from British caves are also Troglophiles, including *Brachydesmus superus* (Latzel) and *Blaniulus guttulatus* (Bosc).

Chilopods or centipedes are much less common than millipedes in caves although a few troglobite species are known from parts of Europe and North Africa. They are not often found in British caves but some nine species have been recorded; it is doubtful whether any of them, with the possible exception of *Lithobius duboscqui* Brolemann, can rank even as a troglophile.

The Class Insecta

There are more species in the Insecta than in the whole of the rest of the animal kingdom, and the list even of those recorded from British caves is quite extensive. Nevertheless, they can be dealt with relatively briefly, as only three orders call for much comment. Of the 30 orders into which the class is usually divided only seven are known certainly to contain troglobites, but even so the number of species involved throughout the world is enormous; in Britain, however, very few of the insects found in caves appear to be troglobites. The seven orders known to contain troglobitic forms somewhere in the world are the Collembola, Diplura, Thysanura, Dictyoptera, Orthoptera, Diptera and Coleoptera; but only the first and the last two are of any great interest in the British context.

Insects belonging to other orders are frequently seen in caves. This applies particularly to aquatic forms, such as water bugs belonging to the Hemiptera, and to those orders such as the Plecoptera (stone flies), Ephemeroptera (May flies) and Trichoptera (caddis flies) which have aquatic juvenile stages. These are carried underground from surface streams, and the adults often emerge in stream passages where they provide food for predators such as spiders. Such forms are accidental trogloxenes as they do not reproduce and establish populations in a cave.

The Collembola or Springtails

This order of aberrant wingless insects, all small and some minute, is a difficult one for the cave biologist. There are two main difficulties; one that many collembola are endogeans, devoid of pigment and blind or with reduced eyes. It is therefore difficult to decide which of the numerous species of such springtails that occur in caves in many parts of the world are likely to be troglobites. The other problem arises from the systematic difficulty of the order; in many cases specimens can only be identified to species groups, and the fact that such a group is represented both underground and on the surface does not rule out the possibility that individual species, if indeed the group really consists of more than one, may be troglobites. Some idea of the problems involved in the case of one group, *Hypogastrura* (*Schaefferia*) *emucronata*, was given by Lawrence (1959, p. 121).

Most of the families of Collembola contain recognized troglobites somewhere in the world, and the number of troglophilic species must be vast. Springtails require dampness, and many feed on moulds, fungal hyphae, bacteria or organic detritus, so that it is not surprising that such species can become established in caves. In their turn they provide food for predatory mites and small beetles.

Collembola are common in most British caves, and the number of species or species-groups recorded is large. These are drawn from some nine families and mostly are the relatively long-bodied forms belonging to the suborder Arthropleona; but some, such as *Arrhopalites* spp., are globular and belong to the Symphypleona. Some of the species recorded may be accidentals, but most are well established underground and are at least troglophiles. How many are troglobites it is impossible to say at present. *Onychiurus schoetti* (Lie Pettersen), the British distribution of which is shown in Fig. 10.2, has only been found underground in this country (Lawrence 1960, 1968) and this is also true of *O. dunarius* Gisin and *Pseudosinella dobati* (Gisin) (H. Gough, private communication). These species can therefore be accepted provisionally as troglobites, and so, rather more firmly, can *Sphyrotheca patritzii* (Cassagnau and Delamare-

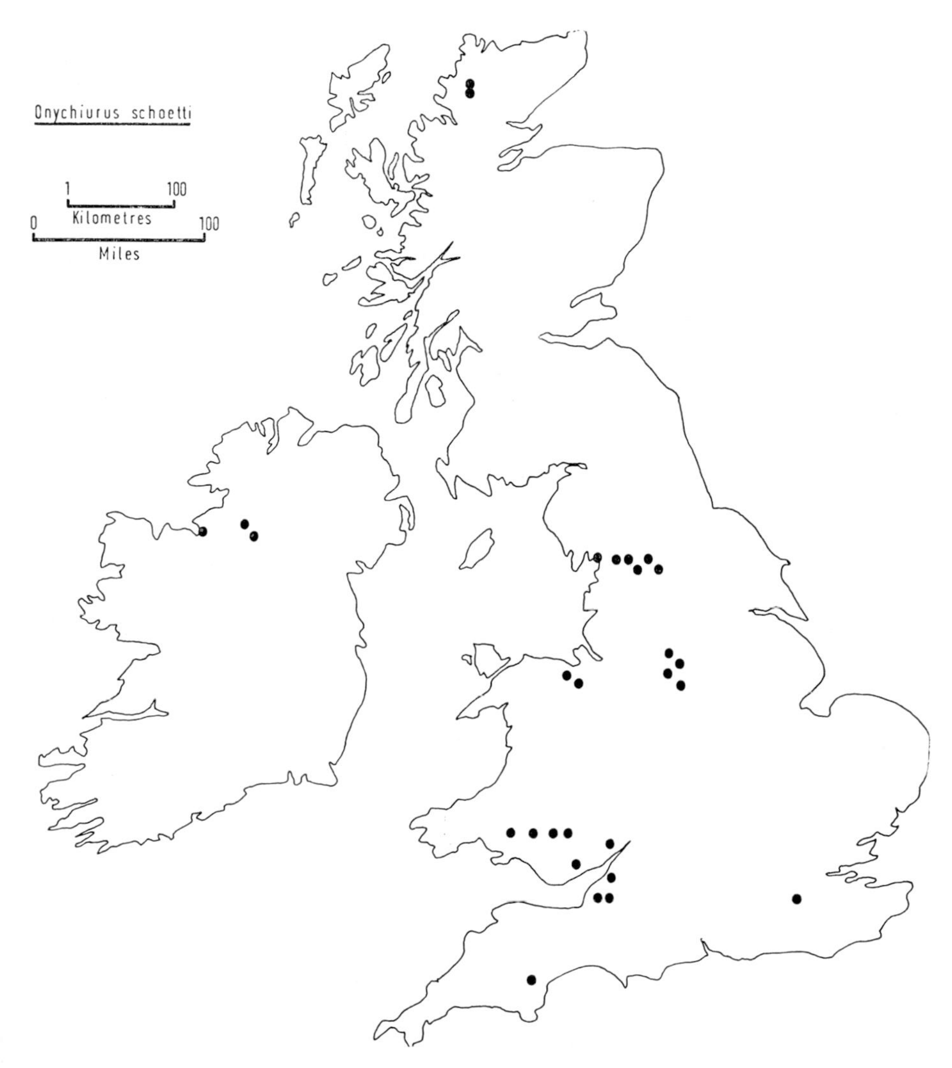

FIG. 10.2. Distribution in the British Isles of *Onychiurus schoetti*.

Deboutteville) (= *Pararrhopalites patritzii*). This symphypleonan is known only from caves, hitherto from further south in Europe, but now also from Devon where W. G. R. Maxwell found several in 1970 (Gough, 1972). Further ecological and taxonomic work can be expected to reveal other troglobites, particularly among some of the "species groups" recorded for most areas.

The Diptera or Two-winged Flies

Many species of this large order are found in caves, sometimes in considerable numbers; but, even on a world scale, surprisingly few of them are troglobites (see Matile, 1970, p. 204). The list of records from British caves is quite extensive, many being adults hibernating in thresholds or accidentally present in the cave, but some are undoubtedly troglophiles with all stages of the life history being completed underground. It is very doubtful whether any rank as troglobites; Mr A. M. Hutson, an authority on this order, says (private communication) that there are no British Diptera modified specifically for the cave environment.

A winter-gnat, *Trichocera maculipennis* Meigen (family Trichoceridae), is frequently found deep in caves and is much more common there than in surface habitats; it is certainly a troglophile. The fungus-gnats or Mycetophilidae are represented underground by several species, the most interesting being *Speolepta leptogaster* (Winnertz), which is often said to be troglobitic. The larva, which occasionally has depigmented ocelli, lives on damp cave walls, but the pupa hangs down freely, suspended from the wall or the roof (Cheetham, 1920; Edwards, 1924, p. 566; Madwar, 1937, p. 56). This fly normally breeds entirely underground, but a few adults have been recorded away from caves. This does not necessarily rule out the possibility of its being a troglobite—it would have to be shown to breed in epigean situations for that—but it casts enough doubt upon it to make it prudent to consider it, provisionally at least, as a troglophile. Flies belonging to the closely related family Sciaridae are also fairly common in British caves, but these are notoriously difficult to identify and little can be said at present about their status as cave animals.

Triphleba antricola (Schmitz), a member of the rather more advanced family Phoridae or coffin flies, occurs widely in caves and is rare elsewhere; it is a troglophile, and in some parts of Europe has been considered a troglobite. Two species belonging to the family Heleomyzidae, *Scoliocentra villosa* (Meigen) and *Heleomyza serrata* (L.) (= *Leria serrata*), are frequently seen underground. Both species are common on the surface, but seem to be well established in caves and are presumably troglophiles. The family Sphaeroceridae is also well represented underground, particularly by species belonging to the genus *Limosina*. One of these, *Limosina racovitzai* Bezzi, seems to be very rare on the surface.

One other family of dipterans deserves mention in the cave context, the Nycteribiidae, a family of highly modified wingless flies which are parasitic on bats. These insects are pupiparous, that is to say the larval stages are completed within the mother, and they do not normally leave the host except when the female deposits her young. Individuals are only rarely found in caves away from their hosts, but one of the three British species of nycteribiids, *Stylidia biarticulata* (Herman) (= *Nycteribia biarticulata*), has been recorded a few times (see also Chapter 12).

The Coleoptera or Beetles

This is an enormous order, in numbers of species much the largest order in the animal kingdom. It also includes a very large number of cavernicolous forms, but in Britain the record is very disappointing. The great majority of cavernicolous beetles belong to just two superfamilies, the Caraboidea and the Staphylinoidea, both well represented in the records of Coleoptera from British caves, but by species which, in the main, seem to be only accidentals. It is doubtful whether the status of troglobite can be claimed for any British beetle, and only a few appear to be troglophiles.

Perhaps the most interesting beetle found in British caves is *Trechoblemus micros* (Herbst). This is a rather rare terrestrial beetle belonging to the family Carabidae, or ground beetles, which has been recorded from several caves, particularly in the Mendip area, but also in Derbyshire and Tipperary. The adult has been found rarely on the surface; but the few records of the larva are from caves (Bagshawe Cavern, Derbyshire, and Pen Park Hole, Bristol). Figure 10.3 shows the total recorded distribution of the species in the British Isles and it has also been found in caves in many parts of Europe.

Although the larva is blind and both larva and adult possess long sensory hairs, these are by no means certainly adaptations to the hypogean habitat and are not comparable with the modifications shown by the closely related North American cave beetles belonging to the genus *Pseudanophthalmus*. *Trechoblemus micros* may be troglobitic (Hazelton, 1963, p. 41), perhaps living mainly in crevices, but it seems likely that it also breeds in some endogean situations and it is probably more realistic to consider it a troglophile. Among the Staphylinidae or rove-beetles

Quedius mesomelinus (Marsham) is certainly a troglophile and the same may be true of *Lesteva pubescens* Mann. and *Ancyrophorus aureus* Fauv., all of which are widely distributed in British and European caves.

The numerous water-beetles recorded from British caves seem to be accidental trogloxenes except for four species belonging to the family Dytiscidae. Two of these, *Agabus guttatus* (Paykull) and *Agabus biguttatus* (Olivier), appear to be able to breed underground and so rank as troglophiles. The other two, *Hydroporus ferrugineus* Steph. and *Hydroporus obsoletus* Aubé, are probably also troglophiles, but it is just

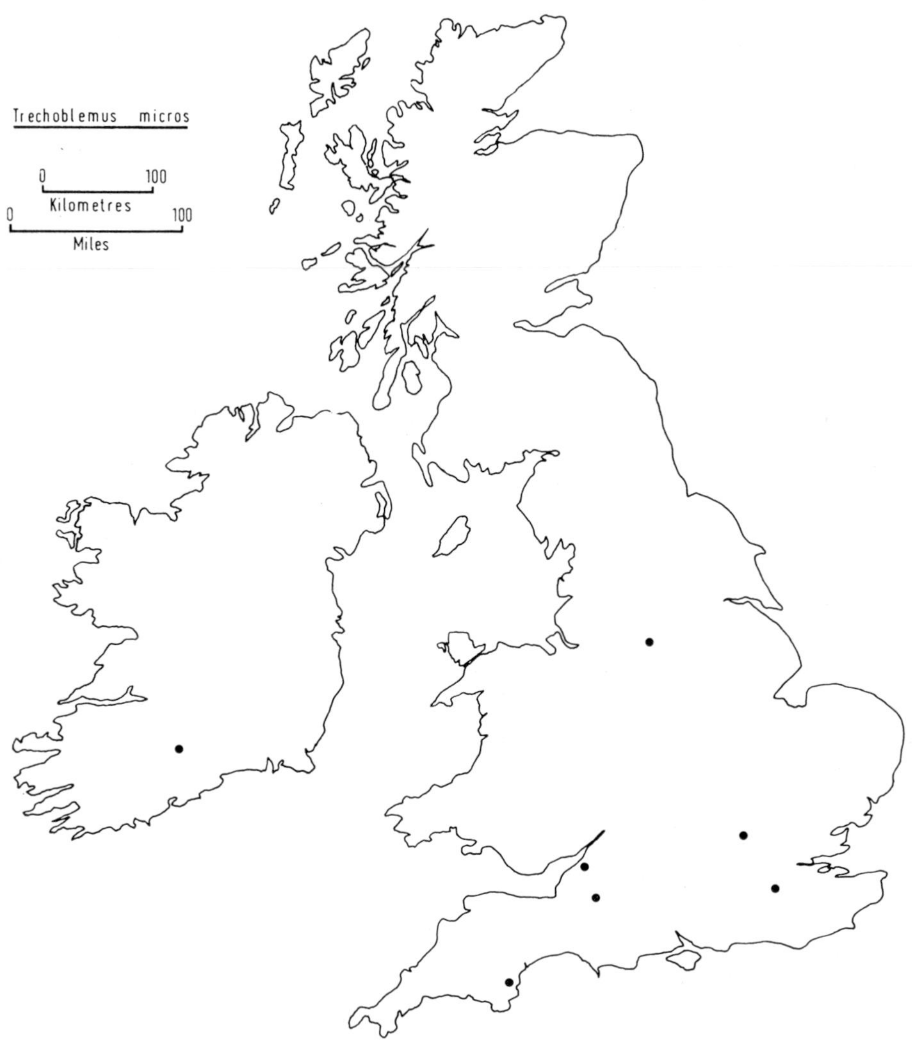

FIG. 10.3. Underground distribution in the British Isles of *Trechoblemus micros*.

possible that either or both may be troglobites; the evidence for this has been reviewed by Hazelton and Glennie (1962, p. 379).

The Class Arachnida

Most of the orders of arachnids are represented in the cavernicolous fauna of some part of the world, but in Britain it is only the spiders (order Araneida) and mites (order Acari) which are of real interest. It is true that a few species of false scorpions (order Pseudoscorpiones) and harvestmen (order Opiliones) have been recorded from British caves (Turk, 1967, pp. 153 and 154) but there is no evidence to indicate that these are anything but accidental trogloxenes. Both groups include troglobites and troglophiles abroad, however, and the possibility cannot be ruled out that some of the pseudoscorpions, at least, could be troglophiles.

The Araneida or Spiders

Some of the threshold spiders have already been mentioned, but this order is also well represented in the dark zone and includes many cavernicolous forms, particularly in the Mediterranean region. The records from British caves include several accidentals; but if these are ignored, only one family, the Linyphiidae, is represented in the dark zone, and of these the genus *Porrhomma* is outstanding. These are very small and unobtrusive spiders which usually spin small patches of webbing in obscure corners. There are ten British species and five of these have been recorded from caves. *Porrhomma pygmaeum* (Blackwall) and *P. pallidum* Jackson, although not often found in caves, seem to be troglophiles, as certainly are *P. convexum* (Westring) and *P. egeria* Simon. The only British troglobitic spider also belongs to this genus. It is *Porrhomma rosenhaueri* (L. Koch) which has been recorded from Tipperary and Co. Clare in Ireland and from South Wales. This species is rather pale in colour and its eyes are either vestigial or completely absent (Locket and Millidge, 1953, p. 332).

Two other linyphiids, *Lessertia dentichelis* (Simon) and *Lephtyphantes pallidus* (O.P.-Camb.), although usually found in the threshold, nevertheless sometimes penetrate deep into caves. These and several other members of the family seem able to establish themselves as dark zone troglophiles where conditions are suitable and the food supply in the form of small insects is adequate.

The Acari or Mites and Ticks

The acarines, with the exception of the ticks, are extremely small animals. There are terrestrial, aquatic and parasitic forms, all three being represented in cave faunas. The list of terrestrial mites recorded from British caves is extensive (Turk, 1967, p. 150; 1972) and although some of these are accidentals, many are undoubtedly troglophiles and several are troglobites.

Among the mesostigmatid mites two species belonging to the genus *Eugamasus* are particularly widespread in British caves. These are *Eugamasus magnus* (Kram.) and *E. loricatus* (Wankel); both seem to be troglophiles, the latter only very rarely having been found anywhere other than underground. *E. magnus* is at the centre of a "species complex" —a whole range of closely related forms, some no doubt specifically distinct, but others merely subspecies or varieties of one grade or another. Two members of the complex which have been given specific status occur in Britain (Turk, 1967, p. 142) and these are both troglobites. One of them, *Eugamasus tragardhi* Oudemans, was first recorded from the Grotte d'Istaürdy in France as a variety of *E. magnus*, but was later accorded specific rank. Although very rare, this species has been found in several British caves. The other is *E. anglocavernarum*, first found by E. A. Glennie in Bagshawe Cavern in 1938 and described as a new species by F. A. Turk (1944). This mite has subsequently been found again in Bagshawe Cavern and in a few other underground localities in various parts of Britain, but the number of records is small. Other mesostigmatid mites are frequently recorded from British caves, particularly species of the genus *Veigaia*, and several of these are undoubtedly well-established troglophiles.

The other group of mites in which cavernicolous forms are particularly well represented is the Prostigmata, and among these the genus *Rhagidia* stands out in Britain. *Rhagidia spelea* (Wankel) is one of the commonest mites in British caves and is a troglophile, as also is *R. gigas* (Canestrini) although this is only rarely found. *Rhagidia longipes* (Trag.), also uncommon in Britain, is a troglobite and this status may also be accorded provisionally to two new species belonging to this genus recently described by Turk (1972, p. 190). These are *Rhagidia odontochela* Turk, described from a single female specimen found by W. G. R. Maxwell in Ogof Cynnes, Breconshire, in 1968, and *R. vitzthumi* Turk, of which two specimens, both females, have been collected, one by Dr Pat Cornelius (*neé* Browne) from Rift Cave, Devon, in 1968 and one by W. G. R. Maxwell from Spratts Barn Mine, Oxfordshire, in 1969. The same two workers also collected during 1968 one specimen each of a third new species of prostigmatid mite which has been named (Turk, 1972, p. 193) *Bonzia brownei* after Dr Browne. Both specimens were found in caves

(Rift Cave and Reed's Cavern, Devon) and this species, too, may be looked upon provisionally as a troglobite.

Many other terrestrial mites have been recorded from British caves. These belong not only to the two groups already mentioned but also to others, particularly the Endeostigmata (often included in the Prostigmata) and the Astigmata. Although some of these, such as the astigmatids, are certainly accidentals, others will no doubt prove to be troglophiles, and as work continues, further troglobitic forms will be found.

Some of the terrestrial mites found in caves are closely associated with water and a few may well prove to be substantially aquatic in habit; they are not, however, water mites in the usually accepted sense as they do not belong to the group often called the Hydracarina. True water mites are seldom found in caves, and this is rather surprising since many hypogean forms which inhabit phreatic waters are known. Several species of such mites have been reported from interstitial and phreatic waters in Britain (Turk, 1967, p. 152; Gledhill, 1971) and some at least of these might be expected to occur in caves; indeed two, *Soldanellonyx monardi* Walter and *S. chappuisi* Walter (family Porohalacaridae), have been found in caves in Yorkshire. Although blind and depigmented, these species occur in lakes as well as in groundwater, and when found in caves should perhaps be looked upon as troglophiles. The sparsity of cavernicolous water mites in Britain is probably more apparent than real and search in the right situations would no doubt yield more, but it is curious that relatively few have been reported from caves in any part of the world.

The parasitic acarines occurring in caves in this country are those associated with bats, and much the most common is the tick *Ixodes vespertiliones* Koch which can infest any of our native bats. Several other species of *Ixodes* have also been recorded as well as four or five species of mesostigmatid mites which are also bat parasites (see also Chapter 12). All of these can be considered to be, like their hosts, trogloxenes.

THE CHORDATA

The protochordate groups, being marine, do not concern us, and among the Craniata or vertebrates it is only in two classes, the fishes and amphibia, that troglobites occur. The three "higher" classes, the reptiles, birds and mammals, all contain cave-dwelling forms in some part of the world, but these are either troglophiles or, more usually, trogloxenes. Pre-eminent among the latter are bats, a group fairly well represented in the British cave fauna and which are dealt with separately in Chapter 12.

Hypogean fishes, of which some thirty-two species are known (Greenwood, 1967, p. 262) all belong to the order Teleostei and occur in many parts of the world, but especially America and Africa. They are, curiously, absent from Europe. These fishes show reduction, or even virtual absence, of the eyes, and the skin is depigmented and often without scales. Most of the fish recorded from underground streams in Britain are accidentals and call for no comment, but mention must be made of the "white cave fish" which have been reported from various regions, but particularly from Durham, Yorkshire and South Wales. The majority of these are brown trout which have become very pale. There is little evidence of any structural modification, and the blanching is presumably a normal chromatophore response to darkness. It is not known whether these fish establish permanent populations underground, in which case they would be troglophiles, or whether their presence there is a purely temporary or accidental phenomenon. There are places where they can almost always be seen, and there is evidence that some individuals have been underground for a long time, but this does not preclude the possibility of interchange with surface parts of the river system.

The Amphibia constitute one of the smallest vertebrate classes but include some of the most remarkable of cave animals. *Proteus anguinus*, the "Olm", was not only the first animal to be recognized as a cavernicole, but must, even today, be the one that most people have heard about. This long slender salamander is, except when very young, eyeless and depigmented —not quite white since its red blood gives it a pale pink hue. It lives in underground streams in Yugoslavia and adjacent parts of Italy (Vandel, 1966, p. 181), but there are seven other known species of cave salamander and these are all North American (Brandon, 1971). The American forms belong to a different family from *Proteus* but resemble it in that all except one of them retain essentially larval characters and remain aquatic throughout their lives. Amphibia are not found in British caves except for the occasional frogs and newts which sometimes find their way into them accidentally and do not seem to survive for long.

CONCLUSION

This brief review of the British cave fauna is clearly incomplete, not only because many animals which have been recorded have not even been mentioned, but also because there must be many more waiting to be found. The fact that any comprehensive review of this fauna is possible is a tribute to the work of both the specialists who have identified the material and those, mostly amateurs, who have collected it underground. It is to be hoped that this work will continue to develop, not only with the aim of finding new underground species but also in order to extend our knowledge of the distribution and ecology of hypogean animals. There are, however, problems. The necessity of collecting specimens for identification is to some extent in conflict with the need for conservation, and this work should only be undertaken by those prepared to take great care with the preservation of specimens, to keep accurate data records and to see that all material collected is properly identified and recorded. Details of the necessary techniques and procedures are given by Hazelton and Glennie (1962, p. 348) and by Driver (1963). Data on fauna and flora collected by the Cave Research Group are not only published by the Group, but are also passed to the Biological Records Centre of the Nature Conservancy where they can be handled by computer. The storage and retrieval of biospeleological information lends itself to computer methods (see Chapter 14), and systems at various levels of sophistication are being developed for this purpose (Wilcock, 1970; Reynolds, 1971).

THE CAVE COMMUNITY

The concept of the community is central to modern ecological thinking. Put very simply, it is that all living things must be members of an integrated community consisting of various kinds of organisms capable of achieving some state of dynamic equilibrium. The relationships between members of a community are complex, but the overriding one is trophic, i.e. concerned with feeding. Food provides an organism with two things: nutrients, the materials out of which it builds its body, and energy; in other words, it is also the fuel which drives it. The old idea of the food-chain was too simple—it is usually a complex web or nexus rather than a chain—but it came close to the fundamental organization of the community: a sequence of "trophic levels" with nutrients and energy passing from one to another.

The first trophic level or link in the chain must consist of *autotrophs* or "producers", organisms which can take up energy and, as it were, "store" it in complex organic substances which they build up from simple inorganic raw materials such as carbon dioxide, water and mineral nutrients. These producers are in general *photo-autotrophs*, using light energy for synthetic purposes, and much the most important are the green plants, whether unicellular algae floating in the sea or higher plants on land. All succeeding trophic levels consist of *heterotrophs* or "consumers":

herbivores, various grades of carnivores and the decomposers. Each consumer level obtains both its energy and its building materials from the preceding one, and all are ultimately dependent on sunlight through the photosynthetic activity of plants. The chemical elements passing along the food-chain are continually being returned for recirculation; carbon and oxygen are passed back to the air as carbon dioxide produced during respiration, and minerals are returned to the soil as a result of the activities of the decomposers, viz. moulds and bacteria. Energy, however, cannot be recycled; it is continually being lost, degraded to heat, all along the food-chain. A community in equilibrium, therefore, requires an external supply of energy; but, given that, the recycling of raw materials means that it is self-sustaining.

All this calls into question the validity of the term "cave community"; the external source of energy, at least in the form of sunlight, is missing, and it must be conceded that the organisms living in caves are, in general, only part of a larger community, starting, as it were, on the surface where solar energy is available. The term "cave community", however, has the merit of convenience and, with this reservation, will be used. Only the situation in the dark zone of caves will be considered, but much of what is said will apply equally to shallow disused mines, although not to deep workings which are only kept free of water by pumping and which, owing to the geothermal gradient, are warm; the situation in such pits is quite different (Jefferson, 1960, p. 72) and will not be discussed here.

The food nexus in a cave is simpler than that of a typical surface community in that the variety of the organisms involved is more restricted. On the other hand it is complicated by the fact that many cave animals seem to be markedly polyphagous, eating a variety of food materials, and generalization is made difficult by the surprising lack of information about the feeding habits of many species. If we start at the end of the nexus, some purely carnivorous forms can be picked out. Among the terrestrial animals in British caves the spiders are predatory, mainly on insects, and so are most of the beetles; some of the mites feed on Collembola and probably on other mites and nematodes as well as on the eggs of any of these which may be available. In other parts of the world the range of carnivores is often greater; there are even troglophilic snakes in some tropical caves, which prey upon roosting bats. The bats themselves, at least most of those found in caves, are insectivorous carnivores; but they feed outside, and their importance in this context is as transporters of organic material into the cave.

In the case of the aquatic cave fauna of this country the blanched trout, like the true cave fishes and amphibia of other parts of the world, are predators, feeding mainly on invertebrates carried in from the upstream surface stretches of the river. The planarians and the water beetles found underground are also essentially carnivorous. Amphipods such as *Niphargus* are actively predatory, but these, like many other cave animals, both aquatic and terrestrial, are catholic in their tastes and better looked upon as omnivores.

Herbivores, in the usual sense of animals feeding on green plants, are clearly not found underground, but there are numerous creatures which live on the dead organic material which, from one source or another, is present in caves. In some cases these animals may be feeding directly on this material, but usually their food seems to consist largely of fungi or bacteria which are growing on the organic matter and breaking it down (see Chapter 11). This organic material is the starting point of the cave food-chain or nexus; it represents the initial source of energy as far as the cave community or ecosystem is concerned. It must in general come into the cave from outside, where initially it was built up by green plants using solar energy. The organic material itself can be of various kinds, but the most important are guano, consisting of the droppings of trogloxenes which feed outside, and detritus derived largely from plant debris carried passively into the cave, especially by floods.

Guano, particularly that produced by bats, but in some cases also by birds and even cave crickets, often supports large cave communities in tropical and warm-temperate regions. These are essentially terrestrial ecosystems, and it is noteworthy that the animals involved, which are often present in great numbers, are almost all either troglophiles or trogloxenes. Many accounts have been given of these remarkable systems, among the more recent being those of Mitchell (1970), Harris (1970), Peck (1971), Richards (1971a, p. 23) and Horst (1972). Poulson (1972) has discussed the differences to be expected between guano ecosystems of various kinds and what he calls deep-cave communities or detrital systems, such as those based either on rotting wood and other plant debris, or on mud with a high organic content. It is only the detrital systems which occur in Britain. That is not to say that none of our caves contain guano; wherever bats roost there will be droppings which will make a contribution to the organic matter of the cave but not in the quantities needed to support a full guano community. The herbivores of the food nexus in British caves, then, are essentially detritus feeders living either directly on dead organic material or on micro-organisms. This applies both to terrestrial and aquatic animals, and all our troglophiles and troglobites other than the pure carnivores and the parasites feed either entirely or, in the case of the omnivores, partly, in this way. Virtually nothing is known about the parasites, other than those of bats, but there can be no doubt that some are present.

The "biomass", usually expressed as weight of living matter per unit area, is extremely small in deep-cave communities although much higher in tropical guano systems. Poulson and White (1969, p. 973) give estimates of biomass in typical North American cave systems; these are of the order of 1 g/ha in drip-fed pools and 20 or 30 g/ha in terrestrial passages; figures are naturally higher for the twilight zone (300) and the underground course of a surface stream (50), but the real basis for comparison is given by the figure of one million g/ha for the entrance. The low biomass reflects not only the scarcity of organisms but also their small size; Barr and Kuehne (1971, p. 49) illustrate this in the case of the relatively numerous and conspicuous cave beetle, *Neaphenops tellkampfii*, in the Mammoth Cave–Flint Ridge system. They estimate that some 750 000 individuals of this species exist in this enormous system; but, with an average weight of 5 mg each, they contribute a total of only 3·75 kg to the biomass, or about 15 g/km of cave passage.

Figures are not available for biomass in British caves, but Jefferson (1969, p. 131) has estimated population densities of *Proasellus cavaticus* on several flowstone slopes in Ogof Ffynnon Ddu in South Wales. These varied greatly; some had none, others 10 or 20 per sq. m and one particular slope averaged 80 per sq. m. Even this last figure, which was exceptional, does not represent a very high biomass in view of the small size of the animal, and, taking the cave as a whole, the biomass must be very low. It seems likely that in British caves the average density of small animals must lie well to the lower end of the range of 0·0001 to 1·0 per sq. m quoted by Poulson (1971, p. 52). There can be little doubt that the low biomass of deep-cave communities is a reflection of the limited supply of energy to the system which, since we are dealing essentially with heterotrophs, is another way of saying that food is the limiting factor (Mohr and Poulson, 1966, p. 97; Barr and Kuehne, 1971, p. 49). On the other hand Husson (1962, pp. 108 and 113; 1964, p. 73) has argued that the underground habitat is often not short of food and has opposed the idea that cave animals, which are sometimes larger than related surface forms, need to be more resistant than the latter to starvation.

Since many of the special features of cave communities are due to the absence of light, it is natural to compare them with those of other regions similarly devoid of an external source of radiant energy. An obvious comparison, often made, is with the ocean depths, an environment of great stability having a constant low temperature and a complete absence of sunlight. Communities in the depths, like those of deep caves, are essentially based on detritus which is derived from a rain of organic matter falling from above and initially built up by microscopic plants in the sun-lit surface layers of the sea. It is not certain whether food is as limited on the ocean bottom as in deep caves; the biomass seems to be much higher, but there is some evidence of a more efficient utilization of what food is available and the diversity of species is greater (see discussion summarized by Poulson, 1971).

Animals living in the ocean depths are modified in various ways, but on the whole the modifications are different from those shown by troglobites. Depigmentation and reduction of eyes are not general features of deep sea forms; the eyes are often very large. This is correlated with the fact that, although devoid of sunlight, the depths of the oceans are not completely dark; many animals have luminescent organs and there is some light present. This is in marked contrast to the situation in caves where bioluminescence is noticeably absent. The ability to produce light might seem to be an obvious adaptation to life in a dark habitat, but it is one which cavernicolous animals have not adopted.

It may be objected that everyone has heard of the "glow-worms" of Waitomo Caves in New Zealand, and there is also a "fire-fly", the beetle *Lychnocrepis antricola*, in the Malaysian Batu Caves (Wycherley, 1967, p. 258). These are certainly luminous animals, the "glow-worms" being the predatory larvae of a midge, *Arachnocampa luminosa*, close relatives of which also occur in Tasmania and Australia (Richards, 1964). They are all, however, troglophiles which also live in surface habitats, and their luminescence is not a particular adaptation to cave life; it is a curious fact that no luminescent troglobites are known.

THE ENERGETICS OF CAVE COMMUNITIES

The problem of the source of the energy utilized by cave communities has often been discussed, and the importance of exogenous organic matter carried in from outside will already be apparent. Whatever the nature of this material and however it is carried into the cave, it all represents solar energy taken up on the surface; the food-chain, as Hawes (1939, p. 3) pointed out, starts outside the cave. This organic matter in the process of being decomposed by bacteria and fungi forms detritus which is the food of so many cave animals. There is little doubt that most of the energy represented by the biomass of caves can be accounted for in this way, and possibly all of it can; if so, there is no real problem; but is this the whole story?

There are, in many cave systems, well-established populations of animals in regions far above the highest flood level and in places without a trace of bat guano. In South Wales, for instance, substantial populations of the troglobitic crustaceans *Proasellus cavaticus* and *Niphargus fontanus* can be found, in just such situations, on flowstone slopes and in gour pools, fed only by seepage water; overt signs of exogenous organic matter are usually completely absent. A certain amount of extraneous organic material is undoubtedly carried inadvertently to such places by the boots of cavers and is presumably utilized, directly or indirectly, by the animals. This, however, can only be a minor supplement to the natural supply of food, since comparable populations have been found in the newly discovered parts of caves. It is the nature of the natural food-supply in areas such as these which poses a problem and leads to speculation about the possibility of an endogenous source of energy from within the cave.

The fact that there may be no visible sign of exogenous food material does not necessarily mean that none is reaching the site. Atkinson (1971, p. 288) has pointed out that in a carboniferous limestone aquifer filtration is minimal, and minute particles of organic matter could no doubt be carried in from the surface, through open joints and bedding planes, by "vadose trickles". There are certainly situations, such as one in Pen Park Hole (Glennie, 1963, p. 35), where this happens; and Holsinger (1966) has shown that unusually large populations of aquatic troglobites in a Virginia cave could be explained by the presence of sewage particles coming from septic tanks situated 13–16 m vertically above the cave passages. Nevertheless there are some sites, presumably fed by "vadose seepage" rather than "vadose trickles" (Atkinson 1971, p. 284), where microscopical examination of the incoming water does not reveal any evidence of extraneous particulate matter, and yet animals are present. Similarly Beatty (1941), working on pool sediments from the Postojna caves, was able to detect, chemically, carotenoids and chlorophyll in one subject to flooding, but not in another where the pool, which contained several *Niphargus* sp., was fed only by roof drips. Beatty considered (p. 150) that water reaching pools in this way is freed of pigments as it percolates through the limestone; if this is so, it must also be devoid of particulate plant debris.

Organic matter could, and some presumably does, reach such sites in the form of airborne particles; pollen grains, spores and bacteria will certainly be carried into the cave by air currents, but it is not clear whether much of this material is carried very far. Although there may be strong draughts in parts of caves, on the whole the air is not very turbulent and particles will tend to settle out fairly near to entrances; this process could be accelerated, wherever the air is saturated, by the condensation of water droplets around airborne particles. On the other hand Mason-Williams and Benson-Evans (1958, p. 67, and 1967, p. 401) found Bryophyte spores and algal cells in soil samples from the dark zone of caves, and these must make some contribution to the organic material available. It seems, nevertheless, most unlikely that the amount of airborne organic matter reaching some of the sites carrying *Proasellus* and *Niphargus* could be adequate to support the number of animals present.

If particulate exogenous material does not appear to be the complete answer to the problem in all cases, there is the possibility that organic matter in solution may be important. Rain water soaking through overlying vegetation and soil could dissolve some organic materials such as carbohydrates and amino acids. Being in solution, these substances would not be removed by any simple filtration occurring as the water seeped through the overburden above the cave. Determination of the small quantities of organic matter likely to be present in solution is difficult, but approximate estimation is possible by using wet oxidation techniques, and there have been several reports of such material having been found in seepage water (Dudich, 1932, p. 198; Ginet, 1960, p. 159; Mason-Williams, 1967, p. 391, 1969, p. 162, and Edington, 1970, p. 33). Concentrations are low, of the order of a few mg/l, and variable, but it is quite possible that this material could be utilized. It has often been suggested that marine animals, particularly planktonic forms such as small crustaceans, might absorb directly appreciable amounts of the organic matter which, in similar concentrations, is present in the sea. There is considerable controversy whether this happens (see Corner and Davies 1971, p. 130, for a brief review of the evidence in relation to nitrogenous organics), but at present the possibility cannot be ruled out that animals such as *Proasellus cavaticus* might be able to take in part of their food requirement directly from solution. Another, and perhaps more likely, possibility is that bacteria, which could be eaten by *Proasellus*, are the actual agents of absorption; they are known to take up organic materials from solution in lake waters, and some of the mechanisms involved have been investigated (Wright and Hobbie, 1965).

In recent years another phenomenon which could be relevant to the present problem has come to be recognized by oceanographers. This is the physical conversion, occurring in the sea, of dissolved organic matter into particle aggregates (see review by Riley, 1970). This particulate material can apparently be utilized as food by animals, since Baylor and Sutcliffe (1963) have shown that the brine shrimp *Artemia salina* will grow on a diet of particles produced in the laboratory by bubbling air through filtered sea water. It is probable that other processes in addition to bubbling also bring about the precipitation of organic matter in nature, and it is at least possible that it occurs in cave water. This has not so far been investigated and the phenomenon may be confined to saline waters, but there are some reasons for thinking that it may happen as water flows over stalagmite surfaces or drips into pools. If so, small animals could no doubt feed on the aggregates, either directly or, again, through the intermediacy of bacteria.

This raises the obvious question of what cave crustaceans such as *Proasellus cavaticus* actually feed upon. Husson (1956, p. 67) states that this species feeds on vegetable detritus, and certainly in the laboratory it will eat dead leaves as will other species (Culver, 1970b, p. 955); it must, however, be able to live on other material since, as we have seen, some of the places where it occurs are substantially devoid of plant debris. It has been suggested (Glennie, 1956b, p. 28; Hazelton and Glennie, 1962, p. 353) that *Proasellus* feeds on a "water fungus", and Pattee (1965, p. 313) found some mycelial filaments associated with it; but there are often no signs of fungal mycelium in places, such as some of the wet calcite slopes, where numbers of the animal are present. The surface of the stalagmite in such situations, however, usually has a slightly slimy feel, and if scrapings are examined under the microscope a rather floccular material is found together with numerous micro-organisms; the gut of *Proasellus* is usually full of what appears to be identical material. Edington (1970, p. 55) has shown that the micro-organisms present in this material include filamentous chlamydo-bacteria and has argued plausibly that the so-called "water fungus" consists of masses of such filaments.

There can be little doubt that, at least in some situations, this slimy material forms a major part of the diet of *Proasellus*; the floccular part of it may be produced by the bacteria or possibly could be the result of the physical aggregation of dissolved organic matter. In any event since bacteria are being ingested by *Proasellus*, the source of energy for those present in the material must be considered. If they are deriving their energy from organic matter, whether particulate or in solution, coming from outside, the energy is clearly exogenous. There is, however, another possibility; there are bacteria, usually called chemo-autotrophs, which, like green plants, build up organic molecules from inorganic materials, but, unlike the plants, do not use radiant energy for this synthesis but energy derived from inorganic chemical reactions (see Chapter 11). As some of these chemo-autotrophic bacteria occur in caves, an endogenous source of energy is, in theory at least, possible, and since this was first pointed out by Dudich (1930, p. 82; 1932, p. 201; 1933, p. 41), there has been considerable discussion about its significance in the cave community.

In the relatively simple situation we are discussing the picture is: if *Proasellus cavaticus* feeds on bacteria whose energy is derived from chemo-autotrophic sources and if in turn it is preyed upon, as it is, by other animals, "we would have an interesting little food nexus independent of solar energy" (Jefferson, 1969, p. 132). This possibility is theoretically very interesting; most ecologists would no doubt find the idea of a food-chain not starting with sunlight quite novel. It must, however, be admitted straight away that the evidence for any significant source of endogenous energy is very slender. The Chlamydobacteria are heterotrophic and so are most of the other bacteria found in the slimy film (Mason-Williams, 1967, p. 393, and, as Edington, 1970, p. 81). They could nevertheless be a second link in the chain, themselves depending on material built up elsewhere by chemo-autotrophs. It is not impossible that the organic matter arriving in solution at the sort of site we are considering, and on which the heterotrophs presumably depend, could be of endogenous origin. There is, however, no real evidence for this and it seems probable that most, if not all, of it comes from the surface.

Niphargus fontanus presents a similar problem to *Proasellus*; it can often be found in pools containing no obvious exogenous organic matter. In such cases examination of gut contents usually reveals only traces of silty material together with larger amounts of the floccular matter which tends to settle on the silt at the bottom of pools; this material is very like that from stalagmite slopes and the same discussion as to its possible origin would apply. Silt has been shown, especially by Ginet (1955, p. 346; 1960, p. 171) and Gounot (1960, p. 515; see also Gounot 1964, p. 179; 1967, p. 121; 1969, p. 103) to play an essential part in the nutrition of some species of *Niphargus*. This is apparently due to accessory growth factors and amino acids produced by micro-organisms living in it (see Chapter 11), and these authors found, as also did Bouillon (1964, p. 545), that the species which they studied also needed exogenous food. Nevertheless, the organic matter present might conceivably be utilized as a source of energy, and silts from pools consistently contain small but by no means negligible amounts of organic material. Total organic contents of 2·3% and 3·3% dry weight have been reported (Jefferson, 1969, p. 132) for silts from two pools in Ogof Ffynnon Ddu; neither pool was subject to flooding and both contained *Niphargus*. These figures are rather higher than some given by Gounot (1960, p. 508; 1967; p. 33; 1969, p. 105) for pool silts from several French caves. If her figures are expressed on the same basis as those just quoted, they range from 0·34% to 1·9%, still indicating an appreciable organic content. Whatever the origin of this organic matter, whether exogenous, endogenous or both, it seems likely that some of it can be used by *Niphargus* as food.

Niphargus is omnivorous and some species at least seem unable to develop satisfactorily unless their diet includes some animal matter (Gounot, 1960, p. 518). It is not known whether this is the case with *N. fontanus*, but, given the opportunity, this species certainly preys on *Proasellus*; the latter seems to be relatively safe in the film of water on stalagmite surfaces, as in this environment it can move more effectively than the larger *Niphargus*; but in pools and small streams it must be vulnerable.

Not only do silts from pools contain organic matter, but so also do the various muds, earths and clays on the floor of the relatively dry parts of caves. The organic content varies widely; a series of determinations by means of a wet oxidation method gave results for earths from two South Wales caves ranging from 0·1% to 3·9%, the average value being 1·5% (Jefferson, 1969, p. 132, and unpublished). Very similar results are quoted by Cavaille (1960, p. 392) for three French caves. Poulson and Culver (1969, p. 156) listed 22 determinations made on earths from part of the Flint Ridge Cave System and these ranged from 0·5% to 4·6% with an average of 3·0%. These results are probably a little inflated as they were obtained by ignition, a method which gives spuriously high figures where carbonates are present. They are, nevertheless, lower than those found by Jubertie and Meštrov (1965, p. 220) for sediments in a Pyrenean cave; these averaged about 4·8% total organics, an unusually high figure. Culver (1970a, p. 474) quotes comparable figures for some American cave muds, but in most cases the amount of organic matter present in cave earths and silts is rather less than in rich surface soils; it must, nevertheless, represent a source of energy available to the community. Whether this energy is entirely exogenous or whether any appreciable part of it is of endogenous origin is still a matter of controversy.

One possible endogenous source of energy has not so far been mentioned. This was suggested by Ford (1965, p. 24), who pointed out that most limestones contain detectable amounts of organic substances such as hydrocarbons and other compounds, the residue of organisms living when the limestone was formed. If there are in caves bacteria which can utilize this material, thereby converting it into a form acceptable to animals, it would represent a source of food continually being liberated at the interface between the rock and the cave. Although endogenous, such food would be based on fossil solar energy. This interesting suggestion has not yet been investigated at all widely, and so far there is little evidence either for or against it. Edington (1970, p. 125) considered the possibility that the so-called "wall fungus" might use this fossil energy, but came to the conclusion that it was doubtful.

In summary, it can be agreed that most of the energy used by cave communities is exogenous. This is clear enough in parts of caves subject to flooding or where guano is present; in other regions some particulate material may be airborne or carried in through open fissures, while in some situations dissolved organic matter could well be important. There is no general agreement whether any significant proportion of the energy is endogenous; the fossil energy possibility has been little discussed, but there are widely divergent views on the importance of chemo-autotrophic bacteria. On the one hand there is the widely held belief that "endogenous cycles . . . play an important role in the subterranean world" (Vandel, 1965, p. 338) while on the other there is the view that the contribution of chemo-autotrophs is either negligible or of only very local significance (Mohr and Poulson, 1966, p. 64; Barr, 1967, p. 475, and 1968a, p. 48; Edington, 1970, p. 134; Barr and Kuehne, 1971, p. 88).

THE ORIGIN AND DISTRIBUTION OF HYPOGEAN FAUNAS

The numerous terrestrial animals found in tropical and sub-tropical caves are nearly all trogloxenes or troglophiles; this is not to say that there are no terrestrial troglobites in such regions, but their numbers are very small compared with those in some parts of the world. The regions having rich faunas of terrestrial troglobites are these: the more southerly parts of Europe, particularly an area from the Pyrenees, around the northern shores of the Mediterranean, to the Caucasus, Japan, New Zealand, the south-easterly parts of the United States; and some highland areas of Mexico. South Africa and Australia, although less richly endowed in this respect, also both have interesting terrestrial troglobites.

The areas with particularly rich terrestrial troglobitic faunas are in parts of the world which were affected by a general drop in temperature during the latter part of the Tertiary, culminating in the Quaternary glaciations, and it is generally agreed that climatic changes played an important part in the establishment of these faunas. It has been suggested that many of the animals involved have been derived from forms which lived in the humus of wet mountain forests, during the Tertiary, where conditions probably resembled those in present-day tropical African forests at altitudes between 2000 m and 2500 m (Vandel, 1964, p. 560, but mistranslated in Vandel, 1965, p. 474). As the climate deteriorated, such forms, to some extent "preadapted" for life in caves by their previous endogean habitat, moved underground and so avoided temperature extremes. Presumably these faunas were wiped out in those regions which, during the Pleistocene, were directly affected by glacial or periglacial conditions. In some cases a progressive increase in aridity associated with climatic changes or caused by karstic evolution may either have driven animals into the Hypogean domain or confined them to it; this could well have been true of South Africa (Vandel, 1965, p. 467) and parts of Australia (Hamilton-Smith, 1969, p. 118; Richards, 1971a, p. 51).

Aquatic troglobites have a somewhat wider distribution. Most of the relatively few troglobitic animals in the tropics are aquatic and some subtropical regions such as Cuba and Florida have well-developed faunas in their underground waters. In general, the regions having rich terrestrial hypogean faunas are also well endowed with aquatic troglobites, but it is noticeable that in Europe the distribution of aquatic forms extends considerably further north to include southern Britain and other areas to about the same latitude. Most aquatic troglobites are thought to date back to the Tertiary, and presumably they were either better able than terrestrial forms to withstand the effects of the Pleistocene glaciations or have been more successful at recolonization since that time. This will be discussed later in relation to the British fauna.

Some aquatic troglobites no doubt evolved from ancestors which entered caves directly from surface habitats, perhaps washed in by flooding; but many, particularly of the smaller species, seem to have entered the hypogean domain through the interstitial medium. The ancestors of such animals were in some cases fresh-water forms, but in others either brackish or marine, the interstitial habitat seeming in some way to facilitate the transition from salt to fresh water. A considerable measure of preadaptation to life underground would again have evolved while the animals were still in superficial interstitial situations and before they became truly hypogean.

An outstanding feature of troglobites, both terrestrial and aquatic, is that so many of them are relicts, rather ancient forms, the ancestors and epigean relatives of which are often extinct. This indicates a long period of isolation in the hypogean habitat, in most cases since Tertiary times. While surface populations of a hypogean species—at this stage a troglophile—still exist in the same area, these will act as a gene bridge allowing genetic interchange between underground populations; but once climatic or other changes have wiped out the surface populations, those surviving below ground will be much more effectively isolated one from another (Barr, 1968a, p. 86, 1968b, p. 191; Poulson and White, 1969, p. 973). Populations within one cave system may be very effectively isolated from those in other systems unless there are crevices or other connections enabling individual animals to pass from one to another.

Isolation of populations leads to divergence and eventually to speciation. This phenomenon is well illustrated on islands where terrestrial and fresh-water forms are effectively isolated and tend to diverge until each island may have its own characteristic species. The situation is similar in the case of caves which, in this respect, resemble islands or, more realistically, groups of islands (Culver, 1971, p. 97). Some troglobitic species are limited to single cave systems and many are restricted to one cave region; terrestrial species tend to have a more restricted distribution than aquatic forms, which seem to have somewhat better opportunities for dispersal through groundwater and interstitial routes (see Holsinger, 1963).

Among cavernicolous animals there are some species which seem to have been established as troglobites for much longer than others. Matile (1970, p. 204), following some ideas of Jeannel (1959, p. 335) and Vandel (1965, p. 465), used the terms "palaeotroglobite" and "neotroglobite" respectively for these. The former are relicts, long established as troglobites, and usually showing profound morphological and physiological modifications, while the latter are recent cavernicoles, less modified and closely related to surface forms. Palaeotroglobitic species are clearly the kind which tend to show restricted distributions.

It is doubtful whether the British cave fauna contains a single terrestrial palaeotroglobite. The uncertainty about the status of many of our terrestrial cave animals is a reflection of their relatively recent occupation of the underground habitat; most are closely related to surface forms and many have not been sufficiently isolated to have diverged to the point at which their specific distinctness is beyond question. Those which show reduction of eyes and pigment are probably at the preadapted stage and little, if any, more modified than endogean relatives. Any terrestrial Tertiary relict troglobites which may have occurred in Britain were presumably exterminated by the Pleistocene glaciations although Turk (1972, p. 187) considers that some may have survived in deep fissures; if any do come to light, the most likely location would seem to be South Devon, the cave area furthest removed from the glaciers. The collembolan *Sphyrotheca patritzii* (= *Pararrhopalites patritzii*) recently found in that region by W. G. R. Maxwell (Gough, 1972) appears to be a troglobite without close epigean relatives and it has a southern distribution in Europe. It is not, however, very markedly modified, and this, together with its wide distribution in Europe, would argue against its being an ancient relict.

Many of the British terrestrial troglobites could well be survivors of a late-Pleistocene periglacial, or at least cold, fauna which occupied tundra-covered areas or perhaps pine forests outside the glacial limits. As the glaciers retreated, this fauna followed in a general northerly direction, but also up mountains, and some elements of it may have found refuge underground from the increasingly high temperatures of summer. Since the last retreat of the glaciers was, geologically speaking, quite recent, any such forms would at most be neotroglobites and would not show the local speciation to be found among the more ancient cave relicts of other parts of the world. The collembolan *Onychiurus schoetti* seems to be an example of this kind of animal (Skalski and Skalska, 1969, p. 216); it is a widely distributed troglobite in Britain (see Fig. 10.2) and is also cavernicolous in Silesia; but in the north there is a record of it from an epigean locality in Norway (Stach, 1954, p. 93). Dr Turk (1967, p. 146) has found little evidence for a really arctic and montane element among British cave acarines but nevertheless recognizes that there is some relationship to a vaguely northern fauna. The genus *Porrhomma*, also, which includes our only troglobitic spider, is a markedly northern group.

It is among the aquatic hypogean animals and particularly the Crustacea that the few British palaeotroglobites are to be found. So little is known about troglobitic ostracods in this country that they cannot be usefully discussed, but all seven hypogean malacostracans seem to belong to ancient or relict groups, and their distribution raises some interesting problems. There are three outstanding features which need to be considered in any attempt to trace the history of the British palaeotroglobitic fauna. These are: (*a*) the marked difference between the hypogean fauna of Devon and that of the rest of the country, (*b*) the similarity between the cave faunas of South Wales and Mendip, and (*c*) the fact that, with few exceptions, the animals involved are, in the British Isles, confined to the south (Jefferson, 1974).

There is general agreement that most hypogean amphipods had a marine origin, but the distribution of the known species of *Niphargus* and of a few closely related genera such as *Niphargellus*, does not indicate a centre of origin in any existing sea but somewhere in a southerly part of central Europe. This origin may have occurred in Miocene times in a sea which became progressively more brackish and from around the perimeter of which various lines of *Niphargus* radiated into fresh water (Ruffo, 1956, p. 33) and spread to many parts of present-day Europe. The *Crangonyx* group of amphipods is much more widely distributed but seems to have been in fresh water for a very long time (Ruffo, 1956, p. 31). Among the isopods the family Asellidae has probably been a fresh-water group since the Mesozoic (Birstein, 1951, p. 17) and the genus *Proasellus* is thought to have been a very ancient inhabitant of Europe with a more extensive range in Tertiary times than today (Birstein, 1951, p. 19). *Bathynella* belongs to a group of "living fossils", the Syncarida, and seems to have been a fresh-water genus for an extremely long time (Chappuis, 1956, p. 52). All these forms must presumably have dispersed through fresh-water routes and have reached Britain at a time when there were land connections.

The British Isles have been connected to Europe in the past and the most recent break is usually put at only about 5000 B.C. It cannot, however, be assumed that the connection was continuous before that; a considerable body of evidence has now accumulated indicating that in the early Pleistocene and perhaps the latter part of the Pliocene, the whole of the lower-lying parts of Britain and Europe were under the sea. Holmes (1965, Chapter 21) has summarized the geological evidence which seems to show that between those times and the present there has been a eustatic fall in sea-level of nearly 200 m. This was not a steady fall; it fluctuated with short-term eustatic and isostatic changes as the glaciers came and went, but the widespread occurrence of a so-called "600 foot wave-cut platform" which, together with several lower platforms, is particularly clear in some parts of Wales (Brown 1960), indicates that there has been a considerable overall lowering of sea-level.

Glennie (1967, p. 132) drew attention to the biological significance of this and postulated direct migrations of various species of *Niphargus* from the sea into underground waters as the sea-level fell. Jefferson, on the other hand, has argued for the existence of a hypogean fauna in Britain during the late Tertiary and has pointed out that the high sea-level would have broken it up into a number of island communities; Devon, parts of Ireland, Wales, Mendip, and the chalk downlands of the south and east of England would all have been separate islands, together with others further to the north (see Fig. 10.4). Differentiation of isolated island faunas would be expected, and because of their limited opportunities for dispersal, the underground faunas, unlike those on the surface, would tend to retain a measure of distinctness after the islands had rejoined. This could explain the sharp distinction between the hypogean fauna of Devon on the one hand, and Mendip together with South Wales on the other.

The distribution of hypogean crustaceans in the British Isles is predominantly southern and is in keeping with the position on the mainland of Europe where such animals are not found any further north. This is usually, and very reasonably, attributed to the destruction of

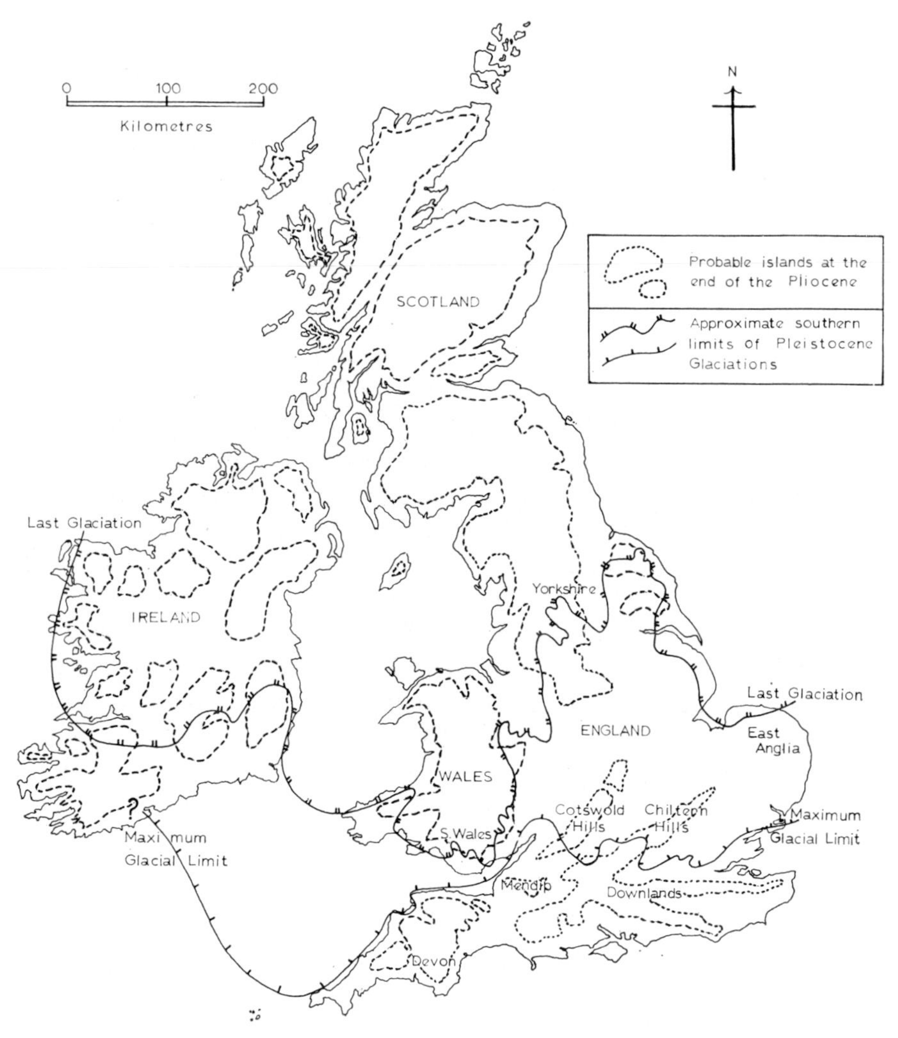

FIG. 10.4. Plio-Pleistocene "islands" and glacial limits in the British Isles.

faunas in those areas affected by ice during the Pleistocene glaciations. Ruffo (1956, p. 34), although sometimes misquoted, argued that the northern limit of the main area of distribution of *Niphargus* spp. agrees with the southern limit of the maximum glaciations, but conceded that there are places where *Niphargus* occurs north of this line. These, he claimed, are still south of the limit of the most recent or Würm glaciation and probably represent a return movement by forms like *N. aquilex* and *N. kochianus*, living in more superficial waters. These and similar arguments concerning other hypogean crustaceans carry conviction in broad terms, but there are difficulties when the details of the distribution in Britain are examined.

In addition to isolated records of other species further north, *Niphargus fontanus* and *Proasellus cavaticus* are common in South Wales in areas which have been repeatedly glaciated, including during Würm times. Again *N. fontanus* and *N. kochianus* occur in the chalk in East Anglia, in areas which have certainly been glaciated, although probably not during the Würm (see Fig. 10.4). This instance may not therefore be completely at variance with Ruffo's views, but *N. fontanus*, far from being a superficial form, is associated with deeper phreatic waters than any other British species. The situation in Ireland is complicated; most of the country, including the places where the Irish subspecies of *N. kochianus* has been found, seem to have been glaciated at some time, but parts of the south-west may have escaped.

There are no firm records of any hypogean crustacea, other than *Bathynella* in Yorkshire, from the main limestone areas of North Wales, northern England or Scotland, and the conclusion seems inescapable that this must be due to the effects of former glaciations. Glennie (1967, p. 134) suggests that the temperature of present-day cave waters in these areas is too far below the optimum for the breeding of these animals, but there is little evidence for this; some species of *Niphargus* live at lower temperatures elsewhere. If aquatic troglobites were able to survive below the ice cover, it seems that this was only possible in areas like East Anglia, South Wales and perhaps parts of Ireland, not too far distant from the edge of the ice sheet. Alternatively some recolonization may have occurred since the retreat of the ice; this can hardly be doubted in the case of superficial interstitial forms like *Bathynella* which has even reached Scotland, and *Niphargus aquilex* which, although not reaching as far north, is widely distributed. Suitable habitats for such animals are widespread, and opportunities for surface transport may occur, so that some northward spread is to be expected. The situation with less superficial forms such as *Proasellus cavaticus*, *Niphargus fontanus* and even *N. kochianus*, is by no means certain.

East Anglia could have been recolonized along the Chiltern Hills from the more southerly chalk downs, underground movement in the chalk possibly being facilitated, as suggested by Glennie (1967, p. 133), by a fluctuating water-table. South Wales might have been recolonized from the Mendips, since in immediately post-glacial times the sea-level was low and the upper part of the Bristol Channel was dry land (Jefferson 1969, p. 130). This would explain the similarity between the two faunas, but so also would some measure of post-glacial interchange along this route if a fauna had survived in South Wales. In Ireland *N. kochianus irlandicus* can probably best be looked upon as the remains of a fauna which differentiated in isolation; but whether it survived beneath the ice or has recolonized from ice-free areas in the south-west is impossible to say.

It may be that recolonization has occurred more extensively than has been considered hitherto. The possibility cannot be ruled out that the whole of the British hypogean fauna was destroyed in glacial times with the exception of *Niphargellus glenniei*, which survived in the relatively favoured south-west, and presumably *N. kochianus irlandicus* which did the same in Ireland. In this event the rest of the hypogean crustacea must have come in from the European mainland in post-glacial times while the south-eastern land bridge survived; their present distribution would represent the extent of their spread to date.

The question of whether animals such as *Proasellus cavaticus* and *Niphargus fontanus* survived beneath the ice or have migrated since cannot be answered with the evidence at present available. Glennie (1967, p. 134) gave reasons for believing that survival beneath an ice-cap might be possible. He pointed out that a cover of snow and ice would act as a protective blanket, and that, in the underlying rock at depths which need not be very great, the Earth's heat would produce temperatures higher than the mean annual temperature at the surface. Further investigation of apparent differences between the populations of *P. cavaticus* in South Wales and Mendip might provide some evidence; if they are in fact sufficiently distinct to be considered as separate subspecies, this would, to some extent, argue against the relatively recent recolonization of South Wales from Mendip and would add a little support to the possibility of survival beneath the ice. It would argue strongly against the idea of recolonization from Europe.

This question of the possible survival of aquatic troglobites beneath the ice was posed by Hazelton and Glennie in *British Caving*. The first edition of that work was published in 1953, and it is appropriate to wind up here with the same question, since we are very little nearer to an answer based on anything more than guesswork. This is unfortunate, not only because of the intrinsic interest of the matter, but because of the light it might throw on another problem we have considered but not resolved. If animals were shown to have survived beneath the ice-sheets, the prolem of their food supply would have to be faced; it is hard to see how it could have come from the surface under those conditions, and it seems that an endogenous source of some kind would have to be invoked. It is impossible to say whether organic matter, which might have come from the surface in pre- or inter-glacial times and been incorporated into cave earths and silts, would persist in sufficient quantities to support animals for the duration of the glacial stages, some of which lasted many thousands of years. If this is admitted as a possibility, however, the question could be argued whether it represented an exogenous or endogenous source. The energy would not have been derived from current receipts of solar energy and could be looked upon as having been a "sub-fossil" endogenous source; but this is largely a matter of semantics and hardly worth pursuing in the absence of factual information. If biospeleology seems to have more than its share of controversy and highly speculative discussion, this is only a reflection of its youth and vigour and should not obscure the considerable body of established knowledge which it also has to its credit.

REFERENCES

Atkinson, T. C. (1971). The dangers of pollution of limestone aquifers, with special reference to the Mendip Hills, Somerset. *Proc. Univ. Bristol Spel. Soc. 12*, 281-290.

Barr, T. C. Jr. (1967). Observations on the ecology of caves. *Amer. Naturalist 101*, 475-491.

Barr, T. C. Jr. (1968a). Cave ecology and the evolution of troglobites. *Evolutionary Biology 2*, 35-102.

Barr, T. C. Jr. (1968b). Ecological studies in the Mammoth Cave system of Kentucky I. The biota. *Internat. J. Speleo. 3*, 147-204.

Barr, T. C. Jr. and Kuehne, R. A. (1971). Ecological studies in the Mammoth Cave system of Kentucky II. The ecosystem. *Ann. Spéléo. 26*, 47-96.

Baylor, E. R. and Sutcliffe, W. H. Jr. (1963). Dissolved organic matter in seawater as a source of particulate food. *Limnol. Oceanogr. 8*, 369-371.

Beatty, R. A. (1941). The pigmentation of cavernicolous animals II: carotenoid pigments in the cave environment. *J. Exp. Biol. 18*, 144-152.

Birstein, Y. A. (1951). *Fauna of U.S.S.R.*, Vol. 7, No. 5, *Crustacea, freshwater isopods* (*Asellota*). Moscow (translated A. Mercado, 1964).

Bouillon, M. (1964). Contribution à l'étude écologique des amphipodes du genre *Niphargus* dans les Pyrénées Centrales. *Ann. Spéléo. 19*, 537-551, 813-818.

Brandon, R. A. (1971). North American troglobitic salamanders: some aspects of modification in cave habitats, with special reference to *Gyrinophilus palleucus*. *Nat. Spel. Soc. Bull. 33*, 1-21.

Brown, E. H. (1960). *The Relief and Drainage of Wales*. University of Wales Press, Cardiff.

Cavaille, A. (1960). Les argiles des grottes. Introduction à l'étude des sédiments souterrains. *Ann. Spéléo. 15*, 383-400.

Chappuis, P. A. (1956). Sur certaines reliques marines dans les eaux souterraines. *Premier Congr. Intern. Spéléo.*, Paris, 1953, *3*, 47-53.

Cheetham, C. A. (1920). *Polylepta leptogaster* Winn., in Yorks. A cave-dwelling dipterous larva. *Naturalist*, *1920*, 189.

Corner, E. D. S. and Davies, A. G. (1971). Plankton as a factor in the nitrogen and phosphorus cycles in the sea. *Adv. mar. Biol. 9*, 101-204.

Culver, D. C. (1970a). Analysis of simple cave communities I. Caves as islands. *Evolution 24*, 463-474.

Culver, D. C. (1970b). Analysis of simple cave communities: niche separation and species packing. *Ecology 51*, 949-958.

Culver, D. C. (1971). Caves as archipelagoes. *Nat. Spel. Soc. Bull. 33*, 97-100.

Delamare-Deboutteville, C. (1960). *Biologie des eaux souterraines littorales et continentales*. Hermann, Paris.

Dixon, J. M. (1973). Biospeleology in north-west England, *in* Waltham, A. (ed.), *Limestones and Caves of Northwest England*. David and Charles, Newton Abbot.

Driver, D. B. (1963). Some simple techniques and apparatus for the collecting and preservation of animals from cave habitats. *Trans. Cave Res. Grp. G.B. 6*, 93-101.

Dudich, E. (1930). Die Nahrungsquellen der Tierwelt in der Aggteleker Tropfsteinhöhle. *Allattani Közlemények 27*, 77-85.

Dudich, E. (1932). Die Biologie der Aggteleker Tropfsteinhöhle "Baradla" in Ungarn. *Speläolog. Monogr.* (Vienna), 13.

Dudich, E. (1933). Die Klassifikation der Höhlen auf biologischer Grundlage. *Mitteil Höhl. Karstf. 3*, 35-43.

Edington, A. (=Mason-Williams, A.) (1970). *Investigations into the microbial populations of caves: their biological and geological significance*. Unpublished Ph.D. thesis, University of Wales.

Edwards, F. W. (1924). British fungus-gnats (Diptera, Mycetophilidae). With a revised generic classification of the family. *Trans. Ent. Soc. London 57*, 505-670.

Efford, I. E. (1959). Rediscovery of *Bathynella chappuisi* Delachaux in Britain. *Nature 184*, 558-559.

Ford, T. D. (1965). An ultimate food source for cave life? *Trans. Cave Res. Grp. G.B.* 7 (No. 3, British Hypogean Fauna No. 1), 24-25.

Fox, H. M. (1964). New and interesting cyprids (Crustacea, Ostracoda) in Britain. *Ann. Mag. Nat. Hist. 13* (7), 623-633.

Fox, H. M. (1967). More new and interesting cyprids (Crustacea, Ostracoda) in Britain. *J. Nat. Hist. 4*, 549-559.

Ginet, R. (1955). Études sur la biologie d'Amphipodes troglobies du genre *Niphargus*. I. Le creusement de terriers; relations avec le limon argileux. *Bull. Soc. zool. Fr. 80*, 332-349.
Ginet, R. (1960). Écologie, Éthologie et Biologie de *Niphargus* (Amphipodes Gammarides hypogés). *Ann. Spéléo. 15*, 127-376.
Gledhill, T. (1971). The genera *Azugofeltria*, *Vietsaxona*, *Neoacarus* and *Hungarohydracarus* (Hydrachnellae: Acari) from the interstitial habitat in Britain. *Freshwater Biol. 1*, 61-82.
Gledhill, T. and Driver, D. B. (1964). *Bathynella natans* Vejdovsky (Crustacea: Syncarida) and its occurrence in Yorkshire. *Naturalist 809*, 104-106.
Glennie, E. A. (1947). Cave fauna. *Cave Res. Grp. G.B. Publication* No. 1, Pt. 1.
Glennie, E. A. (1956a). A brief account of the hypogean Amphipoda of the British Isles. *Premier Congr. Intern. Spéléol.*, Paris, 1953, *3*, 61-63.
Glennie, E. A. (1956b). Cave fauna. *Tenth Anniversary Publication*, *South Wales Caving Club*, 24-32.
Glennie, E. A. (1960). The Hertfordshire Bourne in 1959. *Trans. Hertfordshire Nat. Hist. Soc. 25*, 105-107.
Glennie, E. A. (1962). The Hertfordshire Bourne, 1960-61. *Trans. Hertfordshire Nat. Hist. Soc. 25*, 200-204.
Glennie, E. A. (1963). Chapter IV (The biological investigation) in Tratman, E. K. (ed.), Reports on the investigations of Pen Park Hole, Bristol. *Cave Res. Grp. G.B. Publication* No. 12, 33-38.
Glennie, E. A. (1965). The nature of cave fauna. *Trans. Cave Res. Grp. G.B.* 7 (No. 3, British Hypogean Fauna No. 1), 3-6.
Glennie, E. A. (1967). The distribution of the hypogean Amphipoda in Britain. *Trans. Cave Res. Grp. G.B. 9*, 132-136.
Gough, H. J. (1972). The discovery of *Pararrhopalites patritzii* (Insecta Collembola Sminthuridae) in British caves. *Trans. Cave Res. Grp. G.B. 14*, 199-200.
Gounot, A. M. (1960). Recherches sur le limon argileux souterrain et sur son rôle nutritif pour les *Niphargus* (Amphipodes Gammaridés). *Ann. Spéléo. 15*, 501-526.
Gounot, A. M. (1964). Recherches sur la production de vitamines par la microflora d'un limon argileux souterrain. *Spelunca Mémoires 4*, 178-180.
Gounot, A. M. (1967). La microflore des limons argileux souterrains: son activité. productrice dans la biocoenose cavernicole. *Ann. Spéléo. 22*, 23-146.
Gounot, A. M. (1969). Activité productrice de la microflore de limons argileux souterrains. *Proc. IV Int. Congr. Speleol. in Yugoslavia* (*1965*), *4-5*, 103-108.
Greenwood, P. H. (1967). Blind cave fishes. *Studies in Speleology 1*, 262-274.
Hamilton-Smith, E. (1969). Studies of the Australian cavernicolous fauna. *Proc. IV Int. Congr. Speleol. in Yugoslavia* (*1965*), *4-5*, 113-120.
Hamilton-Smith, E. (1971). The classification of cavernicoles. *Nat. Speleo. Soc. Bull. 33*, 63-66.
Hamilton-Smith, E. (1972). A reply to "The classification of Australian cavernicoles with particular reference to the Rhaphidophoridae (Orthoptera)". *Nat. Spel. Soc. Bull. 34*, 27-28.
Harris, J. A. (1970). Bat-guano cave environment. *Science 169*, 1342-1343.
Hawes, R. S. (1939). The flood factor in the ecology of caves. *J. Anim. Ecol. 8*, 1-5.
Hazelton, M. (1955 *et seq.*). Biological records of the Cave Research Group of Great

Britain. *Biol. Suppl. Cave Res. Grp. G.B.* Pts. *1-5*, *Biol. Rec. Cave Res. Grp. G.B.* Pts. *6-8*, *Trans. Cave Res. Grp. G.B.* 7 (No. 3), 10-19; *9*, 162-241; *10*, 143-165; *12*, 3-26; *13*, 167-197; *14*, 205-230; *15*, 225-253.

Hazelton, M. (1963). Chapter IV (The biological investigation) in Tratman, E. K. (ed.), Reports on the investigations of Pen Park Hole, Bristol, *Cave Res. Grp. G.B. Publication* No. 12, 39-43.

Hazelton, M. (1965 *et seq.*). Fauna collected from caves. *Trans. Cave Res. Grp. G.B.* 7 (No. 3), 26-40; *9*, 242-255; *10*, 167-181; *12*, 75-91; *13*, 198-223; *14*, 231-272; *15*, 203-215.

Hazelton, M. (1973). Biospeleology. In *Limestones and Caves of the Mendip Hills*, Smith, D. I. David and Charles, Newton Abbot.

Hazelton, M. and Glennie, E. A. (1947). Cave fauna preliminary list. *Cave Res. Grp. G.B. Publication* No. 1, Pt. 2.

Hazelton, M. and Glennie, E. A. (1953 and 1962). Cave fauna and flora, in Cullingford, C. H. D. (ed.), *British Caving* (1st edn. 1953, 2nd edn. 1962). Routledge and Kegan Paul, London.

Holmes, A. (1965). *Principles of Physical Geology* (New edn.). Nelson, London.

Holsinger, J. R. (1963). Annotated checklist of the macroscopic troglobites of Virginia with notes on their geographic distribution. *Nat. Speleo. Soc. Bull. 25*, 23-36.

Holsinger, J. R. (1966). A preliminary study of the effects of organic pollution of Banners Corner Cave, Virginia. *Internat. J. Speleo. 2*, 75-89.

Horst, R. (1972). Bats as primary producers in an ecosystem. *Nat. Speleo. Soc. Bull. 34*, 49-54.

Husmann, S. (1968). Ökologie, Systematik und Verbreitung zweier in Norddeutschland sympatrisch lebender *Bathynella*-Arten (Crustacea, Syncarida). *Internat. J. Speleo. 3*, 111-145.

Husson, R. (1956). Considérations sur la biologie des Crustacés cavernicoles aquatiques (*Niphargus*, *Caecospheroma*, *Asellus*). *Premier Congr. Intern. Spéléol.*, Paris 1953, *3*, 65-70.

Husson, R. (1962). Les ressources alimentaires des animaux cavernicoles. *Cahiers d'études biologiques*, Lyon, *8-9*, 103-116.

Husson, R. (1964). Considérations sur la taille des troglobies aquatiques. *Proc. III Int. Congr. Speleol.* (Vienna, 1961), *3*, 71-74.

Hynes, H. B. N., Macan, T. T. and Williams, W. D. (1960). A key to the British species of Crustacea: Malacostraca occurring in fresh water. *Sci. Publ. Freshw. Biol. Ass.* No. 19.

Jeannel, R. (1959). Situation géographique et peuplement des cavernes. *Ann. Spéléo. 14*, 333-338.

Jefferson, G. T. (1960). *In* Brough, J., Matheson, C. and Jefferson, G. T., Chapter IV (Zoology) *The Cardiff Region*, Rees, J. F. (ed.), University of Wales Press, Cardiff.

Jefferson, G. T. (1969). British cave faunas and the problem of their food supply. *Proc. IV Int. Congr. Speleol. in Yugoslavia (1965)*, *4-5*, 129-133.

Jefferson, G. T. (1974). The distribution of hypogean animals in the British Isles. *Proc. V Int. Congr. Speleol.*, Stuttgart, 1969, *4 B8*, 1-6.

Jegla, T. C. (1969). Cave crayfish: annual periods of molting and reproduction. *Proc. IV Int. Congr. Speleol. in Yugoslavia (1965)*, *4-5*, 135-137.

Jegla, T. C. and Poulson, T. L. (1968). Evidence of circadian rhythms in a cave crayfish. *J. Exp. Zool. 168*, 273-282.

Juberthie, C. and Meštrov, M. (1965). Sur les oligochètes terrestres des sédiments argileux des grottes. *Ann. Spéléo. 20*, 209-236.

Ladle, M. (1971). Studies on the biology of oligochaetes from the phreatic water of an exposed gravel bed. *Internat. J. Speleol. 3*, 311-316.

Lawrence, P. N. (1959). Cavernicolous Collembola collections. *Trans. Cave Res. Grp. G.B. 5*, 117-131.

Lawrence, P. N. (1960). The discovery of *Onychiurus schoetti* (Lie-Pettersen, 1896) *sensu* Stach, 1947 (Collembola) in British caves. *Entomologist 93*, 36-39.

Lawrence, P. N. (1968). Stereoscan studies of subterranean springtails (Insecta: Collembola) and their epigean relatives. *Trans. Cave Res. Grp. G.B. 10*, 131-133.

Leruth, R. (1939). La biologie du domaine souterrain et la faune cavernicole de Belgique. *Mém. Mus. R. Hist. Nat. Belgique*, No. 87.

Locket, G. H. and Millidge, A. F. (1953). *British Spiders*, Vol. 2. Ray Society, London.

Lowndes, A. G. (1932a). Occurrence of *Bathynella* in England. *Nature 130*, 61-62.

Lowndes, A. G. (1932b). A new crustacean in England. *Discovery 13*, 285-288.

Madwar, S. (1937). I. Biology and morphology of the immature stages of Mycetophilidae (Diptera, Nematocera). *Phil. Trans. Roy. Soc. Lond. B227*, 1-110.

Maitland, P. S. (1962). *Bathynella natans*, new to Scotland. *Glasgow Naturalist 18*, 175-176.

Mason-Williams, A. (Edington, A.) (1967). Further investigations into bacterial and algal populations of caves in South Wales. *Internat. J. Speleo. 2*, 389-395.

Mason-Williams, A. and Benson-Evans, K. (1958). A preliminary investigation into the bacterial and botanical flora of caves in South Wales. *Cave Res. Grp. G.B. Publication* 8.

Mason-Williams, A. and Benson-Evans, K. (1967). Summary of results obtained during a preliminary investigation into the bacterial and botanical flora of caves in South Wales. *Internat. J. Speleo. 2*, 397-402.

Matile, L. (1970). Les diptères cavernicole. *Ann. Spéléo. 25*, 179-222.

Mitchell, R. W. (1970). Total number and density estimates of some species of cavernicoles inhabiting Fern Cave, Texas (I). *Ann. Spéléo. 25*, 73-90.

Mohr, C. E. and Poulson, T. L. (1966). *The Life of the Cave.* McGraw-Hill, New York.

Moseley, C. M. (1970). The fauna of caves and mines in the Morecambe Bay area. *Trans. Cave Res. Grp. G.B. 12*, 43-56.

Nicholls, A. G. (1946). Syncarida in relation to the interstitial habitat. *Nature 158*, 934-936.

Pattee, E. (1965). Sténothermie et eurythermie les invertébrés d'eau douce et la variation journalière de température. *Ann. Limnol. 1*, 281-434.

Peck, S. B. (1971). The invertebrate fauna of tropical American caves, Part I: Chilibrillo Cave, Panama. *Ann. Spéléol. 26*, 423-437.

Poulson, T. L. (1964). Animals in aquatic environments; animals in caves, in Dill, D. B. (ed.), *Handbook of Physiology*, *4*, American Physiol. Soc., Washington.

Poulson, T. L. (1971). Biology of cave and deep sea organisms; a comparison. *Nat. Speleo. Soc. Bull. 33*, 51-61.

Poulson, T. L. (1972). Bat guano ecosystems. *Nat. Speleo. Soc. Bull. 34*, 55-59.

Poulson, T. L. and Culver D. C. (1969). Diversity in terrestrial cave communities. *Ecology 50*, 153-158.

Poulson, T. L. and Jegla, T. C. (1969). Circadian rhythms in cave animals. *Proc. IV Int. Congr. Speleol. in Yugoslavia (1965) 4-5*, 193-195.

Poulson, T. L. and Smith, M. (1969). The basis for seasonal growth and reproduction in aquatic cave organisms. *Proc. IV Int. Congr. Speleol. in Yugoslavia (1969) 4-5*, 197-201.

Poulson, T. L. and White, W. B. (1969). The cave environment. *Science 165*, 971-981.

Racovitza, E. G. (1907). Essai sur les problèmes biospéologiques. *Biospeologica 1*, 371-488.

Racovitza, E. G. (1926). L'Institut de Spéologie de Cluj. *Trav. Inst. Spéologie, Cluj. 1*, 1-50 and "annexes".

Reynolds, C. F. (1971). Handling cave fauna records on a computer. *Trans. Cave Res. Grp. G.B. 13*, 160-165.

Richards, A. M. (1964). The New Zealand glowworm. *Studies in Speleology 1*, 38-41.

Richards, A. M. (1971a). An ecological study of the cavernicolous fauna of the Nullarbor Plain, Southern Australia. *J. Zool. 164*, 1-60.

Richards, A. M. (1971b). The classification of Australian cavernicoles with particular reference to Rhaphidophoridae (Orthoptera). *Nat. Speleo. Soc. Bull. 33*, 135-139.

Richards, A. M. (1972). Addendum to "The classification of Australian cavernicoles with particular reference to the Rhaphidophoridae (Orthoptera)". *Nat. Speleo. Soc. Bull. 34*, 28.

Riley, G. A. (1970). Particulate organic matter in sea water. *Adv. Mar. Biol. 8*, 1-118.

Ruffo, S. (1956). Lo stato attuale delle conoscenze sulla distribuzione geografica degli Anfipodi – – –. *Premier Congr. Intern. Spéléo* (Paris, 1953), *3*, 13-37.

Schiner, J. R. (1854). Fauna der Adelsberger, Lueger, und Magdalenen-Grotte, in Schmidl, A., *Die Grotten und Höhlen von Adelsberg, Lueg, Planina, und Laas.* Braumüller, Vienna, 231-272.

Serban, E. (1966a). Contribution à l'étude de *Bathynella* d'Europe: *Bathynella natans* Vejdovsky, un dilemme à résoudre. *Internat. Speleol. 2*, 115-132.

Serban, E. (1966b). Nouvelles contributions à l'étude de *Bathynella* (*Bathynella*) *natans* Vejd. et *Bathynella* (*Antrobathynella*) *stammeri* (Jakobi). *Internat. J. Speleol. 2*, 207-221.

Serban, E. and Gledhill, T. (1965). Concerning the presence of *Bathynella natans stammeri* Jakobi (Crustacea : Syncarida) in England and Rumania. *Ann. Mag. Nat. Hist. 13* (8), 513-522.

Sheppard, E. M. (1968). *Trichoniscoides saeroeensis* Lohmander, an isopod new to the British fauna. *Trans. Cave Res. Grp. G.B. 10*, 135-137.

Sheppard, E. M. (1971). *Trichoniscoides saroeensis* Lohmander, an isopod crustacean new to the British fauna. *Internat. J. Speleo. 3*, 425-432.

Skalski, A. and Skalska, B. (1969). The recent fauna of the Polish caves. *Proc. IV Int. Congr. Speleol. in Yugoslavia (1965) 4-5*, 213-223.

Spooner, G. M. (1952). A new subterranean gammarid (Crustacea) from Britain. *Proc. Zool. Soc. Lond. 121*, 851-859.

Spooner, G. M. (1961). *Bathynella* and other interstitial Crustacea in southern England. *Nature 190*, 104-105.

Stach, J. (1954). *The apterygotan fauna of Poland in relation to the world-fauna of this group of insects. Family: Onychiuridae*. Polska Akademia Nauk, Kraków.

Straškraba, M. (1972a). L'état actuel de nos connaissances sur le genre *Niphargus* en Tchecoslovaquie et dans les pays voisins. *Actes du Ier Colloque International sur le genre Niphargus*, 35-46. Museo Civico di Storia Naturale di Verona, Memorie fuori serie N.5.

Straškraba, M. (1972b). Les groupements des espèces du genre *Niphargus* (sensu lato). *Actes du Ier Colloque International sur le genre Niphargus*, 85-90. Museo Civico di Storia Naturale di Verona, Memorie fuori serie N.5.

Sutcliffe, A. J. (1965). Planning England's first cave studies centre. *Studies in Speleology 1*, 106-124.

Tattersall, W. M. (1930). *Asellus cavaticus* Schiodte, a blind isopod new to the British fauna. *J. Linn. Soc. Zool. 37*, 79-91.

Thinès, G. and Tercafs, R. *Atlas de la Vie Souterraine. Les Animaux Cavernicoles*. de Visscher, Brussels.

Turk, F. A. (1944). Some mites from English caves, with the description of a new species. *North Western Naturalist 19*, 146-152.

Turk, F. A. (1967). The non-aranean arachnid orders and the myriapods of British caves and mines. *Trans. Cave Res. Grp. G.B. 9*, 142-161.

Turk, F. A. (1972). Biological notes on Acari recently recorded from British caves and mines with descriptions of three new species. *Trans. Cave Res. Grp. G.B. 14*, 187-194.

Vandel, A. (1964). *Biospéologie. La Biologie des Animaux Cavernicoles*. Gautier-Villars, Paris (English edn. trans. Freedman, B. E., Pergammon Press, Oxford, 1965).

Vandel, A. (1966). The cave salamander, *Proteus*, and its development. *Studies in Speleology 1*, 181-185.

Wilcock, J. D. (1970) Information retrieval for cave records. *Trans. Cave Res. Grp. G.B. 12*, 96-98.

Wolf, B. (1934-1938). *Animalium Cavernarum Catalogus, pars 1-14*. Junk, Berlin and s'Gravenhage.

Wright, R. T. and Hobbie, J. E. (1965). The uptake of organic solutes in lake water. *Limnol. Oceanogr. 10*, 22-28.

Wycherley, P. (1967). The Batu Caves (Malaysia) Protection Association. *Studies in Speleology 1*, 254-261.

11. Cave Flora

B. D. Cubbon

INTRODUCTION

In epigean systems variations in physical, chemical and biological factors cause localized changes of the micro-climate, and micro-habitats arise, self contained entities that reflect in miniature the biological balance of living organisms in general (Cloudsley-Thompson, 1967, p. 3). In the dark zone of caves the botanical problem becomes a micro-biological one. It is essential not to forget this factor of size when dealing with micro-organisms and micro-ecology. For instance, a bacterium measuring 1 μm in diameter occupies a volume of approximately 0·5 μm^3. It would take 2×10^8 of them to fill a volume of 1/10 000 of one cubic centimetre (Burges, 1960, p. 284). This number of bacteria is about normal for 1 g of good agricultural soil, yet this habitat appears almost empty, in spite of the fact that a few million extra organisms may be counted if one estimates the fungi and actinomycetes as well. In relation to their size, however, the distance to which substances produced by these organisms can diffuse is very large. In the same paper Burges also remarked that little regard is paid to the time taken for the completion of ecological succession.

The fauna of caves contain some species, the Troglobites, which are unlikely to be recorded from any other habitat, and which usually stand out clearly from epigean forms in terms of adaptation to their environment. The flora, on the other hand, is generally considered as a representative selection of the species to be found above ground, which have been adapted to living in environments similar to the hypogean by the process of natural selection, and which have developed in the process a number of remarkable characteristics of their own (Tomaselli, 1951, p. 67). Green plants, in order to carry out the photosynthetic reactions summarized by the equation

$$6CO_2 + 6H_2O + \text{energy from sunlight} \rightarrow \underset{\text{(glucose)}}{C_6H_{12}O_6} + 6O_2 \qquad 1$$

must be confined to regions where light penetrates, but the forms found flourishing there are those which in the epigean show some degree of shade tolerance. They are also representative in the main of the species to be seen in the ecosystem adjacent to the cave mouth. Bacteria and Fungi found on particular substrates in the dark zone are often found upon similar substrates above ground, e.g. wood-rotting fungi such as cause problems in wood technology, are frequently met with on decaying timbers in old mine workings.

THE THRESHOLD ZONE

By comparison with the climate outside the cave, the threshold has reasonably stable temperature and humidity, but shows a gradual decrease of light intensity with increased distance from the entrance. This brings about a distinct zonation of the plants found in this region. On a transect taken from the cave mouth, or down a pothole, the more highly evolved plants, flowering plants and ferns, disappear first as the light intensity diminishes, followed by the mosses and liverworts. Algae are most resistant to decreasing light and are therefore found further in. This zonation has been well documented (Douglas, 1938; Tomaselli, 1947, 1953; Dalby, 1966). An early monograph on the various plant species found in caves is that by Morton and Gams (1295), whilst in much later work Morton (1965) paid special attention to the influence of local climate in cave mouths and dolines upon the plant life found therein. The Paradana doline in the Smrekova Draga was cited to show that even small dolines exhibit marked variations in climate between the rim and the bottom of the depression. Examples are also known in caves in the Postojna region (Morton, 1937, 1939).

Reduced light brings about structural changes in organs and tissues of green plants (Dalby, 1966); for instance, the photosynthetic area of leaves is often increased (Fig. 11.1) and dwarfed plants may be encountered as, in the case of the liverwort *Conocephalum*, at the inner limit of growth. Here the poor light restricts vegetative growth reducing the maximum potential size of the plant. Dalby (1966a) has also described the growth of the moss *Eucladium verticillatum* in a poorly illuminated cave. Similar structural changes, and light and shade forms have been reported in the moss genera *Adoxa*, *Phyllitis*, *Asplenium* and the angiosperm *Viola* (Morton, 1958). Growing at the extreme limits of light (either daylight or artificial light) enables the young stages of ferns (prothalli) to grow from the spores but not to develop sexual organs which would normally arise on the prothalli, allowing development to continue to the sporophyte (fern) stage. A similar effect is seen with moss protonemata, which can persist after the development of the moss plant itself (Douglas, 1938, p. 146).

The effects of substrate, pH, humidity and other physical factors, as well as light intensity, on the green plant flora of cave thresholds in South Wales were discussed by Mason-Williams and Benson-Evans (1958). They described the practical methods used for the botanical investigation of caves, and the presence of a potential green plant flora in the dark zone. Spores of Bryophytes, for example, or seeds of Angiosperms, often find their way in through agencies such as flooding, or by cavers, and it is

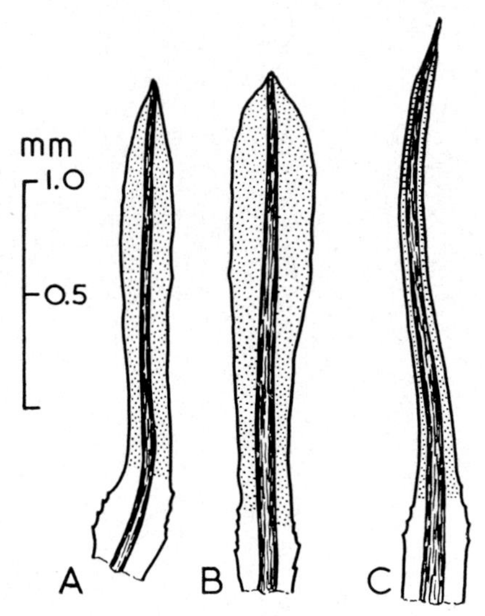

FIG. 11.1. Leaves of the moss, *Eucladium verticillatum*. A and B from plants growing in deep shade, Kimmeridge, Dorset. C from plant growing in the open, Lyme Regis, Devon (Fig. 34 in Dalby, 1966). Reproduced by kind permission Dr D. H. Dalby, Imperial College.

quite common to see etiolated seedlings, which have germinated in the dark and grown long and thin, becoming exhausted in a vain search for light (Plate 11.1). These are usually white or pale yellow, since chlorophyll in higher plants cannot develop in the absence of light. Although they cannot grow, these organisms must contribute to the ecology of the cave environment by providing organic substances which will be employed in the food-cycles within the cave ecosystem.

The German speleo-botanist, Dobat, wrote of certain caves in Germany and France (1966, 1968, 1969, 1969a, 1970), the last paper being devoted to an ecological, morphological and anatomical investigation of the Cryptogams found in the caves of the Swabian Jura. Especially interesting are Dobat's accounts (1969, 1970) of the "Lampenflora"—green plants found growing, concentrically zoned, near to the lights installed in show caves. This phenomenon is a familiar sight to anyone who has been into a show cave, and the managements of these commercial caves are faced with a definite problem, since the plants detract from the beauty of the formations, and must be kept down by periodic chemical treatments. Iwatsuki and Ueno (1959) described the green plant flora of Japan's

PLATE 11.1. Etiolated ash seedlings in Manchester Hole, Yorkshire. (*Photo*: T. D. Ford.)

largest limestone cave, the Akiyoshi-do show cave, the species found growing in the threshold zone, and those around the various types of artificial illumination set up for the tourists. This cave is unusually rich in fauna, and examples are given of animals which are thought to depend on the green plant population.

Another phenomenon sometimes met with in thresholds is the formation of "Eucladioliths" (Dalby, 1966, p. 201; Dobat, 1969a, p. 28), which are tufa-like deposits formed by deposition of lime from calcareous seepage on to growing moss plants (Plate 11.2). As the moss grows (towards the light) more lime is deposited on the older parts, and a formation builds up around the plant. Dobat's examples, termed "Phytogener Excentriques", were taken from the "Lampenflora" of the Nebelhöhle, Germany. He also described a hanging colony of the alga *Gleocystis rupestris* in the form of a small stalactite found growing in the deep threshold region in the Bear's Cave, Erpfingen (Dobat, 1971).

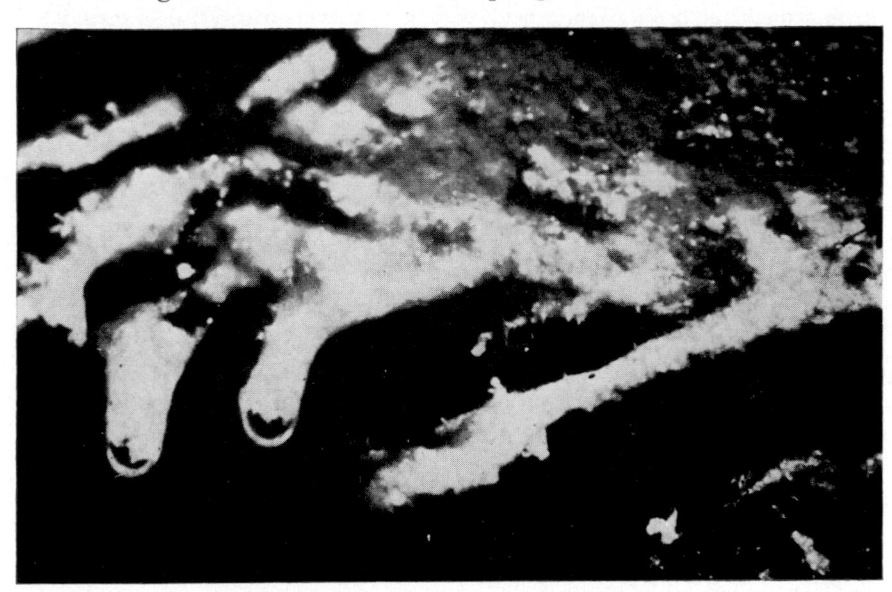

PLATE 11.2. Two short stalactites at left, inclined eucladiolith at right, growing towards the light with living shoots of *Eucladium* at tip. Roof of old mine level, Kimmeridge, Dorset. Natural size (Plate XXXIIB (facing p. 200) in Dalby, 1966.) Reproduced by kind permission Dr D. H. Dalby, Imperial College.

The painted cave of Lascaux in the Dordogne, France, which was discovered in 1940 and became a show cave soon after, was recently the site of a heavy growth of a green alga, *Palmellococcus* sp., that threatened completely to overgrow the prehistoric paintings. The "Maladie Verte" was controlled by application of formalin solutions to kill the algae, following treatment with a mixture of antibiotics to kill the bacteria which were thought to be enriching the nutrients available to *Palmellococcus* on the walls (Lefevre and Laporte, 1969). Green algae growing in the dark can retain their chlorophyll, hence the colour of the "Maladie", though of course the cave was lighted for visiting parties, and the heterotrophic growth was made possible because of very heavy contamination due to the large number of visitors depositing both organisms and organic material derived from the human body and clothing. In time sufficient proliferation of micro-organisms, chiefly bacteria, was brought about to enrich the substrate to the point where *Palmellococcus* could start its spectacularly rapid spread.

THE DARK ZONE

"Caves are able to sustain a complex microbial population in the absence of light. Even though the species represented are not restricted to caves, it is essential to recognize that this varied life exists, and the opinion that bacteria are the only significant micro-organisms found in the cave environment should be rejected."

So said Caumartin (1963, p. 13) and, Mason-Williams (1965, p. 96), fungi are the most obvious flora in undisturbed natural caves.

The cave microflora also includes Actinomycetes (Lovett, 1949; Mason-Williams and Benson-Evans, 1958; Gounot, 1967), a group of organisms intermediate in character between the Bacteria and Fungi, and about which less is known, except in the field of antibiotic production. Actinomycetes are ubiquitous in soil, and it is interesting to speculate what part, if any, their antibiotic-producing capacity has to play in soil microhabitats, including those of the cave. Probably many of the discrete lichen-like colonies frequently noted on walls and formations in the dark zone may be Actinomycetes of the genus *Streptomyces*, since they often have the powdery appearance and characteristic earthy odour common to cultures of this genus, and Streptomycetes have been isolated from wall colonies along with the fungal genus *Fusarium* by Mason-Williams and Holland (1967). It is thought that there is some type of biological association between these organisms in the cave habitat, since mixed cultures in the laboratory produce an effect similar to colonies of wall fungus. It is the odour of Streptomycetes which is probably responsible, at least in part, for the distinctive earthy smell of caving (Caumartin, 1963), known as "Cavers' Perfume" (Picknett, 1967). Many wall colonies of Actinomycetes (and fungi) become covered with drops of condensation or exudate which produce a bright shiny appearance due to reflection on the headlamp beam, making them quite easy to pick out underground. True lichen species have been identified from British and Continental caves, e.g. a *Lecanora* sp., not producing fruiting bodies, was collected from the entrance to Priorsdale Cave, Garrigill, Cumberland (Newrick, 1957, p. 42). It is unlikely that the algae-fungal symbiotic partnership would hold together in the total darkness of the inner cave, though Mason-Williams and Benson-Evans (1958) report a number of lichens including young stages from low light intensities of around 5 lux in South Wales caves.

BACTERIA

Many bacteria are introduced into caves by stream water and seepage, by the air circulation or by the activity of animals, including man. Rarely is any region of a cave completely protected from outside contamination, even in systems not entered by man, though visitation by man evidently leads to much contamination of the environment, as at Lascaux. Bacteria that grow well at 37°C, for example *Micrococcus* and *Streptococcus* spp. and sometimes coliforms may be isolated from caves.

Bacterial action is thought to be responsible for the black deposits on ancient bones found in many of the Mendip caves, Somerset, and also for the black coating sometimes seen on calcite formations in British and Continental caves (e.g. Hölloch, Switzerland). Some of these may be due to manganese-depositing bacteria.

Both heterotrophic organisms (requiring the presence of food in the form of complex organic substances) and autotrophic forms are found, though not usually together in the same population (Vandel, 1964, p. 334). Vandel thought that the chemosynthetic autotrophs were endogenous to caves, and being sensitive to slight variations in the chemical equilibrium were eliminated from environments rich in organic matter where the heterotrophs abound. These bring about the degradation of complex organic debris and the liberation of simpler substances which have a potential food value for other organisms.

The autotrophic group includes Iron Bacteria which get their energy necessary for growth from the simple oxidation of iron compounds. They are not confined to caves, being also found in many soils. *Ferrobacillus ferrooxidans*, which is widespread in caves and mines, particularly in bituminous coal deposits, is a strict autotroph using carbon dioxide as carbon source, and oxidizing ferrous iron to ferric sulphate (2) or ferric hydroxide in acid environments (pH optimum approximately 3·5), to obtain its energy.

$$4FeSO_4 + O_2 + 2H_2SO_4 \rightarrow 2Fe_2(SO_4)_3 + 2H_2O. \qquad 2$$

Another group, the Sulphur Bacteria, both oxidizers and reducers of sulphur, are also found. The former, such as *Thiobacillus thiooxidans* which has been reported from the Grotte Hautecourt, France (Gounot, 1967, p. 56), obtain energy for growth from the oxidation of incompletely oxidized sulphur compounds such as elemental sulphur itself (3), or thiosulphate. As carbon source they employ carbon dioxide or bicarbonate in solution. They are Gram-negative organisms and are widely distributed in soils and mine waters.

$$2S + 3O_2 + 2H_2O \rightarrow 2H_2SO_4. \qquad 3$$

One such organism is *Thiobacillus ferrooxidans* (Temple and Colmer, 1951) which forms acid from the oxidation of sulphur or thiosulphate, and will also oxidize ferrous to ferric iron autotrophically. This ferric iron is capable of oxidizing metal sulphides, like iron pyrites (FeS_2) or Covellite (Copper pyrites, CuS) to the sulphates, and *T. ferrooxidans* is used in concentrating metals in low-grade or inaccessible ores or mine wastes, (Le Roux, 1969)—a large-scale industrial activity for a humble underground micro-organism! The reduction of sulphur compounds can be achieved by most bacteria, but rarely in amounts that yield detectable quantities of sulphide. *Desulphovibrio* spp. (like D. *desulphuricans*) are widespread in caves and play a part in the development of certain minerals. Energy for growth comes from the anaerobic reduction of sulphate accompanied by the simultaneous oxidation of organic matter according to the general equation (4), where C = Organic compound and Me = metal.

$$2C + MeSO_4 + H_2O \rightarrow MeCO_3 + CO_2 + H_2S. \qquad 4$$

A further group of autotrophs metabolise compounds of Nitrogen. Of these, the Nitrifying Bacteria obtain their energy from the oxidation of ammonia to nitrite (Nitrosification, 5) or to nitrate (Nitrification, 6), using carbon dioxide as sole carbon source.

$$2NH_3 + 3O_2 \rightarrow 2HNO_2 + 2H_2O + \text{energy}. \qquad 5$$

$$HNO_2 + \tfrac{1}{2}O_2 \rightarrow HNO_3 + \text{energy}. \qquad 6$$

The first group is represented by *Nitrosomonas* and *Nitrosococcus*, both isolated from South Wales caves, often in the water of small pools (Mason-Williams and Benson-Evans, 1958; Mason-Williams, 1967a). Nitrification is carried out by *Nitrobacter*, also found in South Wales. Kauffmann (1952) attributed corrosion of stone monuments partly to the action of nitrifying organisms, and there would seem to be no reason why the mechanism should not also operate in caves (Caumartin, 1963, p. 9). The organisms oxidize ammonia to nitrous and nitric acids and use carbonate in the stone. Corrosion is caused by leaching out of the calcium nitrite and nitrate so formed. The reduction of nitrate to nitrite (denitrification) has been reported from cave isolates (Gounot, 1967, p. 55), and is said not to be restricted to any particular group of bacteria; however, most would be heterotrophs since *Thiobacillus denitrificans* is the only known denitrifying autotroph, and, though widespread in water, mud and soil, does not seem to have been reported yet from caves. It is an aerobic organism obtaining its energy for the reduction of nitrate to nitrite by oxidizing sulphur and thiosulphate.

A final and important group of possible autotrophs are the Gram-negative aerobes which can fix atmospheric nitrogen, converting it to organic nitrogen compounds within their cells when supplied with a carbohydrate or other energy source. Of these, *Azotobacter* spp. and *Clostridium* spp. have been reported from South Wales (Mason-Williams and Benson-Evans, 1958) and from the French caves (Gounot, 1967, p. 50), whilst *Azotobacter* sp. was found in Borrin's Moor Cave, Yorkshire (Wright, 1965, p. 53). Mason-Williams (1969) has commented on the fact that chemo-autotrophic bacteria are often isolated from small pools. The chemosynthetic autotrophs play a fundamental part in the food economy of the heterotrophs, including other micro-organisms and fauna, modifying the physical environment and nutritional standing of the micro-habitats in the cave.

Some recent research (Tuttle *et al.*, 1968) into the activity of micro-organisms in acid mine waters supports Caumartin's findings of iron and sulphur autotrophs (Caumartin, 1963, pp. 3-6). It was found that mine drainage waters had a different bacterial flora from normal streams, and that iron and sulphur oxidizing autotrophs were always represented. Recently, specimens of rock, water and coal have been examined (Nesterov *et al.*, 1971) as part of a study to determine the factors involved in the generation of methane gas in mines of the brown coal deposits near Moscow. All the mines contained methane-forming bacteria, commonly in the water samples, which generate the gas solely from carbon dioxide and hydrogen under aerobic conditions (7).

$$CO_2 + 4H_2 \rightleftharpoons CH_4 + 2H_2O. \qquad 7$$

There were also saprophytic hydrogen-forming and sulphate-reducing micro-organisms in the coal, rock and water.

An interesting iron autotroph called *Perabacterium spelei*, of curious morphology and uncertain taxonomic position, was isolated from the sediment of certain caves in Southern France (Caumartin, 1957, 1959). A small rod-shaped organism that feeds solely on inorganic compounds and is capable of anaerobic growth, it may prove to be a true cave species.

The botanical studies of caves in South Wales by Mason-Williams and

Benson-Evans (1958, 1967) are the main British papers on the bacterial microflora, and species lists were given for the caves that were investigated. A parallel was drawn with two of the earliest systematic investigations of cave bacteriology: the work of Lovett (1949), who investigated the bacteria found in Gaping Gill, Yorkshire; and of Liddo (1951) in Italian caves. The various habitats from which bacteria were isolated are given, and compared with the corresponding habitats above ground, from the point of view of differing physical factors and the variation of bacteria isolated. A preliminary investigation into the bacteria of Borrin's Moor Cave, Yorkshire, has been carried out (Wright, 1965). Species lists are given in these two papers.

During an investigation of hydrocarbon-oxidizing organisms, a further indication of the relative stability of the cave climate was obtained (Jones and Edington, 1968). They isolated bacteria, actinomycetes and fungi (which made up 10% of the total isolates) from an upland moorland soil in South Wales, and also from a shale band in the cave of Ogof Ffynnon Ddu beneath. A seasonal fluctuation was noted in the numbers of moorland organisms isolated, but this was not seen in the cave.

A magnificent qualitative and quantitative study of cave micro-organisms, particularly the bacteria of cave silts, has been published by the Laboratoire Souterrain, Moulis-Ariège, France (Gounot, 1967); it contains a comprehensive bibliography which relates the speleological field to the study of soil bacteriology generally. Gounot determined the numbers and nutritional requirements of a range of cave bacteria belonging to several genera, and investigated their physiological activities in some detail. She found population densities of the order of several million to several hundred million bacteria per gram of cave silt, approaching the density for a good agricultural soil. With the help of thin layer chromatography on extracts of cave silts and cultures of cave bacteria, she has shown that the organisms enrich the silt by the production of numerous complex organic substances like nitrogenous compounds, amino-acids and growth factors (Vitamins), all of which may be utilized by other forms of life, particularly fauna. She has shown experimentally (1960, 1967) that certain amphipoda are dependent for their survival and growth in caves on a factor produced by bacteria, which may possibly be a B-group vitamin; its discovery serves to underline the fact that this area of study, the interactions between flora and fauna, deserves the greatest possible attention from biospeleologists. The same worker has carried out a bacterial survey of the mud of two caves in the Arctic, finding an abundant active population of cold-loving organisms which presents many analogies with the population of epigean cold biotypes, particularly periglacial soils (Gounot, 1970).

Another topic deserving of further study is the phenomenon of "Moonmilk"—a soft cheesy stalagmitic deposit of widespread distribution, also known as "Rockmilk". A recent American review of the cave environment (Poulson and White, 1969) describes Moonmilk as consisting of a range of complex magnesium carbonates, e.g. hydromagnesite, as well as carbonates of calcium. Some of these substances are alleged to be associated with particular bacteria. In Britain it has been shown (Mason-Williams, 1959) that a variety of organisms can be isolated from Moonmilk deposits, and that one at least—the bacterium *Macromonas* sp.—is capable of depositing crystalline calcium carbonate in the laboratory. Caumartin (1963, p. 7) maintains that organic deposits on dripstone, no matter how small, rapidly become a site of decay, leading to local liberation of carbon dioxide and organic acids, thereby causing impairment of the crystal structure. Algae also may be involved in the formation of Moonmilk. The isolation (Mason-Williams, 1959, p. 137) of a blue-green alga *Synechococcus elongatus* from Moonmilk in the dark zone shows that the organism must have been utilizing alternative metabolic pathways to the normal photosynthetic ones it employs in the light. A related alga *Gleocapsa* sp. was isolated together with certain bacteria (Höeg, 1946) from a Norwegian cave deposit of the Moonmilk type, and it is worth noting that *Gleocapsa magna* has been isolated from the dark zone of Bridge Cave, as well as the threshold zone of Pulpit Hole and Ogof Ffynnon, all in South Wales (Mason-Williams and Benson-Evans, 1958). Moonmilk formation was also observed in Lascaux Cave at the time of the "Maladie Verte" (Lefevre and Laporte, 1969, p. 38).

FUNGI

The larger fungi, which can be obvious in the dark zone, are only a part of the complete fungal population of the cave, which contains in addition a host of microfungi that can be studied only under a microscope. They can be isolated from the soil, air and water of the cave, where they may be present as mycelium or spores, by using a variety of microbiological methods. Organic substances such as plant and animal remains or any matter brought into the caves by man or animals, are likely to be attacked by these fungi, and will probably carry their own microfungal flora when brought in. This flora will proceed to grow and develop if conditions are right. Douglas (1938, p. 143) found moulds on food, orange peel, dung and miscellaneous organic debris from a number of British caves. Fungi Imperfecti of the genera *Penicillium* or *Aspergillus* are frequently found growing saprophytically on all kinds of organic debris. They are often simply referred to as green mould. *Penicillium italicum* was isolated from Read's Cave, Somerset (Bird, 1949, p. 26), and the same worker found *Aspergillus versicolor* in the dark zone of Goatchurch Cavern near by. Moulds forming on splashes of candle grease are a familiar sight, and frequently belong to these genera.

These and other aspects of the growth of fungi in caves were dealt with by Hazelton and Glennie (1962), Caumartin (1963) and Mason-Williams (1965). Tomaselli (1953, p. 30) maintained that the majority of cave fungi are sterile and show a much modified morphology. Caumartin described the isolation of fungi present as spores and as actively growing mycelium within the cave, and also large numbers of mycelial cysts—resting stages—which he believed were formed in response to the presence of sulphide ions and the bacteria producing them, mainly in water-saturated cave clays.

An interesting physiological adaptation of fungi, which could be looked for in suitable caves, is mentioned in a paper by Silverman and Nasa (1969), who demonstrated that fungi were capable of solubilizing rocks, in this case basalt and granite, when growing under conditions deficient in essential inorganic ions.

When fungi enter a cave system, by whatever means, they will most likely begin to grow if they find themselves on a suitable substrate. They may then simply use up that substrate and die or become established members of the cave ecosystem. The outcome depends on their tolerance of the physical conditions within the cave, and the existing types and levels of nutrients in current supply. However, they will certainly in time provide a food source for other members of the biomass already present. This has been discussed at some length in a thoughtful paper written from the viewpoint of an agriculturalist (Newrick, 1957), that makes a good basic introduction, if a little oversimplified in parts, to the problems of cave ecology. Relevant possibly to Newrick's ideas, and also to the production of growth factors by bacteria in caves, is the work of Lenz (1968), who during a study of possible attractive and repellent substances produced by fungi, noted that certain invertebrates could readily complete their life-cycles on a substrate consisting only of fungal mycelium. Lenz was not working with special reference to the cave habitat, but his findings substantiate those of Walton (1943, p. 134) who worked in Read's Cave, Somerset. Using cultures of *Stemphylium botryosum* and *Penicillium* sp.

he found that *Onychiurus fimetarius* and *Heteromurus nitidus* would eat both the conidia and hyphae, and could survive for several months on nothing else. He maintains that *Stemphylium* colonies are difficult to find in the cave because they are grazed down by the population of springtails. *Isotoma bipunctata* and *Tomocerus minor* also could be induced to live for up to eight months on mycelium of the Phycomycete *Pilobolus* sp. *Sciara* larvae were found in association with some mouldy candle wax from which a white monoverticillate *Penicillium* sp. was cultured, which grew successfully on Paraffin wax and was readily eaten by the Collembola. Plate 11.3 shows Collembola in association with fungal hyphae, photographed *in situ* in Lamb Leer Cave, Somerset.

The Phycomycete *Absidia repens*, isolated recently by the author from a soil sample sent via Brigadier Glennie by Mr G. T. Plant, of the 1970 British Speleological expedition to the Himalayas from the State of Bahgal, has been shown in the laboratory to produce an antibiotic active against the bacterium *Staphylococcus aureus*. Hennebert (1960, p. 12), working in the underground laboratory in the caves of Han-sur-Lesse, Belgium, demonstrated the production of an antifungal antibiotic by *Oidiodendron* sp. which was active against *Cladosporium herbarum*, isolated in the same cave. It is not known if the fungi produce these antibiotics in nature, since a number of attempts to isolate antibiotics from soils have been made without success (but see Hill, 1972). Another problem worthy of further investigation stems from the fact that there have been no identifications to date of Nematode-trapping fungi, e.g. *Dactylella* or *Arthrobotrys* spp. from caves. This is perhaps surprising since Nematodes are known to abound, and many of the fungi that prey upon them live an an aquatic environment. It may be that they are being inhibited by soil fungistasis, to which they are known to be sensitive; or that some other factor of the environment is inimical to them, such as a general paucity of rotting vegetable matter. Apparent fungal hyphae have been observed in stream water underground (Hazelton and Glennie, 1962, p. 365), though most are really filamentous chlamydo-bacteria (Mason-Williams, unpublished). Some may possibly be common aquatic Phycomycetes such as *Achlya* or *Saprolegnia* which frequently produce very thick coarse hyphae. The latter genus has been recorded (Dobat, 1966, p. 146) from caves in the Swabian Alps, though normally they prefer shallow sheltered water with abundant organic remains.

An attempt has been made in Fig. 11.2 to summarize the main food cycles met with in the cave habitat, and to show the fundamental nature of the part played by micro-organisms in the recycling of nutrients and

PLATE 11.3. Collembola in association with fungal mycelium. Photographed *in situ* in Lamb Leer Cave, Somerset. Mag. ×20. (*Photo*: A. E. McR. Pearce.)

the provision of other essential substances such as organic nitrogen compounds, vitamins and so forth. Food cycles were also discussed by Vandel (1964, p. 337) and by Gounot (1967, p. 103). Cubbon (1970) gave a list of the fungi and other flora so far isolated by Cave Research Group members. Other lists were given by Mason-Williams (1962, p. 398; 1965, p. 98), mostly from caves in South Wales; Hennebert (1960) for

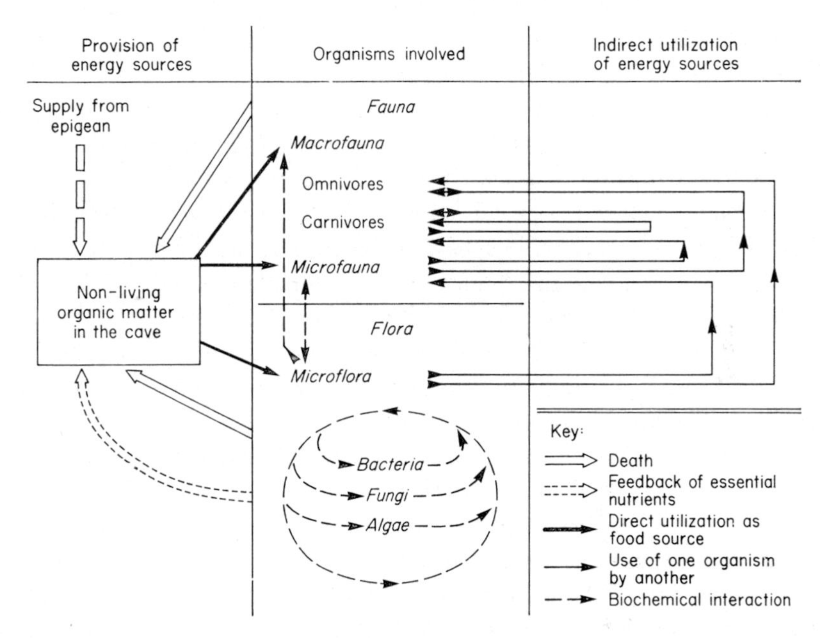

FIG. 11.2. Simplified summary of possible food cycles, showing the part played by flora.

various substrates in the Han Grottoes, Belgium; and Dobat (1966-71), from caves on the Continent, including (1967, p. 16) some remarkable hitherto unpublished notes by Alexander von Humboldt (1769-1859), written in 1794, in which he listed eleven subterranean species of lichens and fungi, some, like *Clavaria nivea*, *Agaricus concentricus* and *Peziza fodinalis*, being described for the first time.

A glance at the species lists shows that a number of fungi are capable of completing their life cycles in the dark, for example *Coprinus* sp. and *Peziza* sp. of which mature fruiting bodies have been found. *Coprinus* spp. particularly seem to be well adapted to living in caves. Their reported occurrence is widespread, even from as far afield as the Batu Caves, Malaysia, where they were growing on bat guano in the dark zone (Wycherley, 1967, p. 257). However, many fungi are affected by light in that some period of illumination is necessary before fruiting will take place, though growth, and in some cases asexual sporulation, can proceed without it. Various aspects of the action of light on fungi constitute current mycological research topics; many fruiting bodies of common wood-rotting saprophytes are found in a deformed and sterile condition in the dark zone, and are thus rendered difficult to identify. Dwarfing of otherwise fully developed fruiting bodies is also frequent, e.g. *Coprinus domesticus* (Mason-Williams, 1965, p. 97)—see Plate 11.4. The causes of this phenomenon are not fully clear, and need further study. Agarics with very long elongated stipes are also seen. Dobat (1971) records a *Mycena* sp. from the Falkensteiner Höhle, Germany, which had stipes up to 20 cm in length, hanging downwards from the substrate under the action of gravity, then turning up at the ends so that the pilei were held in the horizontal position to enable spores to be shed from the gill lamellae. A very similar occurrence has been recorded (Dryden, 1968) in New Dun Iron Mine in the Forest of Dean, Gloucestershire.

The organisms which cause decay of wood in caves and timbers in mines are frequently those met with above ground, and have been well documented (Cartwright and Findlay, 1958; Findlay and Savory, 1960), especially for pit props in coal mines. The main causative agent in British mines is *Poria vaillantii*, which is also found in caves (Newrick, 1957, p. 41), and may be a familiar sight to many speleologists as a white sheet of mycelium spreading fanlike across the substrate, or hanging freely downwards from it in tassels of mycelium which can attain a considerable size. Other wood-rotting fungi recorded from mines and caves include *Lentinus lepideus*, which often forms abnormal cylindrical fruiting bodies; *Fomes annosus*, which is often present in the timber used for pit props when it is cut; *Paxillus panuoides*, and others. *Merulius lachrymans*, the dry rot fungus, is only rarely met with, though it has been collected by Newrick from Hope Level Mine, Stanhope, Co Durham (Cubbon, 1970, p. 63).

Quite a number of fungi are characterized by their growing on the dead bodies of insects and other invertebrates (Plate 11.5). Some of these will be classed simply as saprophytic upon the remains, but others, e.g. *Beauveria* spp. or *Hirsutella* spp. belong to a specialized group, the Entomogenous fungi, which parasitise invertebrates, causing disease and death while completing their life cycles in the tissues of their host. Collections of *Beauveria* spp. from Starbotton Level, Yorkshire, and Fairy Cave, Somerset, exhibit a type of morphological variation that has been described in this genus from other sources (Benham and Miranda, 1953; Macleod,

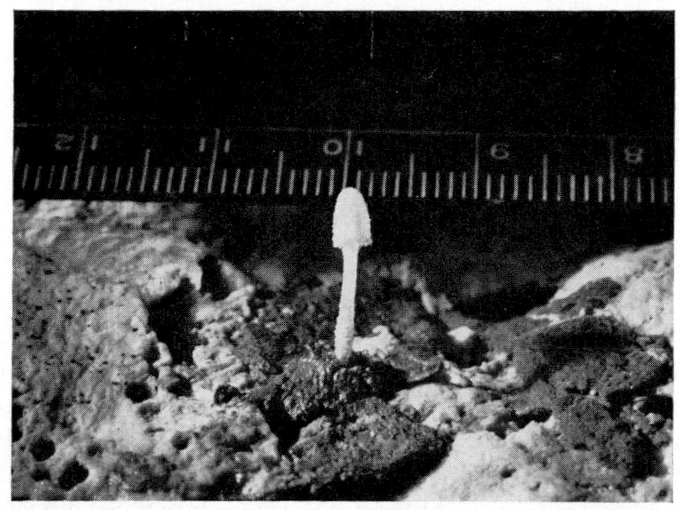

PLATE 11.4. Dwarf *Coprinus* sp. on cave floor. Lamb Leer Cave, Somerset, between cave of falling waters and main chamber. Mag. ×2. (*Photo*: A. E. McR. Pearce.)

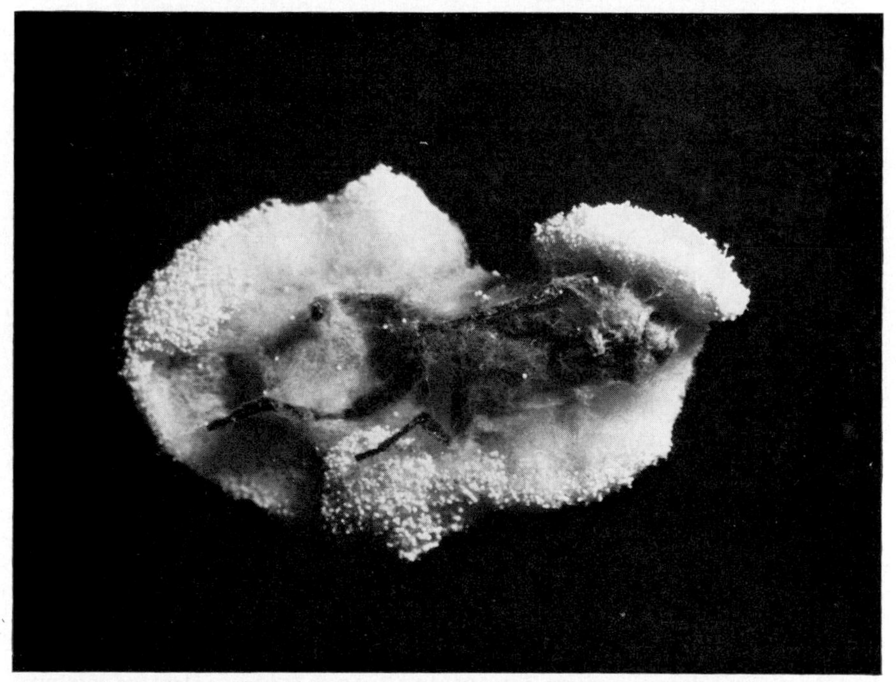

PLATE 11.5. Heavy growth of fungus on dead Dipteron. Mag. × 15. (*Photo*: A. E. McR. Pearce.)

1954) and has led to difficulties in identification. This genus seems to be quite common on dead invertebrates, particularly Dipterous flies, in caves. It is interesting to speculate whether it is endemic as a disease in certain of these animals in the cave habitat, and if so, where they may go to shelter or to hibernate. This possibility is borne out by the fact that *Beauveria Bassiana*, acting as the causative agent of a disease of Codling Moth larvae, requires a low temperature if it is to thrive, and the spread of *B. globulifera* infecting the Chinch Bug is much favoured by cold, wet conditions (Headlee and McColloch, 1913). The fungus is quoted as being unable to grow from spores at less than 90% relative humidity, and the spores have been shown to germinate more readily in total darkness than under normal conditions. Aspects of the infection of insects by these and other fungi generally have been reviewed by Steinhaus (1949). Two other interesting Entomogenous fungi frequently to be seen in caves, which may not yet have been isolated from any other habitat, are *Hirsutella dipterigena* and *Stilbum* (= *Stilbella kervillei*). The latter is often said to be parasitic on the former, which in turn is parasitic on the insect, usually a fly (Hazelton and Glennie, 1962, p. 365). However, the two species are often found separately (Bird, 1949, p. 26; Newrick, 1957, p. 39; Cubbon, 1970), and the apparent parasitism is probably due to a mixed infection. Other species of *Hirsutella* and *Stilbum* are widespread in the epigean on a range of invertebrate hosts. A further genus of Fungi Imperfecti of interest here is *Paecilomyces*, which has been reported on dead flies in Starbotton Level, Yorkshire (Dixon, 1966, p. 10) and from South Wales (Cubbon, unpublished), also on a dead Dipteron (*Rhymosia fasciata* Mg).

Tomaselli (1956, p. 216) describes a number of possibly troglobitic fungi, making an interesting distinction between the possibility of direct troglobites (growing saprophytically on various substrates) such as the ascomycete *Lachnea spelaea*, first described by Barbacovi (1954); and indirect troglobites such as members of the order Laboulbeniales (e.g. *Laboulbenia subterranea*) which are tiny ascomycetes restricted to a parasitic existence on invertebrate hosts. These could be said to be indirect troglobites when growing on troglobitic hosts. He also pointed out that many of the fungi so far recorded only from caves will eventually be found in the epigean domain also, because their substrates are not such as would limit them to the hypogean domain.

The reported occurrence of *Oedocephalum* sp. (Cubbon, 1970, p. 65) as a wall fungus in the Bristol Waterworks Heading, Burrington Combe, Somerset, touches on the relationships between Fungi Imperfecti and their presumed sexual stages, which when they are proved usually turn out to be Ascomycetes or Basidiomycetes. The isolate of *Oedocephalum* agrees well in morphology and cultural characteristics with *O. lineatum* which was described as a new genus by Bakshi (1950), who isolated it from the galleries of wood-boring beetles. He later (1952) described *O. lineatum* as being the imperfect stage of *Fomes annosus*, which is common on decaying wood above and below ground. Cartwright and Findlay (1958, p. 78) refer to *Oedocephalum* (= *Heterobasidion*) as an imperfect conidial stage of *Fomes annosus* (Fig. 11.3), maintaining that the

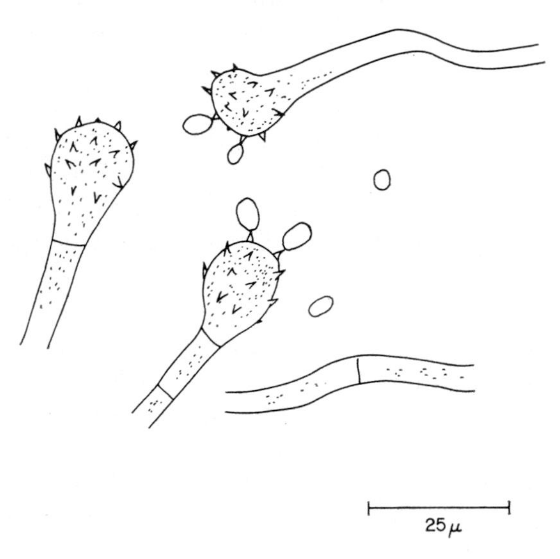

FIG. 11.3. *Oedocephalum* sp. probably *O. lineatum*, the imperfect stage of *Fomes annosus* in culture, formerly known as *Heterobasidion annosum*. Drawn from microscope slides.

conidia, which are formed freely on many media, are an unmistakable guide to the identification of *F. annosus* in culture. (Normal mycological practice is to keep to the original names for the imperfect (asexual) and perfect (sexual) stages, whilst acknowledging that the connection between the two has been proved.) Webster *et al.* (1964), during a study of fungi that colonize burnt ground, described other species of *Oedocephalum* as having certain Pezizales of the Ascomycetes for their perfect stages. The surface of Blackdown Hill overlying the collection site in the Burrington Heading had been burnt extensively about six months before the collection was made. Thus it is not possible at the present time to define with certainty the affinities of this *Oedocephalum* sp., though *F. annosus* is very likely.

The Dermatophytic fungi which cause skin lesions of the "ringworm" type in man and animals, comprise a group frequently isolated from the cave habitat in the U.S.A. and elsewhere, because of their parasitic association with small mammals, especially bats, which frequent the caves. Kajihiro (1965) isolated ten species, all potential human pathogens, belonging to a number of genera, from fresh bat guana in Carlsbad, Sitting Bull, McKittrick and Cottonwood caves in the U.S.A. Lurie and Borok (1955) isolated *Trichophyton mentagrophytes* from the soil of caves, and Lurie and Way (1957) detected this and *Microsporum gypseum* from caves in the Transvaal. The British flora records contain only one notification of a Dermatophyte species, from Starbotton Level, Yorkshire, and it was dubiously identified (Cubbon, 1970, p. 65) as *Trichophyton* sp. because it formed only microspores in culture. A further record of interest is that of *Trichophyton persicolor* which was isolated from the fur of a Pipistrelle bat captured in the Cirencester area (English, 1966). If British bats do prove to be carriers of Dermatophytes, these are quite likely to be recorded from caves, and should be looked for, a start being made on the material from beneath bat roosts.

Shacklette and Hasenclever (1968) in the U.S.A. have studied the effect of flooding in a cave system on the distribution of *Histoplasma capsulatum* within the air, soil and animal life of the cave (including bats). It could be detected in water samples during the flood but was absent for some time afterwards. The problem of Histoplasmosis, "Cave Sickness" (causative agent *H. capsulatum*) was reviewed by Aspin and de Bellard Pietri (1959). Incidences of the disease, an infection of the lungs, have been reported in connection with caves in North and South America, where it is regarded as an occupational hazard of speleologists, and also from parts of Africa.

Another group of organisms which have not yet been mentioned, but which are often found growing saprophytically on wood in the hypogean domain, are the Slime Moulds (Myxomycetes). These possess an amoeboid plasmodial stage, sometimes brightly coloured, which moves slowly about upon the substrate. When mature, the plasmodium gathers itself up into a small area and well-defined capsules (sporangia) are then differentiated. These may be stalked or sessile, and spore liberation is usually aided by some kind of elater mechanism. *Trichia favoginea* (Plate 11.6) is relatively common on old timbers in Box Freestone Mines, Wiltshire.

An effective demonstration that the cave habitat is suitable for the satisfactory and healthy growth of at least some fungi is provided by the fact that caves have sometimes been used for commercial mushroom growing. The first recorded attempt to develop this industry was in France, in the reign of Louis XIV (1638-1715), in limestone caves near

Paris. By the late nineteenth century 1500 miles of cave space was devoted to mushrooms (Smith, 1969, p. 43)! A more recent attempt in Gough's Cave, Cheddar Gorge, Somerset, was less successful; but a fan of mycelium can still be seen spreading across the floor from the old mushroom beds.

PLATE 11.6. Myxomycete, *Trichia favoginea*. Elaters can be seen protruding from ruptured capsules. Specimen from Box Freestone Mines, Wiltshire. Mag. $\times$ 10. (*Photo*: A. E. McR. Pearce.)

COLLECTION AND HANDLING

Green plants from the threshold present no special problems of collection, nor do the larger fungi, e.g. Polyporales or Agaricales, which can simply be put into any convenient container such as a polythene bag, or better, a jar or tin with a light packing of cotton-wool or paper tissue. A fair proportion of these can then be identified at home using one of the well-known guides (e.g. Massee, 1910; Ramsbottom, 1923, or Lange and Hora, 1963). Delicate forms should be packed more carefully, but with many of these the process of decay and autolysis cannot be avoided, and collection is difficult. Some *Coprinus* spp., for example, will scarcely survive the journey back to home or laboratory, which in all cases should be completed without delay, no matter what the type of specimens. Some smaller fungi and microscopic forms can best be collected by taking a small portion of substrate as well, and placing them in a suitable container. For instance, Myxomycetes on timber are easily transported in this fashion in a tin with a slab of cork in the bottom, the specimen being held by a pin pushed through the substrate and into the cork. Other small fungi can be preserved in dilute alcohol or formalin; but when this is done, though identification may well be possible, all hope of culturing the organisms is lost. In handling any material which is to be cultured for identification, or with samples of water, soil or guano from which isolations are to be made, it is advisable to use sterilized equipment and aseptic methods to eliminate chance contaminants.

Samples of water from pools, streams and drips can be collected in sterile screw-capped vials and taken to the laboratory for plating on suitable media. It may be possible for some bacterial investigations to employ filtration methods with subsequent plating of the filter disc. Soil samples can be collected in new polythene bags (which are virtually sterile), or better into small bottles. Soils and clays can also be sampled using sterile vials which can be pushed, after removing the cap, into the soft earth from which the surface layer has been freshly scraped with a sterile implement. Soil samples are usually suspended in a little sterile water or physiological saline in the laboratory, before diluting and plating on agar; or soil particles can be placed directly on to the agar or mixed with cool molten agar before pouring the dishes.

All these methods are standard microbiological procedure, and have been widely employed by workers in all fields. A less common method tried by Wright (1965, p. 53) for sampling the flora of soft stalagmite on the cave wall is to take prepared agar dishes into the cave, remove the lid, and press the agar surface gently on to the surface to be sampled. After returning to the laboratory the plates are incubated, then individual colonies picked off on to nutrient agar slants for subsequent identification. It should be possible to employ in caves some of the other methods devised for the microbiological examination of flat surfaces, such as the agar sausage, or the scotch tape methods, details of which can be found in standard texts.

Air flora are generally sampled by exposing pre-prepared agar plates at suitable sites in the cave for varying lengths of time. There is no record of any of the more sophisticated air-sampling devices being used underground.

Samples of microfungi which can be seen growing as colonies on walls and other immovable objects should be scraped into a small vial and examined by subculturing on to agar, or by dilution and plating as for soil samples. Sometimes, with suitable colonies, time can be saved by aseptically transferring on to agar in the cave (Mason-Williams, 1967, p. 140).

Unless samples are transported to the laboratory within a few hours of collecting, degeneration of organisms may take place, or differential growth brought on by change of conditions may cause one organism to proliferate and swamp the others in the sample, leading to complications in culturing and identification. In large caves with easy access it has been known for the collectors to take an ice chest into which samples can be put to minimize activity during transit (Kajihiro, 1965, p. 721). Collectors without the necessary knowledge or facilities to deal with their own collections should first arrange to have them dealt with by a competent microbiologist, then send (or preferably take) them along immediately after collection.

Details of apparatus and collecting methods are given in Caumartin (1957a), Hennebert (1960, p. 4), Mason-Williams (1967, p. 140) and Cubbon (1969, p. 86). Basic essentials consist of a few robust glass bottles with screw-on lids (2 oz McCartney bottles are ideal), a small scalpel, mounted needle, forceps, a bottle of methylated spirit in which the instruments can be sterilized, and some kind of burner, such as the camping gaz type, for which a bunsen burner attachment nozzle is available; or a spirit lamp will do. A further necessity is a strong box or bag into which everything stows neatly for transport. Pencil and notebook should be included for recording details of collection sites, and pH papers and a thermometer should be included if available, because data on the physical environment can be very significant in ecological work. If no autoclave is available for sterilizing bottles before setting out, this can be done in an ordinary pressure cooker.

Culture media of a host of different compositions are currently in use by microbiologists, and details of methods and media should be sought in a standard textbook, though Hennebert (1960, p. 7) mentioned the use of oatflour, carrot and potato dextrose agars in his study of cave fungi. Problems of contamination can loom very large unless precautions are taken Chance contamination from the air can be ruled out at sampling, but in samples taken from a mixed population it often happens that bacteria swamp the culture dish or slant, being faster growers. Fast-growing fungi may also overgrow slow growers and other types. Media selective for particular types are sometimes used, but a common method for general isolation work is to employ antibiotics in the nutrient agar. A low level of Aureomycin (Cooke, 1954, p. 658) can be used to inhibit bacteria, and is often employed in conjunction with Rose Bengal (Smith and Dawson, 1944; Smith, G., 1969, p. 247) to limit size of fungal colonies, and thereby minimize overgrowth (Plate 11.7). Hennebert (1960, p. 7) used Streptomycin at 35 mg/ml of agar to keep down bacteria.

After growth to a suitable size on the plates, colonies are usually picked off on to nutrient agar slants to be grown up for identification, though a preliminary identification can sometimes be done on the petri dish using a low-power microscope. Except for a few organisms, identification should be done by a skilled microbiologist, as it can be extremely difficult and time-consuming. In general, identification of bacteria is carried out on the physiological characteristics of the cultures (see Bergey, 1957), while in mycology morphological criteria tend to be used, particularly the type of sporing structures, both sexual and asexual, produced by the fungus. The difficulty of identifying micro-organisms has been remarked upon many times (Liddo, 1951; Tomaselli, 1951, p. 61; Mason-Williams and Benson-Evans, 1958, p. 14; Hazelton and Glennie, 1962, p. 366). In the identification of fungi from any habitat, a common cause of difficulty is the complete refusal of the isolate to produce diagnostic features such as spores on whatever medium it is grown. Fungi in the group Mycelia Sterilia have never been observed to form asexual spores. *Ozonium auricomum*, a member of this group, has been reported from many caves and mines. It appears as a mass of rather coarse reddish-orange threads, and has been proved to be a sterile vegetative stage of *Coprinus* sp., groups of which have been observed arising from orange cushions of the *Ozonium* in Wookey Hole, Somerset (Douglas, 1938, p. 144). One organism from Wookey Hole has defied all attempts to classify it. It was isolated from a reel of cable submerged in the stream passage, where it was growing saprophytically on the insulation (Round and Willis, 1956). A detailed investigation of its structure and nutrition was undertaken (Willis, 1961), during which the organism showed a great plasticity of form. It has affinities with both the algae and the fungi, and it did not form any recognizable sporing structures. This organism, however, represents an extreme case, since it is usually possible to effect at least a partial identification.

An entirely different group of methods can be employed to investigate micro-organisms in their habitats where they are to be observed, without any chance of culturing them, and with very little possibility of identification. One such method, the Rossi-Cholodny contact slide technique, makes use of microscope slides coated on one side with gelatin (Cholodny, 1930), which can be pressed against a freshly cut soil profile. Fungal hyphae in the soil stick to the gelatin and can be observed microscopically. The method is useful in estimating the relative amounts of fungus present

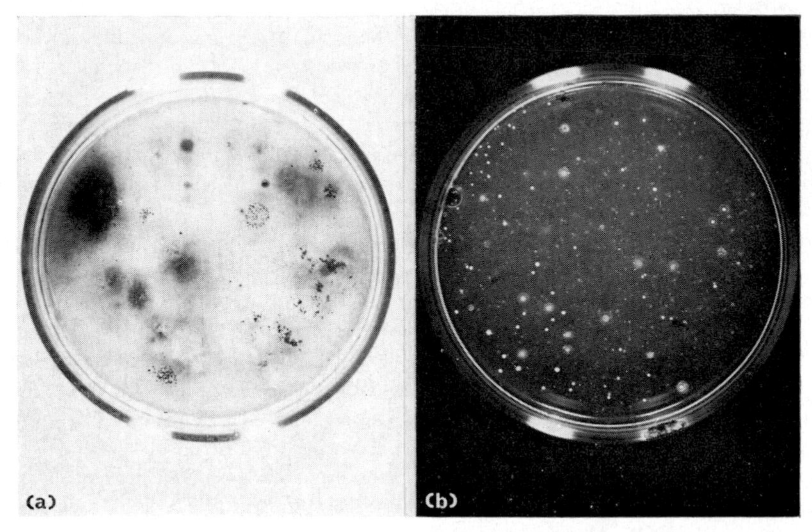

PLATE 11.7. Dilution plates prepared from cave soil. (a) On nutrient agar with the addition of Aureomycin and Rose Bengal, for isolation of fungi. (b) On a medium specific for the isolation of bacteria and actimonycetes. Mag. × $\frac{1}{2}$.

in different soils, and has been employed in the cave of Hautecourt, France (Gounot, 1967, p. 35). A modification not yet tried in caves is to use a film of Colloidin instead of Gelatin (Brown, 1958). During the "Maladie Verte" in Lascaux Cave, direct microscopic examination of organisms on rock samples from the cave walls was tried. It was found that the organisms could be seen, but low-power magnification only could be used (Lefevre and Laporte, 1969, p. 39). An up-to-date technique which has been used to examine organisms on soil particles is stereoscan electron microscopy (Gray, 1967), in which the surface structure of organisms and soil particles (microhabitats) can be seen and photographed *in situ*. The method could equally well be applied to soil samples from caves, and may prove particularly valuable in the investigation of moonmilk phenomena.

REFERENCES

Aspin, J. and de Bellard Pietri, E. (1959). "Cave Sickness"—Benign pulmonary histoplasmosis. *Trans. Cave Res. Grp. G.B. 5* (2), 107-114.

Bakshi, B. K. (1950). Fungi associated with Ambrosia beetles in Great Britain. *Trans. Brit. Mycol. Soc. 33*, 111-120.

Bakshi, B. K. (1952). *Oedocephalum lineatum* is a conidial stage of *Fomes annosus. Trans. Brit. Mycol. Soc. 35*, 195.

Barbacovi, G. (1954). Funghi cavernicoli. Descrizione di una nuova specie di Ascomicete. *Studi Trentini Sci. Nat. 31* (1-2), 50-53. Cited by Tomaselli, R., in *Atti 1st. Bot. Lab. Crittogam.*, Pavia (1956), Ser. 5, *12* (2), 216.

Benham, R. W. and Miranda, J. L. (1953). The genus *Beauveria*, morphological and taxonomic studies of several species and of two strains isolated from wharf-piling borers. *Mycologia*, *45*, 727-746.

Bergey, D. H. (1957). *Manual of Determinative Bacteriology* (7th edn), Breed, R. S., Murray, E. G. D. and Smith, N. R. Baillière, Tindall and Cox, London.

Bird, P. F. (1949). Additions to Fauna list of caves in the Mendip Hills. *Proc. Univ. Bristol Speleo. Soc. 6* (1), 23-26.

Brown, J. C. (1958). Fungal mycelium in dune soils estimated by a modified impression slide technique. *Trans. Brit. Mycol. Soc. 41* (1), 81-88.

Burges, N. A. (1960). Time and size as factors in Ecology. (Presidential address to the British Ecological Society, 3rd Jan. 1959.) *J. Ecol. 48*, 273-285.

Cartwright, K. St. G. and Findlay, W. P. K. (1958). *Decay of Timber and its Prevention* (2nd edn). Min. Tech. (Forest Products Research Laboratory). H.M.S.O.

Caumartin, V. (1957). Recherches sur une bactérie des argiles de cavernes et des sédiments ferrugineux. *Comptes Rendus Acad. Sci., Paris 245*, 1758-1760.

Caumartin, V. (1957a). La microflore des cavernes. *Notes biospéléologiques*, *12*, 59-64. Cited by Hennebert, G. L. in *Ann. Féd. Spéléol. Belg.* (1960), *1*, 3-13.

Caumartin, V. (1959). Morphologie et position systématique du *Perabacterium spelei. Soc. Bot. Nord France Bull. 12* (1), 15-17.

Caumartin, V. (1963). Review of the microbiology of underground environments. *Nat. Speleo. Soc. Bull. 25* (1), 1-14.

Cholodny, N. G. (1930). Über eine neue methode zur Untersuchung der Bodenmicroflora. *Arch. Microbiol. 1*, 620-652.

Cloudsley-Thompson, J. L. (1967). *Microecology*. Institute of Biology—Studies in Biology, No. 6, Arnold, London.

Cooke, W. B. (1954). The use of antibiotics in media for the isolation of fungi from polluted water. *Antibiot.* and *Chemother. 4*, 657-662.

Cubbon, B. D. (1969). The collection of cave fungi. *Mem. Northern Cavern and Mine Res. Soc.* 85-92.

Cubbon, B. D. (1970). Flora records of the Cave Research Group of Great Britain from 1939 to June 1969. *Trans. Cave Res. Grp. G.B. 12* (1), 57-74.

Dalby, D. H. (1966). The growth of plants under reduced light. *Studies in Speleology*, *1* (4), 193-203.

Dalby, D. H. (1966a). The growth of *Eucladium verticillatum* in a poorly illuminated cave. *Revue Bryol. lichen 34*, 288-301.

Dixon, J. (1966). Biological Survey, in *Springswood Level*, *Starbotton*, *Kettlewell*,

Yorkshire. Richardson, D. T. (ed.). Northern Cavern and Mine Res. Soc. Individual Survey Series, No. 1, P. 9.

Dobat, K. (1966). Geschichte und Ergebnisse botanischer und zoologischer Untersuchungen in den Höhlen der Schwabischen Alpen bis zum Jahre 1960. *Jn. Karst-u. Höhlenkde*, 6, 139-158.

Dobat, K. (1967). Ein bisher unveröffentlichtes botanisches Manuskript Alexander von Humboldts—*Plantae subterranae Europ. 1794 cum Iconibus*. Akad. der Wissenschaften und der Literatur. *Abh. der Mathematisch-Naturwissenschaftlichen Klasse Jahr. 1967*, Nr. 6, 16-19.

Dobat, K. (1968). Die Pflanzen- und Tierwelt der Charlottenhöhle. *Abh. Karst-u. Höhlenkde*, Reihe A., *3*, 37-50.

Dobat, K. (1969). Die Lampenflora der Bärenhöhle, in *Die Bärenhöhle bei Erpfingen*— Ed. Wagner, G., 29-35. Erpfingen.

Dobat, K. (1969a). Die Nebelhöhle (Schwabische Alpen). *Abh. Karst-u. Höhlenkde*, Reihe A., *4*, 19-32.

Dobat, K. (1970). Considérations sur la végétation cryptogamique des grottes due Jura Souabe (Sud-Ouest de l'Allemagne). *Ann. Spéléo. 25* (4), 871-907.

Dobat, K. (1971). *Ein biologischer Lehrgang durch die Schauhöhlen der Schwäbischen Alpen*. Reprinted from "*Blatter des Schwabischen Albvereins*", 77 (2), 43-46; (3), 77-80; (4), 117. Stuttgart, slightly altered from Die Schulwarte, *22* (6), 439-456 (1969).

Douglas, E. J. (1938). The speleobotany of Wookey Hole Cave, Somerset. *Caves & Caving 1* (4), 142-154.

Dryden, J. K. (1968). Unusual Fungi—Photographs in *The Speleologist*, *2* (15), 16.

English, M. P. (1966). *Trichophyton persicolor* infection in the Field Vole and Pipistrelle bat. *Sabouraudia 4* (4), 219-222.

Findlay, W. P. K. and Savory, J. G. (1960). *Dry Rot in Wood* (6th edn.). D.S.I.R. Forest Products Res. Bull. No. 1. H.M.S.O.

Gounot, A. M. (1960). Recherche sur le limon argileux souterrain et sur son rôle nutritif pour les *Niphargus* (Amphipodes Gammarides). *Ann. Spéléo. 15*, 501-526.

Gounot, A. M. (1967). La microflore des limons argileux souterrains: Son activité productrice dans la biocoenose cavernicole. *Ann. Spéléo. 22* (1), 23-146.

Gounot, A. M. (1970). Quelques observations sur le micropeuplement des limons dcs grottes arctiques. *Bull. mens. Soc. linn. Lyon 39* (7), 226-236.

Gray, T. R. G. (1967). Stereoscan electron microscopy of soil micro-organisms. *Science*, N.Y. *155* (3770), 1668-1670.

Hazelton, M. and Glennie, E. A. (1962). Cave Fauna and Flora, in *British Caving* (2nd edn.) (ed. Cullingford, C. H. D.), Chap. IX, pp. 347-388. Routledge and Kegan Paul, London.

Headlee, T. J. and McColloch, J. W. (1913). The chinch bug (*Blissus leucopterus* Say.). *Kansas State Ag. Coll. Bull. 191*, 287-353.

Hill, P. (1972). The production of penicillins in soils and seeds by *Penicillium chrysogenum* and the role of penicillin β-lactamase in the ecology of soil bacillus. *J. Gen. Microbiol. 70* (2), 243-252.

Höeg, O. A. (1946). Cyanophyceae and Bacteria in calcareous sediments in the interior of limestone caves in Nord-Rana, Norway. *Nytt. Magasin Naturvid 85*, 99-104.

Iwatsuki, I. and Ueno, S.-I. (1969). The green plants growing in Akiyoshi-dô cave, south-western Japan. *Mem. Coll. Sci. Kyoto Uni*, Ser. B. *26*, 315-322.

Jones, J. G. and Edington, A. (1968). An ecological survey of hydrocarbon-oxidising micro-organisms. *J. Gen. Microbiol. 52*, 381-390.

Kajihiro, E. S. (1965). Occurrence of Dermatophytes in fresh bat guano. *App. Microbiol. 13* (5), 720-724.

Kauffman, J. (1952). Rôle des bactéries nitrifiantes dans l'altération des pierres calcaires des monuments. *Comptes Rendus Acad. Sci., Paris, 234*, 2395-2397.

Lange, M. and Hora, F. B. (1963). *Collins' Guide to Mushrooms and Toadstools.* Collins, London.

Lefevre, M. and Laporte, G. S. (1969). The "Maladie Verte" of Lascaux. *Studies in Speleology 2* (1), 35-44.

Lenz, M. (1968). Zur Wirkung von Schimmelpilzen auf verschiedene Tierarten. *Z. angew. Zool. 55*, 295-374.

Le Roux, N. W. (1969). Mining with microbes. *New Scientist 43* (668), 12-16.

Liddo, S. (1951). Ricerce batteriologiche nell'aria delle grotte de Castellana (contribute allo studio della microflora cavernicola). *Bull. Soc. Ital. Biol. Sper* 17 (3), 496-498.

Lovett, T. (1949). Micro-organisms in caves. *Bull. Brit. Speleo. Assoc. 2* (10), 51-52.

Lurie, H. I. and Borok, R. (1955). *Trichophyton mentagrophytes* isolated from the soil of caves. *Mycologia 47*, 506-150.

Lurie, H. I. and Way, M. (1957). The isolation of Dermatophytes from the atmosphere of caves. *Mycologia 49*, 178-180.

Macleod, D. M. (1954). Natural and cultural variation in Entomogenous Fungi Imperfecti. *Ann. N.Y. Acad. Sci. 60* (1), 58-70.

Mason-Williams, A. (1959). The formation and deposition of Moonmilk. *Trans. Cave Res. Grp. G.B. 5* (2), 133-138.

Mason-Williams, A. (1962). Supplementary Report on Cave Flora, in *British Caving* 2nd edn. (ed. Cullingford, C. H. D.), Chap. IX, pp. 389-391. Routledge and Kegan Paul, London.

Mason-Williams, A. (1965). The growth of Fungi in caves in Great Britain. *Studies in Speleology 1* (2-3). 96-99.

Mason-Williams, A. (1967a). Further investigations into bacterial and algal populations of caves in S. Wales. *Int. J. Speleol. 2*, 389-395.

Mason-Williams, A. (1967b). Ecological methods in caves. *Trans. Cave Res. Grp. G.B.* (3), 140-141..

Mason-Williams, A. (1969). Comments on the microbial population of small pools in caves. *Int. Cong. Speleo.*, Ljubliana, 4.

Mason-Williams, A. and Benson-Evans, K. (1958). *A Preliminary Investigation into the Bacterial and Botanical Flora of Caves in South Wales.* Cave Res. Grp. G.B. pub. 8.

Mason-Williams, A. and Benson-Evans, K. (1967). Summary of results obtained during a preliminary investigation into the bacterial and botanical flora of caves in South Wales. *Int. J. Speleo. 2*, 397-402.

Mason-Williams, A. and Holland, L. (1967). Investigations into the "Wall Fungus" found in caves. *Trans. Cave Res. Grp. G.B. 9* (3), 137-139.

Massee, G. (1910). *British Fungi and Lichens*, Routledge and Kegan Paul, London.

Morton, F. (1937). Monografia fitogeografica delle voragini e doline nella regione carsica de Postumia. Parte I. "*Le Grotte d'Italia*", Ser. 2a, *2*.

Morton, F. (1939). Monografia fitogeografica delle voragini e doline nella regione carsica de Postumia. Parte II. "*Le Grotte d'Italia*", Ser. 2a, *3*.

Morton, F. (1958). Die Pflanzenwelt der Höhlen. *Die Pyramide 6* (3), 87-89.

Morton, F. (1961). Dolinenklima und Pflanzenwelt. *Welter und Leben*, Jahr. 13, Heft 7-8, 155-158.

Morton, F. and Gams, H. (1925). *Hohlenpflanzen.* Speleol. Monograph., Vienna, *5*, 1-227. Cited by Vandel, A. (1964) and other authors.

Nesterov, A. I. *et al.* (1971). Participation of micro-organisms in methane formation in coal mines of the Moscow deposit. *Prikl. Biokhim. Mikrobiol.* 7 (3), 305-309, (in Russian). *Chem. Abs.* (*Rev.*) 75 (9), 60098.

Newrick, J. A. (1957). Some thoughts on cave flora. *Trans. Cave Res. Grp. G.B. 5* (1) 35-51.

Picknett, R. G. (1967). Cavers' Perfume. *Cave Res. Grp. G.B. Nltr. 105*, 13-14.

Poulson, T. L. and White, W. B. (1969). The cave environment. *Science*, N.Y. *165* (3897), 971-980.

Ramsbottom, J. (1923). *A Handbook of the Larger British Fungi.* British Museum, London.

Round, F. E. and Willis, A. J. (1956). A filamentous saprophyte from Wooley Hole caves. *Nature*, London *178* (4526), 215-216.

Shacklette, M. H. and Hasenclever, H. F. (1968). The natural occurrence of *Histoplasma capsulatum* in a cave. Effect of flooding. *Amer. J. Epidemiol. 88*, 210-214.

Silverman, M. P. and Nasa, E. F. M. (1969). *Fungal Solubilization of Igneous Rocks.* (69th Ann. Mtg. Ass. Soc. for Microbiol.), Bact. Proc. A23, 4.

Smith, G. (1969). *An Introduction to Industrial Mycology* (6th edn.). Chap. XI, p. 247. Arnold, London.

Smith, J. (1969). Commercial mushroom production. *Process Biochem. 4* (5), 43-46.

Smith, N. R. and Dawson, V. T. (1944). The bacteriostatic action of Rose Bengal in media used for plate counts of soil fungi. *Soil Sci. 58*, 467-471.

Steinhaus, E. A. (1949). *Principles of Insect Pathology*, Chap. 10 (Fungous Infections), pp. 318-416. McGraw-Hill, New York.

Temple, A. R. and Colmer, K. L. (1951). The autotrophic oxidation of iron by a new Bacterium, *Thiobacillus ferrooxidans*. *J. Bact. 62*, 605-611.

Tomaselli, R. (1947). Notes sur la végétation des grottes de l'Hérault. *Ann. Spéléo. 1* (4), 173-185.

Tomaselli, R. (1951). La vegetazione delle Grotte. *Speleon 2* (1), 63-68.

Tomaselli, R. (1953). Observations de Biospéléologie végétale. *Bull. Féd. Spéléol. Belg. 4*, 28-32.

Tomaselli, R. (1956). Relazione sulla nomenclatura botanica speleologica. *Atti 1st. Bot. Lab. Crittogam. Pavia*, Ser. 5, *12* (2), 203-221.

Tuttle, J. H. *et al.* (1968). Activity of micro-organisms in acid mine waters. I. Influence of acid water on aerobic heterotrophs of normal streams. *J. Bact. 95* (5), 1495-1503.

Vandel, A. (1964). *Biospeleology. The Biology of Cavernicolous Animals.* Translated by B. E. Freeman. Pergamon Press, Oxford.

Walton, G. A. (1944). The Natural History of Read's cavern. *Proc. Univ. Bristol Speleo. Soc. 5* (2), 127-137.

Webster, J. *et al.* (1964). Culture observations on some Discomycetes from burnt ground. *Trans. Brit. Mycol. Soc.* *47*, 445-454.

Willis, A. J. (1961). Further studies on a filamentous saprophyte from Wookey Hole. *Proc. Univ. Bristol Speleo. Soc.* *9* (2), 137-144.

Wright, K. A. (1965). Borrin's Moor Cave, Horton-in-Ribblesdale. Micro-organisms, a preliminary investigation. *Mem. Northern Cavern & Mine Res. Soc.* 52-54.

Wycherley, P. (1967). The Batu caves (Malaysia) Protection Association. *Studies in Speleology* *1* (5), 254-261.

12. Bats in Caves

J. H. D. Hooper

INTRODUCTION

Since bats require shelter, seclusion and an equable temperature during the winter months, when insect food is scarce, it is not surprising that many species choose caves and other underground haunts in which to pass this period of minimum metabolic rate so vital to their survival. Of the fourteen species recognized as natives of Britain, about ten may be described as "cave-dwellers". Some, such as the horseshoe bats, are often seen by the passing caver, since they hang from the open walls and roof; others tuck themselves into crevices and are only found after careful search. Many speleologists, the author included, have developed an interest in bats thus noticed during ordinary caving trips, and it is an easy step to the next stage, a study of the habits of these bats; and this can include the use of banding techniques. Throughout this chapter, emphasis must therefore be placed on the responsibility which such "speleo-naturalists", or indeed any speleologist, must bear in his or her treatment of bats seen in caves.

It is all too easy, when a cluster of Greater Horseshoe Bats is seen, to stand near by and shine lights on the bats and point them out to visitors, and then perhaps to create more disturbance by taking photographs thereby causing many of the cluster to awake—so that when the party comes back along the passage on its way out of the cave, the result will be surprise that most of the bats have gone. But probably no thought is given to the fact that in flying off, those bats have used up reserves of stored fat, and these reserves may not be fully restored before the end of the winter. If bad weather then prevents them from hunting, they may die of starvation. A single disturbance does not automatically result in the death of a bat later in the season, but the risk is there. There should thus be no need to underline the possible consequences of repeated disturbance of bats during the winter, for such activities as banding, weighing and collection of ectoparasites. Stebbings (1971a, p. 34) puts it quite bluntly: "Naturalists studying bats can also be a threat to the continued existence of bats in their haunts and it is only now becoming clear how vulnerable bats are, while in hibernation". He has also pointed out (Stebbings, 1971b, p. 106) that most bat species living in inhabited temperate areas are declining and that the aim of naturalists should be to try to halt this process. This chapter, therefore, in pointing out gaps in our knowledge of bats, is at the same time a warning that attempts to fill those gaps, unless made with care, could do more harm than good.

THE BRITISH BATS

Throughout the world, about 1000 different species of bat are known, many of these being cave-dwelling. The literature on such bats and their biology is voluminous; reprints in my own very modest collection already make a pile over two feet high! So in this following account it has been necessary to be very selective, and to place the emphasis on cave-dwelling bats in Britain, a mere ten species. For details of bat biology, the reader is referred to the standard natural history books, and in particular to: Barrett-Hamilton (1910), Blackmore (1948), Harrison Matthews (1952), Southern (1964) and Yalden and Morris (1975). The fantastic diversity of bat species is best grasped by a study (in the absence of the real thing) of colour photographs, and an outstanding publication in this respect is the book by Leen and Novick (1969). Colour photographs published by Pye (1969, p. 50) and of American bats in a book by Barbour and Davis (1969) are also recommended.

Bats belong to the order Chiroptera, meaning wing-handed; this comprises two sub-orders, Megachiroptera (fruit, pollen and nectar eaters) and Microchiroptera (for the most part, insect eaters, although with some notorious exceptions, e.g. the Desmodontidae, or vampire bats). The Microchiroptera cover 16 families, of which only two are represented in Britain, the Vespertilionidae (12 species) and the Rhinolophidae, or horseshoe bats (2 species). A list of these 14 British species is given in Table 12.1. Those that could be seen in caves are indicated by "(C)". The measurements quoted for *Plecotus austriacus* are based on data published by Corbet (1964, p. 512) and Brink for European bats (1967, p. 60). For detailed measurements of bats in one known British colony of this species, reference should be made to Stebbings (1967a, 291-310). For all the other species, figures published by Southern (1964, pp. 232-248) have been used.

The identification of individual species when seen in a cave or held in the hand can be difficult to those who are unfamiliar with bats. Southern (1964, p. 231) presents a useful key to bat identification; but to the novice,

Table 12.1

THE BRITISH BATS

	Common Name	Measurements (mm)	
		Head and Body	Forearm
Vespertilionidae			
Myotis bechsteini (C)	Bechstein's Bat	50-53	41-44
Myotis daubentoni (C)	Daubenton's Bat	46-51	36-38
Myotis myotis (C)	Mouse-eared Bat	65-79	57-64
Myotis nattereri (C)	Natterer's Bat	43-50	40-43
Myotis mystacinus (C)	Whiskered Bat	46-50	34-37
Nyctalus noctula	Noctule	75-82	51-54
Nyctalus leisleri	Leisler's Bat	60-63	41-45
Plecotus auritus (C)	Long-eared Bat	46-51	38-40
Plecotus austriacus (C?)	Grey Long-eared Bat	46-53	37-41
Barbastella barbastellus (C)	Barbastelle	49-52	38-41
Pipistrellus pipistrellus	Pipistrelle	42-52	29-32
Eptesicus serotinus	Serotine	74-80	50-54
Rhinolophidae			
Rhinolophus ferrumequinum (C)	Greater Horseshoe Bat	65-68	54-58
Rhinolophus hipposideros (C)	Lesser Horseshoe Bat	35-39·5	37-40

such information, however clear it may seem when read in comfort at home, can well appear ambiguous and confusing when he is trying to identify an unknown and furiously struggling bat by flashlight in a cave. The following notes attempt to give some idea of the differences in size and appearance that may be expected for bats seen in caves. The remarks concerning the distribution of the British species are based on the distribution maps compiled as a result of a national scheme put into operation under the aegis of the Mammal Society in 1965 (Corbet, 1971, p. 95).

GREATER HORSESHOE BAT

This species is not found on the American continent, but otherwise is distributed world wide, e.g. in many parts of Asia, in Japan and in south and central Europe. In Britain, although at one time recorded as far east as Kent, it now seems to be concentrated in the south-west e.g. Somerset, Dorset, Gloucestershire and South Wales. It does not occur in the northern counties of England, or in North Wales, Scotland or Ireland. Typical habitats in the south-west are the caves of Somerset and Devon, stone and ochre workings in Gloucestershire, old mine tunnels in the Forest of Dean, underground stone quarries in Dorset and Devon, and the many disused mine adits scattered throughout Devon and Cornwall.

As normally seen in such tunnels, this bat is quite conspicuous, hanging from a roof or against a passage wall, with no attempt at concealment. When it is asleep, the wings are tightly wrapped around the body, folded so that the forearms lie side by side down the middle of the back. This habit is shared by the Lesser Horseshoe Bat, and this feature, a characteristic of the Rhinolophidae, unlike those of the Vespertilionidae, is an aid to recognition at a distance. If seen on a cave wall, vespertilionid bats will have their wings folded down the sides. From feet to nose, the length of a hanging Greater Horseshoe Bat is about 90 mm; the corresponding measurement for a Lesser Horseshoe Bat is about 65 mm. When either species is awake, the folded wings no longer hide the nose and identification is no problem. The extraordinary, plate-like nasal membrane, shaped like a horseshoe, is immediately recognizable, and quite unlike the more mouse-like features of most vespertilionid bats. The ears of the horseshoe bats are long and pointed and—again a point of difference from vespertilionids—have no tragus (a lobe, usually pointed, which stands at the front of the ear passage). The question whether a rhinolophid bat is a greater or lesser horseshoe is resolved simply by size; measurements of body length have been given above and in Table 12.1, and the relative differences in wing span are also very obvious (Greater Horseshoe Bats 330-385 mm and Lesser Horseshoes 228-250 mm). In both species the fur is usually tawny brown, with a mauvish cast (although a ginger-coloured bat may sometimes be seen).

LESSER HORSESHOE BAT

This species has been described above, for purposes of comparison with the Greater Horseshoe Bat. For continental Europe and the rest of the world, the distribution of the two species is very similar, but in the British Isles the Lesser Horseshoe Bat has a somewhat wider distribution, being found in western Eire and in North Wales (e.g. mine tunnels in Denbighshire) as well as in the West Country haunts described for the Greater Horseshoe Bat.

NATTERER'S BAT

This bat is one of the genus *Myotis*, of which there are five species native in Britain, all cave-dwellers. Two of these are both rare and local (Mouse-eared Bat and Bechstein's Bat) but the distribution maps for the other three species (Corbet, 1971, pp. 104, 105, 106) show a general scatter of records throughout England and Wales, although information from Scotland and Eire is still sparse. The other three species (Natterer's, Daubenton's and Whiskered Bats) are also widespread throughout Europe. As they are often confused by the novice, emphasis is given below to some of the more characteristic differences. For example, a feature not commonly mentioned, for purposes of recognition, is facial colouring. As pointed out to me by Dr Pat Morris some years ago, identification of these three bats, on the basis of the following differences, is almost infallible.

Natterer's Bat	Ears	Brownish pink
	Lips and nostrils	Pink
Daubenton's Bat	Ears	Dark brown or black
	Lips and nostrils	Pink
Whiskered Bat	Ears	Dark brown or black
	Lips and nostrils	Dark brown or black

The Natterer's Bat has greyish-brown fur and, like the other *Myotis* species mentioned, the fur on the underside is noticeably lighter, almost white. The ears are fairly long, oval and set far apart. The tragus is slender and more than half the length of the ear. If the ear is folded forward, the tip is long enough to curve round the front of the muzzle (compare with Daubenton's, below). The traditional recognition features for Natterer's Bat are a fringe of hairs along the outer edge of the interfemoral membrane (tail flap) and the fact that the calcar (a spur of cartilage at the edge of this membrane) extends halfway from the ankle to the tail. The calcar of the Whiskered Bat is similar, but for the Daubenton's Bat extends two-thirds of the distance to the tail. The tail fringe clinches identification for Natterer's Bat, but can be difficult to see, as this bat, held in the hand, curls up its tail membrane, and if the bat is wide awake and active, it may not be easy to uncurl this membrane.

DAUBENTON'S BAT

This bat is slightly smaller than the Natterer's Bat (wing span 220-245 mm as against 265-285 mm for the latter). It has no fringe of hairs on its tail membrane (note; a few hairs may sometimes be seen, but not a fringe). The Daubenton's Bat has a very large foot—more than half the length of the tibia. The ear is short, and if folded forward, only just reaches to the end of the muzzle.

WHISKERED BAT

This bat is similar in size to Daubenton's Bat, with a wing span of about 225-245 mm. In other respects, apart from facial colouration and absence of tail fringe, it is similar to the Natterer's Bat. However, although the ear *looks* long, if folded forward it only reaches to the tip of the muzzle.

BECHSTEIN'S BAT

Although recorded from most European countries, this species is generally regarded as rare. For Britain, during the last ten years, reports have been limited to Devon (one individual) and Dorset. Anyone who finds this species in a British cave can thus regard himself as exceptionally lucky. In size and colour it is similar to Natterer's Bat, but has much longer ears (18-26 mm, as against 14-17 mm for Natterer's Bat). These extend for about half their length beyond the muzzle, when laid forward.

MOUSE-EARED BAT

The foregoing remark about luck also applies to any finder of this species. It is very common on the continent of Europe, for example in tunnels in S Limburg, in Holland (Bezem *et al.*, 1960, p. 514) or in the Grotte de Han in Belgium (Anciaux, 1954, p. 170); but in Britain its occurrence appears to be limited to small colonies in tunnels in Dorset (Blackmore, 1956, p. 201) and Sussex (Phillips and Blackmore, 1970, p. 520). These haunts, near the south coast, could be the result of immigration from the Continent. The distinguishing feature of this bat is its large size (see Table 12.1), the wing span being 350-450 mm. This in itself should distinguish it from Bechstein's Bat, with which it could perhaps be confused, because of its long ears. These are 27-28 mm long, but if laid forward only extend about 5 mm beyond the muzzle (Southern, 1964, p. 239).

LONG-EARED BATS

British natural history books published prior to 1964, in referring to the Long-eared Bat, and its common distribution throughout the British Isles, dealt only with *Plecotus auritus*. *Plecotus austriacus* (the Grey

Long-eared Bat) was, however, in that year found to be resident in a small colony in Dorset (Corbet, 1964, p. 515; Stebbings, 1965a, p. 314), and this raises the question of whether some of the earlier records for *auritus* were in fact for *austriacus*. Since it is extremely difficult to distinguish between these two species (with even the experts a little uncertain at times) the same doubt will apply for Long-eared Bats found in the future. The enormous ears (34-37 mm in length), almost equal to the length of the head and body, make it obvious to the finder that the bat is a "long-eared", but all he can then do is to record it as "*Plecotus* species"; to decide which one it is will require inspection and measurement by the expert. One of the many necessary diagnostic features for *P. austriacus* is that the under fur is almost black (grey at the tips), whereas for *P. auritus*, the overall appearance of the fur is more of a sandy colour, with the under fur chocolate brown.

When the Long-eared Bat is seen asleep in a cave tunnel, first appearances may be deceptive. The characteristic ears are folded backwards under the wings, but each tragus still projects upwards, looking rather like ears in themselves, so that a first impression is that the bat is of some other species.

BARBASTELLE

This species can usually be identified without difficulty. The face, flattened in appearance, is black, and can only be described as ugly. The ears are squat, almost square, and—the distinctive feature—are joined together at the base. The fur, and indeed the whole of the bat, is very dark, usually black. Most British records are for the south of England, although it has been found as far north as Yorkshire. Although a cave user, it merits the description of "elusive" given it by Corbet (1971, p. 112), and is rarely seen. In over 20 years of checking Devon caves for bats, only one Barbastelle has been found by the writer.

THE STUDY OF BATS IN CAVES

BANDING, AND OTHER MARKING TECHNIQUES

In attempting to follow the movements of bats from one haunt to another, workers have experimented with a variety of techniques for labelling individual bats so that they can be positively recognized each time they are seen. Such techniques have included colour marking the ears (Ryberg, 1947, p. 63), marking the fur on the back with coloured Indian ink (Kowalski and Wojtusiak, 1951, p. 41), staining the wings of bats with haematoxylin (Mohr, 1934, pp. 26-30), tattooing numbers on the wing membrane (Griffin, 1934, pp. 202-207), and the use of numbered metal tags or bands. Metal bands were first used by Allen (1921, pp. 53-57), who clamped aluminium bird bands round the legs of four bats in Ithaca, New York, in 1916. Griffin (1934, see above) and Mohr, (1934, see above) also experimented with bird bands on the legs of bats, the latter fixing the ring right round the leg by cutting a slit in the interfemoral membrane. He also tried the further method of clipping a "Fingerling" tag (used for marking small fishes) to the ear of a bat, using special pliers. The method that finally gained general favour was that described by Trapido and Crowe (1946, pp. 224-226), who clamped the bands round the forearms of bats. As the band was initially shaped like the letter "C", the two ends could be squeezed gently together so that they touched, but did not puncture, the wing membrane, and care was also taken that the band was loose enough to slide freely along the forearm.

In the U.S.A., after the Biological Survey (later the Fish and Wildlife Service) had authorized the use of its bird bands on bats (in 1937), a growing number of workers started massive banding experiments on the very large bat populations in many of the American caves. Up to about 1932, only 124 bats had been banded in North America; by 1951 the total had risen to over 67 000, and the continued rate of growth of such banding is illustrated by the fact that in a single year (1959), over 75 000 bands were issued (Hitchcock, 1960, p. 277). A more recent paper, by Stebbings (1966, p. 172) gives an estimate for bats ringed in America of over 500 000.

In Europe, bat banding was pioneered in Germany by Eisentraut (1934, pp. 553-560), who used the wing-banding technique and special rings, with the ends folded outwards, so as to minimize the risk of damage to the wing membrane. Other countries soon followed suit, as may be seen from Table 12.2, which gives a selection of some of the banding studies that have been reported.

Table 12.2 giving a total of 137 000 bats, mostly banded in caves, is far from comprehensive and certainly not up to date. However, combined with the figures given for America, it serves to illustrate the very large numbers of bats that have been, and are being, banded world-wide.

In Britain, large-scale banding experiments were first started in 1947, when my wife and I and members of the Devon Speleological Society began a study of the movements of cave-dwelling bats in Devon (Hooper and Hooper, 1956, pp. 1-26). This work covered seven species, but mainly Greater and Lesser Horseshoe Bats. In Somerset, P. F. Bird and members of the University of Bristol Speleological Society began banding bats found in the Mendip caves in 1949. These also were mostly Rhinolophidae (Bird, 1951, pp. 205-207). Work in Somerset has been continued by various other banders, notably Ransome (1968, pp. 77-112), who extended it to include tunnels in Gloucestershire. Other banders of cave-dwelling bats in Britain who have reported on their work include Hesketh (1951, pp. 177-181), who banded over 200 Lesser Horseshoe Bats in mine tunnels in Denbighshire, Pill (1951, p. 7), who studied a colony of Whiskered Bats in Jug Holes, Derbyshire, and Stebbings (1965b, p. 138),

Table 12.2

SOME BAT-BANDING TOTALS REPORTED IN VARIOUS COUNTRIES

Country	No. of Bats banded	Dates	Reference
Australia	59 133	1957-1970	Hamilton-Smith (1971)
Belgium	1 200	1944-1948	Verschuren (1949)
Czechoslovakia	27 411	1948-1967	Gaisler and Hanak (1969)
Denmark	3 828	1955-1961	Egsbaek and Jensen (1963)
France	970	1946-1948	Gruet and Dufour (1949)
Germany	10 887	1934-1943	Eisentraut (1943)
Holland	17 335	1936-1951	Bels (1952)
Italy	1 660	1957-1960	Dinale (1961)
Poland	7 442	1950-1060	Krzanowski (1960)
Spain	3 500	1959-1964	Balcells (1964a)
Sweden	"over 1 000"	1932-1947	Ryberg (1947)
U.K. (Devon)	3 200	1947-1961	Hooper (1962)

who banded 876 bats, mostly *Myotis* species, in chalk tunnels near Bury St. Edmunds between 1949 and 1964.

The bands used in Britain, a variety of designs made to special order, were initially identified by the initials of the workers or Societies concerned (e.g. DSS in Devon, UBSS in Somerset, GEH in North Wales, etc.). However, as the number of workers increased, so also did the permutations of cryptic initials, which were of course meaningless to a member of the public finding a ringed bat. In 1955, therefore, the Mammal Society of the British Isles, feeling the need for some more rational system, proposed a National Bat Banding Scheme. This scheme allowed each worker to apply for blocks of consecutive numbers, unique to himself, which he could use on his bands without fear of confusion with other banders, and also to have his bands stamped with the simple address

"LOND. ZOO". Reports from outside finders of banded bats, if sent into the London Zoo, were then sent on to the appropriate bander.

BAT BANDING—RING INJURY AND OTHER PROBLEMS

The foregoing may perhaps have given the impression that bat banding is all too easy. It is therefore necessary to emphasize most strongly that it is not an activity to be undertaken lightly, or on a short-term basis. In the first place, there is a risk that badly made or incorrectly applied bands will injure the bat. Even experienced bat banders are distressed to find from time to time that, despite all possible precautions, a band is causing trouble. This can vary from simple overgrowth of the metal by skin to acute inflammation. Factors, separately or in combination, which may contribute to this, are:

(1) Deficiencies in ring design or manufacture, if they are too small, too heavy, or have sharp edges.
(2) Incorrect application of the ring, e.g., too tight, too loose, or failure to close it accurately and symmetrically, with the result that the metal is flattened against the forearm, or the ends do not make a butt joint. The latter, a common fault, leaves cutting edges (like scissor blades) to damage the wing membrane.
(3) Possible condensation on the metal in the humid atmosphere of a cave, so that the moisture under the ring could soften the skin and render it more prone to chafing.
(4) The tolerance, or otherwise, of the bat to its ring. Some bats seem to make no effort to bite off their rings, whereas others immediately start to gnaw them, and, if they do, the combination of strong jaws and sharp teeth can flatten the soft metal or produce jagged edges which aggravate the risk of inflammation. We have noticed this many times with bats in Devon. For some individual Greater Horseshoe Bats, the result of banding has been a gnawed and pitted band in a matter of weeks; others we have found ten or even fifteen years after banding, with ring shiny and apparently in mint condition, and with limb and skin beneath it completely unmarked. Lesser Horseshoe Bats seem to be more tolerant of their bands, and in those banded in Devon we have found few cases of injury.

Several workers have commented on the problem of ring injury, notably Hitchcock (1957, pp. 402-405), who strongly criticized the use, in America, of bird bands on bats, and recommended instead the styles used in Holland or Germany (which were constructed so that contact

Plate 12.1. Greater Horseshoe Bat in flight in a Devon cave. Note the numbered aluminium ring on the forearm.

PLATE 12.2. Comparison of size of bats: Greater and Lesser Horseshoe Bats hang side by side in a mine adit in Devon.

PLATE 12.3. A large cluster of Greater Horseshoe Bats in a cave at Buckfastleigh in Devon.

flanged ends. Other banders, and in particular Mr T. J. Pickvance ("Ringing Secretary" of the Mammal Society from 1961) also devoted much thought and experiment to this problem, with the result that by 1968, reasonably general agreement had been reached on four basic designs of Mammal Society bat bands for issue to *bona fide* bat banders under the National Scheme already mentioned. These designs, tailored to British bats, were of 5·5, 4, 3 and 2·5 mm internal diameter, made of an

PLATE 12.6. Aluminium alloy bat rings—4 mm size—as used in Britain, from the design recommended by the Mammal Society for use on Greater Horseshoe Bats.

alloy somewhat harder than pure aluminium, the ends having a 1 mm flange, and the manufacturing process included a "barrelling" stage with abrasive grit to round off all sharp edges. Although it is doubtful if a design that is entirely trouble-free will ever be achieved, experience with the 4 mm version, (weight 100 mg) on Greater Horseshoe Bats has been very satisfactory.

IMPORTANT NOTE. Under the Conservation of Wild Creatures and Wild Plants Act, 1975, it is now illegal in Britain to ring any species of bat without a licence from the Nature Conservancy Council. Greater Horseshoe and Mouse-eared Bats have special protection, and a licence is needed even to handle them.

Bat banding is not an easy and fashionable form of "research" that can be taken up by would-be or dilettante naturalists as a short-term nature project at school or university and then dropped after a year or so. All who undertake bat banding must bear the responsibility of making every effort to ensure that their rings cause no injury throughout the natural life of the bats concerned, which means keeping check on the bats for 10-15 years after the last ring has been applied. They will in any case

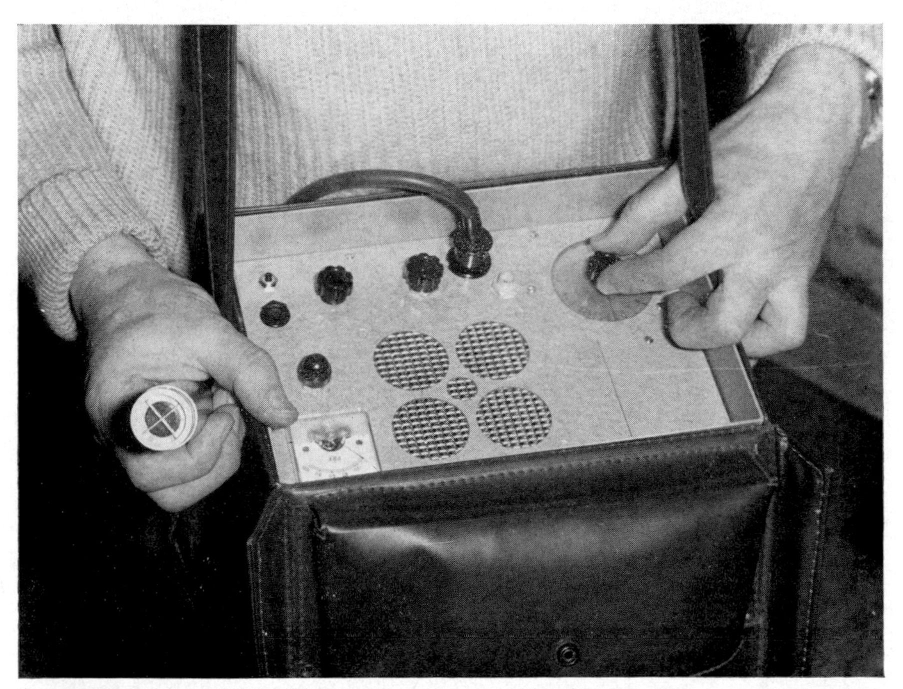

PLATE 12.7. The Holgate Ultrasonic receiver or "Bat Detector" as worn on the chest in a position convenient for use.

have to convince the Nature Conservancy Council that their project is really justified, in order to obtain a banding licence.

Besides direct injury resulting from banding, there may be even more serious harm resulting from the disturbance caused to the bat when it is handled in a cave in winter. This is the time when it is most readily found and handled, if banding is being done; but it is also the time when the bat's metabolic reserves are often at a critically low level. Handling, to put on a band, or to read its number, often causes the bat to awake and fly. At worst, this could cause premature death through starvation later in the winter—which would defeat the object of the banding; at best, the

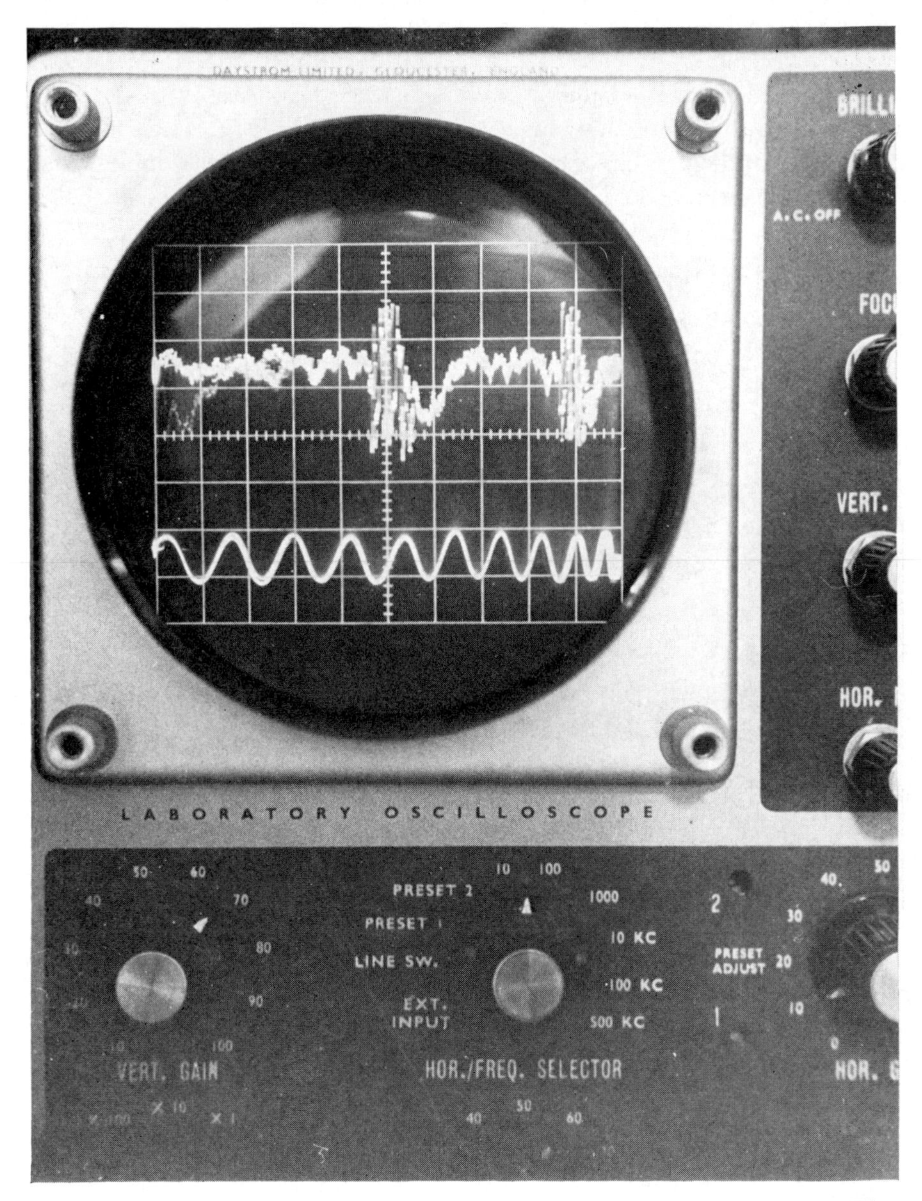

PLATE 12.8. Ultrasonic pulses of a bat displayed on a cathode ray oscilloscope. The lower trace provides the time scale with a peak-to-peak distance of 20 msecs. The two bat pulses on the upper trace are thus separated by a time interval of about 80 msecs.

result is an artificial "movement", a flight which would not otherwise have been made. The naturalist wishing to learn about the movements of bats is thus faced with the vexed problem of trying to decide which movements are natural and which have been induced as a result of subjecting the bat to light, noise, the warmth of human presence and other factors which would normally be absent from a cave environment. It is one thing to record that a bat banded in Cave A in December was then found several miles away in Cave B in January, but how meaningful is this observation? Who is to say whether that movement was natural or whether the flight was made following an unscheduled awakening when the band was put on? There is always the feeling that if the bat had not been disturbed it might still have been in Cave A in January.

A growing awareness of the possible harmful effects of bat banding on population strength has caused a number of workers (including the author) to stop active banding altogether. This was done, for example, in the tunnels of S Limburg, after a statistical analysis of recapture data demonstrated a considerable decline in the populations of at least three species (Sluiter and Van Heerdt, 1957, p. 141). In subsequent population studies in those tunnels, to quote one paper (Daan and Wichers, 1968, p. 262), "One of the requirements of the investigation has been to avoid any disturbance of the hibernating animals. We did not band or even touch any bat, but only silently made notes about their positions and ambient temperatures." The same approach can be commended to bat workers in British caves. A technique which goes halfway towards this is to use colour-coded bands which can be recognized at a distance, so that selected individuals can be followed, without necessarily being handled. This was tried on a limited scale in the Devon caves, where three separate colonies of Greater Horseshoe Bats have been under observation, at Buckfastleigh, in the Chudleigh caves (18 km distant), and in the Virtuous Lady Mine (29 km from Buckfastleigh). A few bats in each area were labelled with coloured celluloid rings (red, yellow and mauve respectively). This meant that if, in a Buckfastleigh cave, a bat with a yellow ring was seen amongst the local inhabitants (red rings), this was a visitor from Chudleigh.

The same idea, but in more elaborate fashion, has been used by workers in the Czechoslovakian caves (Gaisler and Nevrly, 1961, pp. 135-141). This involved a colour code based on the use of plain and painted aluminium rings. By putting two rings on each bat, one painted and one unpainted, either on different forearms or on the same forearm, a set of permutations was built up which enabled 96 bats to be distinctively labelled. Further combinations were devised by using mixtures of plain aluminium and coloured celluloid rings. It was found that, with an ordinary electric flashlight, most bats ringed in this manner could be identified without disturbance and at a distance of up to 3 metres. The short period of illumination needed for this did not waken the bats.

Other Marking Techniques

Brief mention may be made of several more specialized methods that have been employed to follow the movements of bats. Punt and Nieuwenhoven (1957, pp. 51-54) have described the marking of hibernating bats with bands containing radioactive antimony ^{124}Sb. Bats marked in this way could be detected at a distance, even when hidden in deep crevices. The information was limited, since specific individuals were not located, and merely indicated that a banded bat was present.

Griffin (1966) has described two novel methods which he employed for following, or attempting to follow, the movements of homing Spear-nosed Bats in Trinidad. The first was to provide the bat with a small and light-weight tail-light. Miniature lamps, powered by hearing-aid batteries, were taped to the bats, and provided a visible point of light, lasting 15-20 minutes. Unfortunately the low-flying bats were normally lost to sight behind trees or hills before their true flight direction could be established. The tail-lights were therefore replaced by miniature radio transmitters. The emitted signals were picked up by direction-finding receiving equipment mounted on the roof of a car. The transmitters, which were only about one-tenth of the weight of the bat, could function for several days, by which time the bat had worked the radio out of its fur, without damage to itself or to the equipment.

OBSERVATIONS ON CAVE-DWELLING BATS

A SUMMARY OF THE ANNUAL CYCLE

With variation in behaviour between species and differences in habitat and climate, any attempt at generalization is bound to run into difficulties. Nevertheless, no account of cave-dwelling bats would be complete without at least a brief summary of their pattern of existence during the period of the year when they are not normally found in caves, and so it must be made clear that the following picture is based essentially on observations on Greater Horseshoe Bats. These bats leave the caves in the late spring, or as soon as external conditions provide an adequate source of flying insects, and move to other haunts such as the roofs of houses and barns. In these haunts, the young are born, usually in July, and during this month the females collect together into large nursing colonies. Multiple births to a given mother are virtually non-existent, and indeed even a single youngster, which has to be carried on her underside for the first few weeks of its life, soon becomes a cumbersome load, with weight approaching half that of the parent. However, when the adult leaves the roof at dusk to hunt, she normally parks her baby on a rafter to await her return, so that if the roof is visited at this period, one may commonly find an attractive nursery of several dozen juveniles. These grow rapidly over the weeks and by mid-August are capable of flight and of hunting for themselves. By late autumn, when the bats begin to return to the caves, the young bats are virtually indistinguishable from the adults. The foregoing does not, however, mean that the horseshoe bats, and even nursing colonies, are never seen in caves during the summer months, but that when they are, they are wakeful and active, and fly off on human approach.

Copulation takes place in the caves in the autumn, and is promiscuous, with no evidence of pairing off. The male sperm is retained in a hard, jelly-like plug which is formed inside the vagina of the female and remains there during the winter months. This plug can be discharged if the female is disturbed during hibernation, and there is some evidence that mating can also occur in the spring. Normally however, fertilization, which takes place about April, is through the action of the sperm stored since the previous autumn. The period when this occurs seems to be dependent on the weather conditions, since these control the insect activity and hence the resumption of full hunting activity by the bat. A cold spring, therefore, may prolong the bat's stay in the caves and hence mean a late birth. The exact gestation period is uncertain, in view of this process of delayed implantation; but Gaisler (1965, p. 336) has quoted a figure of about 2½ months for Lesser Horseshoe Bats. It is exceptional for a female to bear young until she is at least two years old; this fact, coupled with that of only a single offspring each year, underlines the need for a reasonable number of successful breeding seasons for perpetuation of the species and explains why, as further dealt with below, the natural life expectation of bats is high, in excess of ten years. It follows, however, that the reproduction potential of a bat population is low, and build-up of a population weakened for any reason like human activity can only be a slow process.

WINTER BEHAVIOUR

For British bats, the observations in this section mainly refer to the species most commonly seen in caves, the horseshoe bats. From about October onwards, both Greater and Lesser Horseshoe species tend to be seen in increasing numbers in the West Country caves. The Greater Horseshoe Bats may be found as solitary individuals, in small groups, or even in big clusters of a hundred or more. In such groupings, the bats can be in scattered formation, with individuals not actually touching, or in masses so tightly packed that separate bodies are almost indistinguishable. Lesser Horseshoe Bats, on the other hand, when seen in British caves, are almost invariably found as solitary individuals and never as clusters. On rare occasions we have seen two only a few inches apart, but not in bodily contact. The same behaviour has been noted in Holland (Bels, 1952), Switzerland (Aellen, 1949, p. 50) and France (Jeannel, 1926), although, surprisingly, these workers also found that Greater Horseshoe Bats behaved in the same fashion. This demonstrates the manner in which inter-regional differences render it difficult to make almost any generalization about bat behaviour; further, a photograph of Lesser Horseshoe Bats in a Polish cave (Kowalski, 1953, p. 62), shows about 50 bats in a loose cluster, with several of them in contact with their neighbours.

Populations of Greater Horseshoe Bats are often based on a particular cave, or group of caves. Thus the various caves, quarries and mine tunnels within a radius of a mile or so around Buckfastleigh in Devon normally support, each winter, a colony of about 150 Greater Horseshoe Bats. "Colony" is, perhaps, a misleading word in that it could be taken to imply an integrated group existence such as that of a colony of bees; what is meant is a floating and somewhat restless population, that remains roughly constant in overall total, although its distribution amongst the various tunnels is rarely static for any length of time. The various individuals (their movements followed by banding studies) show no particular constancy of behaviour, and a bat, seen hanging by itself in one cave, could perhaps a week later still be there by itself, or as part of a group in some other cave. In Devon, the large and tightly packed clusters have often been seen during spells of cold weather. This, at first thought, seems a logical observation, but in fact there must be some reason other than to keep warm. For a sleeping bat, the body temperature soon falls to that of the ambient air, and all that clustering would do would be to decrease the rate of cooling for those in the middle. An early belief that bats, in such clusters, segregated themselves into different groups according to sex is now known to be incorrect (cf. Hooper, Hooper and Shaw, 1950, p. 156), and indeed the reverse seems always to be the case, with both sexes present in any given cluster. For a particular population, however, the males often outnumber the females. Although not all workers have found this, and in some cases the samples have been too small to be meaningful, it has nevertheless been the subject of comment for many species. Griffin, for example, (1940, p. 186), although finding a 1 : 1 sex ratio at birth for *Myotis lucifugus*, recorded that out of 3703 of these bats banded in 10 New England caves 72·5% were males. Gruet and Dufour (1949, pp. 138-143) found a preponderance of males for Greater Horseshoe Bats in French caves, and Egsbaek and Jensen (1963, pp. 292-294) noted a figure of 60% for males out of 2505 Daubenton's Bats banded in Denmark during a six-year period. They also found more males than females for three other *Myotis* species. The massive banding experiments in the S Limburg tunnels in Holland showed a preponderance of males for Lesser Horseshoe Bats and also for four *Myotis* species, although not for Daubenton's Bats (Bezem, Sluiter and Van Heerdt, 1960, p. 534). The reason for this excess of males in hibernating bat populations is not known. The Dutch workers mentioned above regarded it as unlikely that more males were born than females and concluded that many of the females spent the winter outside the caves that were being sampled. For British caves, Stebbings (1965b, pp. 138-139) reported an overall male/female ratio of 59 : 41 for five species (Daubenton's, Natterer's, Whiskered, Long-eared and Barbastelles) found in chalk tunnels in West Suffolk. In our own work in the Devon caves, the corresponding ratio for Greater Horseshoe Bats was 52·5 : 47·5 (2001 individuals) and for Lesser Horseshoe Bats was 57·6 : 42·4 (967 individuals).

Myotis bats and other vespertilionid species, in the British caves, are more commonly seen in ones and twos than in large groups, although Pill (1951, p. 7) has reported clusters of Whiskered Bats ("seldom exceeding a dozen") in Jug Holes. This is in marked contrast to caves on the continent of Europe, where *Myotis* bats, of many species, are often present in vast numbers. The Mouse-eared Bat (*Myotis myotis*), for example, hangs in large colonies in the high roof of the river passage in the Grotte de Han, in Belgium, and the population was once estimated at 1000-1500 (Anciaux, 1954, p. 170). Daubenton's Bats also hang in sizeable clusters in this cave and Anciaux has even reported finding single individuals of this species mixed up with a group of Mouse-eared Bats. Such intermingling of species, presumably fortuitous, is occasionally observed in British caves. It is not unknown for a solitary Lesser Horseshoe Bat to hang close to a Greater Horseshoe Bat, or even to a group of the latter, although not in bodily contact. We have also seen a Natterer's Bat hanging only a few inches from a Greater Horseshoe Bat.

During the early part of the winter, cave-dwelling bats maintain their hunting activity, and weighing experiments have demonstrated how the bats gain in weight during this period. Greater Horseshoe Bats found in Devon caves were weighed for four successive winters (Hooper and Hooper, 1956, pp. 16-19), and it was found that the average weight appeared to reach a peak in mid-December (at about 22 g for males and 23·5 g for females). Krzanowski (1961, pp. 249-264), found that for four of the species he was studying (Natterer's, Mouse-eared, Long-eared Bats and Barbastelles) the period of maximum weight occurred rather earlier (late October to early November). This is not surprising, since factors such as differences in altitude, night ambient temperature and hence availability of flying insects, and regional differences in respect of climate and insect populations will obviously affect a number of aspects of bat behaviour. However, as winter progresses and the opportunities for hunting insect prey become less frequent, the bat enters the critical period when it must subsist on its own bodily reserves. Survival then depends on dropping the metabolic rate to a level which makes the minimum demand on such energy reserves as can be held in the form of stored fat. In other words, the periods of torpidity and sleep, when the body temperature drops to ambient—the so-called state of hibernation—become more prolonged. As further noted below, the bat awakes naturally at intervals, and in addition it takes advantage of mild spells of weather to fly out of the cave to hunt. We had confirmation of this one February, in a Buckfastleigh cave, when we found the still kicking remains of a large Dor beetle, its abdomen bitten out, which was obviously the recent victim of a bat. Nevertheless, the balance of existence must, at this stage, be very precarious. A flight from the cave to top up fat reserves could easily use up more energy than was actually gained in terms of caught insects. It should therefore be unnecessary to spell out, yet again, the importance of not disturbing bats seen in a cave during the period December to March. During this period, the weight loss is considerable. The Devon weighing experiments showed that the average weight of Greater Horseshoe Bats decreased (from the December maximum) by about 25% (23% for males and 28% for females) so that, by April, the average weight was down to about 16·5 g. For one particular female, a decrease of 34·3% (from 24·6 to 16·15 g) was noted. Such weighings were carried out, to the nearest 0·05 g, using a simple balance; and although some bats remained torpid on the scale pan, in general less disturbance was caused if the pan was fixed in a vertical plane so that the bat could hang naturally, with its feet gripping the upper edge of the pan.

For horseshoe bats and the other species which use caves in winter, hibernation is by no means continuous. The bats move from one part of a cave to another and even to other caves, throughout the winter. Undoubtedly, some of the movements observed have been artificially induced by human disturbance, but this does not account for the sudden appearance of bats in previously empty passages. The mass of available evidence shows that such winter arousal is natural and indeed is a necessary factor in the bat's winter existence. As long ago as 1907, Coward commented on the light sleep of the horseshoe bats, and Anciaux (1950) pointed out the need of bats to drink at frequent intervals to avoid desiccation. However, more important still is the need for food, and many of the appearances and disappearances of bats in caves, or in parts of caves, seem to be related to hunting activity or preparation for such activity. Many workers have attempted to explain such movements and to correlate them with choice of sleeping quarters (hibernaculum). The picture is not always clear because of the variety of species studied in caves which differ in length, complexity, temperature and external climatic conditions, but undoubtedly cave temperature is the factor which plays a major part in the selection of a hibernaculum and in the frequency of arousal. To deal with this subject in detail would require a chapter in itself and here only a brief outline can be given of "internal migration" within a cave. This concept has been the result, in particular, of many years of study by workers in Holland. The immense underground stone quarries of S Limburg, excavated in cretaceous limestone, form a variety of complex labyrinths that house large numbers of bats. Careful temperature and humidity measurements (Punt and Parma, 1964, pp. 46-59) in one particular system (the Kloostergroeve) established that this set of tunnels could be divided into four temperature zones, the zoning remaining constant throughout the winter. As the winter progressed, the bats tended to migrate towards the entrance, but avoiding places where draughts were evident. In 1961, these workers attempted to modify the temperature and air circulation by boarding up the entrance and warming the chamber inside with a large coke stove. After three nights, with the stoking efforts of biology students at Amsterdam University, nearly a ton of fuel had been burned and temperature changes of up to 5°C had been established in some of the passages. This resulted in a noticeable increase in activity of the bats, but no significant change of zoning, possibly because the experiment was not continued for a long enough period.

Further study in the S Limburg quarries (Daan and Wichers, 1968; Kuipers and Daan, 1970) showed that the pattern of internal movement varied between species in addition to being dependent on temperature zoning. In many of the underground quarries, relatively warm air could flow in in the entrance area, and this part of the system was avoided by the bats at the beginning of the hibernation season. However, in one particular multi-level system, with various ventilation shafts, the internal climate was such that the entrance area of the middle system remained the coldest part throughout the summer, and the bats concentrated in this section on arrival in October. In both types of quarry, individual migration took place from the rear towards the entrance areas as the latter

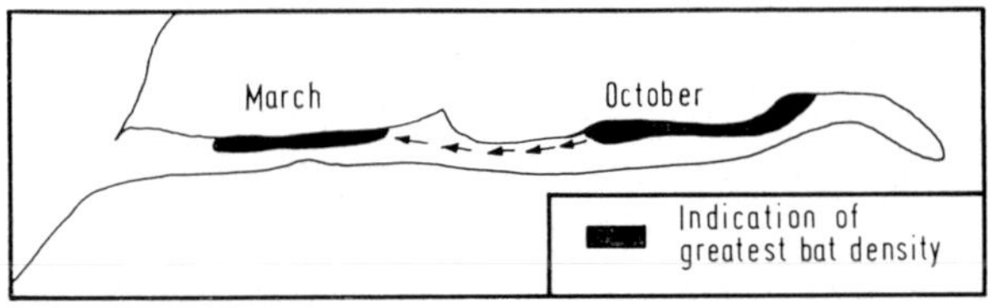

FIG. 12.1. Internal migration of bats within a cave during the winter months. (Based on Kuipers and Daan, 1970.)

cooled down while winter progressed. With regard to variation in behaviour between species, some bats, notably *Myotis emarginatus* showed much less tendency to move from the warmer, inner sections of the quarries. It was speculated that this species, which probably underwent long, uninterrupted periods of hibernation, did not therefore need to hang in a position where the temperature would rapidly reflect external conditions. Some species, believed to spend the shortest time underground like Natterer's Bat, were normally only found in the entrance area, but the bulk of the bat population followed the general pattern already mentioned, of moving progressively from the inner towards the outer tunnels as the winter advanced (Fig. 12.1). Two factors appear to be at work here. The migration towards the colder zones is firstly a sort of protective mechanism which conserves stored body reserves by progressively lowering the metabolic rate and minimizing unnecessary awakening and movement and yet at the same time puts the bat in a position most advantageous for the operation of the second factor. This is the need to react quickly to the onset of favourable hunting conditions, which, occurring as external temperature increases, would become increasingly apparent to the bat as its hibernaculum progressively moved towards the entrance. It follows from this that the blocking of a cave entrance by a solid wall in an attempt to protect the bats that normally use the cave might well have an opposite effect to the one intended, and result in a decrease in bat population through interference with the microclimate inside the cave. A walled-up entrance would prevent the sections inside from reacting rapidly to changes in outside temperature. It would also cause the internal temperatures to increase, and Daan (1967, p. 160), suggested that this latter effect, in particular, was responsible for the depopulation of one cave in Holland.

The importance of temperature as a parameter affecting the habits of bats in winter has been studied by Ransome (1968, pp. 77-112; 1971, pp. 353-371), in caves, mine tunnels and cellars in Somerset and Gloucestershire. In particular, for the Greater Horseshoe Bats hibernating in cellars he was able to introduce a degree of control in his experiments by installing thermostatically controlled heaters. His results confirmed a definite correlation between temperature and arousal frequency, timing of arousals and selection of hibernating site. This correlation varied between the beginning and end of the winter season. In the early winter, the bats woke on average once a day if the temperature was 10·5°C, but only once in six days at a temperature of 8°C. In the spring, a temperature of 8·5°C was sufficient to produce daily arousal, whereas at 6°C the arousal frequency reduced to once in six days. It seems surprising that bats can react so sensitively to such small changes in temperature, but Ransome established that bats normally selected their hibernacula on the basis of a preferred temperature in the range 7-9°C, choosing a temperature about 2°C lower in spring than in early winter. In each case, however, the temperature selected related to the same maximum external air temperature, so compensating for the 2°C seasonal shift mentioned above and hence ensuring the same arousal frequency in relation to the maximum external temperature throughout hibernation.

The foregoing explains why bats are often observed to concentrate in particular zones of a cave. Such choice of zone can be very selective, involving the use year after year not only of a particular passage but of a particular patch of roof in that passage. We have often noticed this for the clusters of Greater Horseshoe Bats seen in the Devon caves, and such highly localized choice seems to be dictated to some extent by shelter from draughts and favourable rock (i.e. adequacy of footholds). At first we thought that location of such a hibernaculum was a matter of trial and error, but this behaviour has been observed so often (e.g. Daan and Wichers, 1968; Ransome, 1971) that undoubtedly memory must play a part in the selection. Indeed, Ransome (1971, p. 365) in his observations on Greater Horseshoe Bats hibernating in a cellar, records several instances of individuals returning to hang from the same brick that they had used on a previous occasion.

In all such studies of bat behaviour in their winter haunts, deductions have to be made from observations as to the presence or absence of bats in a particular haunt at the time when the worker concerned visits it. If a bat is not seen in that haunt, it may just have moved to another part of the cave since it was last observed, or it may have moved to some other cave miles away. The human observer can thus only get an intermittent picture of a bat's activities and never a continuous study of its movements, day and night. Daan (1970) attempted to fill in some of the gaps in the story with the aid of photography and, in particular, of movements in and out of a cave during the hibernating season. This was in one of the underground quarry systems in S Limburg, and he closed the large entrance with a wire grill, all except for a rectangular "window" 0·75 by 1 m in size. A photocell arrangement automatically triggered off a flash each time a bat passed through this window and a motor-driven camera (simultaneously operated) included in its picture not only the flying bat but also a record of date and time from a clock at the top of the frame. The photographs obtained did not allow as precise recognition of bat species as had been hoped, but enabled long-eared bats to be recognized easily and thus distinguished from *Myotis* species. The photographs showed that all the flights through the "window" and out of the cave were made between sunset and sunrise. Movements within the tunnels took place to some extent during the hours of daylight, but the bats did not emerge from the cave until after dark. It would be interesting to carry out a similar experiment in Britain, in a cave inhabited by Greater Horseshoe Bats, in order to obtain a better picture of the winter flights made by this species.

MOVEMENTS OF BATS

The numerous banding experiments already outlined have demonstrated that bats are capable of flying over quite long distances. In Britain, where published results relate essentially to the horseshoe bats, the longest recorded flight still appears to be one made in 1954 by a Greater Horseshoe Bat which was banded at some stone quarries at Beer, in Devon, in February of that year and which was found in October in the roof of a building near Bryanston School in Dorset. The straight-line distance between these two sites is 64 km (40 miles). Six weeks later, this bat was found back again in the Beer tunnels. Another Greater Horseshoe Bat, banded in Dorset, has flown almost as far; and one banded in Somerset flew 43 km (27 miles) to a site in Gloucestershire. Such distances appear, however, to be exceptional. For Greater Horseshoe Bats studied in Devon, some 840 movements greater than 1·5 km (1 mile) have been recorded, but only 5 of these have been in excess of 32 km (20 miles). A breakdown of these Devon results, according to distance, is given in Table 12.3 and demonstrates that flights up to 16 km (10 miles) and even perhaps up to 24 km (15 miles) are commonplace for Greater Horseshoe Bats. For Lesser Horseshoe Bats in Britain, records of long flights are scanty. Even

Table 12.3

MOVEMENTS LONGER THAN 1·5 KM (1 MILE) FOR GREATER HORSESHOE BATS IN DEVON, 1948-1970

Distance in km	Total Movements	Winter Movements (November to March, inclusive)
1·5-3	443	144
3-8	232	39
8-16	97	10
16-24	51	8
24-32	16	1
Above 32	5	—
Totals	844	202

in Devon, where about 970 of this species have been banded, only about 50 movements greater than 1·5 km have been recorded, the longest being about 22 km.

As Table 12.3 shows, movements during the nominal winter period, November to March, are not uncommon, a particularly striking example being a 22 km flight for a Greater Horseshoe Bat from a cave at Buckfastleigh to one at Yealmpton within the short period December 24th to December 29th. Even if it was assumed that such a movement was initiated by disturbance on the 24th, there would appear to be no obvious reason for the bat to fly 22 km, since it could just as well have gone to any one of the several dozen caves and mine tunnels in the Buckfastleigh area, within a radius of approximately 1 km of so. It also seems unlikely that a bat, hunting on a single evening, would end up 22 km away from its starting point. This type of winter movement, and many others like it, e.g. flights over the 18 km distance between the Chudleigh and Buckfastleigh caves in January and February, are difficult to explain. An allied problem is that of selection of destination. Does the bat set out with the intention of flying to a specific cave; and, if that cave is several kilometres away, how does it find it? In any case, it is difficult to explain how a bat, flying after dark and relying on aural rather than visual information, can locate the entrance to a mine or tunnel which may be quite a small hole, and remote from any other similar tunnel. Much obviously remains to be learned about such movements; and, to emphasize this even further, there is the case of a cave in Devon whose entrance, previously unknown and buried beneath earth at the foot of a cliff, was opened up by digging. Within a few weeks of the cave being revealed, it was found to be occupied by several Lesser Horseshoe Bats. How did they know it had been opened?

In summer, many movements of bats can be explained on the basis of a shift from the caves where the winter has been spent, to the building which will be used as a site for a nursing colony. In Buckfastleigh, the roof used for the summer quarters of Greater Horseshoe Bats is only a few metres from one of the caves, and yet the number of bats observed in the nursing colony in that roof always seems to be at least double the total of Greater Horseshoe Bats seen in the local caves during the winter. Similarly, any census of Lesser Horseshoe Bats around the Buckfastleigh caves rarely shows a total of more than about a dozen, and yet in a house roof (known not to be used by such bats in winter) and barely 3 km away, a summer count of at least 60 of this species is a regular event. All this suggests that the nursing colonies are made up of bats drawn from a much wider area than the caves in the immediate vicinity. How great this area is, we do not know, but ringed Lesser Horseshoe Bats from the Beer tunnels have been found amongst a nursing colony, 20 km away, in a house roof near Clyst Honiton. The Greater Horseshoe Bat which flew the 64 km from the Beer tunnels to a roof near Bryanston and back again was a female; perhaps this was a routine movement from a hibernaculum to join the nursing colony which is normally found in that roof each summer.

Apart from such seasonal movements between winter and summer haunts, the Devon banding studies have shown no particular pattern for the movements of Greater Horseshoe Bats except for a tendency for bats born in any one area, e.g. Buckfastleigh, to move out from that area, after a few years, to more distant haunts. Often, though not always, the bat, having made such a move, apparently settles down in its new haunt, and is found there winter after winter and never (at least on the occasion of our checks) back in the caves adjacent to its birthplace. Such moves are occasionally from one major colony to another, as from Buckfastleigh to the Chudleigh caves, but may equally well be to some isolated and quite small mine adit, which then seemingly provides regular winter quarters for only a handful of bats.

Hanak *et al.* (1962, pp. 14-37), commenting on movements of Lesser Horseshoe Bats in Czechoslovakia, also noted that this species dispersed in all directions when leaving hibernating quarters and summer roosts, and that some individuals stopped from time to time in temporary roosts. In view of the relatively short distances flown between summer and winter quarters by the European horseshoe bats, the Czech banders classified them as "stationary" species, as distinct from most European *Myotis* species, which they described as "vagrants", and even stronger flying species such as *Nyctalus* (Noctules and Leisler's Bat) and *Miniopterus schreibersi* (Schreiber's Bat), which they classified as "migrants" (Gaisler and Hanak, 1969, p. 31). Very long flights have certainly been recorded for Noctules in Europe, e.g. 930 km in Czechoslovakia (Gaisler and Hanak, 1969) and 418 km in Poland (Krzanowski, 1960). For the Mouse-eared Bat (*Myotis myotis*), flights of 200-250 km have been recorded by the same two workers, and about the same (230 km) for *M. schreibersi*. Longer flights have, however, been noted for this last-named species. Balcells (1964b, p. 25) recorded a movement of 350 km between caves near Barcelona and Marseilles; and in Australia, where this species is very common, several flights in excess of 500 km have been recorded, and even one of 1248 km (780 miles) (Hamilton-Smith, 1968, p. 1).

ULTRASONICS OF BATS

With the development during recent years of portable ultrasonic receivers, the study of bat ultrasonics, formerly only possible with bulky equipment in a laboratory, has become a recognized field activity for naturalists. Moreover, the ultrasonic receiver, popularly known as a "bat detector", offers scope for gaining information on bat behaviour without causing disturbance to the bats.

The ability of a bat to avoid obstacles, to find its way through tunnels in darkness and to locate its prey by means of echo-location, a natural version of SONAR, is now well known. In brief, the bat in flight emits a continuous series of short pulses of high-frequency sound. If there is an obstacle ahead, these pulses of ultrasound are reflected back to the bat as echoes, and by some mechanism still not fully understood it is able to translate the information received into precise "visualization" of distance and direction and act on it in order to take the necessary avoiding action or even to intercept a flying insect. During the last decade, the ultrasonic receiver has enabled much to be learned about the basic feature of the system, the actual ultrasounds emitted by the bats. According to species, such ultrasounds vary in many respects like frequency range, pulse duration, repetition rate and presence of harmonics, but in general mainly lie within the frequency range 20-120 kHz. (1 kHz = 1 kilocycle/second = 1000 cycles per second.) A definitive account of bat ultrasonics is given in Sales and Pye (1974).

For such high frequencies, a special microphone is required, and use is made of a solid dielectric capacitance microphone (Kuhl *et al.*, 1954, pp. 519-532) in which a thin plastic diaphragm is tensioned over a brass disc with a surface of very fine concentric grooves. This diaphragm, of Mylar, is about 0·00025 in. thick, and acts as the dielectric. A metallic coating on the outer surface provides the earth plate of the condenser, and the other plate, maintained at about 180 volts, is the brass disc. The first really portable bat detector was devised by the Lincoln Laboratory of the Massachusetts Institute of Technology, for use by Dr Griffin, the pioneer of studies on bat ultrasonics (Griffin, 1958). The Lincoln Laboratory instrument (McCue and Bertolini, 1964, pp. 41-49) responded to all ultrasonic signals of short duration within a wide frequency band (20-200 kHz), each ultrasonic pulse picked up producing a click from a loudspeaker. In Britain, a more elegant receiver was subsequently devised by Pye and Flinn (1964, pp. 23-28), based on a tuning system analogous to that of a radio set, with a 5 kHz pass band. With this instrument, helped by the tuning dial, it was possible to select the frequency range at which the signals from particular bats were received most strongly, and indeed to measure that range. Following discussions between members of the Mammal Society and an electronics firm (Holgates of Totton, Ltd.), this firm undertook the construction of a commercial bat detector, based on the Pye and Flinn design. After various prototypes had been field tested (Watson, 1965; Hooper, 1966), a production version was established, known as the Holgate Ultrasonic Receiver, Mark IV. This instrument consists of a metal box, 24 by 16·5 by 13 cm, containing batteries, the necessary circuitry, a tuning dial graduated over the range 10-180 kHz, and a loudspeaker. With its microphone, the total weight is 3·9 kg.

During recent years, much work has been done to establish bat detector responses to the ultrasounds of the various British bats, and to make tape recordings of such responses for subsequent analysis. For some species the combination of tuning setting and type of sound heard through the loudspeaker is sufficiently characteristic to permit positive identification of the species heard (Hooper, 1969, pp. 177-181). For British bats, the ultrasounds emitted are of three main types, for which simplified forms of sound spectra are shown in Fig. 12.2. The sound pattern illustrated by (a) in Fig. 12.2, is characteristic of the *Myotis* species, and comprises a succession of very short pulses, each pulse starting at a high frequency (about 80 kHz) and dropping rapidly through one or two octaves, down to about 30 kHz, in rather less than 5 msecs. Pulse repetition rate, according to species and circumstances, varies between 10 and 35 per second, but can increase to approaching 200 per second as the bat homes

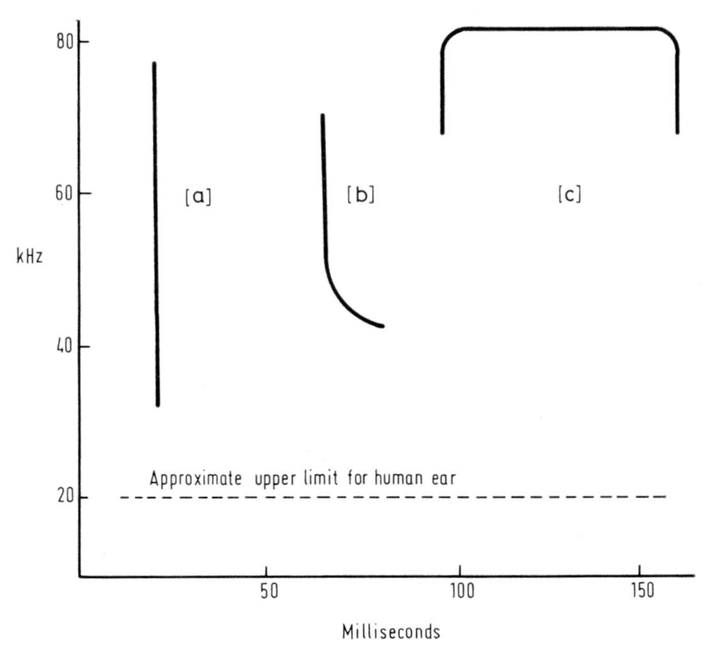

FIG. 12.2. Simplified sound spectra for three types of ultrasonic pulse emitted by British bats. (a) Natterer's Bat; (b) Pipistrelle; (c) Greater Horseshoe Bat.

in on an insect. If the bat detector is tuned to a frequency within the range covered by the sweep, each pulse is heard as a click through the loudspeaker. The second type of pulse ((b) in Fig. 12.2), is also a short duration, descending frequency sweep, but ends with a short period at almost constant frequency. This is typical for Pipistrelles and Noctules. The constant frequency portion of the signal can be combined with a synthetic signal generated within the bat detector circuitry by means of a Beat Frequency Oscillator (BFO). The result of combining the two signals is a "beat note" corresponding to the difference between their two frequencies, and if this difference is low enough, as at 1 kHz, to lie within the "audio" range, this is heard through the loudspeaker as an almost musical bleep or chirrup.

The third type of pulse ((c) in Fig. 12.2) is characteristic of the horseshoe bats. It is relatively long (*ca* 50 msecs) and is essentially at a constant frequency, in effect a sort of trumpet note which is emitted through the nostrils and noseleaf of these species. The frequency is high, *ca* 84 kHz for Greater Horseshoe Bats and *ca* 110 kHz for Lesser Horseshoe Bats, and the repetition rate, when the bat is flying, is about 10 pulses per second. When the bat detector is tuned to the appropriate frequency, with the BFO on, the loudspeaker emits a very characteristic yapping sound.

This summary is somewhat oversimplified, in that for some species distinctive harmonics are also emitted. For the horseshoe bats, the strongest signals are actually second harmonics (at the frequencies mentioned above). The fundamental notes (e.g. at about 42 kHz for the Greater Horseshoe Bat) are normally emitted at much lower intensity.

Caves have provided a useful natural laboratory for much basic study of bat ultrasonics and I have made many tape recordings of bat detector responses for wakeful or flying Horseshoe Bats in the Devon caves, checking various frequencies and comparing relative intensities of the harmonics. An initial hope that a bat detector would provide a useful instrument for locating bats in caves was however soon dispelled; bats at rest and in a torpid state do not emit ultrasounds and are only "on the air", ultrasonically, just before flight, or when disturbed. This latter observation explained why disturbance of only one bat to the point of wakefulness caused any other bats near by to become restless, wake up, and possibly fly off. To human ears, the bat being handled was apparently silent; but if the bat detector was switched on, the ultrasonic shouts of this bat were immediately apparent, and other bats clearly react to such sounds.

Fenton (1970, pp. 847-851), checking for bats emerging from summer roosts in Ontario, has described the use of a Holgate ultrasonic receiver as a means of monitoring such activity without causing disturbance to the bats. By setting up the detector in a suitable position and maintaining a continuous record from sunset onwards, he was able to derive graphs relating degree of activity (based on number of "passes" in a given time) to time of night. Similar experiments could clearly be carried out for bats flying in particular cave passages, or for bats leaving the entrance of a cave to hunt. Although much can be done by visual counting while some daylight remains, the Holgate receiver, set up outside a cave, could provide useful additional data on bats leaving the entrance after dark. Moreover, by appropriate choice of tuning setting, the response could be made selective for particular species. At 50 kHz, for example, only *Myotis* species would be picked up, and as such bats are hard to find in a cave, an ultrasonic count might provide a more reliable method of establishing the true population.

HEALTH AND LONGEVITY

A long-term bonus from banding data is a partial answer to the question "How long do bats live?" Published reports by various workers have made it clear that a life span of 10-15 years is by no means uncommon, and for several species, ages considerably in excess of this have been recorded. This is demonstrated in Table 12.4.

Table 12.4

LONGEVITY DATA FOR VARIOUS SPECIES OF BAT

Species	Country	Age (years) (at least)	Source of Data
Little Brown Bat (*M. lucifugus*)	U.S.A. (Vermont)	24	Griffin and Hitchcock (1965)
Daubenton's Bat (*M. daubentoni*)	U.K. (Suffolk)	18½	Stebbings (1968)
	Holland	15½	Van Heerdt and Sluiter (1961)
Natterer's Bat (*M. nattereri*)	U.K. (Suffolk)	12	Stebbings (1965b)
	Holland	17½	Van Heerdt and Sluiter (1961)
Mouse-eared-Bat (*M. myotis*)	Holland	14½	Van Heerdt and Sluiter (1961)
Whiskered Bat (*M. mystacinus*)	Holland	18½	Van Heerdt and Sluiter (1961)
Lesser Horseshoe Bat (*R. hipposideros*)	Holland	14½	Van Heerdt and Sluiter (1961)
	Czechoslovakia	18½	Gaisler and Hanak (1968)
	U.K. (Devon)	13¼	Hooper (unpublished)
Greater Horseshoe Bat (*R. ferrumequinum*)	Holland	17½	Van Heerdt and Sluiter (1961)
	U.K. (Devon)	20½	Hooper (unpublished)

The figures in Table 12.4 are of course minimum ages, based on the initial date of banding, since the age of the bat at that first finding would not normally be known. It has not yet been established whether an age of 20½ years is exceptional for a Greater Horseshoe Bat. All that can be said is that another individual of this species in Devon was found at an age of at least 19¼ years (Hooper and Hooper, 1967, pp. 1135-1136) and that another 23 bats have been recorded in Devon with ages greater than $15\frac{1}{2}$ years, including three more than $18\frac{1}{2}$ years old.

The fact that bats can live so long emphasizes the point that those who start banding experiments must be prepared to continue for a very long time to check that the bands cause no injury during the life of the bats. It also follows that, except under adverse conditions, bats can be regarded as healthy creatures. In the U.K., they have few natural predators apart from an occasional owl or an optimistic cat, and man. In this last-named category must be included householders who call in Pest Control officers to remove bats from summer roosts (many Devon Greater Horseshoe Bats were killed in this way in a house near Totnes), vicars who resent bats in their churches, those who collect bats for commercial sale to museums or for dissection, and naturalists who in fact have no intention of killing bats but do so by causing overdisturbance and subsequent starvation. Starvation, however caused, would seem to be a major cause of death. The disturbance aspect has been summed up by Stebbings (1967b, p. 520), who stated that "A bat ringer may be his own worst enemy"; but he has also commented on a summer feeding problem, resulting from modern farming practice. Ploughing up of hedges, ditches, ponds and marshy areas is thought to have caused a significant reduction in insect numbers during recent years, thereby reducing available food supplies for bats. Moreover, by virtue of their insect diet, bats are particularly vulnerable to poisoning through accumulation of traces of residual chemicals absorbed from insects that have survived attack by pesticide sprays and then have been eaten. In the U.S.A., Luckens and Davis (1964, p. 948) showed that the Big Brown Bat (*Eptesicus fuscus*) was far more sensitive to DDT than other mammals. They also showed (1965, pp. 879-880) that although this species was not so acutely sensitive to endrin and dieldrin, nevertheless, mortality was caused by quite low doses. More recently (1972, pp. 245-263) Jefferies, of Monks Wood Experimental Station, has reported similar results for British bats.

ECTOPARASITES

Bats in caves may sometimes harbour ectoparasites, in particular the curious wingless fly *Nycteribia biarticulata* Hermann, and ticks such as *Ixodes vespertilionis* Koch. The latter gain access to the bat from the walls of a cave, and usually attach themselves to the back of the neck or under the jaw, so that the bat cannot readily remove them. When gorged with blood, they drop off. Attempts to remove ticks from a bat are usually ill advised, and unless extreme care is taken, all that will be collected is the body of the tick, the head remaining buried in the bat's flesh, a potential source of inflammation. Table 12.5 (based on Thompson, 1961, pp. 131-134) lists some ectoparasites collected from cave-dwelling species of bat in Britain.

Mites are also sometimes found on bats, e.g. *Spinturnix myoti* (Kolenati); and in particular, young horseshoe bats, at the nursery stage, are often

Table 12.5

SOME ECTOPARASITES COLLECTED FROM BATS IN CAVES AND TUNNELS IN BRITAIN

	Host Bat Species	Area
SIPHONAPTERA		
ISCHNOPSYLLIDAE		
Ischnopsyllus s, simplex (Roths.)	*M. nattereri*	Clwyd
Ischnopsyllus hexactenus (Kol.)	*Plecotus auritus*	Clwyd
IXODOIDEA		
IXODIDAE		
Ixodes vespertilionis (Koch.)	*R. ferrumequinum*	Devon
	R. hipposideros	Devon
		Clwyd
		Gloucestershire
ARGASIDAE		
Argas vespertilionis (Latreille)	*R. ferrumequinum*	Devon
INSECTA DIPTERA		
NYCTERIBIIDAE		
Nycteribia (Stylidia) biarticulata (Hermann)	*R. hipposideros*	Devon
Nycteribia (Nycteribia) kolenatii (Theodor and Moscona)	*M. daubentoni*	Surrey
		Kent

found to be heavily infested with *Spinturnix euryalis* Canestrini. As the juveniles get older, such infestation apparently disappears.

When a bat is hibernating, it is difficult to search for and collect ectoparasites without causing considerable disturbance to the bat. Unfortunately, some collectors of such parasites pursue their prey with a single-minded determination that pays no heed at all to the wellbeing of the bat and to the possible consequences resulting from disturbance and unnecessary flight in mid-winter. Excerpts from a letter once received by the writer illustrate this attitude:

> "I had a trip to Box Mines today (Jan. 19th) and there was a chap with the party . . . (who) was armed with a number of tooth brushes and plastic bags, and on finding a bat it was measured not very gently with a 'foot rule'! But worst of all the bat was then scrubbed with the tooth brush to see if it had ring worm! the brush was then put in the plastic bag. . . . When I asked if the treatment was not a little rough for the bat, I was told it was better than killing them to check! Needless to say, after a few strong words, we 'saw' no more bats."

In view of what has already been said in this chapter, further comment seems unnecessary.

CONSERVATION OF BATS

The question "What use are bats?" is commonly asked; and although the answer usually given is to the effect that the insect eaters (e.g. *all* the British bats) are not in any sense pests, do no harm to man, and probably do him some good by helping to control the insect population, the extent of that help is not generally realized. Gould (1955, pp. 399-406; 1959, pp. 149-150) has studied the feeding efficiency of Little Brown Bats (*M. lucifugus*) and showed that these bats (weighing *ca* 8 g) could accumulate insects at the rate of about 1 g per hour. His figures indicate an average catch of about 500 insects per hour, or one every 7 secs; or, put another way, a colony of 100 would consume 19·2 kg in four months of summer activity. On a *pro rata* weight basis, the cave-dwelling Greater Horseshoe Bats in Devon would consume at least a quarter of a ton of insects in the same period. Where the bat population is large, the insect consumption is quite staggering. Davis, Herreid and Short (1962, pp. 311-346), for example, estimated that Mexican Free-tailed Bats in Texas consumed about 6600 tons of insects per annum.

For reasons such as the above, many countries have some degree of protective legislation for bats. Britain has been rather slow in this respect, but since August 1975, under the Conservation of Wild Creatures and Wild Plants Act, legal protection has been given to the Greater Horseshoe Bat and the Mouse-eared Bat. In March 1970 (Stebbings, 1971b), at the Second International Bat Research Conference at Amsterdam, 100 scientists attending agreed upon the following resolution:

> "Bats are world wide and important in our ecosystem and to the human economy.
>
> Delegates representing 20 nations stated that bats are declining, especially in highly urbanized areas.
>
> The main causes are: pollution by insecticides, loss of habitat, killing of bats by man.
>
> Certain countries have recognised the importance of conserving bats and have legislation protecting them.

Since bats migrate internationally the conference recommends that other countries should provide legislation to protect these mammals.

Additionally it was agreed that further investigations in the ecology of bats is essential for conservation, and any necessary control of local populations should be done by qualified persons."

Throughout this chapter, the pressure on bats as a result of direct or indirect human interference has been emphasized. Fenton (1971, pp. 12-19), in a recent plea for conservation of bats in Canada, makes the bitter comment: ". . . man has fostered an increase in the bat population by providing more roosting sites, both in the form of hibernacula and nursing colonies. It is obvious that this was quite accidental and that we are taking steps to rectify the situation." Speleologists and other readers of this chapter may or may not be able to reverse the trend towards a declining bat population, but at least they can refrain from contributing to it. Speleologists and bats both have an interest in caves; for the former, the interest is that of a visitor and often only transient; for the bats, caves represent an environment which may be vital for survival.

REFERENCES

Aellen, V. (1949). Les Chauves-souris du Jura Neuchâtelois et leurs migrations. *Bull. Soc. neuch. Sci. nat.* 72, 23-90.

Allen, A. A. (1921). Banding bats. *J. Mammal 2* (2), 53-57.

Anciaux, F. (1950). *Explorons nos Cavernes*. Dinant, Blaimont.

Anciaux, F. (1954). Observations sur une colonie de Murins (*Myotis myotis* Borkhausen) dans la Grotte de Han-sur-Lesse (Belgique). *Rassegna Speleologica Italiana 4*, 167-183.

Balcells, E. (1964a). Ergebnisse der fledermaus-beringung in Nordspanien. *Bonn. Zool. Beitr. Sonderheft. 1/2*, 36-43.

Balcells, E. (1964b). Datos sobre biologic y migracion del murçielago de cueva (Miniopterus schreibersi, chir vespert.) en el ne de Espana. *Third Internat. Congr. Speleo.* Band III (Section II). Vienna.

Barbour, R. W. and Davis, W. H. (1969). *Bats of America*. University Press of Kentucky.

Barrett-Hamilton, G. E. H. (1910). *A History of British Mammals*. Gurney and Jackson, London.

Bels, L. (1952). Fifteen years of bat banding in the Netherlands. *Publ. Natuurh. Genoots. Limburg 5*, 1-99.

Bezem, J. J., Sluiter, J. W. and Van Heerdt, P. F. (1960). Population statistics of five species of the bat Genus Myotis and one of Genus Rhinolophus, hibernating in the caves of S. Limburg. *Arch. Neerl. de Zool. 13*, 511-539.

Bird, P. F. (1951). The UBSS bat-ringing scheme. *Proc. Univ. Bristol Speleo. Soc. 6* (2), 205-207.

Blackmore, M. (1948). *Mammals in Britain*. Collins, London.

Blackmore, M. (1956). An occurrence of the Mouse-eared bat *Myotis myotis* (Borkhausen) in England. *Proc. Zool. Soc. Lond. 127* (2), 201-203.

Brink, Van Den, F. H. (1967). *A Field Guide to the Mammals of Britain and Europe*. Collins, London.

Corbet, G. B. (1964). The grey long-eared bat *Plecotus austriacus* in England and the Channel Islands. *Proc. Zool. Soc. Lond. 143* (3), 511-515.

Corbet, G. B. (1971). Provisional distribution maps of British mammals. *Mammal Review* 1, 95-142.

Coward, T. A. (1907). On the winter habits of the greater horseshoe, *Rhinolophus ferrumequinum* (Schreber) and other cave-haunting bats. *Proc. Zool. Soc. Lond.* 312-324.

Daan, S. (1967). De Geulhemergroeve; gevolgen van de afsluiting van een Mergelgroeve voor het vleermuizenbestand. *Natuurhist. Maandbl. 56*, 154-160.

Daan, S. and Wichers, H. J. (1968). Habitat selection of bats hibernating in a limestone cave. *Zeit. Saugetierkunde 33* (5), 262-287.

Daan, S. (1970). Photographic recording of natural activity in hibernating bats. *Proc. 2nd Int. Bat Research Conference. Bijdragen tot de dierkunde 40* (1), 13-16.

Davis, R. B., Herreid, C. F. and Short, H. L. (1962). Mexican free-tailed bats in Texas. *Ecol. Monogr. 32*, 311-346.

Dinale, G. (1961). L'Inanellamento di pipistrelli in Liguria negli anni 1959-1969. *Rassegna Speleologica Italiana 13* (2), 52-53.

Egsbaek, W. and Jensen, B. (1963). Results of bat banding in Denmark. *Vidensk Medd. fra Dansk naturh. Foren. 125*, 269-296.

Eisentraut, M. (1934). Markierungsversuche bei Fledermaüse. *Zeit. Morphol. Oekol. 28*, 553-560.

Eisentraut, M. (1943). Zehn Jahre Fledermausberingung. *Zool. Anz. 144*, 20-32.

Fenton, M. B. (1970). A technique for monitoring bat activity with results obtained from different environments in Southern Ontario. *Can. J. Zool. 48*, 847-851.

Fenton, M. B. (1971). Bats—Questions, answers and issues. *Ontario Naturalist 9* (3), 12-19.

Gaisler, J. and Nevrly, M. (1961). The use of coloured bands in investigating bats. *Vestn. Cs. Spol. Zool. (Acta Soc. Zool. Bohemoslov) 25* (2), 135-141.

Gaisler, J. (1965). The female sexual cycle and reproduction in the Lesser Horseshoe Bat (*Rhinolophus hipposideros* Bechstein 1800). *Vestn. Cs. Spol. Zool. (Acta Soc. Zool. Bohemoslov) 29* (4), 336-352.

Gaisler, J. and Hanak, V. (1969). Ergebnisse der zwanzigjahrigen beringung von fledermausen (Chiroptera) in der Tschechoslowakei: 1948-1967. *Acta Sc. Nat. Brno 3* (5), 1-33.

Gould, E. (1955). The feeding efficiency of insectivorous bats. *J. Mammal 36*, 399-406.

Gould, E. (1959). Further studies on the feeding efficiency of bats. *J. Mammal 40*, 149-150.

Griffin, D. R. (1934). Marking bats. *J. Mammal 15* (3), 202-207.

Griffin, D. R. (1940). Notes on the life histories of New England cave bats. *J. Mammal 21* (2), 181-187.

Griffin, D. R. (1958). *Listening in the Dark*. Yale University Press.

Griffin, D. R. and Hitchcock, H. B. (1965). Probable 24-year longevity records for *Myotis lucifugus*. *J. Mammal. 46* (2), 332.

Griffin, D. R. (1966). Homing bats in Trinidad. *New York Zool. Soc. Newsletter*, November, 1-4.

Gruet, M. and Dufour, Y. (1949). Étude sur les chauves-souris troglodytes du Maine-et-Loire. *Mammalia 13* (3), 69-75, and (4), 138-143.

Hamilton-Smith, E. (1968). Banding news. *Australian Bat Research Newsletter* (8), 1.

Hamilton-Smith, E. (1971). Bat-banding in Australia 1967-70. *Australian Bat Research* News (10), 3-6.

Hanak, V., Gaisler, J. and Figala, J. (1962). Results of bat-banding in Czechoslovakia 1948-1960. *Acta. Univ. Carol., Biol., Praha* (1), 9-87.

Heerdt, Van, P. F. and Sluiter, J. W. (1961). New data on longevity in bats. *Natuurh. Maandbl. 50* (3-4), 36.

Hesketh, G. E. (1951). Ringing bats in Denbighshire. *Naturalist*, No. 839, 177-181.

Hitchcock, H. B. (1957), The use of bird bands on bats. *J. Mammal. 38* (3), 402-405.

Hitchcock, H. B. (1960). Bat banding in the United States. *The Ring 2* (25), 277-280.

Hooper, W. M., Hooper, J. H. D. and Shaw, T. R. (1950). Some observations on the distribution and movements of cave-swelling bats in Devonshire. *Naturalist* 835, 149-157.

Hooper, J. H. D. and Hooper, W. M. (1956). Habits and movements of cave-dwelling bats in Devonshire. *Proc. Zool. Soc. Lond. 127* (1), 1-26.

Hooper, J. H. D. (1962). *Horseshoe bats* (*Animals of Britain*, *No. 2*). Sunday Times Publ., London.

Hooper, J. H. D. (1966). The ultrasonic "voices" of bats. *New Scientist*, Feb. 24, 496-497.

Hooper, J. H. D. and Hooper, W. M. (1967). Longevity of Rhinolophid bats in Britain. *Nature 216*, 1135-1136.

Hooper, J. H. D. (1969). Potential use of a portable ultrasonic receiver for the field identification of flying bats. *Ultrasonics* 7, 177-181.

Jeannel, R. (1926). *Faune cavernicole de la France*. Paris, Paul Lechevalier.

Jefferies, D. J. (1972). Organochlorine insecticide residues in British bats and their significance. *J. Zool. Lond. 166*, 245-263.

Kowalski, K. and Wojtusiak, R. J. (1951). Homing experiments on bats. *Bulletin de l'Acad. Polonaise des Sciences*, Serie B (2), 33-56.

Kowalski, K. (1953). Cave dwelling bats in Poland and their protection. *Osobne Odbicie Z'Ochrony Przyrody' 21*, 58-77.

Krzanowski, A. (1960). Investigations of flights of Polish bats, mainly *Myotis myotis* (Borkhausen 1797). *Acta Theriol.* (*Bialowieza*) *4* (2), 175-184.

Krzanowski, A. (1961). Weight dynamics of bats wintering in the Cave at Pulaw (Poland). *Acta Theriol.* (*Bialowieza*) *4* (13), 249-264.

Kuhl, W., Schodder, G. R. and Schroder, F. K. (1954). Condenser transmitters and microphones with solid dielectric for airborne ultrasonics. *Acustica 4* (5), 519-532.

Kuipers, B. and Daan, S. (1970). Internal migration of hibernating bats: response to seasonal variation in cave microclimate. *Proc. 2nd Int. Bat Research Conference. Bijdragen tot de dierkunde 40* (1), 51-55.

Leen, L. and Novick, A. (1969). *The World of Bats*. Edita, Lausanne.

Luckens, M. M. and Davis, W. H. (1964). Bats: sensitivity to DDT. *Science*, N.Y. *146*, 948.

Luckens, M. M. and Davis, W. H. (1965). Toxicity of dieldrin and endrin to bats. *Nature 207*, 879-880.

Matthews, L. Harrison (1952). *British Mammals*. Collins, London.

McCue, J. J. G. and Bertolini, A. (1964). A portable receiver for ultrasonic waves in air. *IEEE Transactions on Sonics and Ultrasonics*, Paper SU-11 (1), 41-49.

Mohr, C. E. (1934). Marking bats for later recognition. *Proc. Pennsylv. Ac. Sci. 8*, 26-30.

Phillips, W. W. A. and Blackmore, M. (1970). Mouse-eared bats *Myotis myotis* in Sussex. *J. Zool. Lond. 162*, 520-521.

Pill, A. L. (1951). Jug Holes Cave and its bats. *Naturalist* 836, 7.

Punt, A. and Nieuwenhoven, Van, P. J. (1957). The use of radioactive bands in tracing hibernating bats. *Experimentia 13*, 51-54.

Punt, A. and Parma, S. (1964). On the hibernation of bats in a Marl Cave. *Publ. Natuurh. Genoots. Limburg. 13*, 45-59.

Pye, J. D. and Flinn, M. (1964). Equipment for detecting animal ultrasound. *Ultrasonics 2*, 23-28.

Pye, J. D. (1969). The diversity of bats. *Science Journal 5* (4), 47-52.

Ransome, R. (1968). The distribution of the greater horseshoe bat, *Rhinolophus ferrumequinum*, during hibernation, in relation to environmental factors. *J. Zool. Lond. 154*, 77-112.

Ransome, R. (1971). The effect of ambient temperature on the arousal frequency of the hibernating greater horseshoe bat *Rhinolophus ferrumequinum*, in relation to site selection and the hibernation state. *J. Zool. Lond. 164*, 353-371.

Ryberg, O. (1947). Studies on bats and bat parasites. *Svensk Natur*, Stockholm, 1-330.

Sales, G. and Pye, D. (1974). *Ultrasonic Communication by Animals*. Chapman and Hall, London.

Sluiter, J. W. and Van Heerdt, P. F. (1957). Distribution and decline of bat populations in S. Limburg from 1942 till 1957. *Natuurhist. Maandbl. 46*, 134-143.

Southern, R. N. (1964). *The Handbook of British Mammals*. Blackwell, Oxford.

Stebbings, R. E. (1965a). *Plecotus austriacus* in Dorset. *Nature 206*, 314-315.

Stebbings, R. E. (1965b). Observations during sixteen years on winter roosts of bats in west Suffolk. *Proc. Zool. Soc. Lond. 144* (1), 137-143.

Stebbings, R. E. (1966). Bats under stress. *Studies in Speleology 1* (4), 168-173.

Stebbings, R. E. (1967a). Identification and distribution of the genus *Plecotus* in England. *J. Zool. Lond. 153*, 291-310.

Stebbings, R. E. (1967b). Conservation of bats. *J. Devon Trust*, No. 13, 517-522.

Stebbings, R. E. (1968). Longevity of vespertilionid bats in Britain. *J. Zool. Lond. 156*, 531.

Stebbings, R. E. (1971a). Bats, their life and conservation. *J. Devon Trust*, *3* (1), 29-36.

Stebbings, R. E. (1971b). Bat protection and the establishment of a new cave reserve in the Netherlands. *Studies in Speleology 2* (3-4), 103-108.

Thompson, G. B. (1961). The Parasites of British birds and mammals XXXVI—New records of bat parasites. *Ent. Mon. Mag. 97*, 131-134.

Trapido, H. and Crowe, P. E. (1946). The wing band method in the study of the travels of bats. *J. Mammal. 27* (3), 224-226.

Verschuren, J. (1949). L'activité et des déplacements hivernaux des chiroptères en Belgique. *Institut royal des Sciences naturelles de Belgique 25* (3), 1-7.

Watson, A. (1965). Observing the natural behaviour of bats in flight. *Studies in Speleology 1* (2-3), 100-104.

Yalden, D. W. and Morris, P. A. (1975). *The Lives of Bats*. David and Charles, Newton Abbot.

13. Cave Palaeontology and Archaeology

A. J. Sutcliffe, D. Bramwell, A. King and *M. Walker*

Part I: CAVE PALAEONTOLOGY

A. J. Sutcliffe

For the palaeontologist looking for fossil remains, caves provide one of the richest fields of research, sometimes with vast concentrations of animal bones packed together in a very small space.

Two factors contribute to the accumulation of such deposits. Firstly, caves are places where remains tend to become concentrated by natural processes. Some caves are dangerous, with shafts in their roofs. The remains of animals and vegetation falling down shafts become mixed with earth and rocks which build up as conical heaps on the cave floor below, perhaps filling the shafts to the level of the ground surface. Some animals use caves as breeding places, for eating or sleeping shelters and they may die there or leave the bones of their prey there. In European caves great quantities of bones of bats, of brown and cave bears and of spotted hyaenas accumulated in this way. The hyaena remains are commonly associated with fossilized hyaena droppings, known as coprolites, and with gnawed bones of other animals, which the hyaenas had carried underground to eat. Human habitation débris and the food remains of birds of prey may accumulate in a similar manner.

Secondly, though not invariably, caves are places where remains are very likely to survive as fossils, once they have been deposited there. Under normal conditions, in the open, animal and plant remains are not preserved at all. The body of an animal is usually quickly destroyed by scavengers; and bacterial decomposition and the effects of sun, rain and frost soon destroy both animal and plant remains. Their preservation is abnormal and can occur only where the process of decomposition is prevented, for example, by rapid burial. In the cave environment this frequently happens. Remains are protected from processes of weathering by the cave roof; and the alkaline conditions prevailing in limestone caves favour the preservation of bone. In non-calcareous caves, on the other hand, bone is less likely to survive. In a lava cave on Mt Suswa, Kenya, the bones of a rhinoceros, probably dating from the present century, are already flaking and in an advanced state of decomposition. The first bone remains found in the Cotte de St Brelade, Jersey (a granite cave), were so decomposed that they were at first mistaken for kaolin. In caves with flowstone floors, impressions of insect and plant remains are sometimes found between the calcite layers. Impressions of caddis fly wings and of a maple leaf, found in Elderbush Cave, Staffordshire, by members of the Peakland Archaeological Society, provide a notable example (Bramwell, 1964).

For part of his information, the cave palaeontologist must intrude on the terrain of the archaeologist. Since early times Man has portrayed the things he has seen around him, sometimes on the walls of caves and rock shelters where protection from the weather has allowed their survival to the present day. Subjects depicted have included such diverse items as the woolly mammoth, woolly rhinoceros, giant ox, reindeer and ibex, so frequently portrayed in Western European cave paintings, to motor cars still being drawn today in the rock shelters of East Africa. For the palaeontologist contemporary illustrations of extinct Pleistocene animals are of special interest and have been extensively used in artists' reconstructions of such species as the woolly mammoth (Fig. 13.1) and woolly rhinoceros, otherwise known only from fossil remains.

When a palaeontologist sets to work excavating a cave, how does he proceed and what does he hope to discover? His principal object is to reconstruct the history of the fauna and flora of the neighbourhood, from which he may incidentally obtain evidence of past fluctuations of climate and other information. The interpretation of a sequence of deposits in a cave is based on the assumption that the lowest deposit was laid down first and is the oldest, those above being successively younger up to the present-day cave floor where sedimentation may still be taking place. In geological nomenclature this is known as the law of superposition. If changes of fauna (which may indicate changes of climate) occur while a cave is becoming filled, this will be reflected in the fossil remains, which may vary from layer to layer. It is essential, therefore, that excavation is conducted in such a way that the remains from the different layers are kept separate, so that faunal changes can be recognized. The earliest crude method of attempting to record stratigraphic division of faunal remains in caves was based on the excavation of the deposits in horizontal layers, the deeper layers being regarded as older than those overlying them. Such a method is reliable, however, only where stratification is horizontal and where there has been no disturbance of the deposits, a situation which very rarely occurs.

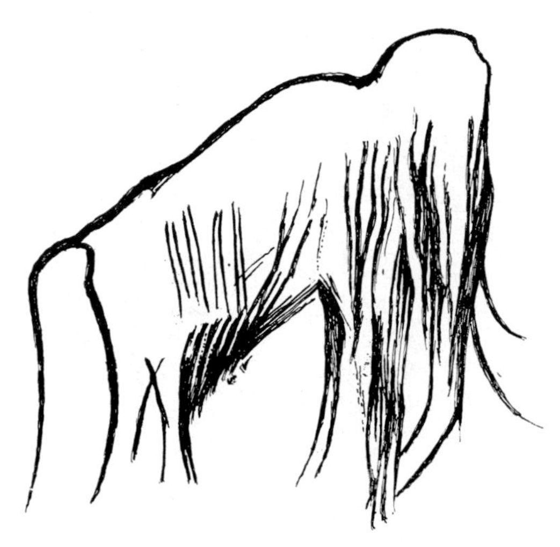

FIG. 13.1. Upper Palaeolithic painting of a woolly mammoth in the cave of Pech Merle, France. Length 60 cm. The portrait, like many other cave paintings of mammoths, shows an animal with a high domed head, a deep neck, steeply sloping back and a long hairy coat. Such contemporary illustrations are of great assistance to the palaeontologist trying to reconstruct the appearance of animals otherwise known only from fossil remains. Drawn by Rosemary Powers from a photograph.

Fig. 13.2 shows a section along an imaginary bone cave in which the commonest types of cave deposit and disturbance are represented.

The confusion that can result from the excavation of such deposits in horizontal layers is obvious. The layers of the talus cone are sloping and the burial lies at the same level as deposits older than itself. The burrow is a source of double confusion, for not only have remains been thrown from depth on to the surface, but a burrowing animal has also died in the burrow. In stream caves deposits may become partly washed out, leaving cemented remnants suspended on the cave walls high above younger deposits accumulating on the cave floor below.

The excavation of a bone cave must be treated like a dissection; each horizontal layer, each talus cone, each burial, each burrow being examined separately, if the full history of the deposits is to be reconstructed successfully. We will return to this topic in the last section of this chapter.

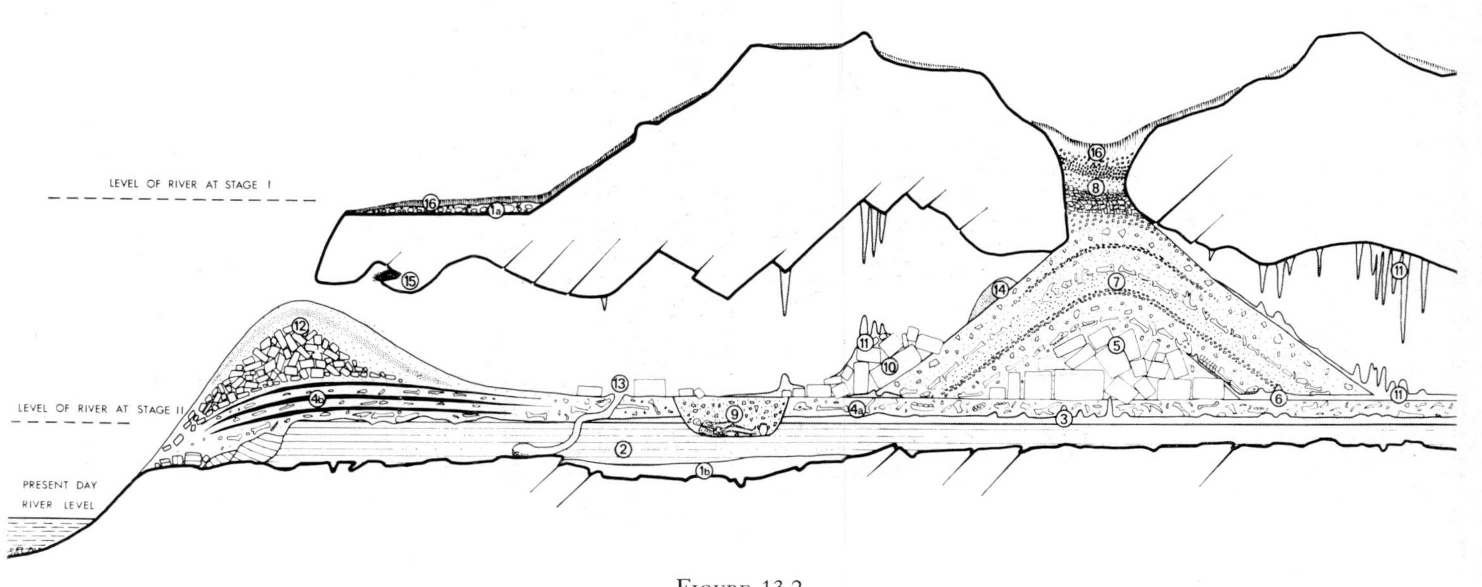
LEVEL OF RIVER AT STAGE I
LEVEL OF RIVER AT STAGE II
PRESENT DAY
RIVER LEVEL
1a
1b
2
3
4a
4b
5
6
7
8
9
10
11
12
13
14
15
16

FIGURE 13.2

HISTORY OF THE SCIENTIFIC EXCAVATION OF PALAEONTOLOGICAL REMAINS IN CAVES

Although Man has entered caves for various purposes since Palaeolithic times, the scientific investigation of cave deposits for palaeontological (and archaeological) remains is a recent development probably going back little more than two centuries. Often the earliest finds were regarded as the remains of mythical animals (in Germany, unicorns; in China, dragons) and it was not until the advent of their scientific study that their true identity as remains of cave bears and other animals was recognized.

FIG. 13.2. Section of an imaginary bone cave. Drawn by Una Sutcliffe. Reproduced, with permission, from *Studies in Speleology 2* (2), 1970.

The section shows the principal types of deposit found in caves and emphasizes some of the problems associated with reconstructing the sequence of events leading to their accumulation.

The section shows how, in general, the lowest deposits in a cave are the earliest; those above later, but that deposits may be discontinuous or disturbed, so that their relationship is not always easy to interpret. The earliest deposit shown (1a) is a river-terrace deposit laid down on the hill above the cave at a time when a river was flowing at level I and the entire cave lay beneath the water-table. While this deposit was being laid down, cave formation was still taking place under phreatic conditions and insoluble residue from the limestone was settling on the cave floor (1b).

Valley downcutting then caused the river to fall to a lower level, II, draining all but the lowest part of the cave. A vadose stream flowing out of the cave laid down sand and silt (2) in its bed. Animals remains sometimes occur in such deposits, but are uncommon. Further valley deepening caused the river to fall again, leaving the cave dry and causing layer 2 to be truncated by erosion at the cave mouth. A layer of flowstone (3) then accumulated on the floor of the inner part of the cave. Carnivorous animals, entering the cave, left the débris of their food and other remains (4a). Man, sheltering at the cave mouth, left the ashes of his fires (4b).

A rock fall (5) then opened a shaft in the cave roof and another layer of flowstone (6) formed. Earth and rock fragments and the remains of animals which had fallen into the cave by accident formed a talus cone (7), beneath the shaft, which finally became completely filled by sediment (8).

A human body (9) was then buried in the cave floor, disturbing deposits 2, 3 and 4, but subsequently being covered by part of another rock fall (10) and by further flowstone associated with stalagmites and stalactites (11). A talus cone at the cave mouth (12) accumulated directly upon layer 4 but cannot be related to layers 5-11 of the cave interior, because it is isolated from them.

A burrowing animal dug a burrow (13), deflected by flowstone layer 3, in the cave floor, throwing up fossil bones from depth onto the surface and subsequently dying in the burrow. A pile of bat dung (14) accumulated beneath a bat roost inside the cave and a pile of rodent bones beneath a nest of bird of prey (15) at the cave mouth. A layer of soil (16) formed on the hill top above the cave. Two further common types of disturbance not shown in the section are slumping and washing out by water.

Man has nevertheless excavated in caves for materials of economic value since early times. In North America, for example, pre-Columbian miners sometimes penetrated into caves for more than two miles in search of mirabilite and gypsum; other caves have been excavated for guano; and some interest in fossil remains may have arisen. The ammonites found in Aveline's Hole, Somerset, are thought to have been taken there by Upper Palaeolithic Man, though these are surface finds and not from a cave (Donovan, 1968). Human agency has sometimes been invoked to explain cave fossils in surprising stratigraphic contexts. Buckland, for example (1823), considered that the rods and rings of ivory found with the skeleton at Paviland Cave, Wales, had been fashioned from a fossil antediluvian tusk excavated from the floor of the same cave. Only more recently has it been shown that both ivory and skeleton are of Upper Palaeolithic age and probably contemporary. Collecting of cave fossils by Upper Palaeolithic Man has been similarly postulated to explain the occurrence of teeth of sabre-toothed cat, in deposits believed to post-date the extinction of this animal, in Kent's Cavern, Devon. In instances such as these it is unfortunately generally not possible to prove Man's agency in the collecting of fossil material in caves.

One of the earliest European references to animal remains *in situ* in a cave is that of Grebner (quoted in Kurten, 1969) who wrote a poem about bones which he had seen in 1748 in the famous cave of Gaelenreuth, Germany. Although the bones were, in fact, of cave bears, he supposed them to be human. A new epoch in the literature of caves began with Esper's investigations of fossil remains discovered in European Caves, continued subsequently by Rosenmuller and Goldfuss. In a publication of 1774 Esper described and illustrated remains of bears and other animals from Gaelenreuth. More detailed summaries of early cave excavation in Europe have been given by Buckland (1823) and Dawkins (1874).

Although animal remains were already known to occur in caves in Britain before the end of the eighteenth century (Buckland, 1823, for example, recorded remains of elephant, rhinoceros, hyaena and other animals being found in a cave near Swansea in 1792) it was not until 1816, when some rhinoceros remains were found in a cave at Oreston, Devon, that the first scientific palaeontological cave investigation took place in this country (Home, 1817).

The first comprehensive work on fossiliferous British caves is Buckland's *Reliquiae Diluvianae* (1823) with its account of the hyaena den in Kirkdale Cave, Yorkshire, and of Paviland Cave, South Wales, with its famous human skeleton. From the 1820s the excavation of caves for palaeontological remains mushroomed in Britain and has continued to do so there and throughout the world until the present day. Excavations which have attracted most attention were those where remains of Man and extinct fauna were found together, especially between about 1820 and 1860, during which time Man's antiquity was under greatest dispute. Among the more notable excavations of this period in the British Isles were those of MacEnery in Kent's Cavern, 1825-1841 (MacEnery, 1859); Pengelly in the Brixham Cave, 1858; Dawkins in Wookey Hole Hyaena Den in 1859; Colonel Wood and Falconer in the Gower Caves from 1858-1861; followed by the British Association excavations in Kent's Cavern, Devon, directed by William Pengelly, commencing 1865, and in Victoria Cave, Yorkshire, in the early 1870s.

Three stages of excavation technique can be identified since the search for palaeontological remains was begun. The earliest excavators were naturally so astonished by the animal remains which they found that their attention had to be given to such matters as whether the animals concerned had actually lived in the neighbourhood where they had been found, or whether the remains had been carried there from distant parts by Diluvial waters. There was relatively little attention to stratigraphy at this stage. As the significance of stratification came to be realized, excavation began to be conducted in layers. The most recent development has been a great increase in the number of specialists necessary at every excavation. This will be discussed further in a later section of this chapter.

SOME OVERSEAS FOSSILIFEROUS CAVES

Although remains of plants and animals have accumulated by similar processes in caves throughout the world, fauna and flora differ regionally so that each country has its own characteristic assemblage of fossil remains dating from each stage of time. The diversity of these assemblages is richly apparent in the fossil cave material. Only a few examples can be mentioned here.

Of all the animals known from fossil remains in caves, probably the bear (especially the great cave bear of western Europe) has attracted the greatest attention. The Drachenhöhle, a cave near Mixnitz in Austria, is estimated to have contained the remains of over 30 000 cave bears which accumulated as a result of a small number of animals continuing to occupy the cave over a long period of time (Kurten, 1969). Outside Europe true cave bears are unknown, but other species of bear have been found to make their dens in caves in North Africa, Asia and America. The great cave bear thrived on the continent of Europe during late Pleistocene times. Other Continental cave deposits are of much earlier date. A fissure at Egerkingen, Switzerland, produced mammalian remains of Middle Eocene age; and the richly fossiliferous Upper Eocene-Oligocene cave deposits of Quercy, France, were formerly worked commercially for phosphate. Mammalian remains from Quercy include the creodont *Hyaenodon*, the artiodactyl *Cainotherium*, anthracotheres, early bats, rodents, and many other mammals and birds.

Islands often have unusual faunas which have evolved in isolation from those on the nearest mainland, sometimes in absence of competition from other species, sometimes under adverse conditions of food supply. In cave deposits of Pleistocene age on some of the Mediterranean islands there have been found remains of dwarf and giant and other peculiar mammals, which evolved in this way. Remains represent pigmy hippos and pigmy elephants on Malta, Sicily and Cyprus; a pigmy deer on Crete; a giant dormouse on Malta; the peculiar goat-like *Myotragus balearicus* on Majorca; and many other species.

Among the most important African cave sites are the famous Lower Pleistocene australopithecine localities of South Africa; Makapansgat (Dart, 1957) and Sterkfontein (Cooke, 1969). The Sterkfontein fauna consists almost exclusively of extinct animals including a giant pig, *Notochoerus*, almost twice the size of a warthog; *Libytherium*, a giraffid with antler-like horns; *Hipparion*, the three-toed horse; carnivores; *Dinopithecus*, a giant baboon; and many other genera.

Madagascar, long separated from the mainland of Africa, has its own distinct fauna, reflected in fossil remains found in its caves. Of special interest are the fossil lemuroids, first described from the Cave of Androhomana, near Fort Dauphin in the southern part of the island. These remains are probably of relatively recent date.

There are many fossiliferous caves in China, most notable being the Lower Pleistocene site of Choukoutien, where Peking Man was found (Kowalski, 1965). Many other Chinese caves have been dug for "dragon's teeth" and the finds from them have been sold for medicinal purposes in apothecaries' shops, from where they have sometimes been rescued by palaeontologists. Colbert and Hooijer (1953) have described some excavations for this purpose in Pleistocene deposits in limestone fissures in the province of Szechwan, at which the best specimens were acquired from the workmen for scientific study. Mammals represented included monkeys, porcupine, panda, hyaena, tiger, *Stegodon*, elephant, tapirs, a chalicothere, rhinoceros, pig, deer and buffalo. In recent years there have been many scientific excavations in Chinese caves.

In the S.E. Asian region, one of the most important cave excavations has been that in Niah Cave, Borneo (Plate 13.1 from Harrisson, 1964; Medway, 1964). A rich fauna of late Pleistocene and post-Pleistocene age has been found associated with important archaeological remains. Mammalian species include various bats and monkeys, orang-utang, pangolin, rodents, bear, tapir, rhinoceros and deer. Most of these animals still survive in the region at the present day.

PLATE 13.1. Excavation in progress in Upper Pleistocene deposits (50 000 years old) in Niah Great Cave, Borneo. Note the orderly arrangement of the work in spite of the apparent complexity of the scene. A carefully prepared section of the deposits is exposed in the balk in the foreground; the layers are individually labelled; and everything has been carefully swept so that there is no loose sediment anywhere. The Director of the excavation, Tom Harrisson, is seen taking records which will be needed later for publication. (*Photo* by courtesy of the Sarawak Museum.)

Fossil remains have been found in many Australian caves, some of the most important sites being situated in the Nullabor region of Western Australia. The mammalian fauna of Australia is peculiar in that this continent became separated from the rest of the world in Cretaceous times, probably about 100 million years ago, and became a place of radiation of marsupial and not of placental mammals. With the exception of some remains of dingos and rodents (which reached Australia more recently), the fossil mammalian remains are all of marsupials. Lundelius (1966) has shown that the carnivorous marsupial *Sarcophilus*, the Tasmanian Devil, probably used caves as dens during the past, leaving there the fragmentary bone remains of bandicoots, kangaroos, wallabys, rats, reptiles and other animals. The Tasmanian Devil still survives in Tasmania, although it is extinct in Australia. A dried carcass recently discovered in Thylacine Hole, Western Australia, was shown by Carbon-14 dating to be about 4600 years old (Lowry and Merrilees, 1969).

New Zealand has no indigenous mammalian fauna. Remains of the extinct flightless Moa have been found in caves there.

An unusual deposit of bird remains was found in fumaroles on the Island of Ascension in the South Atlantic. Olson (1973) described an accumulation of bones of the recently extinct Ascension Island Rail, apparently resulting from these flightless birds having climbed up the slopes of the fumaroles, where they fell into the vents, from which they could not escape. He considered that the bones accumulated over some period of time, though the latest specimens (judging from their condition) were probably not more than a few hundred years old.

Palaeontological remains have been found in many caves in North and South America. Turnbull (1961) described a supposed carnivore den deposit of Lower Tertiary age from Colorado. A map of the more important North American Pleistocene localities and a list of the principal North American publications was given by Hibbard *et al.* (1965). Most of the cave deposits are of Upper Pleistocene age, though some earlier sites are also known, including Port Kennedy Cave, Pennsylvania, and Conard Fissure, Arkansas, which are probably Middle Pleistocene. A mammal of special interest from the former locality is the early sabre-toothed cat, *Smilodon gracilis*, an ancestor of the Upper Pleistocene sabre-tooths found in some later cave deposits. Port Kennedy Cave also provides the earliest North American record of bear. Anderson (1968) gave a list of the mammals from twelve of the most important North American late Pleistocene cave sites. Included are ground sloths, porcupine, bears, sabre-toothed cat, mammoth, tapir, peccary, camel, reindeer, bison and other animals. This assemblage, which includes forms with South American affinities (porcupine, ground sloths) and Eurasian affinities (mammoth, reindeer, bison), vividly illustrates the mixed origin of the North American Pleistocene mammalian fauna. Some of the North American cave deposits are, by their nature, most unusual. Some of the Florida bone caves, for example, have been flooded to depth of 25 m by subsequent rise of sea-level and can be excavated for their mastodon and other remains only with the aid of diving equipment. Wells and Jorgensen (1964) described an unusual cave deposit in the Mohave Desert, composed mainly of plant remains carried there by wood rats. The earliest part of the deposit was shown by Carbon-14 to go back 40 000 years. A study of the plant remains showed evidence of climatic change during the time of their accumulation.

One of the best known South American caves is that of Ultima Esperanza in Patagonia. So favourable were the conditions for preservation (probably the combined effects of aridity and mineral salts) that large pieces of skin, with hair, of the extinct Pleistocene ground sloth *Mylodon* still survived in the deposits of the floor of the cave.

FOSSILIFEROUS CAVES OF THE BRITISH ISLES

Few parts of the world can rival the British Isles for the variety of fossil remains found in caves. The earliest finds date back to Triassic times, probably about 200 million years ago; there is an abundance of Pleistocene cave deposits; and remains of plants and animals are still accumulating in some caves at the present day.

Throughout this time there were many changes in the fauna and flora of the British Isles. There were slow long-term evolutionary changes and there were also short-term local changes resulting from fluctuations of climate during the Pleistocene. Throughout the world the Pleistocene was a time of great climatic change, with northerly species of plants and animals being displaced southwards in the British Isles as the ice sheets advanced, southerly species moving northwards during the interglacials when the climatic became more temperate. In England and Wales the southerly hippopotamus and straight-tusked elephant alternated in Upper Pleistocene times with such northern forms as the woolly mammoth, woolly rhinoceros, reindeer and wolverine. Their remains are commonly preserved in caves, sometimes in stratified sequences within a single cave, providing information about the climatic history of the British Isles.

Organic remains tend to accumulate in caves principally near entrances and to become progressively less common towards the cave interior until empty chambers are reached. Most cave entrances are vulnerable to denudation so that any fossiliferous deposits situated there are likely to be relatively quickly destroyed, exposing deeper parts of the cave system where fresh organic remains can accumulate before being destroyed in turn themselves. The survival of fossil material in caves becomes progressively more uncommon with age. In the British Isles deposits with Upper Pleistocene and Holocene remains are numerous. Earlier cave deposits are very rare.

Some of the more important fossiliferous caves of the British Isles may now be considered, commencing with the oldest. The probable stratigraphic range of their deposits is shown in Table 13.1. The rarity of any cave deposits earlier than about 100 000 years old is clearly shown. (For a more detailed list, see Jackson, in Cullingford (1953), Chapter 8.)

Caves with Mesozoic Fossils

The earliest fossiliferous cave deposits in the British Isles are the Mesozoic infillings in the Carboniferous Limestone of Somerset and South Wales. A valuable summary of the Somerset sites had been given by Halstead and Nicoll (1971). Fissures at Holwell contained great quantities of teeth and scales of fish, including the shark *Acrodus*, and also teeth of the reptile *Oligokyphus*. Remains from a Triassic cave infilling at Cromhall Quarry included lacertilian lizards and a small dinosaur. Lizard and dinosaur remains were also found in a cave deposit at Durdham Down, Bristol. Recently Kermack, Mussett and Rigney (1973) and others have investigated Rhaeto-Liassic fissure infillings in South Wales with remains of early mammals including *Morganucodon* and *Kuehneotherium*. These are regarded as among the earliest known records of mammals anywhere in the world.

Cave Deposits of Tertiary Age

Tertiary cave infillings in the British Isles are unfortunately rare. A series of sand- and clay-filled solution pockets in the Carboniferous Limestone, near Brassington, Derbyshire (Ford and King, 1969; Walsh *et al.*, 1972), were found to contain plant remains, including *Scadiopitys* and *Tsuga*, which are probably of Pliocene age (Boulter and Chaloner, 1970). Other areas have been searched unsuccessfully for fossiliferous Tertiary cave deposits. Savage (1957) drew attention to the occurrence of potentially fossiliferous swallets in the surface of chalk covered by Tertiary basalt in Co Antrim, Ireland. Near Bovey Tracey, in Devon, Oligocene lake deposits with plant remains rest unconformably on cavernous Devonian limestone, but no fossils have yet been found in any of the cavities.

It is to be hoped that cave deposits of Tertiary age with vertebrate fossils may yet be found in the British Isles.

The Lower Pleistocene Fauna of Dove Holes, Derbyshire

The earliest Pleistocene cave deposit in the British Isles is Dove Holes, Derbyshire (Dawkins, 1903), with a mammalian fauna including mastodon, sabre-toothed cat and horse. It is probably contemporary with the

YEARS AGO			IRELAND	
	/ERTEBRATE FAUNA		SITES	VERTEBRATE FAUNA
Present day	ɪr, brown ɪlf, ?lynx.			
10,000	mammoth, giant deer, r, reindeer, woolly ros, horse, spotted hyaena, bear, cave lion, wolf, fox, rrow-skulled vole.		➡ Castlepook Cave, Ballymintra Cave, Kilgreaney Cave, Castletownroche Cave, Edenvale Cave.	Woolly mammoth, giant deer, reindeer, spotted hyaena, brown bear, wolf, lemmings.
70,000	-tusked elephant, eer, red deer, fallow ppopotamus, pig, arrow-nosed rhino- potted hyaena, brown ve lion, wild cat, x, badger, hare.			
	r, horse, ne, brown amster, lemming, le.			
	ear.			
	n rhinoceros, horse, sabre- l cat, Deninger's bear, wolf, desman, leopard (?jaguar).			
	on, horse, oothed cat.		Ireland almost complely glaciated.	
2-3 million				
7 million				
26 million				
38 million				
65 million	als *anucodon* *neotherium* s *kyphus* rtilian lizards saurs *lus*			
225 million				

facing page 506.

Table 1.1. Distribution of bone-bearing deposits in British Caves.

Upper Cave Earth

Lower Cave Earth

Black Mould

Granular Stalagmite

Stratum

Hyaena Stratum

Crystalline Stalagmite

Stratum

Breccia

Westbury-sub-Mendip Fissure, Somerset

Crag deposits of Suffolk, and was formerly regarded as being of Pliocene age. The fauna was recently revised by Spencer and Melville (1974).

Westbury Fissure

Although highly fossiliferous deposits of Cromerian age have long been known to occur along the Norfolk coast, until recently no cave deposits of this age had been found in the British Isles. The discovery of a cave deposit with abundant mammalian remains and rare artifacts (see the detailed reports in Heal, 1970, Bishop 1974, 1975) of approximately Cromerian age, or probably slightly later, is of the greatest importance. Two main deposits are present in the cave; a lower water-laid deposit with extensively rolled teeth and bone fragments; and an upper deposit which appears to represent a bear den. The fauna includes Etruscan rhinoceros, Mosbach horse, sabre-toothed cat, Deninger's bear, Mosbach wolf, hyaena, a medium-sized felid, possibly related to a jaguar, and desman (water mole). This exceptional fauna is unlike that of any other British cave. The possible existence of other Cromerian cave infillings in the Mendip area must not be overlooked. For a long time the so-called leopard remains from Bleadon Cave, figured by Dawkins and Sandford (1872, plate 24), have presented a problem. These remains closely resemble those of the medium-sized felid from Westbury, suggesting a possible early date for part of the deposit in this cave.

Kent's Cavern, Devon

The long stratified sequence of deposits in Kent's Cavern has recently been reinvestigated by Campbell and Sampson (1971). The principal deposits in the cave, in descending order, were as follows:

(5) Black mould
(4) Granular stalagmite
(3) Cave earth
(2) Crystalline stalagmite
(1) Breccia.

The precise stratigraphic range of these deposits is uncertain, though there is evidence that the Breccia, which has a lower Palaeolithic industry, may be relatively early. The association of cave bear, sabre-toothed cat and the rodent *Pitymys gregaloides* suggest a horizon somewhere in the cave which may be of approximately the same age as Westbury. The Crystalline Stalagmite appears to represent a period when the cave was sealed so that no animal remains or sediment could reach it from outside. The fauna so typical of the Last Interglacial (including hippopotamus, narrow-nosed rhinoceros, straight-tusked elephant and fallow deer), commonly found in cave and river terrace deposits throughout England and Wales, is lacking from Kent's Cavern. The Cave Earth has a typical Last Glaciation fauna including woolly mammoth, woolly rhinoceros, horse, red deer, reindeer, giant deer, spotted hyaena and cave lion, associated with Upper Palaeolithic artifacts. The Granular Stalagmite and Black Mould have Mesolithic and later industries and may be Post-Pleistocene.

Tornewton Cave, Torbryan, Devon

The most important sequence of Upper Pleistocene cave deposits in the British Isles is that of Tornewton Cave (Sutcliffe and Zeuner, 1962). The main deposits, in descending order, were as follows:

(4) "Diluvium"
(3) Reindeer Stratum
(2) Hyaena Stratum
(1) Glutton Stratum.

The Glutton Stratum apparently accumulated at a time when the cave was occupied as a lair by brown bears. Associated remains of reindeer and glutton (wolverine) suggest a cold climate. This deposit is referred to the penultimate (Wolstonian) Glaciation. By the time of accumulation of the Hyaena Stratum the cave had been taken over as a lair by spotted hyaenas. Fragmentary remains of at least 300 hyaenas, juvenile and adult, were found in a deposit measuring only 10 m long, 2 m wide and $\frac{1}{2}$ m deep. Associated remains of hippopotamus, narrow-nosed rhinoceros, red and fallow deer show that by this time the climate had become interglacial (Last Interglacial). The Reindeer Stratum produced remains of reindeer, woolly rhinoceros, horse and hyaena, associated with a sparse Upper Palaeolithic industry (Last Glaciation). The so called "Diluvium" is probably Holocene.

This site is of very great importance in the study of the Upper Pleistocene mammalian faunas of the British Isles, being the only known locality where deposits of Wolstonian and Last Glaciation age are separated by an interglacial stratum with hippopotamus. On first inspection the mammalian faunas of the two cold horizons are indistinguishable and the two faunas could be confused if found in isolation. Fortunately the deposits contained abundant rodent remains. These have been studied by Kowalski (1967), who found the two faunas to be different. The lower cold stage is characterized by, among other rodents, two forms of hamster, steppe lemming and snow vole, all absent from the upper cold layer, where they are replaced by large quantities of narrow-skulled vole. Other rodents, including the collared lemming, occur in both levels. This key for the separation of penultimate and Last Glaciation cave faunas in Southern Britain is of great stratigraphic importance. All the above rodent species are now extinct in the British Isles.

Other Caves with pre-Last Interglacial Deposits

In light of the faunal sequence demonstrated at Tornewton Cave, it would be surprising if other cave deposits dating from the time interval between the Last and Hoxnian interglacials, (the exact age of which had not previously been recognized) did not exist. Clevedon Cave, Somerset (Hinton, 1926) with a fauna of snow and root voles, and Hutton Cave, with hamster, are caves with possible faunal remains of this age.

Last Interglacial Cave Deposits

Although earlier fossiliferous cave deposits are rare, those of Last Interglacial age are relatively common in the British Isles as far north as Yorkshire. The cave richest in species is Joint Mitnor Cave, Buckfastleigh (Plate 13.2), with straight-tusked elephant, narrow-nosed rhinoceros, hippopotamus, pig, giant deer, red deer, fallow deer, bison, hare, wolf, fox, wild cat, cave lion, spotted hyaena, badger and bear. This faunal assemblage is typical of the warmest part of the Last Interglacial. A similar assemblage has been found in deposits of the same age in the Upper Floodplain Terrace of the River Thames at Trafalgar Square, London.

Another important Last Interglacial site is Victoria Cave, near Settle, Yorkshire. The sequence of deposits, excavated a century ago, has been discussed by Warwick (1956) (see also King, 1974), who regarded a lower cave earth with hippopotamus, straight-tusked elephant, narrow-nosed rhinoceros and hyaena as being of this age. A long sequence of later deposits suggests that the area was subsequently glaciated, with later occupation by Upper Palaeolithic, Mesolithic and Romano-British Man. There can be few British caves which demonstrate in such a striking manner the climatic fluctuations of the Pleistocene. The cave is situated on craggy moorland at nearly 450 m altitude. It seems incredible, at the present day, to think of hippos walking across this wild part of the Yorkshire moors. That the climate was then more temperate than at the present day is confirmed by palaeobotanical studies at other Last Inter-

PLATE 13.2. Pleistocene mammalian remains exposed by excavation in the Last Interglacial talus cone in Joint Mitnor Cave, Buckfastleigh, Devon. A tooth of a straight-tusked elephant lies close to a crushed humerus of a bison. An upper molar of bison can also be seen. (*Photo*: A. E. McR. Pearce.)

glacial sites. The subsequent glaciation of the area (whilst Tornewton Cave and Kent's Cavern lay south of the ice-sheet) is no less astonishing.

A series of old sea caves with raised beach and fossiliferous deposits of Last Interglacial and later age occurs along the coast of the Gower Peninsular, South Wales. Three sites are of special importance, Ravenscliff Cave, Bacon Hole and Minchin Hole (Murchison, 1868; Allen and Rutter, 1948; Sutcliffe and Bowen, 1973) (Fig. 13.3). In each of these caves

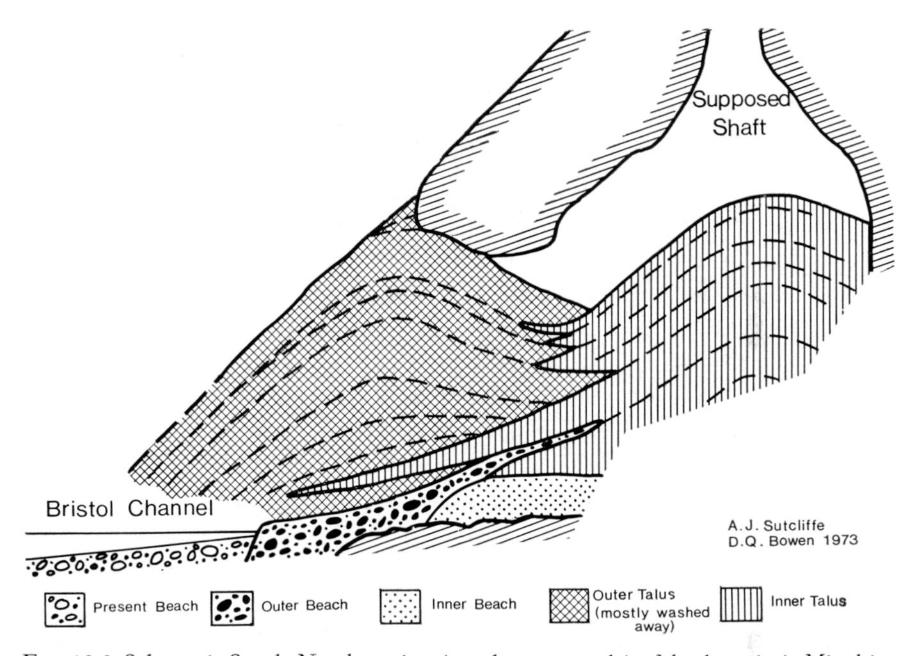

FIG. 13.3. Schematic South-North section (not drawn to scale) of the deposits in Minchin Hole, an old sea cave on the Gower Coast, South Wales. Fossiliferous raised beach deposits, believed to be of Last Interglacial age, are overlain by two interfingering talus cones. (Reproduced, with permission, from *Newsletter 21* of the William Pengelly Cave Studies Trust, 1973.)

the lower part of the sequence consists of old beach deposits with marine shells, indicating a sea-level a little higher than at the present day, followed by interglacial cave earths with remains of narrow-nosed rhinoceros and straight-tusked elephant and (in Ravenscliff Cave only) hippopotamus. Other deposits include blown sand, suggesting the falling sea-level of the Last Glaciation; and breccias of limestone fragments, probably representing a period of frost shattering. The ice of the Last Glaciation is known to have reached Langland Bay, only 6 km from the caves, so that evidence of intense frost shattering is to be expected.

Other British caves believed to be of Last Interglacial age, with hippopotamus, straight-tusked elephant and narrow-nosed rhinoceros include Eastern Torrs Quarry Cave, Yealmpton, Devon; Milton Hill Cave, Somerset; Durdham Down Cave, Bristol; Cefn Cave, North Wales; Hoe Grange Cave, Derbyshire (this cave lacked hippopotamus); Raygill Fissure, and Kirkdale Cave, Yorkshire.

Last Glaciation and Holocene Cave Deposits

Nearly all fossiliferous cave deposits in the British Isles are of Last Glaciation and Holocene age. Relatively few excavators ever encounter anything earlier. Sites of this age are so numerous that only a few of them can be mentioned here. At the height of Last Glaciation the northern part of the British Isles and most of Wales and Ireland were glaciated whilst the southern part of England remained free of ice and continued to support plants and animals. This fauna was not uniform but shifted with the advances and melting of ice. A fairly typical mammalian fauna can nevertheless be distinguished for this period, including woolly mammoth and woolly rhinoceros, horse, reindeer, cave lion, wolf and other animals. The spotted hyaena, so abundant during the preceding Last Interglacial, apparently adapted itself to colder conditions and continued to flourish in great numbers. In addition to the Cave Earth of Kent's Cavern and Reindeer Stratum of Tornewton Cave, already mentioned, the deposits of the following caves are probably of Last Glaciation age: The Ightham Fissures, Kent; Brixham Cave, Levaton Cave, Torcourt Cave, Cow Cave, Chudleigh Fissure, Yealm Bridge Cave, Oreston Caves, Devon; Wookey Hole Hyaena Den, Badger Hole, Soldiers Hole, Aveline's Hole, Uphill Cave, Dulcote Hill Cave and Bleadon Cave, Somerset; Great Doward Cave and King Arthur's Cave, Herefordshire; Bosco's Den, Paviland Cave and other caves, Glamorganshire; Langwith Cave, Dowel Cave, Etche's Cave, Fox Hole Cave, Pin Hole Cave and Dream Cave, Derbyshire; Dog Holes Cave, Lancashire, and many other caves.

Some excavations recently conducted by members of the Peakland Archaeological Society in Derbyshire caves (Bramwell, 1960; Pernetta, 1966) are of special interest for the information they have provided about the transition from the Late Pleistocene to the early Holocene.

Many caves contain deposits of Holocene age. Bramwell, for example (1957), has recently described a Holocene deposit in a cave in North Staffordshire. Among the large mammals which survived into Holocene times were the brown bear and wolf and possibly lynx and reindeer.

Bone Caves in Scotland

Remains of lynx, brown bear, reindeer and lemming have been found in a small cave near Inchnadamph, Sutherland (Newton, in Peach and Horne, 1917). The age of these remains is uncertain but the caves are situated in wild mountainous countryside and the possibility that they are of Post-Pleistocene age needs to be taken into consideration. Some bones of brown bear found in a nearby cave have recently been dated by Carbon 14 as only about 2700 years old (Burleigh, 1972).

Bone Caves in Ireland

The colonization of Ireland by mammals appears to have occurred very late during the Pleistocene so that the long sequence of mammalian faunas known from England and Wales is absent there. Ireland was extensively glaciated during the penultimate and Last Glaciations. The earliest mammals known probably crossed from England at some stage during the Last Glaciation. The most important Irish Cave is Castlepook Cave near Cork with remains of mammoth, giant deer, reindeer, two species of lemming, hyaena, brown bear and wolf. All these animals were abundant in England during the Last Glaciation. Other species such as the woolly rhinoceros and bison failed to reach Ireland. It is perhaps significant that the two lemmings are extreme northern species which, at a time of amelioration of climate, would be expected to reach Ireland before other rodent species. Other important caves with Pleistocene fauna are Ballynamintra, Kilgreany, Castletownroche and Edenvale Caves. For discussion of Irish Pleistocene mammals see Coleman (1965), Savage (1966) and Mitchell (1969).

METHODS OF CAVE EXCAVATION; AND THE FUTURE

Although few cavers set out with the intention of conducting a palaeontological or archaeological excavation in a cave, some encounter bone remains or artifacts at one time or another and there are many who enjoy taking part in excavations organized on a broader basis than their own time and other resources permit. What should the caver do who has made such a discovery (which may be holding up his own work in another field) or who would like to help at an excavation or to conduct an investigation of his own? There is much which the caver and the palaeontologist can offer one another.

Let us first look at an excavation such as might be conducted by a large institution with sufficient resources of men and equipment. If reconstruction of the past history of the fauna and flora, climate and geomorphology of an area is to be the main objective of an excavation in a cave, then as many different lines of evidence as possible need to be investigated. In consequence, as new techniques for study are devised, excavation becomes even more and more a task for a team of specialists working together. The fossil mammal specialist, for example, might interpret a cave deposit with remains of hippopotamus as dating from a warm period, but can he really be sure that the British hippopotamus did not have a long woolly coat and live here under cold conditions? If, on the other hand, its remains were to be associated with those of southerly species of birds and molluscs and plants, then several independent lines of evidence would all point to southerly conditions and it is likely that the climate was indeed warmer than at present. Similarly, the specialist who can carry out Carbon-14 dating can give an absolute age for associated palaeontological remains up to about 50 000 years old; and the sedimentologist can tell the palaeontologist how the various deposits accumulated.

It is the task of the director of an excavation to co-ordinate the various lines of study up to the time of their final publication. The extent to which he takes responsibility for the various parts of the field work himself and the extent to which other specialists take part in an excavation varies, from site to site, according to the circumstances. Usually the director will draw his own plans and sections and take his own photographs but, especially if an excavation is extensive, he may delegate this work. Sometimes, material for study (for example sediment samples to be studied for pollen; charcoal or bone for Carbon dating) can be collected, with suitable care, during an excavation and sent to the appropriate specialist, but it is better if the specialist can visit the site himself and collect his own samples. Other specialist tasks include sieving for remains of small mammals, reptiles, amphibia and fish and for shells of molluscs. If a cave has any unusual features, these must be taken into consideration. Flowstone, if sufficiently pure, can be dated by chemical means; and an old sea cave may contain marine sand with sea shells, ostracods and foraminifera. All lines of evidence must be taken into consideration.

One of the first tasks of the director of an excavation, after drawing a plan and obtaining a photographic record of the appearance of the unexcavated cave, is to cut a trial trench in order to determine the sequence of strata beneath the cave floor, in preparation for more extensive excavation. If the cave has previously been disturbed, then any earlier trenches must be located and, if they lie within the excavation area, all spoil must be removed from them. The stratification of the undisturbed deposits can then be studied in the side of the trench, which acts as a substitute for the trial trench. In either case the section must be carefully prepared, using a builder's trowel or a triangular paint scraper so that the various layers can be distinguished. Trial trenches should, where possible, be laid out along the direction of slope of a deposit and not across it, where a misleading picture of horizontal stratification may be obtained. This point is illustrated in Fig. 13.3, a longitudinal section of Minchin Hole showing two interfingering talus cones. A transverse section, halfway along the cave, would misleadingly show apparently horizontal stratification with no differentiation between the fingers of the two cones, which are of different origin and source.

The course of further excavation, after the completion of a trial trench, depends on the nature of the cave in question. Adequate sections must be recorded; disturbed deposits must be recognized; and finds from the various layers must be kept separate.

In general, no greater part of the deposits should be excavated than is necessary for the study in hand and as much as possible should be left for further investigation in the future, when the development of new techniques opens new lines of study not available at the present time. In a cave which can be protected, a standing section of the deposits is of constant interest to visitors. In Joint Mitnor Cave, at the Pengelly Cave Studies Centre, Buckfastleigh, Devon, for example, a complete section of the basal water-laid and overlying Last Interglacial ossiferous deposits is permanently preserved for demonstration purposes. Where such protection cannot be given to the remaining deposits it may be necessary to refill the excavation in order to provide protection from disturbance by casual visitors. At the conclusion of the excavation in Minchin Hole in 1973 the trenches were lined with a large sheet of polythene before refilling, so as to facilitate distinction between disturbed and undisturbed deposits at the time of any future excavation in the cave.

Finally, after completion of an excavation the director must assemble all the plans and sections and specialist reports and prepare these and a discussion of the results for publication. No excavation is justified unless it is followed by publication, and much valuable evidence has been destroyed in the past when caves have been excavated from wall to wall and from ceiling to floor, with no subsequent report on the findings. For a model report see Guilday *et al.* (1964), whose detailed palaeontological account of New Paris Sink Hole 4, Pennsylvania, leaves few lines of enquiry unexplored.

Let us finally consider the role of the caver who, although he does not wish to become involved in a palaeontological excavation himself, has made an accidental discovery about which he needs guidance; that of the part-time palaeontologist; that of the cave landlord who has a palaeontological cave site on his property and who may have difficulty in knowing how to deal with the requests to excavate there; and the circumstances under which palaeontological excavation in caves is or is not justified.

The caver who has made an accidental discovery should seek specialist advice. Maybe he has only encountered the skeleton of a cow, buried there by the farmer, and his own project can continue without hindrance. If fossil material has been discovered and a specialist visits the site, a well-prepared section of the deposits in the side of the excavation would help him to assess its importance. Whenever possible, a cave should not be excavated from wall to wall at any part since it then becomes impossible to correlate deposits outside the cave mouth with those inside. In newly opened caves a look-out should be kept for unusual features such as footprints of bears on the cave floor. Many important excavations in the past have followed accidental discoveries by cavers.

For the would-be excavator, who would like to take part in the excavation of a cave but who has no previous experience, I urge that the first move should be to join a well-organized excavation at a Roman or Mediaeval open site. Here it is easier to learn about the laying out of trenches, the preparation and recording of sections and the meticulous tidiness which must go with every excavation. Only when preparing sections has become more fascinating than finding pots or bones is he ready to start work in a cave.

For the cave landlord who has received a request to conduct a palaeontological or archaeological excavation on his property three points need to be taken into consideration. Firstly most caves, especially stream caves with sink-hole entrances, probably do not contain palaeontological or archaeological remains and any general restriction on excavation would be a great disappointment to those working in other fields of speleology. Rockshelter-type cave openings along valley sides, on the other hand, are more likely to contain remains and should be treated with greater reserve, especially if finds have already been made in other caves in the neighbourhood. Secondly, consideration needs to be given as to whether the cave is safe or whether it is imminently liable to be destroyed by quarrying or some other process. Some of the most important British bone caves (the South Wales and Mendip Triassic Fissures; Dove Holes, Derbyshire; Westbury Fissure, Somerset; Eastern Torrs Quarry Cave, Devon, and many others) would never have come to light had there been no quarrying, and their investigation has been made possible by the kindness of the quarry owners. Such rescue excavations merit the most urgent attention with fewest reservations. Lastly, where a cave is in no danger and where it is likely to contain palaeontological remains or is already known to do so, the landlord may find himself faced with a less urgent request to excavate. He should satisfy himself that the person making the request is not planning to excavate with only a single line of investigation in mind, but has sufficient specialist help; and that he has the determination to see the excavation through to publication. In the latter respect some of the most eminent professional palaeontologists are probably the greatest villains. Many are drastically overcommitted and should be writing up past work instead of embarking on new field excavations.

Bearing in mind the improved aids which will be available to future generations of excavators, there can be no case for disturbing known palaeontological cave deposits except under the most controlled conditions or where the site is in danger of destruction by quarrying or other causes.

APPENDIX

Names of Pleistocene mammals from mainland Europe and the British Isles mentioned in the text. (For a more comprehensive list of European Pleistocene mammals, see Kurten, 1968.)

Insectivora

Desman	*Desmana* cf. *moschata* (Linn.)

Carnivora

Wolf	*Canis lupus* (Linn.)
Mosbach wolf	*Canis lupus mosbachensis* (Soergel)
Fox	*Vulpes vulpes* (Linn.)
Cave bear	*Ursus spelaeus* (Rosenmüller and Heinroth)
Brown bear	*U. arctos* (Linn.)
Deninger's bear	*U. deningeri* (Reichenau)
Wolverine	*Gulo gulo* (Linn.)
Badger	*Meles meles* (Linn.)
Spotted hyaena	*Crocuta crocuta* (Erxleben)
Cave lion	*Panthera spelaea* (Goldfuss)
Sabre-toothed cat	*Homotherium latidens* (Owen)
Wild cat	*Felis silvestris* (Schreber)
Lynx	*Felis lynx* (Linn.)

Proboscidea

Mastodon	*Anancus arvenensis* (Croizet and Jobert)
Woolly mammoth	*Mammuthus primigenius* (Blumenbach)
Straight-tusked elephant	*Palaeoloxodon antiquus* (Falconer and Cautley)

Perissodactyla

Woolly rhinoceros	*Coelodonta antiquitatis* (Blumenbach)
Narrow-nosed rhinoceros	*Dicerorrhinus hemitoechus* (Falconer)

Etruscan rhinoceros	*D. etruscus* (Falconer)
Horse	*Equus caballus* (Linn.)
Mosbach horse	*E. mosbachensis* (Reichenau)
Artiodactyla	
Wild pig	*Sus scrofa* (Linn.)
Hippopotamus	*Hippopotamus amphibius* (Linn.)
Giant deer	*Megaceros giganteus* (Blumenbach)
Fallow deer	*Dama dama* (Linn.)
Red deer	*Cervus elaphus* (Linn.)
Reindeer	*Rangifer tarandus* (Linn.)
Giant ox	*Bos primigenius* (Bojanus)
Bison	*Bison* cf. *priscus* (Bojanus)
Ibex	*Capra* cf. *ibex* (Linn.)
Lagomorpha	
Hare	*Lepus* sp.
Rodentia	
Hamster	*Cricetus cricetus* (Linn.)
Hamster	cf. *Allocricetus bursae* (Schaub)
Steppe lemming	*Lagurus lagurus* (Pallas)
Collared lemming	*Dicrostonyx torquatus* (Pallas)
Norwegian lemming	*Lemmus lemmus* (Linn.)
Vole (extinct)	*Pitymys gregaloides* (Hinton)
Snow vole	*Microtus nivalis* (Martins)
Root vole	*M. oeconomus* (Pallas)
Narrow-skulled vole	*M. gregalis* (Pallas)

REFERENCES

Allen, E. E. and Rutter, J. G. (1948). *Gower Caves*. Thomas, Swansea. (Previously published in *Proc. Swansea Scient. Fld. Nat. Soc. 2* (6-7), 1943, and *2* (8-9), 1947).

Anderson, E. (1968). Fauna of little Box Elder Cave, Converse County, Wyoming. The Carnivora. *Univ. Colo. Stud. Earth Sci. 6*, 1-59.

Bishop, M. J. (1974). A preliminary report on the Middle Pleistocene mammal-bearing deposit of Westbury-sub-Mendip, Somerset. *Proc. speleol. Soc. Bristol 13*, 301-318.

Bishop, M. J. (1975). Earliest record of Man's presence in Britain. *Nature, Lond. 253*, 95-97.

Boulter, M. C. and Chaloner, W. G. (1970). Neogene fossil plants from Derbyshire (England). *Rev. Palaeobot. Palynol. 10*, 61-78.

Boylan, P. J. (1970). An unpublished portrait of Dean William Buckland, 1784-1856. *J. Soc. Biblphy nat. Hist. 5*, 350-354.

Bramwell, D. (1957). Report on work at Ossum's Erie Cave, season 1957. *Newsl. Peakland Archaeol. Soc. 14*, 8-15.

Bramwell, D. (1960). The vertebrate fauna of Dowel Cave—Final Report. *Newsl. Peakland archaeol. Soc. 17* (pages unnumbered).

Bramwell, D. (1964). The excavations at Elderbush Cave, Wetton, Staffs. *North Staffs. J. Field Studies 4*, 46-60.

Brown, B. (1908). The Conard Fissure. *Mem. Am. Mus. nat. Hist. 9*, 157-208.

Buckland, W. (1823). *Reliqiae Diluvianae; or observations on some organic remains contained in Caves, fissures and Diluvial gravel, and on other geological phenomena attesting to the action of a Universal Deluge*. London.

Burleigh, R. (1972). Carbon 14 dating with application to dating of remains from caves. *Stud. Speleol. 2*, 176-190.

Campbell, J. B. and Sampson, G. G. (1971). A new analysis of Kent's Cavern, Devonshire, England. *Univ. Ore. anthrop. Pal. 3*, 1-40.

Colbert, E. H. and Hooijer, D. A. (1953). Pleistocene mammals from the limestone fissures of Szechwan, China. *Bull. Am. Mus. Nat. Hist. 102*, 1-134.

Coleman, J. C. (1965). *The Caves of Ireland*. Anvil Books, Tralee.

Cooke, H. B. S. (1969). The Sterkfortein ape-man cave site, South Africa. *Stud. Speleol. 2*, 25-34.

Cullingford, C. H. D. (ed.) (1953). *British Caving*. Routledge and Kegan Paul, London.

Dart, R. A. (1957). The osteodontokeratic culture of *Australopithecus prometheus*. *Trans. Mus. Mem. 10*, 1-105.

Dawkins, W. B. (1874). *Cave Hunting*. London.

Dawkins, W. B. (1903). On the discovery of an ossiferous cavern of Pliocene age at Dove Holes, Buxton, Derbyshire. *J. Geol. Soc. Lond. 59*, 105-129.

Dawkins, W. B. and Sanford, W. A. (1872). British Pleistocene Felidae. *Palaeontogr. Soc. (Monogr.) 1* (4): 177, plate 24.

Donovan, D. T. (1968). The ammonites and other fossils from Aveline's Hole, Burrington Combe, Somerset. *Proc. Univ. Bristol Speleo. Soc. 11* (3), 237-242.

Ford, T. D. and King, R. J. (1969). The origin of the silica sand in the Derbyshire limestone. *Mercian Geol. 3*, 51-69.

Guilday, J. E., Martin, P. S. and McCrady, A. D. (1964). New Paris No. 4. A Pleistocene Cave deposit in Bedford County, Pennsylvania. *Bull. Nat. Spel. Soc. 26*, 121-194.

Halstead, L. B. and Nicoll, P. G. (1971). Fossilized caves of Mendip. *Stud. Speleol. 2*, 93-102.

Harrisson, T. (1964). Borneo caves. *Stud. Speleol. 1*, 26-32.

Heal, G. J. (1970). A new Pleistocene mammal site, Mendip Hills, Somerset. *Proc. speleol. Soc. Bristol 12*, 135-136.

Hibbard, C. W. and others (1965). Quaternary mammals of North America. *In* Wright, H. E. and Frey, D. G. (eds), *The Quaternary of the United States*. Princeton University Press.

Hinton, M. A. C. (1926). *Monograph of the Voles and Lemmings Living, and Extinct*, Vol. 1. British Museum (Nat. Hist.), London.

Home, Sir E. (1817). An account of some fossil bones of rhinoceros discovered by Mr Whitby. *Phil. Trans. R. Soc. 107*, 176-182.

Kermack, K. A., Mussett, F. and Rigney, L. (1973). The lower jaw of *Morganucodon*. *Zool. J. Linn. Soc. 53*, 87-175.

King, A. (1974). A review of archeological work in the caves of North West England, Chap. 10 *in* A. C. Waltham, *Limestones and Caves of N.W. England*. David and Charles, Newton Abbot.

Kowalski, K. (1965). Cave studies in China today. *Stud. Speleol. 2*, 75-81.

Kowalski, K. (1967). *Lagurus lagurus* Pallas (1873) and *Cricetus cricetus* (Linnaeus 1758) (Rodentia, Mammalia) in the Pleistocene of England. *Acta Zool. Cracov. 12*, 114-124.

Kurten, B. (1969). Cave bears. *Stud. Speleol. 2*, 13-24.

Kurten, B. (1968). *Pleistocene Mammals of Europe*. Weidenfeld and Nicolson, London.

Lowry, J. W. B. and Merrilees, D. (1969). Age of the desiccated carcase of a Thylacine from Thylacine Hole, Nullabor Region, Western Australia. *Helictite* 7, 15-16.

Lundelius, E. L. (1966). Marsupial carnivore dens in Australian caves. *Stud. Speleol. 1*, 174-180.

MacEnery, J. (1859). *Cavern Researches* . . . Ed. E. Vivian. London.

Medway, Lord (1964). Post Pleistocene changes in the mammalian fauna of Borneo. *Stud. Speleol. 1*, 33-37.

Mitchell, G. F. (1969). Pleistocene mammals in Ireland. *Bull. Mammal Soc. Br. Isl. 31*, 21-25.

Murchison, C. (1868). *Palaeontological Memoirs and Notes of the late Hugh Falconer*. Vol. 2. Hardwick, London.

Olson, S. L. (1973). Evolution of the Rails of the South Atlantic islands (Aves Rallidae). *Smithson. Contr. Zool. 152*, 1-53.

Peach, B. N. and Horne, J. (1917). A bone cave in the valley of Allt nan Uamh (Burn of the Caves) near Inchnadamph, Assynt, Scotland. *Proc. R. Soc. Edin. 37*, 327-349.

Pernetta, J. C. (1966). Report on the findings at Etches Cave, Dowel Dale, Derbyshire. *Newsl. Peakland Archaeol. Soc. 21*, 11-16.

Savage, R. J. G. (1957). Fossiliferous fissures in the chalk of north-east Ireland. *Irish Nat. J. 12*, 1-7.

Savage, R. J. G. (1966). Irish Pleistocene mammals. *Irish Nat. J. 15*, 117-130.

Spencer, H. E. P. and Melville, R. V. (1974). The Pleistocene mammalian fauna of Dove Holes, Derbyshire. *Bull. Geol. Surv. Gt. Br.* No. 48. 43-49a, Pls. 2-5.

Stringer, C. B. (1975). A preliminary report on new excavations in Bacon Hole Cave. *Gower*, Swansea, *26*, 32-37.

Sutcliffe, A. J. and Kowalski, K. (1976). Pleistocene rodents of the British Isles. *Bull. Br. Mus. nat. Hist.* (*Geol.*) *27* (2).

Sutcliffe, A. J. and Zeuner, F. E. (1962). Excavations in the Torbryan Caves, Devonshire, 1. Tornewton Cave. *Proc. Devon Archaeol. Explor. Soc. 5*, 127-145.

Sutcliffe, A. J. and Bowen, D. Q. (1973). Preliminary report on excavations Minchin Hole, April-May 1973. *Newsl. Pengelly Cave Stud. Trust. 21*, 12-25.

Turnbull, W. D. (1961). A fossil carnivore den. *Bull. Chicago Nat. Hist. Mus. 32* (11), 4-5.

Walsh, P. T., Boulter, M. C., Ijtaba, M. and Urbani, D. M. (1972). The preservation of the Neogene Brassington Formation of the southern Pennines and its bearing on the evolution of Upland Britain. *J. Geol. Soc. London 128*, 519-560.

Warwick, G. T. (1956). Caves and glaciation—1. Central and Southern Pennines and adjacent areas. *Trans. Cave Res. Grp. G.B. 4*, 127-158.

Wells, P. V. and Jorgensen, C. D. (1964). Pleistocene wood rat middens and climatic change in the Mohave Desert; a record of juniper woodlands. *Science, N.Y. 143*, 1171-1174.

Part II: BIRDS AS CAVE FOSSILS

D. Bramwell

The bones of birds have become incorporated into cave sediments usually as a result of one or more of the following circumstances:

(a) As remnants of the food brought to the cave by late Palaeolithic hunters and by later prehistoric human groups.
(b) From pellets regurgitated in the cave by certain species of owls and falcons.
(c) From remains of food brought to their dens by mustelid, felid and canid species of mammals.
(d) Through the natural deaths of cave-roosting bird species.

Birds are usually much more mobile than mammals and also show greater adaptability to climate, for example raven and eagle owl range from the Arctic to tropical Africa, and so would hardly be reliable indicators of past climatic conditions. Other species from the warbler, swallow, flycatcher and cuckoo families, for example, are migratory, spending the breeding season in Europe and then wintering in tropical Africa. Thus the most useful indicators of past climates and environments are those birds of a more sedentary habit, particularly those which spend a large part of their lives feeding on the ground. The most important order is thus the Galliformes or game birds which includes the various grouse species and partridge. The habitual food of willow grouse and ptarmigan is mainly the shoots and berries of such low-growing shrubs as heather and crowberry, so these two bird species are indicative of a moorland or tundra-type environment. Partridge prefers a little more shelter with more grassland and the fact that it occurs alongside the grouse, in cave remains, indicates that the landscape was more varied than is sometimes conjectured. In addition there were undoubted occurrences of remains of woodpeckers in British and European caves, alongside the grouse forms, indicating trees with reasonably thick trunks, probably birch species, forming thickets in some more sheltered situations. Birds of coarse grasslands also appear fairly consistently with the above-mentioned species, corncrake being noteworthy, while marsh-haunting plover and dunlin are not uncommon. Finally, water birds, including swans, geese, ducks and divers testify to the numerous tarns and lakes lying over the late glacial landscape.

The bones of forest-type birds are not too well known from caves at present but some caves in the Peak District have shown an interesting rise in forest forms as the late glacial gave way to the post-glacial. Such conditions were marked by the capercaillie in the early pine forests; but owls, doves, tits and warblers, blackbird and robin mark the later mixed oak forest which dominated central Europe till human clearance for agriculture modified it.

Part III: CAVE ARCHAEOLOGY

A. King

Prehistoric people left behind no writings, so the archaeologist attempts to reconstruct their ways of life from the material evidence that has survived. Both evolutionary and actual physical steps were taken by early man away from the forest environment of the other primates; in both the glacial and peri-glacial conditions of the north and in the deserts and semi-deserts of the south, caves provided shelter and some strange bedfellows.

The caves that welcomed man subsequently harboured the evidence that he left; it is for this reason that the greatest respect must be paid to all cave deposits. The chance of survival, as of becoming a fossil in a specific geological horizon, varies, like the skill of the team doing the excavation. Cave archaeology is a multi-disciplinary subject, and chances can no longer be taken. The directors of excavations must be able to call upon many specialists for their help; specific questions must be answered; and he or she must be able to understand the implications of these replies.

The lines above may convince the reader that the subject is a science, but anyone who has walked along the passage of Lascaux or turned the pages of a picture book of cave art will know this is not the whole truth, and that many facets of art history are to be found underground. Gombrich (1972) has stated that there is no art, only artists; without doubt our study is one of people, and our aim is a more complete understanding of them.

Many textbooks will widen a student's understanding, like W. E. Le Gros Clark's (1964) *Fossil Evidence for Human Evolution* which sets the stage for the prehistorians; F. Bordes' (1968) *The Old Stone Age*; Clark and Piggotts' (1965) *Prehistoric Societies*; G. Clark's (1967) *Stone Age Hunters*, and J. Coles and E. S. Higgs' (1969) *The Archaeology of Early Man* all add detail. Breuil, H. and Windels, F. (1952) *Four Centuries of Cave Art* needs no comment, but Ucko and Rosenfeld (1967) *Cave Art* may be easier to obtain in Britain. Chapter VIII of *British Caving* (Cullingford, 1962) remains the source book for the contents of British Cave Archaeological sites; Valdemar (1970) has added detail of Welsh sites.

For readers interested in the excavations of the nineteenth century Boyd Dawkin's *Cave Hunting* (1874) has been reprinted recently. Those striving for more and more reliable age determinations should consult the Council for British Archaeology's Archaeological Site Index to Radio Carbon dates, the source of the uncorrected ^{14}C dates in this chapter. American work on the growth rings of the Bristlecone pine has shown that radiation has varied with time and that a correction must be applied to many ^{14}C dates.

HISTORICAL BACKGROUND

The excavated artefact, whether bead, bowl or inscription, may have some intrinsic value; but the find, its provenance and horizon, taken with the other finds of the assemblage, have archaeological significance.

Intellectual freedom followed in the wake of Darwin's *Origin of the Species* (1859). Previously the Church allowed only one interpretation of the book of Genesis, and whilst fossils may have been found in unlikely spots they were taken as evidence for the Flood. Fossils were bought, sold and collected throughout Europe in the eighteenth century, and one of the first clerical palaeontologists who discovered fossil ivory in Gailenreuth Cave, Bavaria, Pastor Esper, found human bones in association with animals bones and identified the bear skulls correctly. The finds of "unicorn's horn" or "dragon's teeth" were due for reappraisal in the early part of the nineteenth century. Guettard argued, as early as 1746, against a single Flood in favour of periodic incursions by the sea and gradual day-to-day change. Summoned by the censor to the Sorbonne, Guettard published a recantation. Buffon avoided the censors, and saw that the Catastrophic theory become widely held, enabling his successor Lamarck to talk of "boundless ages . . . millions of centuries". The history of science can be read elsewhere, but it is important to realize the intellectual climate of the time when considering the early cave explorations.

One concept fundamental to cave science was to spring from the self-trained mind of William Smith. As a surveyor he drew geological maps and sections; he recognized that some finds were index fossils of particular strata, and this understanding became the key to cave deposits, though not until much later. Buckland's excavation of Kirkdale Cave in the Yorkshire Corallian Limestone in 1821 appeared to break no new ground, but he showed that the animal material had been carried in mainly by hyenas; he dispensed with the idea of the deepening waters sweeping northwards carrying with them the bones of tropical species. But he was unable to forget the Flood, considering that the clay deposit has been laid down by those waters and that the finds were "antediluvian", i.e. before the Flood.

EARLY MAN

Zoologically man and anthropoid apes are now joined in one superfamily, Hominoidea. Early man had a brain size no greater than the large apes; but Oakley (1951) has argued that the term *man* may be used from that stage when he was intelligent enough to make tools.

Nuclear research has given archaeologists a technique of assessing time by measuring the decay of the isotope of ^{14}C but it cannot be used with any accuracy for periods of time greater than 60 000 years. Geologists on the other hand can measure much greater lengths of time by using the Potassium-Argon decay method. The potassium isotope ^{40}K has a half-life of 1330 million years, and has been used to date many ages of rocks, particularly granites in Britain, but may be as inaccurate for short spans of time as ^{14}C is for more lengthy ones. The close dating of Late Pliocene-Early Pleistocene finds is not yet possible and, although the faunal correlation of the African Tertiary and Quaternary still has to be worked out in detail, it is there that limestone extraction has drawn attention to cave deposits.

In 1936-7 Broom found hominoid fragments at three sites near Johannesburg-Sterkfontein, Krondaai and Swartkrans; as this evidence was similar to the juvenile skull previously known from Taung, he considered them Australopithecines. In 1947 Dart, working in breccia in the Makapansgat Cave, found further evidence, which with earlier material showed that the Australopithecines had a cranial capacity of 430-600 cc, an upright stance and dentition similar to that of present man.

Further north is the Tanzanian Rift Valley, where the Olduvai Gorge section in the Serengeti plain shows a Pleistocene sequence 40 km long and up to 100 m high, discovered by Kattwinkel in 1911. Louis Leakey organized his first field trip there in 1931 with Professor Reck, and that year found evidence for an Oldowan industry; but it was left to Mrs Leakey in July 1959 to find the skull of what they called *Zinjanthropus*, possibly a contemporary of the Taung skull; but both were Australopithecines.

It used to be accepted that man, using pebble or core tools, advanced culturally to use the flakes struck off the cores and later developed the technique of flaking long parallel-sided blades off nodules usually of flint. But even in the Early Oldowan pebble culture some flakes are seen to have been utilized.

Homo erectus

The name *Pithecanthropus* has been in use for over seventy years, but it has now given way to the term *Homo erectus*. With a cranial capacity of about 1000 cc, the more famous of this species are Peking and Java Man. The evidence for Peking Man has come from a group of caves in Silurian limestone near the village of Chou-Kou-tien. The hominid remains were associated with occupation debris including not only mammal remains but pebble and flake tools and well-defined hearths. An age of about 400 000 years has been suggested for these remains whilst the Javan deposits in the alluvia of the Solo River could be older; and there is little doubt that these hominids were game-hunters.

During the Middle Pleistocene the hand axe developed in Africa, for in Olduvai the pointed tool gave way to an artefact with a broader cutting edge. Taking their name from the French site at St Acheul, these Acheulean axes and the folk using them were spread over the whole of Africa, a good deal of western Europe and south-east Asia. Only one Acheulean axe has been found in Wales, at Cardiff, and none north of Scarborough in England. This suggests that hunting groups were not attracted to Highland Britain, and there must have been more than enough game on the riverine lands of south-east England. Some Acheulean evidence has been gleaned from caves such as Le Lazaret at Nice and Combe Grenal in the Dordogne, where fifty Mousterian layers and nine Upper Acheulean levels of the Penultimate glacial period showed a very cold fauna of reindeer and saiga antelope. Artefacts included cores, flakes and scrapers, blades, points and burins.

Homo sapiens neanderthalensis

At some time during the Hoxnian or Great Interglacial (Fig. 13.4) the hand axe Acheulean industries became of less significance when compared with the industries that had a higher proportion of flake elements in their make up. Geographically, this change of emphasis occurred in north Africa, southern Europe and south-west Asia. The hominoid remains show skulls with more refined features than the Lower Palaeolithic types but still with protruding brow ridges. The size of the brain (1350-1600 cc) has convinced modern anthropologists that Neanderthal Man, first

	TODAY	IRON AGE
POST GLACIAL		BRONZE AGE
		NEOLITHIC
		OBANIAN MESOLITHIC AZILIAN
ALLEROD		CRESWELLIAN
GLACIAL	10	LATE UPPER PALAEOLITHIC
	20	EARLY UPPER PALAEOLITHIC
GLACIAL		
UPTON WARREN INTERSTADIAL	30	HOMO SAPIENS SAPIENS
	40	
GLACIAL	50	LOWER PALAEOLITHIC (MOUSTERIAN & ACHEULEAN)
CHELFORD INTERSTADIAL	60	
	thousand years B.C.	DEVENSIAN BASE AT 75,000

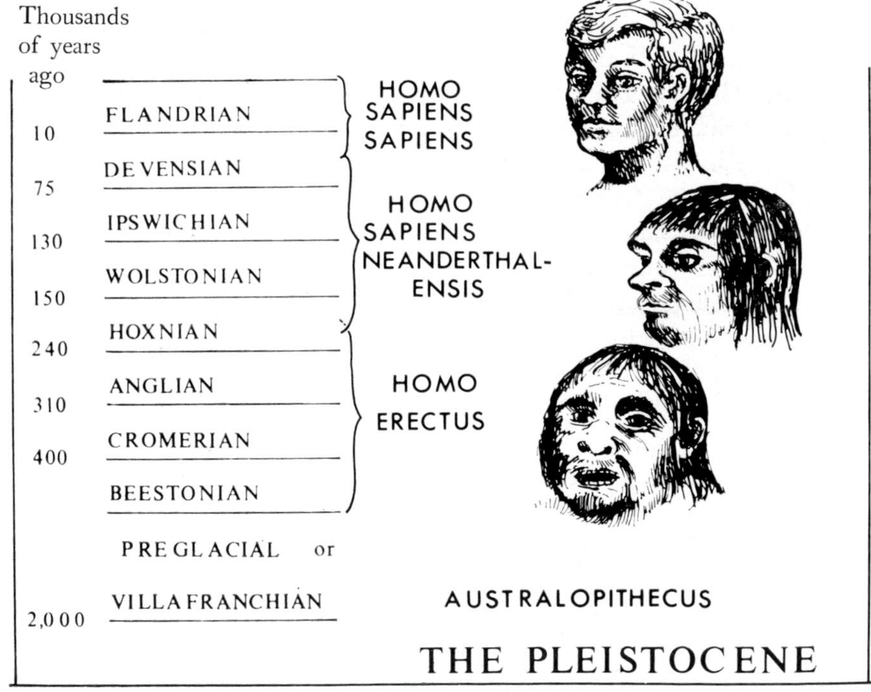

FIG. 13.4. Charts of the chronological distribution of prehistoric man and some of his cultures in relation to divisions of Pleistocene time.

identified from remains found in Neanderthal Cave near Dusseldorf in 1856, is not a species distinct from modern man (Fig. 13.4).

In the later Mousterian culture, the earlier part of which is contemporary with Acheulean, it is possible to find evidence illustrating for the first time that we are without doubt dealing with people with feeling.

In Shanidar Cave, Baradost Mountain, Iraq, pollen analyses indicate a climate warmer than today during the Last Interglacial. Seven Neanderthal men were buried in the cave and soil samples taken close to the skeletons show not a broad ecological picture but concentrations of pollen from a few flower species which, it is suggested, must have been picked and placed in the cave with the body.

A more advanced blade-using Neanderthal community gave a child a cave burial at Tesik Tas, Uzbekskaya, U.S.S.R. The boy's head was partially surrounded by a halo of ibex horns stuck into the ground; whilst one cannot hope to interpret the ritual or other significance of this sort of evidence, it is generally easier to build models of the ways of life of the earlier hunting communities than it is to suggest explanations for technologically more advanced urban societies.

The evidence from the Mount Carmel Caves sheds some light on rapidly evolving man. The basal layers yield an Acheulean industry passing upwards into *Levallois-Mousterian* associated with a fauna possibly representative of the warm dry Last Interglacial, or more probably of an early Last Glacial interstadial as the top of that horizon has been dated between 45 000 and 41 000 B.P. This is followed by a wetter, colder climate during which time at least ten Neanderthals were buried. The adjacent cave Mugharet el-Wad continues the sequence showing a transition from Upper Palaeolithic intensive hunting, mainly of deer and wild ox, to intensive plant collecting shown by the Natufian sickle blades and hafts.

BRITAIN

In Britain Father Burdo suggested a Pre-Mousterian data for the material excavated from his lower levels at the sea cave of La Cotte de St Brelade in Jersey and this has been substantiated by the recent work of McBurney and Callow (1971). A small number of cordiform hand axes and African form cleavers were obtained from loess which appears to have been laid down upon an Eemian (Riss-Würm) inter-glacial beach deposit. A concentration of mammal bones of a cold/arctic fauna is considered food debris; it included crania of three rhinoceroses and remains of at least five mammoths piled together or placed in a pit.

The notes which MacEnery and Pengelly made at Kent's Cavern, in particular Pengelly's diary, together with the faunal collections in various museums, have recently been re-analysed by Campbell and Sampson (1971). In their composite section given in their Fig. 2, the B1 Breccia had yielded a lithic assemblage considered by Campbell to contain thick asymmetric hand axes that are possibly early Acheulean. The presence of seven teeth considered to be from a sabre-tooth cat are now thought to be Lesser Scimitar Cat and the deposit later than Hoxnian Interglacial. "Generalization about the origins of the Acheulean complex in Africa and its dispersal into western Europe may require revision in the light of the Kent's Cavern evidence" (*ibid.* p. 23). Attention is also drawn to the association of artefacts and the cave bear remains that make up 90% of the animals represented in that horizon. Possibly hibernating bears were the target of the hunters; there is no suggestion here of the bear cult found in some European caves where skulls and bones were found in stone cists. One Alpine "bear cave" 2500 m above sea-level shows the spread of Neanderthal Man into European mountain areas. The subsequent deposit B/A2 on Campbell and Sampson's Fig. 2 has yielded the largest British Mousterian sample, and as it passes upwards into Upper Palaeolithic proto-Solutrean levels a number of relationships may well be deduced if excavation of the deposits remaining in the cave could be allowed. Other caves associated with Mousterian occupation include Coygan Cave, South Wales, and Plas y Cefn Cave in the Vale of Clwyd. Prehistorians are now able to appreciate the potential of reinvestigation on the lines of the above painstaking research.

Homo sapiens sapiens

It is not possible to continue the British sequence of events without returning to the Continental caves and rock shelters which have given us the abundance of Upper Palaeolithic type sites. The journey to Britain had to be made in the increasingly severe conditions marking the onset of the Last Glaciation, and it is not unlikely that at least one ice advance geographically isolated Neanderthals from the communities bordering the Mediterranean by Pyrenean and Alpine ice.

The advanced Upper Palaeolithic people *Homo sapiens sapiens* (modern man) appeared about 40 000 years ago, emerging in the zone peopled by Neanderthals. The expanding population, dependent upon the migrating herds of mammoth, bison, horse and deer, spread across Eurasia and into the Americas. Australia was settled rather later but still well before the end of the Pleistocene. The warmth of the interstadials made settlement of the northern parts of Russia possible if the burial site at Sungir is accepted as evidence. In 1964 O. N. Bader found two burials of *Homo sapiens sapiens* below the permafrost and dated them by ^{14}C to between 33 000 and 40 000 years B.P. In addition to the covering of red ochre, 1500 bone ornaments were found in the oval grave (Bacon, 1971, pp. 44-45).

In the Mediterranean region one of the finest stratigraphical columns occurs at Hava Fteah in Libya and covers a period of possibly 100 000 years. The Levallois-Mousterian culture appears to be submerged below an intrusive Pre-Aurignacian horizon which gives way to a fully developed Upper Palaeolithic level with backed blades, burins and end scrapers. Though named from Aurignac in France, this advanced culture seems to have spread from SE Asia, entering Europe via the Balkans, where it is succeeded by the Gravettian, another culture that spread from the east but named after La Gravette in the Dordogne. The Gravettian hunters moved to the caves of France by way of the more northerly loess lands, and these are the people who may be credited with developing the art of carving figurines, initially in the mammoth ivory that lay scattered around their tended camp sites. Animals represented include horse, mammoth, bison and ibex and the many "Venus" figures, over 60 of which have now been found. New concepts continued to be introduced into art; bas-reliefs in friezes along cave walls probably came with the Solutreans; the later beautifully sensitive polychrome mural work is considered Magdalenian. More radio-carbon dates are needed, but the problem of dating graffiti on cave walls is not easy. If their purpose was known it might be possible to say at what height above the cave floor the work would have been executed; but the paintings had many functions, not least amongst them being a creative recreational one. Powell (1966) discussing the Krapova Cave, an outlier of cave art in the Ural Mountains, has suggested that painted tent walls may be the common origin of this type of decoration.

The sophistication of the Upper Palaeolithic art is matched by the beauty of the flints worked by the Solutreans. Their technique of pressure-flaking blades off cores, presumably with a wooden or bone punch, and of their skilled shallow secondary surface finishing, is unique. Readers may be excused for imagining a lack of continuity in the archaeological record when faced with the proliferation of cultures, but this is not so; just as the Neanderthals gave way to modern man so the Solutrean horizons merge upwards into the evidence of the final Upper Palaeolithic people, the Magdalenian.

This culture was adapted to the late Glacial environment centred on the Franco-Cantabrian region, and their life-style was based on the products of specialized reindeer hunting. Working antlers by cutting deep grooves before splintering they produced their "type-fossil", the harpoon. Initially an antler point with a basal facet, the projectile developed into uni-serially and finally bi-serially barbed harpoons (see Fig. 13.6). The acme of cave art exemplified at Lascaux and Altamira is early Magdalenian. Decadence comes with the ameliorating climate.

THE UPPER PALAEOLITHIC IN BRITAIN

Whilst there may be genetic changes resulting from increasing cold elsewhere, Britain, during the greater part of the Devensian (Würm) Glacial, was concealed by ice. Interdisciplinary speleological research should provide information for both geomorphologists and archaeologists.

THE UPPER PALAEOLITHIC OF BRITAIN

The most recent Devensian (Würm) glacial epoch is characterized by three periods of refrigeration, the most recent from about 16 000 to 12 000 B.C., which was the period of maximum cold. The limit of this Devensian Newer Drift was to the north of the suggested edge of the earlier glacial advances. Fig. 13.5 shows that whilst the highland areas of Wales and north Britain were buried some of the coastal areas of South Wales and southern England were free of ice. It seems doubtful whether permanent settlers could have occupied the caves of South Wales, because sea-level would have been so much lower and the Mendips a comfortable tundra environment two days' walk away.

McBurney (1965) has drawn attention to the problem of Upper Palaeolithic nomenclature. For too long the term "Proto-Solutrean" has been used for a culture found both before and after the last full glacial; it is intended here to use the self-explanatory "Early" and "Late" Upper Palaeolithic periods. Without this simplication future parallels with cultures in Belgium, Germany and Poland will become increasingly difficult. It is also apparent from recent work that the finds from the Early Upper Palaeolithic have a homogeneity suspected in the south-west, but now shown to be more widespread by the appearance of bifacial and unifacial leaf-shaped points elsewhere.

At Kent's Cavern (Campbell and Sampson, 1971) the Early Upper Palaeolithic (henceforth E.U.P.) material from above the Mousterian horizon and from the Loamy Cave Earth shows a more intensive use of flint instead of Greensand Chert. In addition to the bifacial and unifacial spearheads mentioned above, the typical assemblage contains scrapers and strong bone sewing-needles. At Wookey, Tratman (1971) considered the

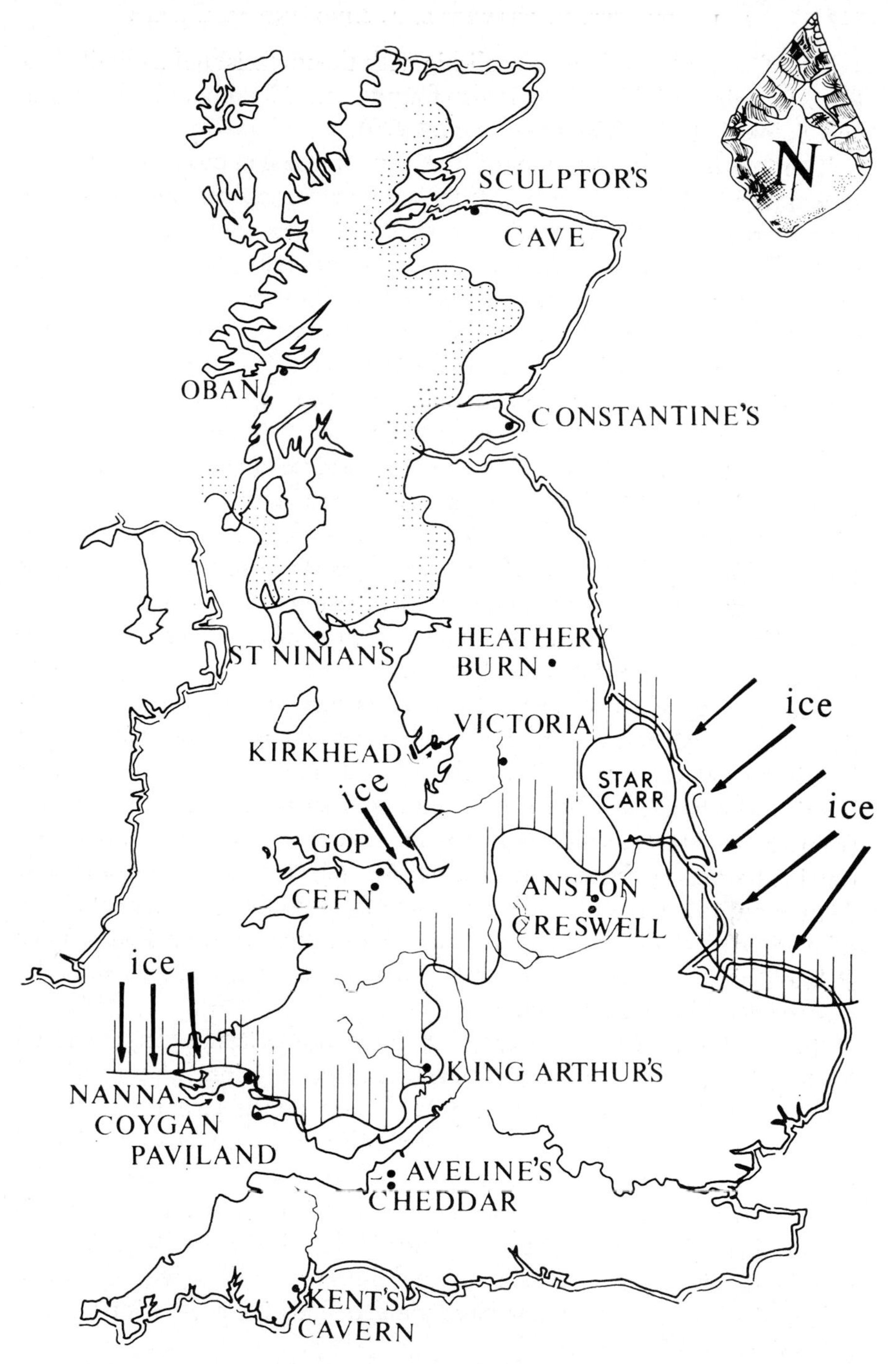

FIG. 13.5. Important Upper Pleistocene sites in relation to glacial limits in Britain.

small bifaces to be Middle Palaeolithic with the unifacial points limited to E.U.P. horizons. This culture is also found in the Mendips at both Badger and Soldier's Holes (Campbell *et al.*, 1970).

The original identification of the sex of the famous "Red Lady" skeleton from Paviland Cave was incorrect; also, as Buckland excavated the material in 1832 it seems that too much emphasis should not be placed on the date of 16 460 B.C.±340. The similarities with the Sungir permafrost body suggest that it may well be an underestimate; in addition to the mammoth ivory bracelet and finds of perforated animal teeth, both burials were situated in oval graves and covered in red ochre. A short discussion between Drs Bowen and John on the geomorphological implications of the date has appeared in the 1970 and 1971 issues of *Antiquity*. The Paviland finds have to be considered unstratified but those from Cae Gwyn and Ffynnon Beuno, caves east and west respectively of Trefnant in the Vale of Clwyd suggest occupation in North Wales before the final Devensian ice advance, the end scraper from Cae Gwyn being found between two layers of boulder clay. In Derbyshire at Robin Hood's Cave, Creswell, the E.U.P. horizon lies between a basal red silty sand and a thermoclastic scree (Campbell, 1970).

The wasting of the ice of the final Devensian glacial and the colonization of the north and west by both plants and animals meant that a larger hunting province was available for exploitation by specialist reindeer hunting communities. The Isle of Man appears to have been free of ice from about 17 000 B.C. although in Cumbria a date of 10 650 B.C. has been obtained from an organic fill in a hollow left by the ice (Evans and Arthurton, 1973). Rising sea-level about 12 000 years ago meant the loss of a good deal of coastal fringeland, and from that time Late Upper Palaeolithic (L.U.P.) people spread northwards, and lived on into the post-glacial Flandrian period.

The L.U.P. in Britain comprised the indigenous Creswellian, identified by its triangular and trapeziform backed blades, burins, borers and end scrapers of flint, and the Magdalenian typified by the javelin points and harpoon heads of bone or more usually antler. Basal faceted and grooved points comparable with the early part (Magdalenian O, I, II, III) of the French sequence cannot be expected on climatic grounds. The later uniserially barbed harpoons (Mag. V) and the biserial forms (Mag. VI) are readily identified (Fig. 13.6) but cultural links may well be with Germany where sites such as Petersfels have yielded finds identical with English examples.

Campbell and Sampson (1971) were able to relocate 144 of the 522

L.U.P. items found by Pengelly, including the two uni-serially barbed and the one bi-serial harpoon head. In the Mendips over 7000 artefacts are known to have been found in Gough's Cave. Other sites in Somerset include Sun Hole, Soldier's Hole, Flint Jack's Cave and Aveline's Hole,

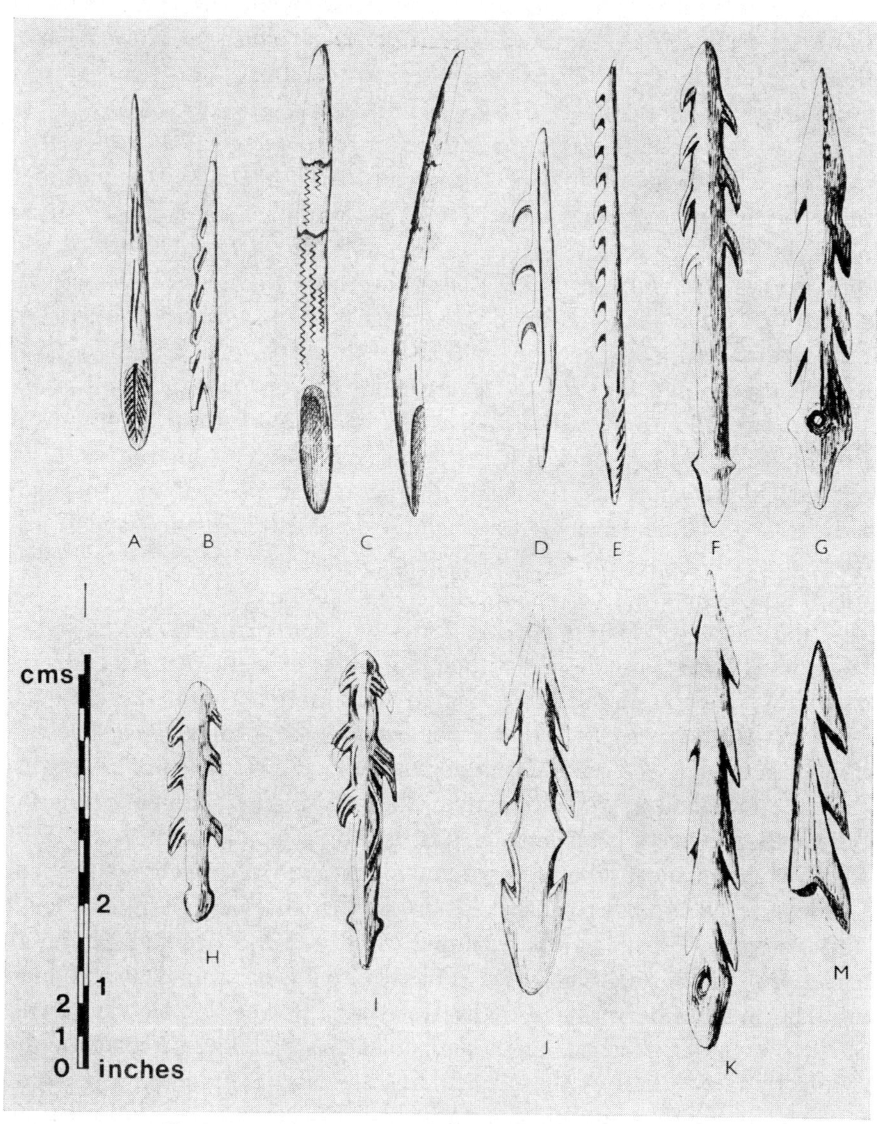

FIG. 13.6. Late Upper Palaeolithic barbed harpoons.
A, B, D, E, F, G, I are Classic French forms (after Bordes, F.): C, J, from Victoria Cave, Yorkshire; H from Aveline's Hole, Somerset; K from MacArthur Cave, Oban; M from Druimvargie Rock Shelter, Oban.

where a bi-serially barbed antler point was discovered. In Wales, Cefn, Cathole, Hoyle's Mouth, Little Hoyle, Lynx, Nanna's Cave, Caldey and Priory Farm Cave are the better known Creswellian locations whilst King Arthur's Cave near Monmouth should be mentioned.

The type sites are the Creswell Caves flanking the east-west valley in the Magnesian Limestone outcrop on the Derbyshire-Nottinghamshire border. Here woolly rhinoceros appears to join horse and reindeer as a common meat (Dawkins, 1876; Campbell, 1970).

In the north-west the reindeer antler rods from Victoria Cave have been figured by the Abbé Breuil (1922) and considered Magdalenian: an earlier projectile point, matching some Petersfels examples perfectly, has been noted by Jackson (1945); this is at least as early as Magdalenian V and may belong to group IV (Fig. 13.6c). To these L.U.P. finds should be added the point from Kinsey Cave (Jackson and Mattinson, 1932). Excavations in 1968, 1969 and 1970 at Kirkhead Cavern, Cumbria, have pushed the known limit of the Creswellian further north (Wood *et al.*, 1969; Ashmead and Wood, 1974). A serious research team commitment to the archaeological and other Pleistocene problems in the north is needed. A few dates are available for the L.U.P. period; one from Sun Hole at Cheddar, 10 428 B.C. ± 150, and the three from Anston Cave, South Yorkshire, of 7800-7990 B.C., are close to, if not later than, the spread of the Mesolithic culture across Britain.

With a date of 7607 B.C. ± 210 for a wooden platform amongst the lakeside dwellings of Mesolithic Star Carr it is evident that no clear-cut chronological boundaries are going to be found between the cultures. Nor does the writer believe that groups of sedentary hunters had become housed in the regions centred on the caves. The distributions of cavernous areas is controlled by geology and is widespread. There are no significant differences between the assemblages of the Mendips, of South Wales or of Creswell; it is more likely that communications were well established.

There is an almost total lack of art in British Palaeolithic caves; no mural art is known, and the examples of carved and decorated bone or antler are very rare. A rib bone from Gough's Cave found by Tratman and Hawkes has been described by Rosemary Powers (in Hawkes *et al.*, 1970). With its notched edges and cross-hatched face it matches the enigmatic tally sticks of the French U. Palaeolithic and should be compared with the one published by Marshack (*in* Sieveking, 1971). In the same volume Sieveking, referring to an assemblage of finds from Kendrick's Cave, Llandudno, focussed attention on to the incised zigzag decoration on a horse mandible symphysis which is considered to be upper Palaeolithic.

THE MESOLITHIC IN BRITAIN

Attention has already been drawn to ^{14}C dates for the two Yorkshire sites of Anston Cave and Star Carr (see Fig. 13.2). The former L.U.P. horizon has reindeer present, but in Mesolithic Star Carr red deer were the common food source. The hunting and gathering communities appear, from the evidence of their microlithic flints, to have spread over the greater part ot upland southern Britain and continued to occupy some lowland areas. It is likely that they began to "manage" some of the herbivores they had previously chased. Using this practice and by burning vegetation, they possibly had some impact upon the edge of the woodland zone, particularly in the uplands. Mesolithic evidence in caves is slight, but some attention should be given to the groups living on the raised beaches of western Scotland.

In Ireland there is no upland evidence for a Mesolithic occupation; communities lived in riverside locations as salmon fishers, especially in the valley of the river Bann (Estyn Evans, in press). A number of dates for occupation levels in Ireland are found in the sixth millennium B.C., not much older than some of the shell middens of Oronsay. The finds from these middens and from the caves along the raised beaches have been termed "Obanian"; but probably the Mesolithic culture had developed some regional specialization (Lacaille, 1954). Mesolithic levels in MacArthur Cave, Oban, are covered by two layers of shingle showing that storm conditions could drive beach material into the cave which is now at 10·4 m O.D. Stone finds include "limpet hammers" and a few flint flakes, together with bi-serially barbed antler points which continue the late Palaeolithic tradition (Fig. 13·6k, m). The Azilian harpoon from Victoria Cave was an intermediate type (Fig. 13.6j). In addition to the dietary evidence from mollusc shells, fish and animal bones, Mercer (1970) found carbonized seeds of *Rubus* (Bramble), *Potentilla sterilis* (Barren Strawberry), acorns and quantities of hazel nuts at Lussa River, North Jura.

Away from the coast, near Melfort, a cave close to the top of An Sithean has produced a quantity of worked flint of this period, as has a neighbouring rock shelter (see *Discovery and Excavation in Scotland* 1959, 4; 1963, 9). The conservatism of the aboriginal fishing collecting communities in Scotland may be seen in the series of dates in the third millennium B.C. from their middens, which postdate a number of Neolithic monuments and occupation levels also from western Scotland.

THE NEOLITHIC IN BRITAIN

Possible human interference with the upland type cover has been mentioned, but man was not solely responsible. Increasing amounts of rain fell and evaporation in a "maritime" regime became more difficult. It is suggested that the first people to penetrate into the lowland forest and practise arable farming would have exploited the better drained land. The calcareous soils of the Cretaceous chalk and Jurassic limestone belts were colonized in England; valleys, coastal plains and the islands of western Britain were similarly occupied. The initial "landnam" (Viking, land taking) phase is seen in pollen analytical diagrams by a decrease in elm pollen followed by a decrease in pine, oak and even birch. The polished stone axes traded as "roughs" from the many axe factories are the type fossil of the settlements and their earthen and stone burial mounds.

Regional variations exist amongst the community-built stone monuments but usually a passage leads to a burial chamber. The passage of New Grange in Ireland is 20 m long, but at Maes Howe in Orkney it is 12 m long. Some passages are very short and not all are built above ground; rock-cut tombs are known around the Mediterranean as in Sicily and Malta (Daniel, 1962; Piggott, 1965). The Dwarfie Stone on Hoy, Orkney, appears to be the best British example. It is difficult from some reports of excavations to judge whether a passage or rock shelter has been occupied by the living, the dead, or both, since a layer of charcoal may be interpreted in many ways. Manby (1970) has noted that grave goods are notably absent from the burial deposits in the earthen long barrows of northern England, but that occupation refuse, such as broken pottery and animal bones, appear to have been imported for some ritual purpose.

The largest burial cairn in Wales at Gop in Gwaenysgor parish, Clwyd, is 100 m in diameter; beneath this, in a series of shafts and passages, a confused series of finds has been made. In the adjacent rock shelter a fireplace upon which rested a thick layer of charcoal was removed, exposing a chamber 1·35 m $\times$1·5 m $\times$1·20 m containing the remains of 14 contracted inhumations. The three cist walls were made up of regular courses of limestone (Davies, 1949, p. 278). Subsequent excavations opened up the north-west passage, at the entrance of which an unpolished Craig Llwyd axe was found stuck upright in clay, and where one inhumation in a passage recess was contained inside a rubble wall.

In the Peak district, caves which have been used for sepulchral purposes in the Neolithic appear to be a simple underground extension of the distribution of chambered tombs (Manby, 1958) as exemplified by Dowel

Cave, a narrow passage 7 m long, sealed by portal slabs and two internal blockings. Ten individuals had been buried; one child's skull was held up by a limestone plate in a sloping fissure (Bramwell, 1959). A restorable Peterborough bowl was found amongst a mass of disarticulated human bones in the drystone walled cist of Church Dale rock shelter (Piggott, 1953). The sepulchral use of natural passages and shafts is also to be found to the north-west in Yorkshire (Gilkes, 1973) and Lancashire (King, 1974). Elbolton Cave in Wharfedale is a 25 m shaft containing skeletal remains, pottery and six curved bone pins, each with a side swelling; this possibly foreshadows the digging of grave shafts by later metal-using communities.

Apparently burials are not to be found in Mendip caves. Rowberrow, Chelm's Combe Cave and Tom Tivey's Hole have yielded good pottery bowls, while the bowl of a ceramic spoon has come from Sun Hole. Many cave sites have produced limited amounts of Late Neolithic Beaker pottery; that from the 25 m fissure of Antofts Windypit, Helmsley, Yorkshire, was associated with hazel charcoal from which a ^{14}C dating of 1800$\pm$150 has been obtained.

THE BRONZE AND IRON AGES IN BRITAIN

Grinsell (*in* Campbell *et al.*, 1970) has counted over 300 round burial mounds in the Mendips, and the Peak district is liberally strewn with similar monuments, but caves in the limestone have continued the earlier underground burial tradition, Ogof yr Esgryn, Glyntawe, being noteworthy. A bronze rapier, razor awl and a gold biconical bead were on the stalagmite floor, and over 2000 human bones were crowded into sand pockets, but they may belong to the Romano-British period as finds of that period were also obtained and as cremation had virtually replaced inhumation in the Bronze Age (Mason, 1968).

There can be no doubt that continuity of belief persisted from the Bronze Age to the Iron Age and even later. The metal-using Celtic people venerated springs, wells, caves and burial mounds as entrances to the other world (Ross, 1968). Votive offerings were made at these sites, and the author has previously suggested that speleologists should recognize the potential sanctity of the site (King, 1969). The head too was worshipped in a real or carved state and this cult may be represented amongst some cave deposits but will be difficult to identify. Cases are recorded where skulls attached to one or two vertebrae, which show marks of cutting, some being sliced, indicate beheading. The Hay Wood cave burials were equally interesting, the young males having their incisor teeth mutilated (Everton, 1972).

Some Bronze Age caves have contained late bronzes; the rather exotic finds and flat-rimmed pottery from the Sculptor's Cave in Moray, Scotland, have given their locality name to the Covesea phase of bronze-working, but in quantity these pale to insignificance when compared with the material excavated from Heathery Burn Cave, Stanhope, Co. Durham. Exposed by quarrying, it was about 150 m long; and between 1859 and 1874 a fine selection of bronzes was collected mainly by Greenwell but figured by Briton (1968). The four pairs of nave bands give the first evidence of the wheel in a British cave context, and since bones of at least three individuals were found, including an entire skeleton, some sepulchral use must be accepted. Both the above caves, Ogof yr Esgryn and many other caves in South Wales have produced Romano-British metalwork and pottery (Valdemar, 1970). In Yorkshire, Dowkerbottom Hole, Attermire and Victoria Caves are noteworthy (King, 1970). In Lancashire, Dog Hole at Haverbrack, a 10 m shaft with a mouth 1·5 m × 1·3 m reopens the case for ritual or votive use of shafts.

Where a cave has a clear terminal date consideration must be given to its sealing. Frequently the seal or blockage is considered insignificant and removed; often it is not shown on plans and part of the story of that cave is thereby lost. Caves continued to be used in the Migration and even Medieval periods. Early Christian sculpture has linked St Ninian to the sea cave at Glasserton, Galloway; crosses were found carved on a sandstone slab in Constantine's Cave, Fife (Wace and Jehu, 1915), and Pictish carvings gave the name to Sculptor's Cave; migration period metalwork too has been found showing a number of interesting trade links, especially around the Irish Sea.

Recognizing the law of superposition and the importance of all finds, cavers must not accept the cult of the antique; the oldest find cannot be claimed as the best. Clearly the upper levels should not be the province of the amateurs. The Anglo-Saxon penny is as relevant as the Palaeolithic flake to any serious worker. Too much has already been sacrificed in the race to bottom the shaft and clear the passage. It is to be hoped that established links between the B.C.R.A. caving clubs and regional Council for British Archaeology groups will become more widely established.

This chapter was not intended to be a gazeteer, for a fine one already exists in Dr J. Wilfred Jackson's chapter in *British Caving* (Cullingford, 1962). Rather it has attempted to draw cavers and archaeologists more closely together. It was made a great deal easier by the magnificent help given by Dr John Campbell.

REFERENCES

Ashmead, P. and Wood, R. H. (1974). Second report on Archaeological Excavations at Kirkhead Cavern. *North-West Speleology 2*, 24-33.

Bacon, E. (1971). *Archaeology, Discoveries in the 1960s*. Cassell, London.

Bader, O. N. (1963). Krapova Cave. *Soviet Archaeology*, 125-134.

Bordes, F. (1968). *The Old Stone Age*. Weidenfeld and Nicolson, London.

Bramwell, D. (1959). The Excavation of Dowel Dave, Earl Sterndale. *Derbys. Archaeol. J.* 77, 97-109.

Breuil, H. and Windels, F. (1952). *Four Hundred Centuries of Cave Art*. Montignac.

Britton, D. (1968). *Inventaria Archaeologica*. Great Britain 9th Ser., British Museum, London.

Buckland, W. (1823). *Reliquae Diluvianae*. Oxford.

Campbell, J. B. (1970). Excavations at Creswell Crags, *Derbys. Archaeol. J. 89*, 47-157.

Campbell, J. B. *et al.* (1970). *The Mendip Hills in Prehistoric and Roman Times*. Bristol.

Campbell, J. B. and Sampson, C. G. (1971). *A New Analysis of Kent's Cavern, Devonshire*. Univ. of Oregon Anthropol. Papers, No. 3.

Clark, G. (1967). *Stone Age Hunters*. Thames and Hudson, London.

Clark, G. and Piggott, S. (1965). *Prehistoric Societies*. Hutchinson, London.

Clark, W. E. Le Gros (1964). *Fossil Evidence for Human Evolution*, 2nd edn. Chicago.

Coles, J. M. (1960). Scottish Late Bronze Age metalwork, typology, distributions and chronology. *Proc. Soc. Antiq. Scot. 93*, 16-134.

Coles, J. M. and Higgs, E. S. (1969). *The Archaeology of Early Man*. Faber, London.

Daniel, G. (1962). *The Megalith Builders of Western Europe*. Penguin, London.

Davies, E. (1949). *Prehistoric and Roman Remains of Flintshire*. Cardiff.

Dawkins, W. B. (1874). *Cave Hunting*. Reprinted 1972 by S. R. Publications, Wakefield.

Dawkins, W. B. (1876). On the mammalia and traces of man in Robin Hood Cave. *Quart. J. Geol. Soc. London 32*, 245-258.

Evans, W. B. and Arthurton, R. S. (1973). North West England *in* G. F. Mitchell *et al.*, *A Correlation of Quaternary Deposits in the British Isles*. Geol. Soc. London, Special Report No. 4.

Everton, A. and R. (1972). Hay Wood Cave Burials, Mendip Hills. *Proc. Univ. Bristol Spelaeo. Soc. 13*, 5-29.

Gilkes, J. (1973). Neolithic and Early Bronze Age pottery from Elbolton Cave, Wharfedale. *Yorks. Archaeol. J. 45*, 41-54.

Gombrich, E. H. (1972). *The Story of Art*. Phaidon, London.

Hawkes, C. J., Tratman, E. K. and Powers, Rosemary (1970). A decorated rib-bone from the Palaeolithic levels at Gough's Cave, Cheddar. *Proc. Univ. Bristol Speleo. Soc. 12*, 137-142.

Jackson, J. W. (1945). Note on a javelin point from Victoria Cave. *Antiq. J. 25*, Pl. XII.

Jackson, J. W. and Mattinson, W. K. (1932). A Cave on Giggleswick Scars. *Naturalist*, 5-9.

King, A. (1970). Romano-British metalwork from the Settle district of West Yorks. *Yorks. Archaeol. J. 42*, 410-417.
King, A. (1974). Chapter 10 *in* Waltham, A. C., *The Limestones and Caves of Northwest England*. David and Charles, Newton Abbot.
Lacaille, A. D. (1954). *The Stone Age in Scotland*. Oxford Univ. Press.
McBurney, C. B. M. (1965). The Old Stone Age in Wales *in Prehistoric and Early Wales*. Ed. I. Ll. Foster and Glyn Daniel. London.
McBurney, C. B. M. and Callow, P. (1971). The Cambridge excavations at La Cotte de St Brelade, Jersey. *Proc. Prehist. Soc. 37*, 167-207.
Manby, T. G. (1958). Chambered tombs of Derbyshire. *Derbys. Archaeol. J. 78*, 25-39.
Manby, T. G. (1970). Long barrows of Northern England: Structural and dating evidence. *Scot. Archaeol. Forum*, Edinburgh, 1-27.
Marshack, A. (1971). Upper Palaeolithic engraved pieces in the British Museum *in* G. Sieveking's *Prehistoric and Roman Studies*. British Museum, London.
Mason, E. J. (1968). Ogof yr Esgyrn, Dan yr Ogof, Brecknock, Excavations. *Archaeologia Cambrensis*, 18-71.
Mercer, J. (1970). The Microlithic succession in N Jura, Argyll. *Quaternaria 13*, 177-185.
Oakley, K. P. (1951). A definition of man. *Sci. News*, *20*, 69.
Piggott, S. (1953). Editorial Note. *Proc. Prehist. Soc. 19*, 229.
Piggott, S. (1965). *Ancient Europe*. Edinburgh Univ. Press.
Powell, T. G. E. (1966). *Prehistoric Art*. Thames and Hudson, London.
Ross, A. (1968). Shaft pits and wells—Sanctuaries of the Belgic Britons *in* Coles, J. M. and Simpson, D. D. A., *Studies in Ancient Europe*. Edinburgh.
Tratman, E. K. *et. al*. (1971). The Hyaena Den (Wookey Hole) Mendip Hills. *Proc. Univ. Bristol Spelaeo. Soc. 12*, 245-279.
Ucko, P. J. and Rosenfeld, A. (1967). *Palaeolithic Cave Art*. Weidenfeld and Nicolson, London.
Valdemar, A. E. (1970). Preliminary Report on the Archaeological and Palaeontological Caves and Rock Shelters of Wales. *Trans. Cave Res. Grp. G.B. 12*, 113-116.
Wace, A. J. B. and Jehu (1915). Cave excavations in East Fife. *Proc. Soc. Antiq. Scot. 44*, 233-255.
Wood, R. H., Ashmead, P. and Mellars, P. A. (1969). First report on the archaeological excavations at Kirkhead Cavern. *Northwest Speleology*, *1* (2), 19-24.

Part IV: CAVE ART

M. Walker

It was only in the last and coldest phase of the ice-age, 32 000-10 000 years ago, with the development of modern-looking human types, that recognizable art was executed in the caves of the French Pyrenees and Dordogne and Vézère valleys further north, as well as the coastal belt of the northern

Spanish Cantabro-Asturic mountains. Other cave paintings of the same period are found elsewhere in France, Spain, Sicily, mainland Italy, and even in Russia and Turkey. All the paintings were executed by *Homo sapiens*, and none can be attributed to Neanderthal or earlier human species.

Paintings, often quite large and many-coloured, depict mammoth, woolly rhinoceros, bison, brown bear, cave lion, ibex, reindeer, red deer, wild horse and oxen, glutton, wolf, and so on. Human outlines and also outlines of human hands are sometimes painted, as well as less clear "signs". There are also incised outlines of animals, and even bas-relief sculptures. There are a few anthropomorphic figurines or statuettes, and also small stone plaques and pieces of bone and antler engraved, and sometimes painted, with animal and other motifs. These, together with bone and stone tools, have been excavated from earth deposits in cave floors, sometimes from caves with wall paintings, sometimes from caves otherwise devoid of paintings. Associated organic material can sometimes be dated by radiocarbon determination. The differences in successive tool-kits have led to the identification of successive stages in particular cultures named after French locations (e.g. Breuil and Windels, 1952).

Mousterian-Aurignacian transition	35 000-30 000 B.C.
Aurignacian	32 000-25 000 B.C.
Perigordian or Gravettian in France*	25 000-19 000 B.C.
Protomagdalenian	10 000 B.C.
Solutrean	18 000-15 000 B.C.
Magdalenian	15 000-9 000 B.C.

* Tool types of this kind linger until about 10 000 years ago in Eastern Europe and some Mediterranean countries.

Out of about 120 decorated caves only a score can be definitely assigned to particular cultures, since not all have an earth floor in which art pieces became lodged with other dateable objects, and conversely not all caves with archaeological deposits and tools have artistic artefacts in them. The most famous painted caves, Lascaux in France and Altamira in Spain, belong to the Magdalenian culture. However, each of the major archaeological cultures is represented by cave art of one form or another. Human figurines seem to be especially common in Gravettian cultural deposits; monochrome black paintings are said to be characteristic of the Solutrean; and carved spear-throwers typify the Magdalenian, and reindeer portrayals are said to be mainly from that culture. Various schemes have attempted to define styles of art corresponding to particular cultures in more detail, but objections can be put forward against many of these (for bibliography and criticism, see Ucko and Rosenfeld, 1967).

Recent work suggests that the art might be grouped into three broad categories: bas-reliefs, engravings and other sculpture at cave mouths, perhaps associated with habitation débris; art which begins at cave entrances but continues somewhat deeper into the cave; and thirdly, art which is very far underground indeed. For this latter category there are no modern ethnographic parallels, and many theories have been proposed to explain it. An explanation allowing for the heterogeneity of the content of the paintings might be that these obscure situations were used for initiation ceremonies and for the execution of totemic paintings. The one-time popularity of hunting and/or fertility sympathetic magic as an interpretation of the paintings has many strong objections, as pointed out by Ucko and Rosenfeld (1967). Perhaps the artistic endeavours of early man found in or near to the daylight zone, however, represent little more than secular, aesthetically inspired efforts. At some sites it would seem that there were many throw-away practice sketches on small slabs of stone. There seems to have been a good deal of preference for depicting particular animal species, and neglect for some of the more common species which may have been easily acquired food sources (e.g. rabbits, plants, birds). The possibility that the existing record of cave art is far from complete and not necessarily representative remains none the less. Certainly, the technical accomplishment in some cases is amazing, considering the uneven surfaces of cave walls and roofs and the primitive materials for painting and lighting presumably used. They show a mastery of the use of colour and of perspective drawing which excites our attention even today.

Only recently, however, have cave paintings received rigorous critical study (Ucko and Rosenfeld, 1967). To claim that prehistoric man painted bisons that they might flourish and lions for the opposite reason, not only attributes inconsistency to his motives, but really represents no more than dressing out our own subjective thoughts and feelings in the guises of fertility magic or sympathetic magic. Such fanciful speculation leads nowhere. There is no way of testing such hypotheses.

Paintings in shallow caves and rock shelters were executed by illiterate artists, whether contemporary with our era (as in South Africa or Australia) or somewhat earlier in Neolithic cultures *ca.* 6000-1500 B.C. (northern Africa, eastern and southern Spain). African and Australian rock artists can tell us why they paint, but different groups may have several different and conflicting reasons, making it impossible to extrapolate to ancient paintings. Moreover, modern rock artists do not work in deep caves. However, a scientific examination of cave art can throw important light on the environment of early man and his responses to it, when taken together with remains from archaeological excavations. Such an approach takes the art at its face value, and does not search for hidden meanings. It assumes that man painted objects in the real world around them more or less faithfully.

Clearly, this approach demands attention to find details of the paintings and is hindered if they are faded. They must never be fingered nor must water be thrown on them to make them stand out for photography. Fine water sprays from aerosol dispensers must be used. Ultraviolet flash photography with the appropriate colour film, and infrared black and white film with red filters can also enhance contrast. Skilled artists can execute valuable copies, such as those shown here kindly made for the writer by Mrs R. de Méric. Digging in caves with art must be supervised by experienced archaeologists according to the local legal requirements, but speleologists can provide useful information by observations on cave sediments, speleogenesis and cave archaeology at other caves in the area.

To illustrate what can be learned from this approach, brief mention will be made of the writer's survey of cave art sites in south-east Spain. The art is later than, and has a different fauna from, that of earlier northern Spanish and French cave art, and unlike the latter is confined to daylight zones and rock shelters. There seems to have been an early phase of naturalistic paintings depicting usually man, red deer, horse, cattle, and goat or ibex, and less often boar or pig, dog, chamois, fallow deer, bear, sheep, birds and insects. Archaeological bony remains testify to their occurrence in eastern Spain between 7000 and 3000 B.C., embracing the later Upper Palaeolithic, Neolithic and beginnings of copper-using society. More schematic paintings then became popular, lasting till 1500 B.C. (Walker, 1971). Archaeological evidence distinguishes the cultures from contemporary rock painting North African cultures, and limb and body proportions of the painted human figures are also significantly different.

Many animal figures are only a few inches high, and so careful preservation is essential to extract maximum information. Most naturalistic art caves are near springs or streams on south-facing hillsides where they receive the full blast of the sun. These have high incidences of humans with bows and arrows and animal depictions. In high mountains deer and ibex predominate (see Plate 13.3), but in open country cattle and horse are also painted (Plate 13.4). Three caves show cattle whose horns have been overpainted with stag antlers (perhaps they were used as hunting decoys?).

PLATE 13.3. Red deer, ibex and archers: Cañaíca del Calar (El Sabinar, Moratalla, Murcia). (Courtesy: Mrs R. de Méric.)

Possibly the sites were close to those waterholes where hunters drove and/or killed animals. The wide variety of painted bow and arrowpoint types (Plates 13.5, 13.6, 13.7) may reflect that variation in Neolithic stone arrowpoints known from excavations. Agricultural activities are not portrayed, but the close proximity between humans and animals suggests

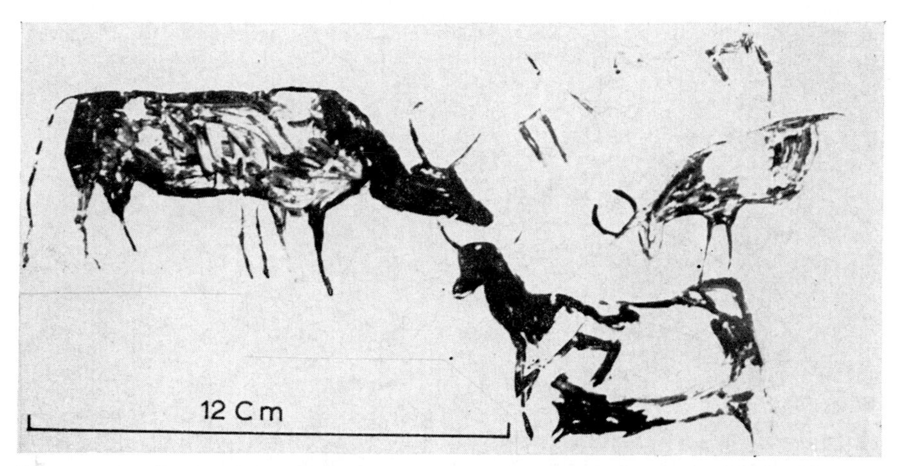

PLATE 13.4. Cattle: detail from Cantos de la Visera (Monte Arabí, Yecla, Murcia). (Courtesy: Mrs. R. de Méric.)

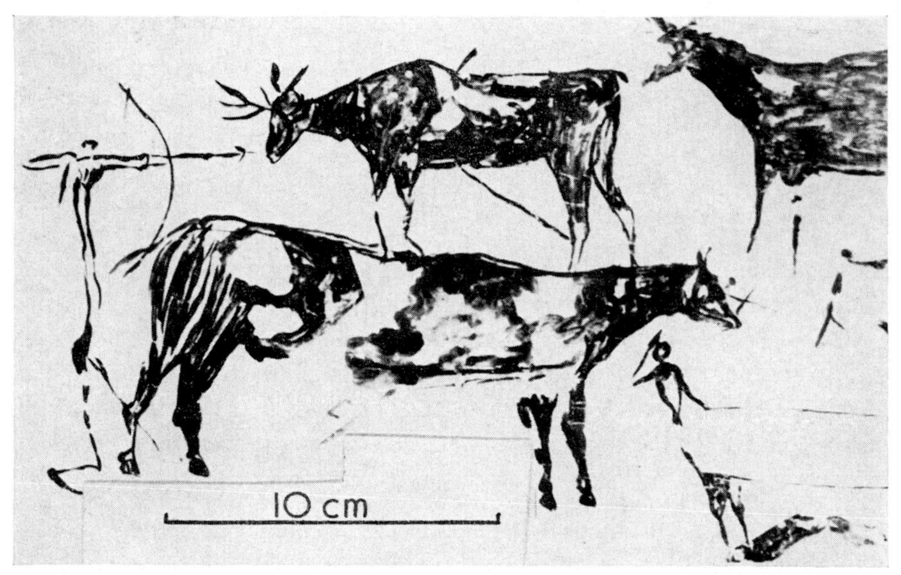

PLATE 13.5. Archer in close proximity to quadrupeds: detail from Solana de las Covachas (Nerpio, Albacete). Scale in centimetres. (Courtesy: Mrs R. de Méric.)

incipient herding. Humans wearing skirts and without weapons (possibly women) are less often painted, occurring especially at one group of north-facing, shaded caves (Abrigo Grande, etc., see Walker, 1972) where there are only a few painted animals (mainly red deer). This may have been used for tool-manufacture or domestic purposes. Studies of artefact types

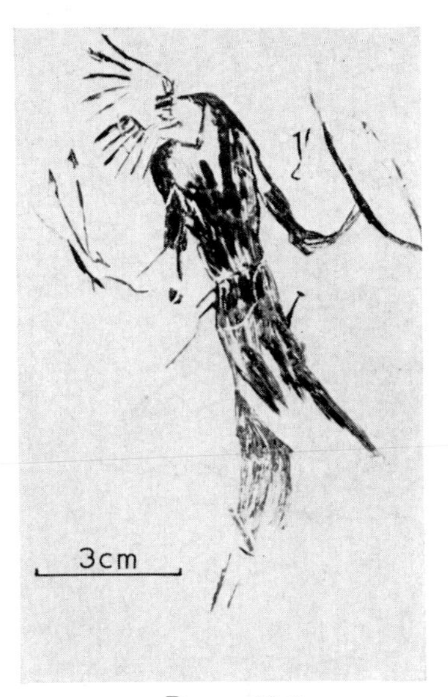

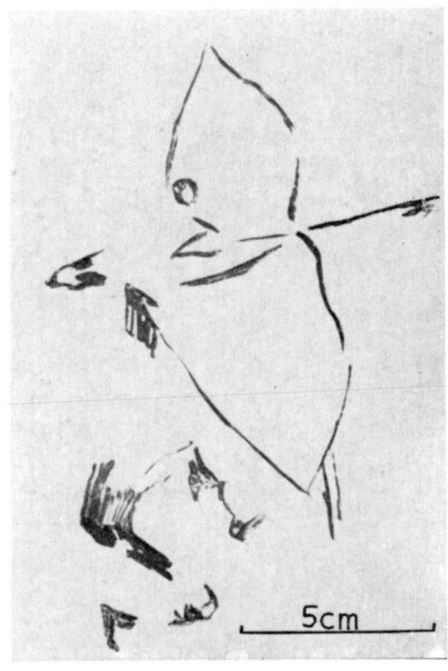

PLATE 13.6 PLATE 13.7

PLATE 13.6. Archer with arrows: detail from Cueva de la Vieja (Alpera, Albacete). As in Plate 13.5, the arrowpoints appear to be barbed and tanged in type. (Courtesy: Mrs R. de Méric.)

PLATE 13.7. Archer with notched arrow: detail from Cueva de la Vieja (Alpera, Albacete.) The arrowpoint appears to be a single-barbed or else a "microlithic" type. (Courtesy: Mrs R. de Méric.)

from different neothermal period caves in the region seem to indicate functional differentiation in site-utilization.

Cave art sites follow the upper and middle reaches of the valleys of SE Spain. Yet there are caves near the coast which seem not to have been occupied at this period. Some north-facing caves (e.g. Abrigo Grande) have later Upper Palaeolithic and early Neolithic artefacts in deposits of

FIG. 13.7. Site Exploitation Territory of Abrigo Grande. This map indicates types of land within 1, 5, and 10 km of site (solid circles) and within 1 hour and 2 hours' walk from site (broken lines). The site is a rock shelter which has now been excavated and represents an occupation of later Upper Palaeolithic-Early neolithic age (see Walker, 1972) and is next to another rock shelter with rich naturalistic paintings.

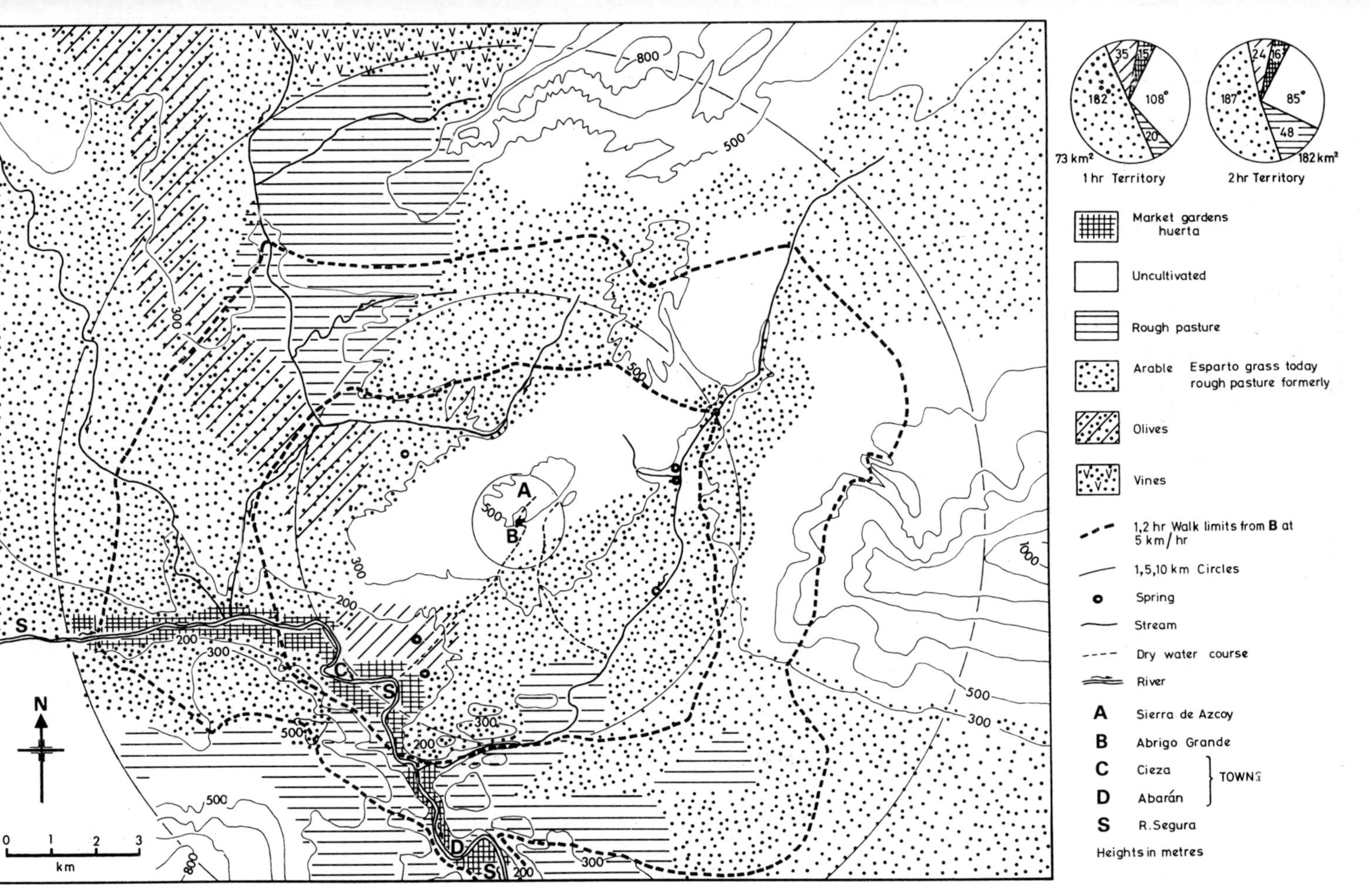
73 km²
1 hr Territory
182 km²
2 hr Territory
Market gardens huerta
Uncultivated
Rough pasture
Arable
Esparto grass today rough pasture formerly
Olives
Vines
1,2 hr Walk limits from B at 5 km/hr
1,5,10 km Circles
Spring
Stream
Dry water course
River
A Sierra de Azcoy
B Abrigo Grande
C Cieza
D Abarán
TOWNS
S R.Segura
Heights in metres
N
km

FIGURE 13.7

windblown or aeolian sand. Neolithic and copper age material is often found in later thermoclastic layers (Cuenca, 1973). In the valleys, riverine sediments show that at a time perhaps corresponding to the close of the last glaciation the rivers were inactive, and palaeosols and clays seem to indicate ephemeral swamps. A phase of intermittent torrential erosion and rising temperature followed, responsible for the lowest of a series of valley bottom sediments eventually 30 feet thick. Much of this thickness was due to aeolian sands and occasionally loess. It is reasonable to correlate these with the cave sands, and to suggest a subdesert, dustbowl phase in the neothermal period, say 7000-5000 B.C. (confirmed by ^{14}C dates). Thermoclastic deposits then follow, as in the caves, and with similar archaeological contents. Some consolidation of valley soils may thus have occurred between 4000 and 2000 B.C., allowing large-scale arable cultivation for the first time. (Later much of the 30-foot thick alluvia was removed by erosion between 500 B.C. and A.D. 1000 and redeposited nearer the coasts where the modern agricultural centres are situated.) Radiocarbon dates tell us that copper-using townships were established before 4000 B.C., and Neolithic sites with cereal grains are known from earlier still. Yet there was not enough arable land to support such communities by large-scale cereal cultivation. Interestingly, new townships appear in the middle reaches of the valleys when such land was becoming consolidated *ca* 4000-2000 B.C.; but animal bones are still found in enormous quantities.

Where does cave art come into the picture? It suggests that animal-based activities, rather than agriculture, were predominant during the neothermal period, in contrast to much archaeological opinion. Making allowances for changes in soil type and utilization, and assuming a higher protein and fat intake than we are accustomed to, it can be calculated from a site catchment analysis at one group of excavated cave art sites (Fig. 13.7) that there is no reason why there should not have been enough venison alone to support a hundred people all the year round. There are about thirty skirted figures (women) painted, suggesting some such population; and other sites have up to 150 human figures. A meat-based diet is subject to less fluctuation than a cereal-based one, and expansion of a neothermal period population throughout the area would (a) have overgrazed the land, causing a dustbowl, and (b) led to setting up of towns and villages from which transhumance would have been necessary to maintain the same dietetic standards and population expansion. This assumes that hunting led to partial and finally full domestication of wild animals on a local basis. Only at an advanced date could townships rely on cereals or other crops to supply more than a trivial element of the diet. Future projected searches will indicate whether this model needs revision or can be substantiated. Such an approach to cave art tells us more than speculation about what the artists "really" tried to tell us. It could profitably be extended to other countries with cave paintings.

REFERENCES

Breuil, H. and Windels, F. (1952). *Four Hundred Centuries of Cave Art*. Montignac.

Cuenca Payá, A. (1973). El cuaternario reciente en la cuenca del Vinalopó (Alicante). *Estudios Geológicos*.

Ucko, P. J. and Rosenfeld, A. (1967). *Paleolithic Cave Art*. Weidenfeld and Nicholson, London.

Walker, M. J. (1971). Spanish Levantine rock art. *Man 6* (4), 553-588.

Walker, M. J. (1972). Cave dwellers and artists of the neothermal period in south-eastern Spain. *Trans. Cave Res. Grp. G.B. 14* (1), 1-22.

14. The Computer in Speleology

J. D. Wilcock

The computer . . . "has no pretensions whatever to originate anything. It can do whatever we know how to order it to perform."

Ada Augusta, Countess of Lovelace, daughter of Lord Byron the poet, Cambridge mathematician and the first computer programmer (to Charles Babbage's Analytical Engine)

WHAT IS A COMPUTER?

Many people tend to think in extremes about the computer; there are those who think it is an infallible electronic "brain", while others believe it is merely a machine which is capable of doing a large number of calculations in a very short time. It has been better described as an obedient, accurate, but unintelligent clerk with an exceptionally good memory who never gets tired, and who can work 24 hours a day, 7 days a week. In spite of recent sophisticated developments it cannot think or originate anything. It is, however, capable of learning and modifying its own *program* (i.e. set of instructions which direct its operation) and it can also indicate inherent order within a set of *data* (i.e. information given to the computer) which might have been quite unsuspected by a human research worker. Basically a computer can add, subtract, multiply and divide. It can also compare two numbers, make simple logical decisions of the Yes/No variety, and "remember" numbers. But where the computer really scores is that it can do all this at very high speeds, e.g. an addition typically takes 1 μs (one millionth of a second).

The basic layout of a *digital* computer is shown in Fig. 14.1; there is a second variety of computer, the *analog* computer, which receives mention below.

Information is fed into the system through the *input* devices (e.g. punched card reader, punched paper tape reader, magnetic tape unit, magnetic disc, teletypewriter) and is placed in named locations in the *store*. The store may be likened to a sorting frame of named or numbered pigeon holes. Each item of data has its own pigeon hole, and when it is required for a calculation it is referred to not by its *value* but by the *name* or *number* of the pigeon hole. This name or number is known as the *address* of the storage location.

Calculations and manipulations of the data are carried out by the *arithmetic unit* and the answers are placed back in the store.

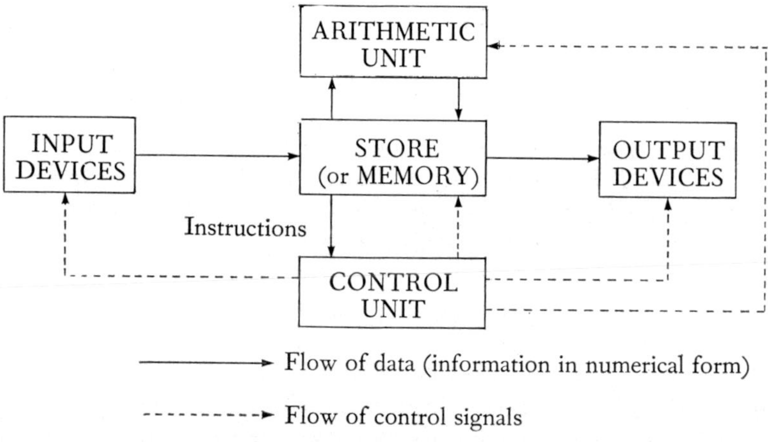

FIG. 14.1. Block diagram of a digital computer.

Results are communicated via the *output* devices (e.g. card punch, paper tape punch, magnetic tape unit, teletypewriter, line printer).

Each process within the machine, whether an arithmetic operation or the transfer of information from one part of the machine to another, is supervised by the *control unit*. Electronic switches are actuated by the control unit as directed by the coded instructions of the program, which it retrieves from the store and deciphers. Having extracted an instruction from its storage location and obeyed it, the control unit usually takes the next instruction in sequence from the store, unless a *jump* instruction is decoded which allows the next instruction to be taken from any place within the program.

Formulating the instructions and sequencing them correctly is the job of the programmer. The completed list of instructions is known as a *program*. It is the programmer's responsibility to ensure that the program performs in the machine as intended. Most computer blunders of the "million pound gas bill" type seized on by the press in the silly season are the result of human error, either the inadequate checking of the program or the use of corrupt data; naturally, if the input data is faulty, the output can be expected to be meaningless, or, as the Americans have it, "garbage in—garbage out".

The form in which instructions are presented to the computer control unit is known as the order code or "low-level" assembly language, and this is peculiar to each make of machine. It can take a considerable time to learn, and there is the disadvantage that the programs are only acceptable by the particular machine for which they were written. Accordingly, several international "high-level" computer languages have been developed which are acceptable by most machines. Those which are most suitable for speleological use are ALGOL, FORTRAN, PL/I and BASIC. All speleologists hoping to use the computer in their research are advised to learn one of these languages, which can be accomplished in a few days, or even hours, of study from standard texts. BASIC is particularly useful since it is designed to be used from a remote terminal (see below). These high-level languages are very simple to learn and programs may be written very quickly in them. The instructions are powerful in the sense that a single instruction can accomplish an involved mathematical procedure. A special program called a *compiler* is necessary to convert each instruction in the high-level language into a corresponding instruction or group of instructions in the order code of the particular machine in use, and so a considerable loss of speed is experienced. This disadvantage, which is not serious for infrequently-used programs, must be weighed against the advantages of simplicity in use and ease of learning of the high-level language.

Before any problem can be put on a computer it must be broken down into simple steps. The whole process of writing a program is susceptible to errors, both of the "spelling mistake" and "logical error" varieties. For this reason coding sheets are very carefully checked, the program is followed through manually with specimen data, and before even the first instruction is written down a *flow-chart* is constructed. Each step is enclosed in a diagrammatic box and the boxes are joined in logical sequence by *flow-lines* with direction of flow indicated by arrowheads.

The basic operations available are:

(a) input and output of data;
(b) replacement of one variable by another (a variable is an item of data which may take several values);
(c) addition, subtraction, multiplication and division;
(d) testing the value of a variable against some known criterion, making a simple decision and following some consequent course of action;

(e) counting the number of times a loop of operations is to be carried out on different data.

A typical flow chart is illustrated in Fig. 14.2.

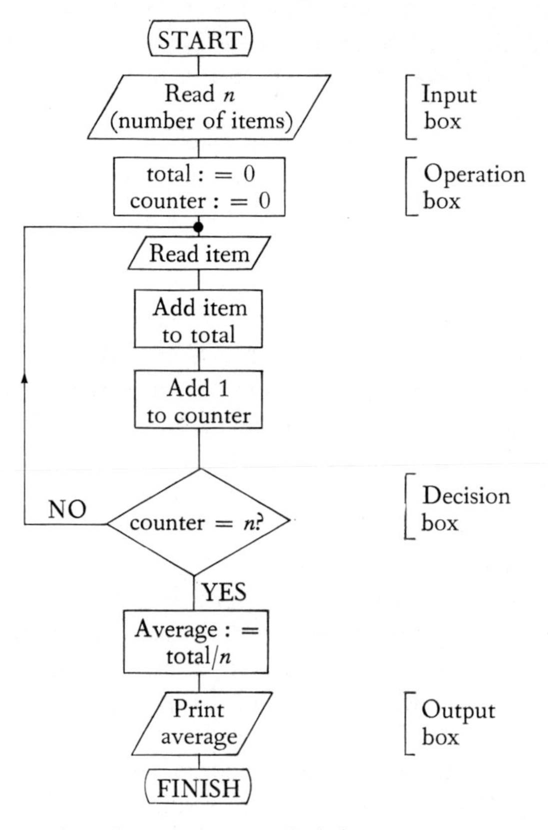

FIG. 14.2. Flow-chart for a program to find the average of a list of numbers.

The computer is but a tool, the use of which can be mastered by intelligent people. When speleologists realize that computing is a technique within their capacity the development of the science of speleology will rapidly accelerate.

APPLICATIONS OF THE COMPUTER IN SPELEOLOGY

The uses of the computer in speleology may be grouped under several headings:

(a) information retrieval, e.g. for cave archaeological records;
(b) routine processing of data, as in cave survey applications; and the production of cave location maps from geophysical observations;

(c) graphics, i.e. the production of cave plans, sections and other diagrams by computer for direct publication;
(d) the use of remote terminals to pass data to a computer from any location in the field or laboratory;
(e) the use of statistics and complex computation as in the application of the computer to hydrological studies and morphometric analysis.

A general survey of information retrieval, routine processing, graphics and the use of remote terminals is given by Wilcock (1971).

INFORMATION RETRIEVAL

CAVE BIOLOGY

The technique of storing large bodies of specialist information within a computer library, usually on magnetic tape, and the recall of sections of data in response to specific requests, have been used in many disciplines; but there have been few attempts to use it for speleological purposes. For this reason techniques derived from related fields with applications in speleology are described below as well as specific speleological studies.

Dr Perring, of Monks Wood Experimental Station, the Nature Conservancy, has collaborated for several years with Miss M. Hazelton, Hon. Biological Recorder of the British Cave Research Association, in the setting up of cave biological records. The computer has been used in the drawing and publication of maps for the Atlas of the British Flora (Perring and Walters, 1962; Perring, 1963), and the computer system may be used to process BCRA collections when the volume of data justifies it. Perring described the computer requirements of the research worker in biology in two articles (Perring, 1967; Soper and Perring, 1967), being of the opinion that electronic data processing should be a by-product of the initial requirement to label all items accurately. At present (1972) the Monks Wood Experimental Station has about 5000 CRG records on punched cards, a large number of which are 40-column, since the Biological Records Centre did not change to 80-column cards until 1969. All the records are hand-written and are punched with species number and vice-county number for preliminary sorting. The data from the hand-written cards is transferred to a second set of 80-column cards to form the *data bank*. Selections can be made from the data bank using an ICL 302/0 Group Select Sorter, which will sort 8 columns of a card simultaneously, and the punched information on the cards may be printed out on an IBM 870 Document Writing System (this same machine will also plot distributions on pre-printed maps). The biological distribution maps which appear in Chapter 10 of this publication were made by selecting cards for the appropriate species, sorting them into grid reference order, and tabulating on the IBM 870. The punched cards carry data in fixed *fields* (groups of columns) and the contents may readily be printed by a tabulator.

A specimen tabulation of biological records is shown in Fig. 14.3. The columns from left to right list the order number; genus and species; grid reference; vice county number; locality; altitude in tens of metres; date of collection (e.g. 10/957 = October 1957); source of reference (e.g. 3 indicates literary source); location of voucher material (e.g. CRG for *Trans. Cave Res. Grp. G.B.*); and expert's initials (e.g. TBR = T. B. Reynoldson). Most of the Nature Conservancy card records have now been transferred to magnetic tape at the Atlas Computer Centre in Cambridge, and the computer is accessed by remote terminal from Monks Wood. The results of the searches are punched on cards which are later used *off-line* (i.e. not connected to the computer) to plot distribution maps on the IBM 870.

CAVE ARCHAEOLOGY

The computer has lately been used for the recording of cave archaeological finds. Cave records from all parts of the country are fed to the BCRA Hon. Recorder for Archaeology and Palaeontology for entry into the computer system (Wilcock, 1970b). The associated problem of the recording of museum catalogues has been studied by the Information Retrieval Group of the Museums Association (IRGMA). This is a British body which aims eventually to place all museum records on computer file. The Chairman is Geoffrey Lewis, Director of the City of Liverpool Museums. Lewis (1965) gave a general description of information retrieval methods and presented some thoughts on the national museum index. A series of papers presented at the 1967 Colloquium on information retrieval for museums (Lewis *et al.*, 1967) contains many points of interest, including a summary of the computer requirements of the research worker in archaeology by Renfrew (1967). Renfrew made the point that archaeological codes of description can never have the finality of species names in biology. This raises the problem of archaeological "types" of object and how they are defined. The computer has been of great assistance in the classification of objects in archaeology, often revealing groupings of objects which were unsuspected before the analysis. IRGMA published draft proposals for an interdisciplinary museum cataloguing

Order no	Species	Grid ref.	Vice County	Locality	Altitude	Date	Source	Expert
2503	PHAGOCATA VITT	20/742665	3	BAKERS PIT	7	10/960	3 CRG	4
2503	PHAGOCATA VITT	20/743665	3	REEDS CAVE BUCKFASTL	7	05/964	3 CRG TBR	4
2503	PHAGOCATA VITT	20/776837	3	MORTON HAMPSTEAD	21	09/969	3 CRG TBR	
2503	PHAGOCATA VITT	20/838631?	3	BAKERS PIT CAVE	7	/960	3 CRG	4
2503	PHAGOCATA VITT	22/838160	42	DAN YR OGOF	24	05/966	3 CRG	4
2503	PHAGOCATA VITT	31/484593	6	RICKFORD FARM CAVE	6	04/966	3 CRG TBR	4
2503	PHAGOCATA VITT	34/481738	60	ARAGONITE BAND MINE	1	10/969	3 CRG TBR	
2503	PHAGOCATA VITT	34/430738	60	MOSS HOUSE MINE	1	09/969	3 CRG TBR	
2503	PHAGOCATA VITT	34/483735	60	CRAG FT MINE WALTON	6	02/966	3 CRG TBR	4
2503	PHAGOCATA VITT	34/483735	60	CRAG FT MINE WALTON	6	06/965	3 CRG TBR	4
2503	PHAGOCATA VITT	34/500775	69	HALE MOSS CAVE	3	01/965	3 CRG TBR	4
2503	PHAGOCATA VITT	34/801778	64	BROWGILL CAVE NORTON	33	06/966	3 CRG	4
2503	PHAGOCATA VITT	35/987403	66	WATERFALL CAVE	25	08/956	3 CRG	4
2503	PHAGOCATA VITT	52/007061	20	BOURNE GUTTER		07/967	3 CRG TBR	4
2503	PHAGOCATA VITT	42/966075	20	MARLINS FM BKHAMSTED	15	04/964	3 CRG TBR	4
2503	PHAGOCATA VITT	43/291580	57	TEMPLE MINE MATLOCK	0	10/957	3 CRG	4

FIG. 14.3. Specimen Tabulation from 80-column record cards. The species *Phagocata vitta* (Duges) is a carnivorous, slug-like aquatic troglophile from the Order Tricladida of the Phylum Platyhelminthes (Flatworms). By courtesy of Miss D. W. Scott, data processing officer, Monks Wood Experimental Station. (Note. The Cave Research Group has now merged with the British Speleological Association to form the British Cave Research Association. The above records were taken from *Trans. Cave Res. Grp. G.B.*)

system in April 1969 (Lewis *et al.*, 1969). This is a full list of codes used in the IRGMA pilot scheme, but contains little guidance about how the system may be adapted to cover archaeological objects or site records, mainly because the experts concerned were more familiar with geological, fine art and other museum objects. Since 1969, IRGMA has held discussions with the international communication organization for museums ICOM, set up by UNESCO, with the result that ICOM has agreed to accept the IRGMA system, now described as the ICOM-IRGMA format. The idea behind this system is to provide a common language for communication between local systems designed for different computers and thus necessarily incompatible. Just as the high-level languages mentioned previously are accepted by most computers, the common data language proposed by IRGMA can be translated into the local data format and vice versa using a translator specially designed for the local system. If all local systems have these individual translators, transfer of data can be effected; efficient and economic cataloguing, however, can only come when computer manufacturers standardize computer equipment.

The initial software for the IRGMA system has been developed by Cutbill (Sedgwick Museum of Geology, Cambridge) and Williams (British Museum (Natural History)) for the IBM 370 and ICL System 4 computers under a grant from OSTI. Under its development title of "Cambridge Geological Data System" it has now completed trial catalogues for the Sedgwick Museum. There is evidence that the cost per record, about £0·10, is comparable to the cost of traditional cataloguing methods. The full system is described by Lewis (1970/71). Interest has recently (1972) revived in the representation of archaeological objects and site records, and a new sub-committee is to be formed to consider the requirements, on which the British Cave Research Association will be represented. The British Museum (Natural History) already uses the ICOM-IRGMA system for its Mammal Collection and also for animal remains from archaeological sites. The Nature Conservancy has also been represented on the IRGMA working parties and will probably adopt the scheme for the Biological Records Centre at Monks Wood.

Roads (1968) of the Imperial War Museum described the use of 80-column punched cards and *aperture cards*, which are 80-column cards with a space for the insertion of a microfilm photograph of an archaeological object.

There have been many museum information retrieval exercises carried out in the U.S.A. using the computer. Ellin (1968a and b) reported the initial stages of the Museum Computer Network project, a consortium of

25 east-coast museums, and he also reviewed internationally similar work (1968c), including studies by Perring, Roads, Chenhall, Gaines and Schneider reported in this chapter. Vance (1970) of the Museum of Modern Art, New York, examined GRIPHOS, a general language for information retrieval, and commented that it remained to be seen whether the various museum projects would proceed independently or under the auspices of the Museum Computer Network. Chenhall (1967), editor of the *Newsletter of Computer Archaeology*, described a system for the computer cataloguing of archaeological items by culture, location, material and method of manufacture. He discussed plain-language description, recording and listing (rather than cryptic codes), a feature which is most important from the archaeologist's point of view; but his actual implementation used complex machine codes. Chenhall (1970, 1971) continued to advance his ideal of a world-wide computer-oriented data bank for archaeologists. Computers have only recently become adequate, and archaeological theory sufficiently defined for such a project to be feasible. He described four operational projects: the Smithsonian, the Museum Computer Network, the Arkansas Archaeological Survey and the University of Oklahoma systems. The Smithsonian system has the useful feature of a *Global Reference Code*, based on latitude and longitude, an extension of National Grid References. Chenhall is of the opinion that standardization is necessary before progress can be made, and has made the plea that this be achieved before large bodies of material are converted for computer input. Schneider (1971) expressed the contrary opinion that it may not be necessary or even desirable to implement a standard system for all archaeological data banks because a large system designed to cover all the differing requirements of various users may be too unwieldy for convenient use. It has been pointed out above that local systems can be used if translators are available to allow them to communicate with other systems via a standard language. Vance (1971) made several points in answer to Schneider: transcription of data is expensive, and translation systems are generally inefficient and costly. He urged groups of workers to collect data uniformly. Lately (1975) the Museum Computer Network has been launched commercially. The Archaeological Data Bank Conference, held May 1971 in U.S.A., set up a committee to assemble the details of a minimal information system for archaeology. Recent developments (1972) suggest that this committee may accept the ICOM-IRGMA format, and if this is achieved an international data communication system for archaeological purposes will at last come into existence, a noteworthy development.

AN EXAMINATION OF SOME PRACTICAL INFORMATION RETRIEVAL SYSTEMS

As an aid to practical information retrieval, Reynolds (1969, 1970, 1971a, b, c and d), of the Department of Computer Science, Brunel University has formulated a radically new approach to man-computer systems, the COntext Dependent Information Language (CODIL). The limitations of conventional information retrieval techniques are found to be a consequence of the concept of "stored-program computer", i.e. the limitations experienced in setting up a practical program and designing data formats. Usual solutions are the simple system with arbitrarily fixed data lengths (which often requires distortion of data) and the complex system which caters for all exceptional conditions but which is expensive, difficult to alter and awkward in use. The CODIL system has variable-length data items of the form NAME = SMITH, YEAR = 1972 and DISEASE NOT MEASLES, i.e. each has a set name, an operator and a value. A statement is a list of items separated by commas and terminated by a full stop. The level of an item is a number representing the distance of the item from the beginning of the list. Successive statements may be combined by eliminating common leading items from one of them. This system allows rapid input, output and processing. During a retrieval run the CODIL system compares each *facts statement* with a set of *criteria.* If all facts in the facts statement are true in the criteria statement currently being processed, then any remaining items in the criteria file statement are transferred to the facts statement. This is a process which is akin to human logical thought processes. The use of this language effectively changes the "stored-program computer" into the "information language computer" which is flexible and without restrictions. Reynolds (1971e) applies the CODIL system to cave fauna records, an example of which is shown in Fig. 14.4, and the system could also be applied to cave archaeological records. Referring to Fig. 14.4, if the criteria statement is

CAVE = BAKERS PIT
YEAR <1965
PRINT FACTS

then a search of the facts file yields the following references to worms:

CR 50 Enchytraeidae, Bakers Pit, 1960
A 51 Enchytraeidae, Fredericia, Bakers Pit, 1948

Walker (1970a and b) and Wilcock (1970a) have worked together on an information retrieval system for general archaeological purposes. Items

```
*CODIL PAGE    221,  31/07/70,  FILE IDENTIFIER = BBSCODIL.TRIALS

221    1 FILE = CAVE RESEARCH GROUP BIOLOGICAL RECORDS,
221    2      NOTE = TAKEN FROM CAVE RESEARCH GROUP PUBLICATIONS.
221    3      PHYLUM = ANNELIDA,
221    4           CLASS = CHAETOPODA,
221    5                ORDER = OLIGOCHAETA,
221    6                     CAVE = PIXIES HOLE,
221    7                          YEAR = 1960,
221    8                               REF = CR 20.
221    9                     CAVE = PRIDHAMSLEIGH,
221   10                          YEAR = 1961,
221   11                               REF = CR 68.
221   12                               REF = CR 76.
221   13                     FAMILY = ENCHYTRAEIDAE,
221   14                          CAVE = BAKERS PIT,
221   15                               YEAR = 1960,
221   16                                    REF = CR 50.
221   17                               YEAR = 1968,
221   18                                    REF = FBA 20.
221   19                          CAVE = PRIDHAMSLEIGH,
221   20                               YEAR = 1961,
221   21                                    REF = CR 73.
221   22                                    REF = CR 81.
221   23                               YEAR = 1968,
221   24                                    REF = FBA 22.
221   25                          CAVE = REEDS CAVE,
221   26                               YEAR = 1956,
221   27                                    REF = MR 49.
221   28                          GENUS = FREDERICIA,
221   29                               CAVE = BAKERS PIT,
221   30                                    YEAR = 1948,
221   31                                         REF = A 51.
221   32                     FAMILY = LUMBRICIDAE,
221   33                          CAVE = BICKINGTON POT,
221   34                               YEAR = 1960,
221   35                                    REF = CR 29.
221   36                                    REF = CR 32.
221   37                                    REF = CR 59.
221   38                     FAMILY = NAIDIDAE,
221   39                          GENUS = PRISTINA,
221   40                               SPECIES = ACQUISETA,
221   41                                    CAVE = BAKERS PIT,
221   42                                         YEAR = 1968,
221   43                                              REF = FBA 20.
221   44                               SPECIES = FORELI,
221   45                                    CAVE = BAKERS PIT,
221   46                                         YEAR = 1968,
221   47                                              REF = FBA 20.
221   48                     FAMILY = TUBIFICIDAE,
221   49                          GENUS = CLITELLIO,
221   50                               SPECIES = AREMARIUS,
221   51                                    CAVE = BAKERS PIT,
221   52                                         YEAR = 1968,
221   53                                              REF = BM 578.
221   54                          GENUS = TUBIFEX,
221   55                               SPECIES = TUBIFEX,
221   56                                    CAVE = PRIDHAMSLEIGH,
221   57                                         YEAR = 1968,
```

FIG. 14.4. A specimen CODIL print out.

are recorded on magnetic tape file, and during an information retrieval run all items are compared with a set of criteria joined by logical operators, of any desired degree of complexity, e.g. cave archaeological records may be examined for items satisfying the following criteria:

$$(\text{Iron Age} \vee \text{Roman}) \wedge \overline{\text{Hyena remains}} \wedge \text{Bronze brooch} \wedge 2/(\text{cave entrance facing South} \vee \text{more than 200 m above sea level} \vee (\text{less than 10 Km from a main land route} \vee \text{less than 5 Km from navigable water}))$$

This particular logical function means that the request is for any cave records satisfying the following:

Date Iron Age OR Roman (or both);
AND NO Hyena remains;
AND a bronze brooch was found;
AND two out of the following three criteria are satisfied:
(1) cave entrance faces south;
(2) cave entrance more than 200 m above sea level;
(3) less than 10 Km from a main land route
OR less than 5 Km from navigable water (or both).

Besides a printed listing of the items which satisfy the request, the computer provides as a by-product a distribution map. The map outlines are carried in numerical form on magnetic tape, and any desired map may be drawn to any scale, complete with border, north point and scale marker. The locations may be plotted using a range of symbols, and a printed title may be added automatically. The finished distribution maps are ready for publication in all respects. Several such distribution maps are illustrated in the above references.

SOME PHILOSOPHICAL CONSIDERATIONS

An important factor in the design of any information retrieval system concerns the choice of *keywords*. Keywords are used to describe the subsets of the classification into which objects may be distributed. Three main options are available:

(a) an autocratic classification system, for which the permissible keywords are set up before the data is classified;
(b) a restricted classification system, which starts without any control on keywords, but for which a list of preferred keywords is gradually built up, a list of synonyms compiled, and the preferred keywords added to the synonyms;

(c) a totally free classification system, which allows any keyword and which uses the computer to cross-correlate between synonyms.

Of these, the totally free system is very expensive to operate, while the autocratic system is difficult to design since the requirements for the classification are often not accurately known at the outset. The restricted system is sensible in operation and economic to run, and is the system most suited to speleological requirements.

Each record in an information retrieval system should carry some indication of the reliability of the data, usually the name of the expert who has identified the specimen, and also the name of the classifier and the date of classification. These details enable the computer to assess quality of data and, for example, to amend all records originated by a particular classifier before a certain date.

The usefulness of negative data should be appreciated. There is a distinction between *no* entry in a particular classification, which is ambiguous, and a *negative* entry, which is a positive statement that something is not present.

Distribution maps, a by-product of an information retrieval run, should be interpreted with care. It is found that hypogean fauna distribution maps reflect not only the caving areas, but such factors as the distribution of collectors, of caving clubs, transport routes and access rather than a strict distribution of species. Glennie (1968) has shown that hypogean species inhabit regions which are quite different geologically from the caving regions. Also collectors probably tend to look for rare rather than common species, the result being the sparse recording of very generally distributed species.

Security of the information in any collection of data is becoming an increasingly severe moral problem. The mere publication of the fact that a certain rare species inhabits a certain cave may lead to extinction of the species in this habitat by over-collecting. As another example, the publication of locations of promising cave archaeological deposits may lead to the unauthorized use of metal detectors on the site by coin collectors and unscrupulous dealers. The owners of private collections may also be most reluctant for it to become known generally that they have important objects in their collections.

The dimension of objects may have been recorded in Imperial measure, and the computer may be employed in conversion to metric. If this is done, attention must be paid to the accuracy of the original measurements, and the probable accuracy of the corresponding metric conversion. It would be absurd, for example, to convert a dimension of 6 inches, probably measured to the nearest inch, to 15·24 cm.

ROUTINE PROCESSING OF DATA

CAVE SURVEY

The computer application in speleology which is likely to be regarded as the most practical by the majority of speleologists is the routine calculation of cave survey co-ordinates (see Chapter 1). Full details are to be found in *Surveying Caves* by B. M. Ellis (British Cave Research Association, Bridgwater, 1976).

GEOPHYSICAL SURVEY FOR THE LOCATION OF CAVES

The use of resistivity meters in the location of subterranean cavities is a well-known technique (see Chapter 9). A variety of probe configurations may be employed. High resistance readings are expected over the caves. For a large site a considerable number of readings must be taken and require experience in interpretation. The ideal is a map of the site indicating where to dig for the cave, and the computer is very useful both for reducing the data and for plotting the site map. Gravimetric and electromagnetic methods may also be used in the location of caves, and both require elaborate corrections and calculations, for which the computer is a help.

Eve and Keys (1929, 1930) carried out a series of tests using radio and electromagnetic waves at the Mammoth Caves, Kentucky. Since the first modern computer appeared in 1944, and computers were not commercially available until the 1950s, the computer could not be used at that time. Palmer (1949, 1951, 1954, 1959, 1960) used resistivity surveys specifically for both speleological and archaeological purposes, and he also designed a new probe configuration, in which each potential electrode is physically linked to its nearest current electrode. The two pairs of electrodes are moved independently and the potential probe spacing is typically about 80% of the current probe spacing. Because the potential probe spacing is a variable proportion of the current probe spacing, reduction formulae become necessary for the calculation of the resistivity. A slide-rule method for the calculation of resistivity readings in the field using Palmer's formulae was evolved by Wilcock (1963, 1965) with advice from Palmer for the Oxford University Expedition to Northern Spain in 1961 but the computer may be employed to better advantage for this purpose. Palmer and Glennie (1963) described the use of the resistivity meter in the location of Pen Park Hole. Typical meters and techniques were evolved by Evershed and Vignoles Ltd. (1961a and b), Nash and Thompson Ltd. (1960) and Tagg (1956, 1957, 1965). Tagg produced a large number of curves based on complex calculations for use in the interpretation of resistivity readings. Direct use of the computer for this interpretation would now be more appropriate. Further coverage of these methods is given in Chapter 9.

It is often necessary in geophysical work to "filter" the readings to remove unwanted noise and background effects, and a computer is invaluable for such work. Scollar and Linington have worked for a long time in this field (Scollar and Linington, 1966; Linington, 1970). The filtering process involves complex computation in which each survey station is allowed to be influenced by all its neighbour stations. Filtering techniques often employed are the *simple differencing filter*, *high-pass*, *low-pass* and *band-pass* filters. More complex filtering methods have also been used (Black and Scollar, 1969; Linington, 1972/3; Gubbins, Scollar and Wisskirchen, 1971), where the use of computers is essential.

It is important to pay attention to the man-computer interface when entering data into the computer and presenting computed geophysical results. Wilcock (1969b) described the design of an 80-column form for the recording of geophysical observations in the field, ready for direct conversion into computer input. The transmittion of results by communication line to a remote terminal was also described.

Gardner (1972, unpublished) has written a simple program for the construction of depth tables for use in the location of caves by magnetic induction, based on an algorithm by Glover.

GRAPHICS

The term *graphics* is applied to all those computer techniques which are concerned with the production of diagrams, graphs, plans, etc. The finished diagrams are ready for publication in all respects. Most of the relevant devices have been described by Wilcock (1970c). The most common devices are the *digital incremental plotter* (graph plotter) and the *line-drawing display unit* (Cathode ray tube) with *light pen*. The graph plotter has a continuous roll of paper which may be stepped up or down by the movement of a drum. A pen may also be moved across the paper on a pen carriage, and so by a combination of pen and drum movements the pen may reach any point on the paper. The pen may be lowered, when it draws a line on the paper, or raised and moved to another location for the start of a new line. Large plotters may have several pens with different coloured inks. Red ink is most suitable if the diagrams are to be duplicated by Xerox methods. A more sophisticated type of plotter is the *automatic drafting machine*, which uses a photographic plate in place of paper and a photo-exposure head in place of a pen. Typical automatic drafting machines and a program for controlling them are described by the Gerber Scientific Instrument Company (1969). The *line-drawing display unit* has a cathode ray tube with associated brightness and focus controls, a *light pen*, *function switches* and often a keyboard. The light pen is used to define points on the screen, and the function switches control the action of the computer program in the drawing of lines, circles, curves, etc., as necessary for the building up of the diagram. Captions may be added by using the keyboard. Any parts of the diagram which are undesirable may be deleted. When the operator is quite satisfied with the presentation, the diagram may be produced in permanent form by the digital incremental plotter and is ready for publication. Three dimensional figures may also be viewed on the screen by programming the computer to calculate perspective. The computer may also display a series of groups of characters along the bottom of the screen, such as

XYROT YZROT ZXROT PNRT PNLFT TCIN TCOUT ZMIN ZMOUT.

These are called *light-buttons* and can control the action of the computer program when recognized by the light pen. Thus the light buttons can rotate a perspective figure about the Z axis (XYROT, which signifies the rotation of the X and Y axes around the Z axis), X axis (YZROT) or Y axis (ZXROT), can pan the sight line to the right (PNRT) or to the left PNLFT), can track the observer's viewpoint in towards the object (TCIN) or out, away from the object (TCOUT) and can zoom in (ZMIN), i.e. enlarge the figure uniformly or zoom out (ZMOUT) to contract the figure uniformly. The three-dimensional figure can be manipulated at will until the presentation is satisfactory.

These techniques can be useful in the manipulation of cave survey plans and elevations, in the temporary display of information retrieval data, and in the creation of hydrological graphs, to mention but a few applications.

The control of these devices is by complex computer program, a typical example of which is discussed by Notley (1970). Popular accounts of graphic devices are given by Taylor (1971) and Pitteway (1972).

It is often necessary to create numerical information inside the computer which represents a diagram, e.g. the outline of a map or a cave passage. Such information can be created by the *d-Mac pencil follower*, a device with a table to hold a diagram and crosswires which may be moved over existing points and lines. A hand or foot-switch is available which causes the position of the crosswires to be recorded to an accuracy of 0·01 cm in two axes. Curves may be built up from a series of such points, and a keyboard is available for the addition of captions. The device produces a punched paper tape for input to the computer, or in some later models is connected directly to the computer.

THE USE OF REMOTE TERMINALS

It is now becoming fairly commonplace for the computer to be operated from any location which is convenient for field research. This could be a caving club headquarters or even a hotel, the only requirements being a telephone and a 13-amp power point. The necessary equipment consists of a *standard ASR-33 teletype* (a special sort of typewriter) and *a modulator-demodulator* (MODEM) which converts the signals into a suitable form for transmission over the telephone line. The most convenient form of modem to use is the acoustic coupler; this has a box into which the telephone handset is placed and the connection is made by sound (at normal audio frequencies) rather than by electrical connection. The procedure is to dial the number of the selected computer bureau in the normal way, wait for the ringing tone to be replaced by the high-frequency carrier wave, then place the handset in the modem. Within a few seconds the computer will send a message to the teletype requesting identification of the user. Provided that the correct passwords are typed in (this is to prevent unauthorized persons using the computer) the teletype then effectively becomes part of the computer, although the main installation may be tens, perhaps hundreds of kilometres away. The computer asks the user what he wishes to do; the options open to him are the running of an existing program with fresh data, or the creation of a new program. The international language BASIC is the most suitable for routine operation of the computer from a remote terminal. The first use of a remote terminal in the field for archaeological purposes is described by Newman (1969). With characteristic American flair this was from a field location in Hawaii to a computer in California via Satellite link. However, the normal telephone system is quite adequate for transmitting data, except for some small country exchanges with antiquated equipment. A similar British exercise was described by Wilcock (1969a). A more recent exercise allowed persons with little or no computer training to operate the remote terminal, producing worthwhile results (Gaines, 1971a and b). Wilkins (1972, private communication) has used a remote terminal for the processing of survey data, using the standard least-squares method for the closing of traverses in complex networks such as St Cuthbert's Swallet, Mendip. Badly fitting traverses are identified by the algorithm; the exercise of an option enables them to be ignored, and the network is closed by matrix transposition and inversion. An additional routine analyses each leg in bad traverses for possible bearing errors, and will also apply leg length corrections. Wilkins has also used the remote terminal to experiment with "random walks" in simulated blocks of limestone to test the action of variable constraints such as dip and faulting.

STATISTICS AND COMPLEX COMPUTATION

HYDROLOGICAL STUDIES

The main application of the computer to hydrological studies has been in the investigation of the flood-pulse technique (see Chapter 6 of this volume). The technique is described by Ashton (1965, 1966, 1967). Ashton employs the method of representing the pulses observed at the resurgence as a sequence of *binary digits*, a representation which is highly suitable for computer analysis (since computers calculate in the *binary*, or power of 2, system). In looking for a repetition of a basic sequence of pulses, caused by alternative passages within the system, it is then necessary to find all possible factors of the binary number. Ashton (1966) does not consider the possible superimposition of pulses, and Wilcock (1968) has extended the analysis to allow for this possibility. The computer analysis reveals all possible timings for the input pulses at the sinks and all possible superimpositions of pulses by passage divisions, junctions and oxbows which could produce the observed output pulse train at the resurgence. A brief coverage of the use of the *analog computer* in the simulation of cave hydrological systems is also given in this article. The analog computer represents values by electrical voltages rather than arithmetic quantities, and the electrical circuits are designed to simulate the action of various parts of the system; e.g. it is found that the electrical analog of the phreas and epiphreas of a cave system is simply a capacitor (a detailed mathematical analysis of this phenomenon is given by Wilcock, 1968).

MORPHOMETRIC ANALYSIS

Williams (1966) has applied morphometric analysis in the analysis of Karst landforms in the Ingleborough, Whernside and Chapel-le-Dale district. Morphometric analysis consists of the allocation of quantitative values to geomorphological features. The analysis is assisted by graphs and diagrams of various sorts, e.g. plots of swallet order against the logarithm of the number of swallets in each order, of swallet order against the logarithm of the mean area of swallet catchments, and of swallet order against the logarithm of the mean distance to the nearest swallet of the same order. It is obvious that data has to be collected from many sampling points to enable this analysis to be carried out, and Haggett (1965) suggests how hexagonal sampling grids may be set up and distribution maps plotted. The computer is an invaluable aid in the storage and analysis of all this data. The general theory of quantitative studies in geography, geology and geomorphology is given in a work edited by Berry and Marble (1968), which contains chapters by Harbaugh and Preston on the use of Fourier Series analysis in geology, and by Kao on the use of computers in the processing and analysis of geographic information.

Hanna and High (1970) have applied Fourier analysis to meanders in underground streams, using a FORTRAN IV computer program.

CAVE CHEMISTRY

(with acknowledgments to R. G. Picknett)

Calcium carbonate solutions contain many different ions, and a fuller understanding of the behaviour of such solutions requires that the concentrations of all ions present should be calculable. An example of this need is the study of the rates of solution and deposition of limestone, discussed in Chapter 7, where it is shown that the concentrations of the calcium and carbonate ions control both rates. The chemical theory of limestone solutions is given in Chapter 7. The theoretical equations which are derived there involve both ion concentrations and ion activities. This gives rise to a classical difficulty in making calculations: concentrations cannot be evaluated until activities are determined, but activities cannot be determined until concentrations are known. The dilemma is only soluble by an iterative technique which, fortunately, is well suited to computer usage.

In considering first non-saturated solutions of calcium carbonate, it is necessary to explain the method of solution as simply as possible, so the complex algebraic expressions needed in the calculation will not be used here. Instead, the following statements will be made, the first two defining the terms used.

(a) There are six different ions involved: Ca^{2+}, H^+, $CaHCO_3^+$, HCO_3^-, CO_3^{2-} and OH^-.

(b) Activities are represented thus: (Ca^{2+}), and concentrations thus: [Ca^{2+}]. For any given ion, activity and concentration are related by the activity coefficient, γ.

(c) From chemical theory all ion activities and concentrations can be calculated from (H^+) and (Ca^{2+}), i.e. for any ion J:

$$(J) = f((H^+), (Ca^{2+}), \gamma) \quad 1$$

$$[J] = f((J), \gamma). \quad 2$$

(d) For all ions, values of γ can be obtained from the ionic strength I, which in turn is obtained from the ion concentrations $[J]$:

$$\gamma = f(I) \quad 3$$

$$I = f([J]). \quad 4$$

There are four equations of type 1 (one for each of the ions other than Ca^{2+} and H^+), six equations each of types 2 and 3 (one for each of the ions), and one equation of type 4, giving a total of 17 equations. In these equations there are 19 variables: 6 concentrations, 6 activities, 6 activity coefficients and the ionic strength,

$$I = \tfrac{1}{2}(2[Ca^{2+}]+2[CO_3^{2-}]+[CaHCO_3^+]+[HCO_3^-]+[H^+]+[OH^-]).$$

It follows that all the equations can be solved provided values are given for any two variables. In this example values for (Ca^{2+}) and (H^+) are assumed to be known. Further coverage of the theory is given in Chapter 7.

For the method of solution, the flowchart for the computer program is shown in Fig. 14.5. A reasonable approximation to be made in the second step is to equate I with 3(Ca^{2+}), which results in a fairly rapid convergence of I and I' values. For calcium carbonate solutions which are nearly saturated, only five or six circuits of the loop are needed to obtain I correct to 1 part in 10^7, while for solutions far from saturation, up to 20 circuits may be needed.

Although a value of (H^+) may well be available from pH measurement, (Ca^{2+}) is often not so well known, and so it is more convenient to input the total calcium concentration $[Ca]_T$, instead of (Ca^{2+}). This can be done by modifying the flowchart to that of Fig. 14.6. The unusual feature of relaxing the values of both I' and (Ca^{2+}) in the same loop works well. Taking the criterion for both I' and $[Ca]_T$ as agreement to 1 part in 10^7 with I' and T respectively, only three to eight extra iterations are required

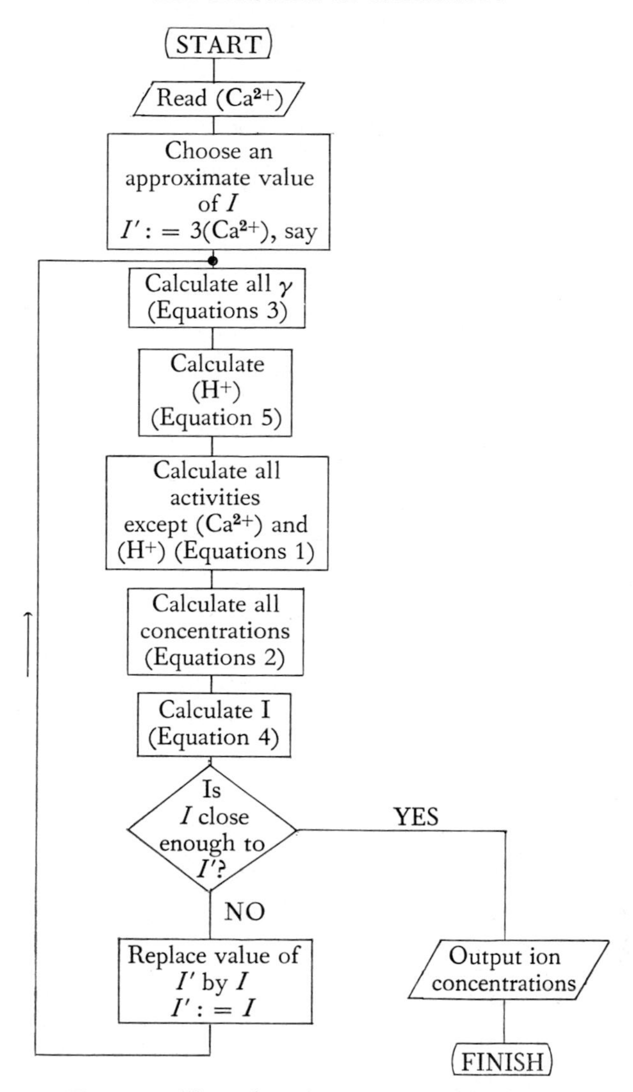

FIG. 14.5. Flow-chart for non-saturated solutions.

over and above those required by the algorithm of Fig. 14.5. Other pairs of input parameters may be used in similar flowcharts; the principles are the same. The parameter T (total calcium in solution at any stage of the iteration) is given by

$$T = [Ca^{2+}] + [CaHCO_3^+] + [CaCO_3^0].$$

This is compared with $[Ca]_T$ as the criterion for terminating the iteration.

As a final example of computer usage, consider calculations for saturated

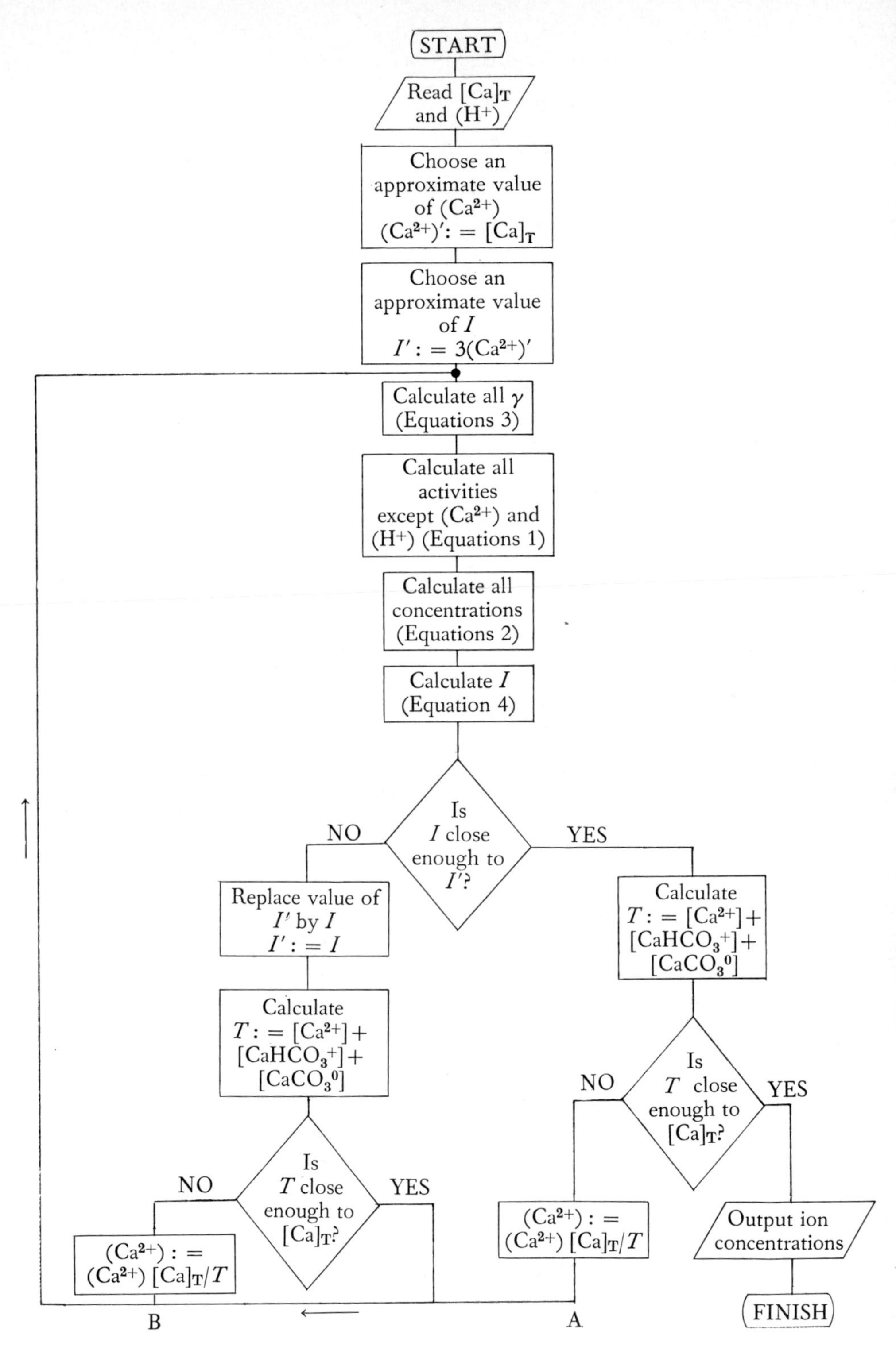

FIG. 14.6. Improved flow-chart for non-saturated solutions.

solutions. An extra theoretical equation becomes available in this case (Eqn 7, p. 226), and it is possible to calculate (H^+) from (Ca^{2+}) and γ:

$$(H^+) = f((Ca^{2+}), \gamma). \quad 5$$

Furthermore, only one variable need be known to allow a complete

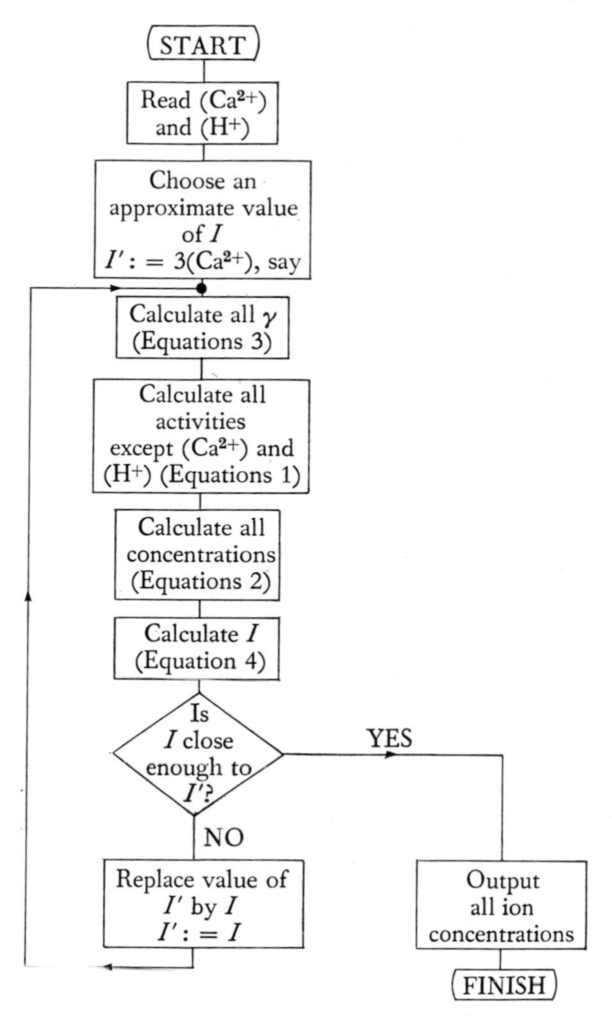

FIG. 14.7. Flow-chart for saturated solutions.

solution of the problem. Take this variable as (Ca^{2+}). The flowchart for the calculation is shown in Fig. 14.7. With the same approximation as before for the second step, less than six circuits of the loop yield values of the ionic strength, I, accurate to 1 part of 10^7. It is, of course, possible to arrange to take $[Ca]_T$ as input, the modifications being the same as those used in Fig. 14.6.

Before computers became widely available, the theoretical study of limestone solutions was seriously retarded by the amount of calculation necessary for even a simple analysis. The situation has now changed and many computer programs are in use, some being sufficiently complex to cope with all the dissolved materials to be found in natural waters. Computer calculations have already been used in reaction rate studies of the solution of limestone (Reddy and Nancollas, 1971), in the evaluation of degrees of saturation (Hostetler, 1964; Wigley, 1972) and as a test of the accuracy of analytical data (Thrailkill, 1971). Wigley (1972) considers not only the ions mentioned above, but also magnesium, sodium, potassium, sulphate, chloride and nitrate ions, together with alkalinity, pH, temperature, and carbon dioxide concentration. His method is suitable only for relatively dilute solutions, but could be modified to suit other cases. The use of computer calculations in cave chemistry will almost certainly increase as the reactions become better understood.

A CAVE ARCHAEOLOGICAL APPLICATION

L. R. and S. R. Binford (1966) have applied a statistical technique called *cluster analysis* to reveal unsuspected correlations between cave archaeological finds, producing groups of objects which form the basis for future classification.

CONCLUSION

This brief summary of the use of computers in speleology has been intended as an appetizer. The computer can perform anything that we can tell it to do, and the limitation is on human ability rather than computer inadequacy. It is hoped that speleologists, particularly cave surveyors and those who have need of routine reduction of field data, will be inspired to write simple computer programs. The procedure is quite simple and within most people's ability. It is true that only those speleologists in educational establishments and a few others in research institutions have regular access to a computer at present; but by the turn of the century computer terminals will be as common in homes as the present-day telephone. Perhaps the sight of a computer terminal in the club laboratory will become commonplace within the next few decades.

REFERENCES

Ashton, K. (1965). Preliminary report on a new hydrological technique. *Cave Res. Grp. G.B. Newsletter 98*, 2-5.

Ashton, K. (1966). The analysis of flow data from Karst drainage systems. *Trans. Cave Res. Grp. G.B.* 7 (2), 161-203.

Ashton, K. (1967). The University of Leeds Hydrological Survey Expedition to Jamaica 1963. *Trans. Cave Res. Grp. G.B. 9* (1), 36-51.

Berry, B. J. L. and Marble, D. F. (eds) (1968). *Spatial Analysis. A Reader in Statistical Geography*. Prentice-Hall, Englewood Cliffs, New Jersey.

Binford, L. R. and Binford, S. R. (1966). A preliminary analysis of functional variability in the Mousterian of Levallois facies. *American Anthropologist 68* (2), 238-295.

Black, D. I. and Scollar, I. (1969). Spatial filtering in the wave-vector domain. *Geophysics 34* (6), 916-923.

Chenhall, R. G. (1967). The description of archaeological data in computer language. *American Antiquity 32* (2), 161-167.

Chenhall, R. G. (1970). S 100. The Arkansas archaeological data bank. *C. Hum. 4* (5) in Directory of Scholars Active 1970, 351.

Chenhall, R. G. (1971). The archaeological data bank: a progress report. *C. Hum. 5 (3)*, 159-169.

d-Mac Ltd. (1967). *d-Mac Pencil Follower Type P.F. 10,000 Mark 1B General Description and Specification*. d-Mac Ltd., Glasgow.

Ellin, E. (1968a). Information systems and the humanities: a new renaissance. *Metropolitan Museum Conference on computers and their potential applications in museums*. New York.

Ellin, E. (1968b). *Report of the Museum Computer Network Project*. New York.

Ellin, E. (1968c). An international survey of museum computer activity. *C. Hum. 3* (2), 65-86.

Eve, A. S. and Keys, D. A. (1929). Radio and Electromagnetic Waves at Mammoth Caves, Kentucky. *Canada Dept. of Mines geol. Surv. Mem. 165*, 89-104.

Eve, A. S. and Keys, D. A. (1930). Geophysical investigations at the Mammoth Cave, Kentucky. *Canada Dept. of Mines geol. Surv. Mem. 170*, 1-17.

Evershed and Vignoles (1961a). *Resistivity Prospecting with Megger Earth Testers*. Evershed and Vignoles Ltd., London. Publication 245/2.

Evershed and Vignoles (1961b). *A Pocket Book on Resistivity Prospecting*. Evershed And Vignoles Ltd., London.

Gaines, S. W. (1971a). Note [on data transmission from archaeological site to computer] in *Newsletter of Computer Archaeology 6* (4), 1.

Gaines, S. W. (1971b and c). Computer application in an archaeological field situation. *Newsletter of Computer Archaeology* 7 (1). 2-4; *MASCA Newsletter* 7 (2), 2-3.

Gerber Scientific Instrument Company (1969). *Gerber Graphics Generator Program (3G)*. Prepared for the Gerber Scientific Instrument Company, Hartford, Connecticut, by Applied Programming Technology Corporation, Sudbury, Massachusetts, 27 September 1969.

Glennie, E. A. (1968). The discovery of *Niphargus aquilex aquilex* Schiodte in Radnorshire. *Trans. Cave Res. Grp. G.B. 10* (3), 139-140, with distribution map.

Gubbins, D., Scollar, I. and Wisskirchen, P. (1971). Two dimensional digital filtering with Haar and Walsh transforms. *Ann. Geophys. 27* (2), 85-104.

Haggett, P. (1965). *Locational Analysis in Human Geography*. Arnold, London.

Hanna, K. and High, C. (1970). Spectral analysis of meanders in underground streams. Panel on computer applications in cave survey. Symposium on Cave Surveying. *Trans. Cave Res. Grp. G.B. 12* (3), 219-223.

Harbaugh, J. W. and Preston, F. W. (1968). Fourier series analysis in geology in Berry, B. J. L. and Marble, D. F. (eds), *Spatial Analysis: A Reader in Statistical Geography*, 218-238. Prentice-Hall, Englewood Cliffs, New Jersey.

Hostetler, P. B. (1964). *The Degree of Saturation of Magnesium and Calcium Carbonate Minerals in Natural Waters*. International Association of Scientific Hydrology, Commission on Subterranean Waters (General Assembly at Berkeley) Publication 64, 34-49.

Kao, R. C. (1968). The use of computers in the processing and analysis of geographic information, *in* Berry, B. J. L. and Marble, D. F. (eds), *Spatial Analysis: A Reader in Statistical Geography*, 67-77. Prentice-Hall, Englewood Cliffs, New Jersey.

Lewis, G. D. (1965). Obtaining information from museum collections and thoughts on a national museum index. *Museum Journal*, *65*, 12-22.

Lewis, G. D. (1970/71). An interdisciplinary communication format for museums in the United Kingdom. (*UNESCO*) *Museum*, *23* (1), 24-26. (Museums and Computers.)

Lewis, G. D. *et al.* (1967). Information retrieval for museums. *Museums Journal 67*, 88-120.

Lewis, G. D. *et al.* (1969). *Draft Proposals for an Interdisciplinary Museum Cataloguing System*. Information Retrieval Group of the Museums Association (IRGMA), London. 23rd April 1969.

Linington, R. E. (1970). A first use of linear filtering techniques on archaeological prospecting results. *Prospezioni Archeologiche 5*, 43-54.

Linington, R. E. (1972/73). Topographical and terrain effects in magnetic prospecting. *Prospezioni Archeologische 7/8*, 61-84.

Nash and Thompson (1960). *Operating Instructions for the Tellohm Range of Soil Resistance Meters*. Nash and Thompson Ltd., Chessington.

Nature Conservancy (1971). *Monks Wood Experimental Station Report for 1969-1971*. The Nature Conservancy.

Newman, S. (1969). Letter to the Editor [concerning the first transmission of archaeological data by satellite circuit]. *Newsletter of Computer Archaeology 5* (2), 1.

Notley, M. G. (1970). A graphical picture drawing language. *Computer Bulletin 14* (3), 68-74.

Palmer, L. S. (1949). Preliminary report on some earth resistance measurements made near Tynings Farm on the Mendip Hills, Somerset. *Univ. Bristol Proc. Spel. Soc. 6* (1), 27-36.

Palmer, L. S. (1951). Earth electrical resistance measurements near the Bath Swallet, Mendip Hills, Somerset. *Univ. Bristol Proc. Spel. Soc. 6* (2), 208-221.

Palmer, L. S. (1954). Location of Subterranean Cavities by geoelectric methods. *Mining Magazine 91*, 137-141.

Palmer, L. S. (1959). Examples of geoelectric surveys. *Proc. Inst. Elec. Eng. 106A*, 231-244.

Palmer, L. S. (1960). Geoelectric surveying of archaeological sites. *Proc. Prehist. Soc. 26*, 64-75.

Palmer, L. S. and Glennie, E. A. (1963). The geoelectric survey and excavation in Tratman, E. K. (ed.), *Reports on the Investigations of Pen Park Hole, Bristol.* Cave Res. Grp. G.B. publication 13, 15-24.

Perring, F. H. (1963). Data processing for the Atlas of the British Flora. *Taxon 12*, 183-190.

Perring, F. H. (1967). The [computer information retrieval] requirements of the research worker in biology. *Museums Journal 67*, 108-111.

Perring, F. H. and Walters, S. M. (1962). *Atlas of the British Flora.* Thos. Nelson and Sons, London.

Pitteway, M. L. V. (1972). The impact of computer graphics. *Nature, London 235*, 83-85.

Reddy, M. and Nancollas, G. H. (1971). The crystallization of calcium carbonate I. Isotopic exchange and kinetics. *J. Colloid and Interface Science 36* (2), 166-172.

Renfrew, C. (1967). The [computer information retrieval] requirements of the research worker in archaeology. *Museums Journal 67*, 111-113.

Reynolds, C. F. (1969). *An Introduction to CODIL.* International Computers Limited, London.

Reynolds, C. F. (1970). Meeting on CODIL. Advanced Programming Study Group. *Computer Bulletin 14* (7), 244-245.

Reynolds, C. F. (1971a). CODIL, Part 1. The importance of flexibility. *Computer Journal 14* (3), 217-220.

Reynolds, C. F. (1971b). CODIL, Part 2. The CODIL language and its interpreter. *Computer Journal 14* (4), 327-332.

Reynolds, C. F. (1971c). CODIL—a new concept in information languages. *Computer Weekly*, May 13, 1971, 6.

Reynolds, C. F. (1971d). An interactive CODIL interpreter. *CODIL News 4* (Dec. 1971), 1-4.

Reynolds, C. F. (1971e). Handling cave fauna records on a computer. *Trans. Cave Res. Grp. G.B. 13* (3), 160-165.

Roads, C. H. (1968). Data recording, retrieval and presentation in the Imperial War Museum. *Museums Journal 68*, 277-283.

Schneider, M. J. (1971). Archaeological data banks. Letter to the Editor of *C. Hum. C. Hum. 5* (4), 239-241.

Scollar, I. and Linington, R. E. (1966). Data processing of geophysical measurements on archaeological sites. *Proceedings of the International Symposium on Mathematical and Computational Methods in the Social Sciences.* International Computation Center, Rome.

Soper, J. H. and Perring, F. H. (1967). Data processing in the herbarium and museum. *Taxon 16* (1), 13-19.

Tagg, G. F. (1956). Megger Earth Tester used in search for King John's Treasure. *Evershed News 4* (5), 3-9 and 12. Evershed and Vignoles Ltd., London.

Tagg, G. F. (1957). A resistivity survey in the Wash area. *J. Int. Elec. Eng. 3*, 5.

Tagg, G. F. (1965). Three dimensional resistivity maps. *Evershed News 8* (2), 10-13. Evershed and Vignoles Ltd., London.

Taylor, F. E. (1971). Computer driven displays and the [British Computer Society] display group. *Computer Bulletin 15* (1), 4-11.

Thrailkill, J. (1971). Carbonate deposition in Carlsbad Caverns. *J. Geol. 79*, 683-695.

Vance, D. (1970). Museum data banks. *Inform. Stor. Retr. 5*, 203-211. Pergamon Press, Oxford.

Vance, D. (1971). Museum Computer Network. Letter to the Editor of *C. Hum. C. Hum. 6* (1), 47-48.

Walker, M. J. (1970a). An analysis of British petroglyphs. *Science and Archaeology 2/3*, 30-61.

Walker, M. J. (1970b). An analysis of British petroglyphs [further remarks]. *Science and Archaeology 4*, 26-29 N4/1.

Wigley, T. M. L. (1972). *A Computer Programme for Water Quality Analysis*. Department of Mechanical Engineering, University of Waterloo, Ontario. Technical Note 15.

Wilcock, J. D. (1963). Cave [resistivity] Surveying in Northern Spain, *in* Milner, R. E. (ed.), *Evershed News* 7 (7), 12-14. Evershed and Vignoles Ltd., London.

Wilcock, J. D. (1965). Geophysical survey and results, *in* Wilcock, J. D. (ed.), *Oxford University Expedition to Northern Spain 1961*. Cave Res. Grp. G.B. publication no. 14, 31-38.

Wilcock, J. D. (1968). Some developments in pulse-train analysis. *Trans. Cave Res. Grp. G.B. 10* (2), 73-98.

Wilcock, J. D. (1969a). Computers and Camelot. South Cadbury, an exercise in computer-archaeology. *Spectrum*, *British Science News 60*, 7-9. Central Office of Information, H.M.S.O., London.

Wilcock, J. D. (1969b). Computer analysis of proton magnetometer readings from South Cadbury 1968—a long-distance exercise. *Prospezioni Archeologiche 4*, 85-93.

Wilcock, J. D. (1970a). Petroglyphs by computer. *Science and Archaeology 2/3*, 27-29.

Wilcock, J. D. (1970b). Information retrieval for cave records. *Trans. Cave Res. Grp. G.B. 12* (*2*), 96-98.

Wilcock, J. D. (1970c). A review of computer hardware of particular use to the cave surveyor. Panel on computer applications in cave survey. Symposium on Cave Surveying. *Trans. Cave Res. Grp. G.B. 12* (3), 201-210.

Wilcock, J. D. (1971). Non-statistical applications of the computer in archaeology, *in* Hodson, F. R., Kendall, D. G. and Tautu, P. (eds), *Mathematics in the Archaeological and Historical Sciences*. Edinburgh University Press, 470-481.

Williams, P. W. (1966). Morphometric analysis of temperate Karst landforms. *Irish Speleology 1* (2), 23-31.

Cave and Fissure Index

Subject Index

D